D1226646

ANAEROBIC DIGESTION AND CARBOHYDRATE HYDROLYSIS OF WASTE

Proceedings of the information symposium under the EEC programme on recycling of urban and industrial waste, Luxembourg, 8–10 May 1984

Secretariat of the symposium and publication arrangements:

P. P. Rotondó
Commission of the European Communities
Directorate-General Information Market and Innovation

ANAEROBIC DIGESTION AND CARBOHYDRATE HYDROLYSIS OF WASTE

Edited by

G. L. FERRERO,

M. P. FERRANTI

Commission of the European Communities

and

H. NAVEAU

Catholic University of Louvain

ELSEVIER APPLIED SCIENCE PUBLISHERS
LONDON and NEW YORK

ELSEVIER APPLIED SCIENCE PUBLISHERS LTD
Ripple Road, Barking, Essex, England

Sole Distributor in the USA and Canada
ELSEVIER SCIENCE PUBLISHING CO., INC.
52 Vanderbilt Avenue, New York, NY 10017, USA

British Library Cataloguing in Publication Data

Anaerobic digestion and carbohydrate hydrolysis of waste.
1. Recycling (waste, etc.) 2. Bacteria, Anaerobic—Industrial applications
I. Ferrero, G. L. II. Ferranti, M. P.
III. Naveau, H.
628.3′54 TD794.5

ISBN 0-85334-324-1

WITH 100 TABLES AND 140 ILLUSTRATIONS

Organisation of the Conference by the Commission of the European Communities, Directorate-General Science, Research and Development, Brussels, in co-operation with the Directorate-General Information Market and Innovation, Luxembourg.

EUR 9347

Printed in Great Britain by Galliard (Printers) Ltd, Great Yarmouth

PREFACE

This seminar was organized in the framework of the EEC R&D programme on Recycling of Urban and Industrial Waste, which started in November 1979, and will be terminated by the end of 1985.

Besides coordination activities in the different research areas of the programme, about 40 shared-cost contracts have been attributed in the field of anaerobic digestion and carbohydrate hydrolysis of waste (Area III of the programme); some of them were concluded and some were entering a final phase to complete the results obtained. At that stage it seemed appropriate to organize a seminar (open also to scientists not directly involved in the EEC programme) to provide a wider forum for an open and free discussion on the validity of the work done and to indicate the needs in research and development for future actions. The response to this seminar, to which over 200 scientists from 18 countries participated, the quality of the papers presented and the numerous special meetings on particular topics which spontaneously took place parallel to the official programme, showed the significance of the subjects treated.

The present volume includes the full texts of the survey papers, summaries of the contractor's reports and descriptions of the posters presented as well as reports on discussions.

The editors express their gratitude to all those who with their contributions and willingness to actively participate in discussions and special meetings have permitted a better understanding of the present developments on anaerobic digestion and carbohydrate hydrolysis of waste.

Brussels, May 1984.

M.P. FERRANTI G.L. FERRERO H. NAVEAU

CONTENTS

CONCLUDING SESSION

OPENING SESSION

INTRODUCTION AND GENERAL PRESENTATION OF THE RESEARCH AND DEVELOPMENT PROGRAMME ON RECYCLING OF URBAN AND INDUSTRIAL WASTE

Ph. BOURDEAU and G.L. FERRERO

Directorate-General for Science, Research and Development

Commission of the European Communities

SUMMARY

The research and development programme in the field of recycling of urban and industrial waste is described in broad outline, and the specific activities carried out in the fields of both Anaerobic Digestion and Hydrolysis, together with the results so far obtained, are briefly presented.

The aims of the Seminar and the expected effects on future CEC activities are briefly outlined.

For some years the European Economic Community has been engaged in various sectorial research and development programmes in priority areas, namely energy, raw materials, environment, agriculture, some aspects of medical research and industrial research, etc. These various actions have now been consolidated into a "Framework Programme" based on essential Community goals.

The research programmes take three different forms :

- direct action, i.e. research carried out directly by the Commission at its Joint Research Centre;

- indirect action, i.e. research which is contracted out with a proportion of the expenditure being borne by the Community, up to a maximum of 50 % of the total cost of the research. Contracts are signed with industry, university laboratories, regional, provincial or municipal authorities, national research centres, etc.;

- concerted or coordinated action, i.e. the Community coordinates publicly-funded research projects being carried out in the individual Member States with the research being performed under contract to the EEC.

The R&D programme on the recycling of urban and industrial waste is carried out entirely by means of contracts and coordination activities.

Over the last few years the problem of waste in general has become increasingly important, both because of its environmental impact and because of the energy and raw materials which could be saved if the products contained in the waste could be recovered.

According to recent studies, the quantities of waste arising in the European Economic Community alone have been estimated at some 2 300 Mio tonnes a year which amounts to 6.3 Mio tonnes a day. The annual total consists of :

120 Mio tonnes of household waste
950 Mio tonnes of agricultural waste
160 Mio tonnes of industrial waste
300 Mio tonnes of sewage sludge
250 Mio tonnes of waste from the extractive industries
170 Mio tonnes of demolition waste and debris
120 Mio tonnes of consumer waste (discarded vehicles, used tyres, etc.)

These quantities increase by an average of some 3 % a year - although there are variations from country to country - in spite of the low rate of growth in industrial production and in gross national product.

It is obviously worthwhile therefore to develop technologies which can use these waste materials to the full and reduce the volume of waste arising from today's industrial processes, in particular because these recycling processes can to some extent help to solve increasingly serious environmental problems.

It is extremely important, however, that as well as developing methods of recycling we should also develop markets for the secondary materials which recycling produces so that we can make full economic use of the results of these recycling operations.

Waste can therefore be seen as a reserve - in some cases a considerable one - of unused or partially used resources which can and should be exploited more fully. If properly organized, the recycling of waste can be tantamount to conserving rare and essential resources (energy and raw materials) while at the same time yielding economic benefits.

It is estimated that at present only some 35 Mio tonnes of the Community's total municipal and commercial waste stream are recovered :

paper and board	9 - 10 Mio tonnes
ferrous scrap	20 - 22 Mio tonnes
non-ferrous scrap	1 - 2 Mio tonnes
glass	1 - 2 Mio tonnes
rubber	1 Mio tonnes
tin-plate	1 Mio tonnes

In addition to the 35 Mio tonnes which are already being recovered each year, it would be possible to recover another 60 Mio tonnes under prevailing technical and economic circumstances (25 Mio tonnes are considered non-recoverable).

As regards industrial waste, between 64 and 96 Mio tonnes are already being recovered, representing between 40 and 60 % of the annual total arising.

It would be technically and economically feasible at present to recover a further 25 to 35 Mio tonnes . This extra potential represents a value of some 500 or 600 Mio ECU.

In the medium and long term, it might be possible to recover more than the estimated 25-35 Mio tonnes depending on the following factors :

- shortages of raw materials;
- increases in the prices of primary raw materials;
- increases in waste disposal costs;
- new recovery technologies.

In 1981 it was estimated that, if all materials could be recovered, savings in the costs of imported raw materials would be in the region of 12 000 Mio ECU. It now seems, however, that these savings would probably be considerably higher.

Over the last few years, the public authorities, industry and scientists in the various Member States have shown increasing interest in the recovery of useful materials from waste, chiefly because of the potential for saving raw materials and energy but also in order to reduce dependence on imported resources and to improve protection of the environment, which is seen as our inalienable heritage and of prime importance to society.

At Community level, the problem of supplies of raw materials has resulted in research and development programmes and in the preparation of other measures to guarantee security of supplies.

These Research programmes deal not only with primary raw materials but also with the recycling of urban and industrial waste (including waste from agriculture and forestry) and with materials substitution.

They were based on the findings of a number of special studies carried out by national experts to identify those areas of common interest where research at Community level would be preferable, in order to achieve greater efficiency by pooling national research activities and eliminating unnecessary duplication, and with the aim to meet the common needs of society in all Member States.

The programme consisted of four research areas, which were subdivided in turn into various research topics and were tailored to Community requirements and the priorities worked out from preparatory studies :

Research Area I - Sorting of household waste
Research Area II - Thermal treatment of waste
Research Area III - Fermentation and hydrolysis
Research Area IV - Recovery of rubber waste

Each of these research areas made provision for coordination of research projects being carried out in the Member States with the help of finance by the public authorities, and a number of specific research contracts part-financed by the Community in the form of indirect actions. Table 1 lists all these activities, giving a breakdown by research area.

On 25 May 1982 a non-Community country, Sweden, signed a cooperation agreement with the EEC concerning this programme and is thus participating on an equal footing with the other Member States in the indirect action and coordination activities.

On 12 December 1983 the Council, on a proposal from the Commission, has extended the programme for two years (1984-1985) merging it with the subprogramme on the recycling of non-ferrous metals of the research and development programme in the raw materials sector.

This seminar aims at presenting the activities undertaken in the fields of anaerobic digestion and hydrolysis in the framework of the CEC programme, together with the results so far achieved. It also aims to provide a forum for discussion in order to improve contact between researchers and to disseminate information. It is intended also to assess the research and development requirements in these two fields and to draw up guidelines for future actions on fermentation and hydrolysis. In fact this research area has aroused far more widespread interest than was anticipated when the programme was decided. It therefore seems justified for activities in these fields to continue in the future within a Community framework as a direct consequence and extension of the results so far obtained.

After more than three years' activity the positive outcome of the research performed under the various contracts of this programme confirms the topical interest of these areas. It has also demonstrated the value of the Community activity, which has brought together researchers of various countries in a common effort to resolve existing problems.

All these results will be presented in detail during the course of the seminar.

The results of the coordination activity will also be presented. This constitutes one of the most positive aspects of the programme.

It is possible for us at this stage to express satisfaction at the considerable interest which has been aroused among such a large Community of researchers by the activities undertaken under this programme.

We are convinced that the aims of this seminar will be attained, thus contributing substantially to a deeper knowledge of the countless on-going actions in the fields of Anaerobic Digestion and Hydrolysis.

TABLE I

RESEARCH AND DEVELOPMENT PROGRAMME

RESEARCH TOPICS	Activities to be coordinated	Indirect Actions
RESEARCH AREA I		
Sorting of household waste		
1. Assessment of waste sorting projects	x	-
2. Methods for sampling and analysis of household waste	x	-
3. Evaluation of health hazards	x	-
4. Technology for the sorting of bulk waste	x	x
5. Materials recovery		
5.1 Paper	x	x
5.2 Plastics	x	x
5.3 Non-ferrous metals	x	-
6. Energy recovery	x	x
7. New collection and transport systems	x	-
RESEARCH AREA II		
Thermal treatment of waste		
1. Firing of waste derived fuel	See L6.	See L6.
2. Pyrolysis and gasification	x	x
3. Recovery of metal and glass from residue	x	-
RESEARCH AREA III		
Fermentation and hydrolysis		
1. Anaerobic digestion	x	x
2. Carbohydrate hydrolysis	x	x
3. Composting	x	-
RESEARCH AREA IV		
Recovery of rubber waste		
1. Retreading	x	-
2. Size reduction	x	-
3. Reclaiming and recycling of rubber powder	x	-
4. Pyrolysis	x	-

WASTE MANAGEMENT ACTIVITIES

L. KLEIN

Directorate-General for Environment,
Consumer Protection and Nuclear Safety
Commission of the European Communities

Waste management is an important part of the overall environment policy. This policy has been defined in 3 successive programmes, in 1973, 1977 and 1982. The first programme dealt mainly with pollution; the second, with prevention and economy of resources; the third, with the necessity of integrating the environment policy with other Community policies.

These different aspects of the environment policy are reflected in the waste management policy. This policy is orientated around three axes, from which stem the Community activities in this area :

- prevention, with the development of the so-called "clean technologies",
- elimination of wastes and recycling.

The elimination of toxic and dangerous wastes has a high priority. A Directive to control the flow of wastes was adopted in 1978. A proposed regulation, dealing with the transfrontier shipping of dangerous wastes, was presented to the Council in June 1983 with a view to complementing the Directive of 1978. Around 30 million tonnes per annum (tpa) of toxic and dangerous wastes are produced, which 3 million tpa are shipped across frontiers.

Recycling remains one of the main aims of the waste management policy. Not only does it reduce pollution but at the same time saves costly resources. Since 1975 a Directive on waste oils has been in force with a view to a better use of these oils, either by regeneration or by burning with energy recovery. In 1981, the Council adopted a Recommendation to develop the use of recycled paper, especially in the public administrations.

Since 1981, a proposal for a Directive, which aims to improve the re-use and/or recycling of drinks containers - which represent 10 % of domestic waste - has been under discussion by the Council.

Finally in 1982 the Commission presented to the Council a proposal on the use of sewage sludge, an important problem from both a quantitative as well as from a qualitative aspect. Expressed in dry matter, the total sludge actually produced is 6 million tpa. In 1990 this will have risen to 15 million tpa.

These sludges contain fertilizer elements (phosphates and nitrates), which must either be imported because the Community has no rock phosphates or, in the case of nitrates, produced with a high energy cost. At the moment, only 29 % of these sludges are used in agriculture. On the other hand, these sludges also contain harmful elements, thus the Commission's proposal sets rules for use and criteria for quality (for sludges and for the soil). This proposal is based on the results of the research project "Cost 68".

SESSION I

CARBOHYDRATE HYDROLYSIS

ACID HYDROLYSIS REVIEW

H.E. Grethlein

Thayer School of Engineering
Dartmouth College
Hanover, NH 03755

Summary

Cellulosic biomass in forest and agricultural residues or in municipal refuse is a renewal resource for carbohydrates and phenolic compounds via acid hydrolysis. While concentrated acid hydrolysis processes give over 90% conversion of the cellulose to glucose, acid recovery is required, and difficult, for economic operation. Dilute acid hydrolysis processes do not require acid recovery, but by lower yields of glucose - 50% to 65%. Past industrial practice for dilute acid hydrolysis featured a perculation process with about a 3 hour cycle time at 180°C. A continuous hydrolysis is possible at 240° to 260°C with 3 to 20 seconds reaction time, which gives a high reactor volumetric productivity. The kinetics of cellulose and hemicellulose hydrolysis are developed using a laboratory scale bench scale continuous plug flow reactor. The rate of various sources of cellulosic materials is fairly similar. The kinetic model gives the glucose, xylose and furfural yield as a function of acid concentration, temperature and reaction time.

1. INTRODUCTION

There are numerous studies that show a large variety of biomass is available and more can be produced (1, 2). Forest and agricultural residues and the paper fraction from municipal refuse, which at one time represented a disposal problem, and new biomass plantations become the resource base for biomass conversion if an economic conversion technology is developed. The major problem on the front end is that the dispersed nature of the biomass, with its high moisture and low density, requires labor, energy and capital for collection and transportation to the processing plant. On the process end there are the problems of how conversion yields and products and by-products mix effect the capital and energy requirements. While many conversion processes are possible, this review will focus on dilute acid hydrolysis.

2. THE SUBSTRATE

Since there are good reviews available on the structure and chemical composition of cellulosic materials (3), we will not go into great detail here. In acid hydrolysis the kinetics are much less dependent on the physical structure of the cellulosic material than in enzymatic hydrolysis.

With reasonable size reduction to say 1 mm or less in one of the dimensions of the substrate, acid hydrolysis is not diffusion controlled. This is evident by the magnitude of the activation energy for hydrolysis, discussed later, which is above 25 Kcal/g mol. This is well above the apparent activation energy range of 3 to 5 Kcal/g mol for diffusion controlled reactions.

The biomass contains primarily cellulose, hemicellulose and lignin where the relative amounts vary from 35:50:15 in corn stover to 50:25:25 in wood to 70:15:10 in newspaper, respectively.

Cellulose, which occurs in both crystalline (or perhaps a better term is resistant) and amorphous (or available) forms, is a linear homopolymer of anhydroglucose units - a glucan which is very resistant to hydrolysis. A good review of the detailed changes in cellulose during acid and enzymatic hydrolysis is given by Chang (4).

For our purpose hemicellulose is defined as a mixed, branched polymer of hexans and pentosans which are relatively easy to hydrolyzate to pentose and hexose. The pentose is mostly xylose with some arabinose, and the hexose consists of mannose, galactose and glucose.

The lignin is a branched polymer of an indefinite structure consisting of propylphenol moieties joined together by carbon-carbon bond or carbon-oxygen bonds. While most of the lignin remains insoluble in dilute acid, a small fraction is acid soluble, with the amount increasing with the temperature. The residual lignin first decreases in average degree of polymerization and then increases with longer reaction time due to recondensation (5).

3. ACID HYDROLYSIS PROCESSES

Acid hydrolysis of cellulose is a well-known phenomenon and can be carried out with concentrated or dilute acid - usually H_2SO_4 or HCL (6). The major industrial operating plants are now in the USSR, where wood is hydrolyzed to sugar for either ethanol or fodder yeast production. At various times plants have operated in the U.S. (6), Europe (7, 8), and Japan (9), but the economics of ethanol production via ethylene since World War II has ecllipsed these activities. With the increase in oil prices since 1973, the hydrolysis and fermentation to ethanol and other chemicals is obviously being reexamined.

Although concentrated acid (72% H_2SO_4 or 42% HCL) gives 90% or more conversion of potential glucan in biomass to oligomers which are recovered as glucose by a dilute post hydrolysis, the large among of acid used per unit of glucose produced must be be recovered efficiently in order to have an economic process. The Bergius process (7), operated in Germany in the 1930's and improved by Schoenemann (10) at Rheinau in the 1950's, used concentrated HCl, which was recovered by vacuum distillation.

These plants are no longer in operation. However, Battelle Geneva (11) has studied improvements of the HCl processes in recent years while Tsao and co-workers at Purdue (12) have tried to work out a practical recovery system for a modified concentrated H_2SO_4 process. Because of the high glucose yield potential, there is a great insentive to continue inventive work on ways to make acid recovery practical. In this spirit, work on concentrated HF processes are under study (13). In the laboratory, hydrolysis of cellulosic materials with concentrated H_2SO_4 is used in quantitative saccharification schemes to measure the potential sugars in biomass sample (14).

On the other hand, hydrolysis with dilute acid, usually about 1% H_2SO_4 in the water phase, is low enough so that acid recovery is not necessary. This simplifies the process, but now to get reasonable rates of reaction the temperature must be raised to above 150 °C and the glucose yield is reduced to 50 to 65% of the potential glucan with decomposed sugar compounds as by-products. Since the substrate is a major cost item for any hydrolysis process, the final process economics are sensitive to the glucose yield and the utility of the by-products.

Dilute acid hydrolysis can be carried out in batch, semi-batch and continuous processes. In the semi-batch process the liquid phase is removed from the biomass mass solids periodically as in the Scholler Process (8) or continuously as in the Madison Process (6) as a way to reduce the sugar decomposition. The latter, known as a perculation process, is used in the USSR and has recently been studied in the pilot plant scale in New Zealand (15, 16) and Brazil (18).

In the perculation process hot dilute acid is started at 145 to 150°C and then the temperature is programmed to rise to 180 to 190°C. The low temperature part of the cycle removes the readily hydrolyzable hemicellulose and the 180° or more is needed for the cellulose hydrolysis. The total perculation time is between 1 to 3 hours depending on the final temperatures. The yields of glucose are as high as 68% of the potential value in poplar and 63% in pine from the New Zealand pilot plant (15). There are good yields for a dilute acid process. When using hardwood, the pentose can be separated from the hexose by segregating the initial hemicellulose hydrolyzate from later cellulose hydrolyzate. The total sugar concentration in the accumulated hydrolyzate is 3 to 5%. It has been recognized that in the perculation process the material must be in the form of wood chips or sawdust which has the required permeability to insure plug flow of the liquid through the bed with good perculation rates in order to remove the sugars from the reactor in a timely manner. If the hydrolysis is carried out so that only lignin is left in the residue the bed can collapse and plug the reactor (16). One advantage of the perculation process is that the reactor is loaded at atmospheric pressure by gravity so no high pressure solids feeding is required. However, some biomass forms such as waste paper or agricultural residues may not form a suitable perculation bed. The scale-up of the perculation reactor tends to increase the residence time of the perculate and lower the yields (17).

The reason to consider a continuous dilute acid hydrolysis is to reduce the capital investment and operating cost compared to a perculation process (19). The operating temperature for the cellulose hydrolysis is in the range of 240 to 260°C with 5 to 20 seconds reaction time. This results in a much smaller reactor per unit of product than for a perculation reactor. In contrast to the perculation process, the solid and liquid phase have the same reaction time. However the front end of the continuous process is more complex because a pumpable slurry of cellulosic biomass is used which requires the size reduction of the substrate to about 1 mm. Hammer milling, hydropulping, or steam explosion are ways of making pumpable slurries. Special solids handling pumps developed by Church and Woolridge (20) or continuous solids feeders used in the NYU Process (21) or the Stake Process (22) can handle particle greater than 1 mm and slurry concentration of about 20% to 40% solids.

4. BENCH SCALE CONTINUOUS PLUG FLOW REACTOR

Over the last seven years the work at the Thayer School of Engineering at Dartmouth College has focussed on developing a continuous plug flow reactor for dilute acid hydrolysis. A bench scale reactor has been developed that is suited for isothermal kinetic studies with known reaction times in the range of 3 to 30 seconds (23, 24). A schematic diagram of the system is given in Figure 1 which consists of the following major components:

A. A continuous moving cavity positive displacement pump with an open throat which can handle up to 20% solids feed slurry over a flow range of .5 to 1.5 L/min.

B. An acid injection pump which handles 98% H_2SO_4 with flow rate adjusted to give 0 to 2% acid in the mixing tee.

C. A live steam injection zone in which feed slurry is heated to reaction temperature in less than .7 sec.

D. A reactor section made of a 1.27 cm Zircornium tube with a 0.09 cm wall, which has numerous thermocouples attached along the outside walls and is well insulated.

E. An orifice with a diameter of 1 mm to maintain the reactor pressure above the vapor pressure of water and allows for flash cooling of the reactor effluent which quenches the reaction almost instantaneously.

F. A self-contained 60 KW electrically heated steam boiler with an adjustable operating pressure to 800 psig (5.6 MPa).

Any substrate under investigation is Wiley-milled to pass through a .25 or .40 mm screen and prepared into a water slurry with a solids concentration of 5% to 15%. When the narrowest solids particle dimension is 0.5 mm or less the time to heat the interior of the solid by conduction to 95% of the external fluid temperature is of the order of 0.5 sec or less. Moreover, the thixotropic nature of the cellulose slurry promotes plug flow in the reactor tube which facilitates kinetic studies.

5. KINETICS OF ACID HYDROLYSIS OF BIOMASS

It is desirable to have a kinetic model that is useful to simulate the hydrolysis reactions for process design studies. Then the selection of the temperature, acid concentration, reaction time, and number of stages of hydrolysis can be made to satisfy some process optimization criterion.

For our purpose, the carbohydrate in hardwood or agricultural residue can be thought to consist of glucan and xylan which yields glucose and xylose respectively as the first products by acid hydrolysis in sequence of reactions shown in Figure 2.

Because hemicellulose hydrolyzes much easier than cellulose, the relatively mild conditions that give high xylose yield only give modest glucose yield and little decomposition of xylose and glucose. However, the more severe reaction conditions needed to achieve the optimum glucose yield also give significant glucose decomposition products such as hydroxymethylfurfural (HMF), levulinic and formic acid and almost complete decomposition of xylose.

Thus, it is possible to achieve high yields (about 90%) of xylose in the initial period of the perculation run (15) while the temperature is in the range of 150° to 160°C or in the first pass of a substrate through a flow reactor at about 200°C (34). The residual solids can then be hydrolyzed for glucose yield in the later part of the perculation cycle at 180 to 190° or in a second pass through a flow reactor at 240 to 260°C.

First, we will consider the kinetics of cellulose which has been studied by Saeman (25) in batch reactors. The same model will also apply to a continuous reactor although the temperature is 60° to 80° higher (24). Less work has been done on the rapid kinetics of hemicellulose but some preliminary results from the flow reactor for mixed hardwood are discussed. Saeman (25) found that the hydrolysis of cellulose in dilute acid can be modelled as a homogeneous system of consecutive pseudo-first order reactions in series by expressing the cellulose as potential glucose. The reaction sequence I in Figure 2 is represented by:

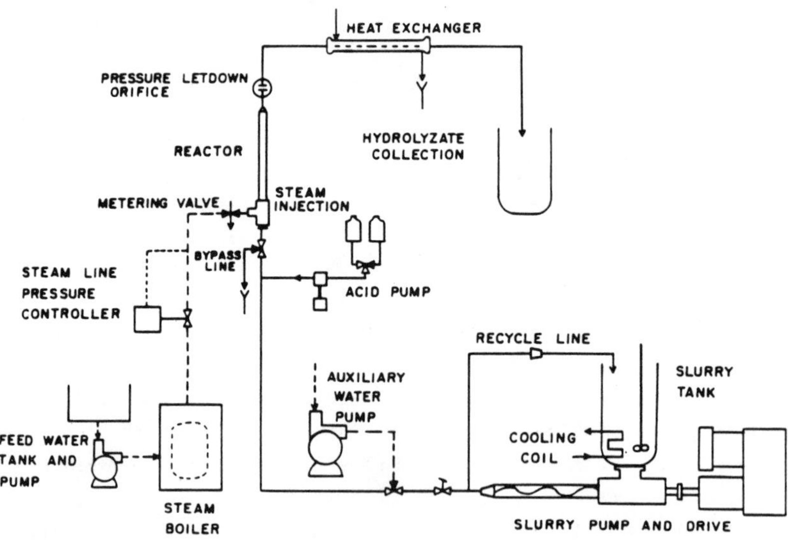

Figure 1: Schematic Diagram of Laboratory Scale Continuous Plug Flow Reactor for Dilute Acid Hydrolysis of Cellulosic Biomass.

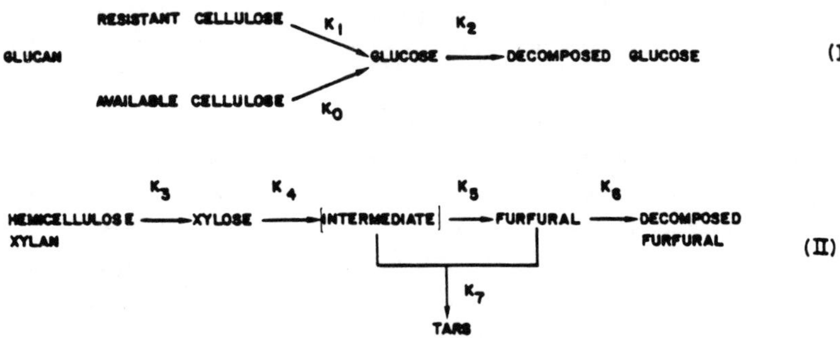

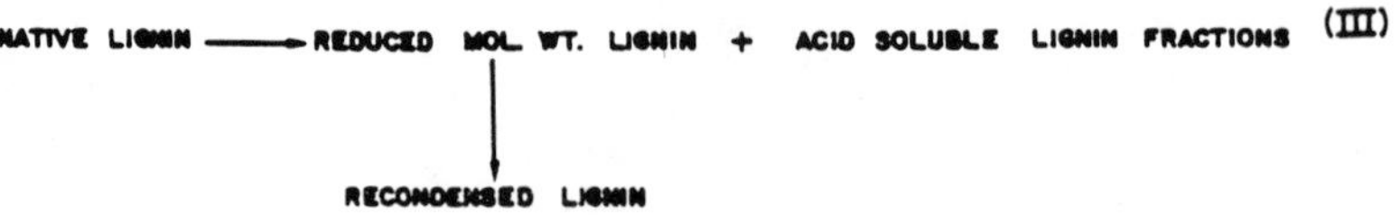

Figure 2: Reactions of Biomass in Dilute Acid.

$$\frac{dC}{dt} = -k_1 C \tag{1}$$

$$\frac{dG}{dt} = k_1 C - k_2 G \tag{2}$$

See Nomenclature at the end of chapter for definition of symbols.

In practice the rate k_o is large compared to k_1, so the glucose from available glucan is taken as free glucose at the initiation of the hydrolysis. Each rate constant k_i has the Arrhenius form with a modification for the effect of the acid as shown in equation (3):

$$k_i = P_i \; A^{n_i} \exp(-E_i/RT) \tag{3}$$

The parameters are the pre-exponential factor P_i, the acid exponent n_i, and the activation energy E_i, which are estimated from isothermal reaction studies. Since the glucose decomposition is a homogeneous reaction, and independent of the cellulose hydrolysis reaction, it can be studied separately and the parameters of the rate constant k_2 can be determined accurately. These parameter values can then be imposed when evaluating the parameters in k_1. While McKibbins (26) determined these parameters from batch reaction condition, Smith (27) found that in the continuous flow reactor a larger rate constant for k_2 was needed to predict the rate of glucose decomposition than given by McKibbins. In fact, the first order glucose decomposition model is an over simplification which is adequate for process design studies when the focus is on glucose.

For a given temperature and acid concentration the integration of equation (2) gives the predicted glucose at time t.

$$\hat{G}\,(t) = G_1 \;\left[\frac{k_1}{k_1 - k_2}\right] [\exp\,(-\,k_2 t) - \exp\,(-k_1 t)] + G_o \exp\,(-k_2 t) \tag{4}$$

Note that G_o is the glucose fraction at t = o from the available glucan and is another parameter to be estimated for a given cellulosic material. The value of $\hat{G}$ (t) depends only on the parameters assigned to k_1 and k_2. The parameters of k_2 are assigned from prior work of Smith (27) and the parameter P_1, n_1, E_1, and G_0 are found by a non-linear search technique that minimizes the sum of squared differences between the observed and predicted glucose yield; namely:

$$\sum_{i=1}^{n} (G(t_i) - \hat{G}(t_i))^2 \quad (5)$$

where i is summed over the number of data points.

The kinetic parameter and the evaluation of the rate constant at 200° and 240°C for 1% acid are given in Table I for a number of substrates which include hard and softwood, paper and corn stover studied in the Dartmouth plug flow reactor. Note the last entry is for glucose and relates the parameters for k_2.

While at first there appears to be a wide range of values for P_i and E_i for the various substrates, in fact the parameters P_i and E_i are highly correlated. Hence, the rate constant, k_1, for a specific acid concentration and temperature (as given as in the last two columns of Table I) is remarkably similar for all these cellulosic substrates. Moreover, since the activation energy $E_1 > E_2$ and the acid exponent $n_1 > n_2$, the selectivity, defined as k_1/k_2, increases as the temperature or the acid concentration are increased. A typical set of glucose yield curves is shown in Figure 3 which illustrates the increased selectivity with temperature for Solka Floc BW-200 using the parameters from Table I.

The kinetic parameters in Table I are for the first pass through the plug flow reactor. It is of interest to know whether the kinetics are altered for the residual cellulose on a second pass through the reactor. Data plotted in Figure 4 show that the glucose yield (based on the potential glucose in the reactor input) for steam exploded poplar is essentially the same when it is run through the flow reactor for a one or two passes (28). Thus, one can use the kinetic parameters in Table I to simulate processes with one or more hydrolysis steps, or where residual cellulose is recycled to the inlet of a reactor. These variations achieve higher glucose yield than one step of hydrolysis.

From the work of Thompson (23) it was demonstrated that the kinetic parameters do not depend on the solid concentration in the reactor over the range of 5% to 14%. This is because the reaction is first order in potential glucose and the acid is a catalyst in which the hydrogen ion concentration fixes its activity, not the acid to solid ratio. When biomass such as wood has a certain ash content which neutralizes some of the acid (17), the acid concentration used in the model is the net acid concentration after the ash neutralization. We expect the kinetic model used here to hold up to 20% if not 30% solids.

Table I: Kinetic Parameters for Acid Hydrolysis of Various Substrates in Equations 1 and 2 and Instantaneous Glucose Yield

Substrate	G_0	P_i	n_i	E_i	k_i min^{-1} for 1% acid		Ref.
		min^{-1}		cal/gmol	at 200°C	at 240°C	
Corn Stover	.146	9.62×10^{14}	1.40	32,800	0.670	10.2	24
Solka-Floc BW200	.030	5.15×10^{16}	1.14	37,000	0.428	9.22	23
Poplar	.011	6.12×10^{15}	0.987	35,150	0.350	6.47	
White Pine	.043	7.80×10^{13}	0.963	30,170	0.880	10.78	
Mixed Hardwood 90% Birch 10% Maple	.006	1.45×10^{15}	1.16	33,720	0.380	6.24	34
Newsprint	.038	1.18×10^{17}	0.696	38,300	0.237	5.68	28
Computer Paper	.039	3.86×10^{15}	0.763	34,850	0.306	5.51	28
Causted Extracted Steam Exploded Poplar	0	2.875×10^{17}	0.816	39,090	0.246	6.36	28
Kraft Pulp Solka Floc BNB-100	.015	1.68×10^{15}	1.20	33,860	0.381	6.33	28
Douglas Fir Prehydrolyzed	0	1.73×10^{19}	1.34	42,900	0.260	9.15	25
Glucose		3.96×10^{8}	0.569	21,00	0.775	4.43	27

Table II: Rate Constants for Hemicellulose Hydrolysis from Mixed Hardwood (34)

	k_3	k_4	k_6	k_8
E cal/gr mol °K	27826	27130	15279	23943
lnP ln(min^{-1})	31.76	18.48	13.23	26.74
n	1.17	0.69	0.58	0.52
k_i at 200°C and 1 wt% acid, min^{-1}	8.64	0.67	0.05	3.55
k_i at 240°C and 1 wt% acid, min^{-1}	86.95	6.51	0.17	25.66

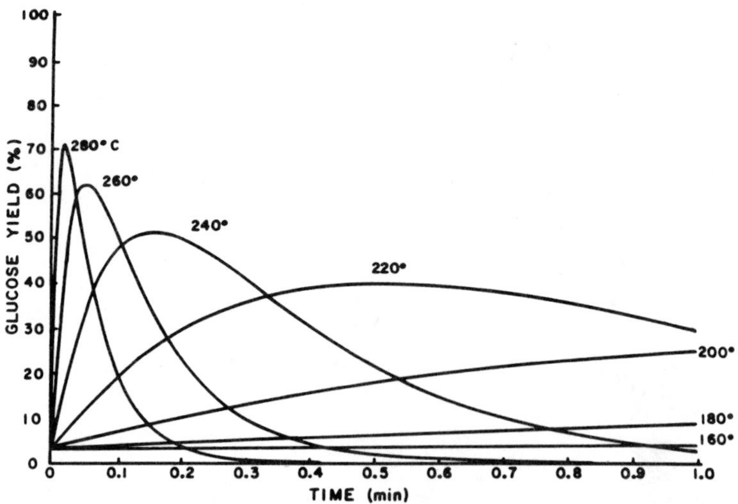

Figure 3: Glucose Yield Predicted by Equation 4 for the Parameters from Table I for Solka Floc BW-200 for 1% Acid.

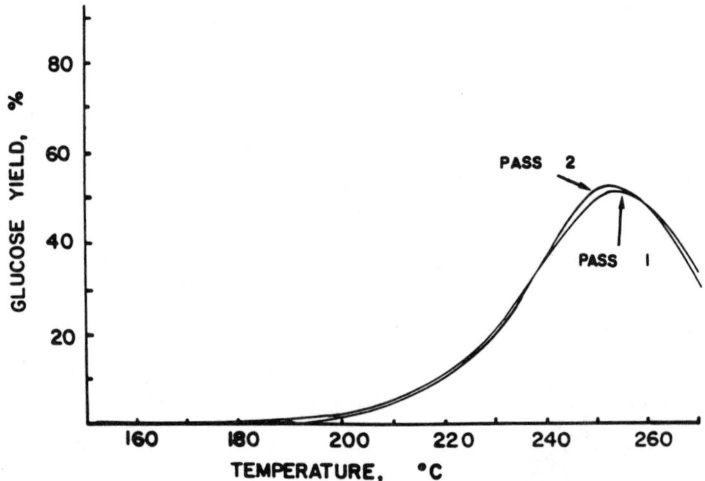

Figure 4: Glucose Yield vs. Temperature for Steam-Exploded Wood and Second Pass Residual Solids Reacted at 1% H_2SO_4, .1 minute Reaction Time.

The hydrolysis of hemicellulose in sweet gum and Douglas fir with dilute acid was studied by Harris, et al (29) using glass ampules from 170 to 190°C but no kinetics were given. Springer and Zock (30) found that the xylan removal in aspen, birch, elm and maple follows two kinetic regions - one hydrolyzing faster than the second. Scarringelli, et al. (31) recently found the same two region behavior in the hydrolysis of xylan in sweet gum wood using 20 to 30% HCl at 30 to 60°C in batch hydrolysis. The work of Trickett, et al. (32) on xylan removal from bagasse in the range of 80° to 100°C with 1 to 4% H_2SO_4 also observes the two kinetic regions. From the work on the hydrolysis of mixed hardwood hemicellulose reported below, we did not find evidence of two kinetic regions when using the flow reactor.

Since a significant fraction of biomass is hemicellulose, the reaction sequence II on Figure 2 was also modelled for the plug flow reactor.

The hydrolysis of the hemicellulose is taken to follow the analogous sequential pseudo first-order reactions as cellulose to give xylose and decomposed xylose. The xylose decomposition is modelled after the work of Root (33) in order to predict the furfural yield. Each rate constant in reaction sequence II has the form of Equation (3) with the usual three parameters P_i, n_i and E_i. The rate equations are given as:

$$\frac{dQ}{dt} = -k_3Q \qquad (6)$$

$$\frac{dX}{dt} = k_3Q - k_4X \qquad (7)$$

$$\frac{dI}{dt} = k_4X - k_5I - k_7I\,FQ_o \qquad (8)$$

$$\frac{dF}{dt} = k_5I - k_6F - k_7IFQ_0 \qquad (9)$$

where Q, X, I, and F are the fraction of potential xylose as xylan, xylose, intermediate, and furfural and Q_o, is the initial xylan in mol/l.

By assuming that $\frac{dI}{dt} = o$, (the pseudo steady state hypothesis) and taking $k_5 >> k_7Q_oF$, the Equations (8) and (9) can be reduced to Equation (10):

$$\frac{dF}{dt} = k_4X - F(k_6 + k_8XQ_o) \qquad (10)$$

where $$k_8 = \frac{k_4\,k_7}{k_5} \qquad (11)$$

While Equation (6) and (7) have analytical solutions and are analogous to Equation (1) and (2), the solution to equation (10) must be done numerically and simultaneously with Equation (6) and (7).

Since all the reactions are homogeneous except for xylan hydrolysis, independent estimates from the plug flow reactor were obtained for the decomposition of furfural, k_6 and xylose, k_4 by following the loss of pure furfural and pure xylose respectively over the temperature and acid concentration range of interest. The ratio k_8 was obtained by measuring the appearance of furfural from pure xylose. Finally, k_3 was obtained by measuring the appearance of xylose from mixed hardwood (90% birch and 10% maple). The parameter estimation technique is the same as used in equation (5) except the experimental xylose yield and the predicted xylose yield given by integrating equation (7) are used. The xylose is measured for the same reaction conditions used to find k_1 for the glucose appearance. To date mixed hardwood has been studied for the complete hydrolysis of cellulose and hemicellulose. The kinetic parameters and values of the rate constants for 200° and 240°C for 1% acid are given in Table II (34).

A typical set of observed data for glucose, residual glucose, xylose and furfural are given in Figure 5 for mixed hardwood along with the predicted values from the integration of Equations (7) and (10). Note that at temperatures above 240° the furfural yield is 80% of the potential xylan, while the glucose yield is 55% of the potential glucose.

A set of parametric curves of the xylose and furfural yield is given in Figures 6 and 7 using the parameters of Table II for mixed hardwood. Note that the furfural yields depend on the solids concentration feed to the reactor since Equation (10) is second order in xylose and furfural. Glucose and xylose yields are independent of solids concentration. When the particle size is below 1 mm, diffusional effect can be neglected.

Together the kinetic models for cellulose and hemicellulose can give a useful prediction of the glucose, xylose and furfural yields as a function of acid concentration, temperature and time. We view the results of the hemicellulose model as preliminary since only one substrate has been studied. Moreover, the model under predicts the xylose yield in the region of the maximum by about 10%. Thus, actual xylose yields have been has high as 90% when the model predicts 80%. Work is in progress to improve the parameter estimates.

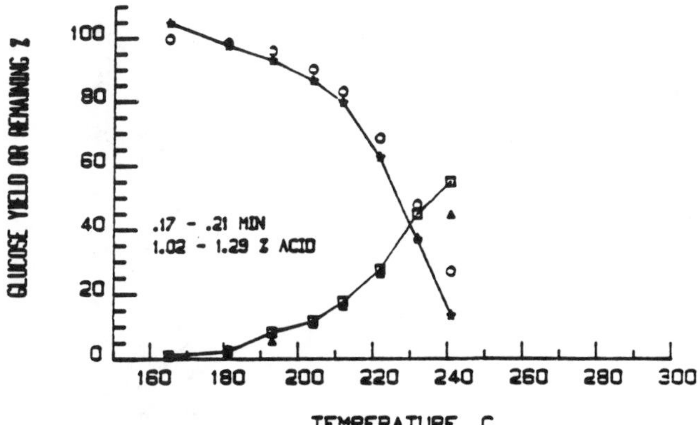

Key: □ Observed Glucose Yield
△ Predicted Glucose Yield
* Observed Glucose Remaining
○ Predicted Glucose Remaining

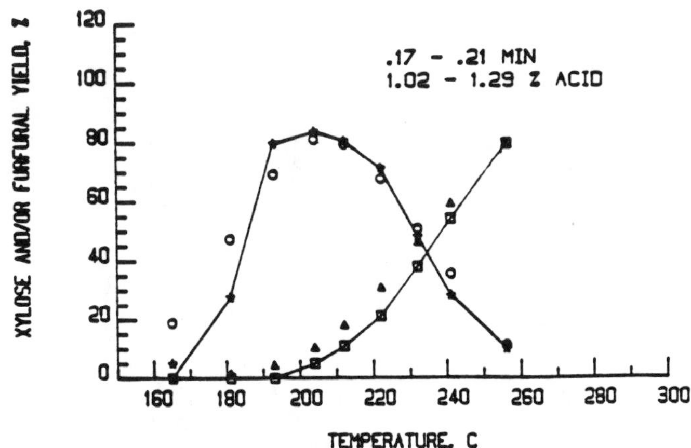

Key: □ Observed Furfural Yield
△ Predicted Furfural Yield
* Observed Xylose Yield
○ Predicted Xylose Yield

Figure 5: Observed and Predicted Yields as Percent of Potential values in Plug Flow Reactor for Feed Solids of 6.6%.

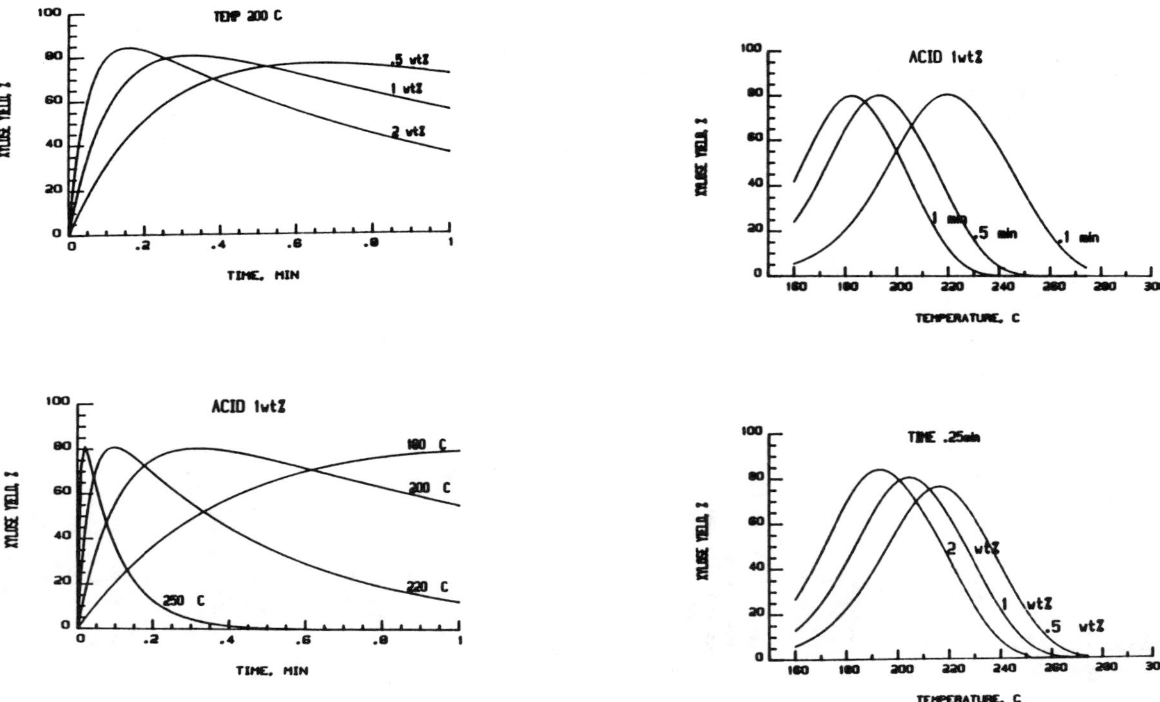

Figure 6: Xylose Yields (% of Potential Xylose) as a Function of Reaction Conditions for Acid Hydrolysis of Mixed Hardwood in a Plug Flow Reactor.

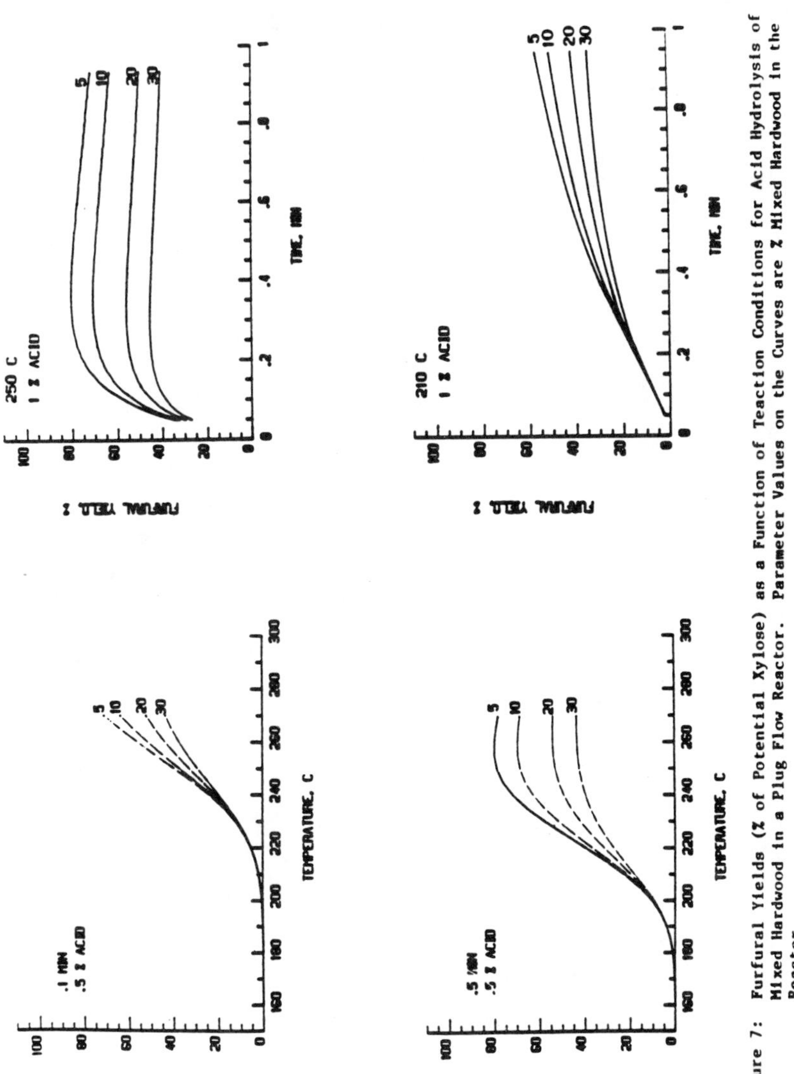

Figure 7: Furfural Yields (% of Potential Xylose) as a Function of Teaction Conditions for Acid Hydrolysis of Mixed Hardwood in a Plug Flow Reactor. Parameter Values on the Curves are % Mixed Hardwood in the Reactor.

Nomenclature

A	weight % H_2SO_4 in the aqueous phase
C	fraction unreacted cellulose (remaining potential glucose)
E_i	activation energy of reaction i cal/gr mol °R
F	fraction potential xylose as furfural, F x 100 is percent furfural yield
G, G(t)	fraction cellulose as glucose at time, t, G x 100 is percent glucose yield.
G_1	fraction cellulose as unavailable at t = 0
G_0	fraction cellulose converted to glucose at t = 0
$\hat{G}$	fraction cellulose as glucose at time, t, predicted by model in Equation 4
I	fraction potential xylose as intermediate xylose decomposition product
k_i	rate constant for reaction i, min^{-1}, reactions 0, 1, 2, 3, 4, 5, 6, and 7 defined in Reactions I and II in Figure 2
k_8	k_4k_7/k_5
n_i	dimensionless power on acid concentration in Equation (4) for reaction i
P_i	preexponential factor in Equation (3) for reaction i, min^{-1}
Q	fraction potential xylose as unreacted
Q_o	concentration of initial potential xylose in feed to reactor, g mol/L. This is equal to slurry concentration g/L x fraction potential xylose X (1/150)
t	time
X	fraction potential xylose as xylose, X x 100 is present xylose yield

References

1. Coté, W.A., editor "Biomass Utilization" NATO ASI Series Plenum Press, NY pp. 1-241, (1983)
2. Palz, W., P. Chartier and D.O. Hall eds. "Energy from Biomass" Applied Science Publ. pp. 159-275, (1981)
3. Cowling, E.B. and T.K. Kirk Biotech & Bioeng Symp No 6 95-125 (1976)
4. Chang, M. and G.T. Tsao Cellulose Chem. Tech., 15 383-395, (1981)
5. Lora, J.H. and M. Wayman TAPPI, 61 (6) 47-50 (1978) and J. Appl. Polymer Sci. 25 589-596 (1980)
6. Hajney, G.J. Res. Paper FPL 385 Mar 1981, For. Prod. Lab, Madison WI
7. Bergius, F., Ind. Eng. Chem. 29 247 (1937)
8. Scholler, H. German Patents 676,967 and 704,109, US Patent 2,188,192
9. Oshima, M. "Wood Chem. Proc. Engg Aspects" Noyes Dev Corp, NY (1965)
10. Schoenemann, J., Chim. Ind. (Paris), 80 140 (1958)
11. Battelle Memorial Institute, Geneva, Switzerland
12. Tsao, G.T., et al. Chapter 1, Annual Reports on Fermentation Processes, Vol. 2 (1978) Academic Press
13. Selke, S.M., et al. Ind. Eng. Prod. Res. Dev. 21 11-16 (1982)
14. Saeman, J.F., J.L. Bubl and E.E. Harris; Ind. Eng. Chem. Anal. Ed. 17 35 (1945)
15. Burton, R.J.; Proc. Int'l Symp. Ethanol from Biomass", Royal Society of Canada, Ottawa, p. 247 (1983)
16. Machle, K., K. Deverell and I. Challander, Ibid, p. 271
17. Saeman, J.F., Ibid, p. 231
18. Araujo Neto, H.J-S.; Studies on Wood Hydrolysis", Foundation for Ind. Tech. Av. Venezulla 82, Rio de Janero (1980)
19. Wright, J.D.; "Evaluation of Acid Hydrolysis", SERI/TR-231-2074, Sept. 1983, SERI, Golden, Co.
20. Church, J.A. and D. Wooldridge, Ind. Eng. Chem. Prod. Res. Dev. 20 371-378 (1981)
21. Rugg, B.; "Opt. NYU Continuous Cell. Hydrolysis Process", SERI/TR-9386-1, Dec. 1982, SERI, Golden, Co.
22. Stake Technology Ltd., Oakville, Ontario Canada.
23. Thompson, D.R. and H.E. Grethlein Ind Eng Chem Prod Res Dev 18 166 (1979)
24. McParland, J.J., H.E. Grethlein and A.O. Converse, Solar Energy 28 55-63 (1982)
25. Saeman, J.F.; Ind. Eng. Chem. 37 45 (1945)
26. McKibbins, S.W., J.F. Harris, J.E. Saeman, and W.K. Neil; Forest Product Jr. 12 17-33 (1962)
27. Smith, P.C., H.E. Grethlein, and A.O. Converse, Solar Energy 28 41-48 (1982)
28. Currier, P.M.; Bachelor of Engg. Thesis, Sept. 1982, Thayer School of Engg, Hanover, NH
29. Harris, J.F., J.F. Saeman, and E.G. Locke; "Wood as a Chemical Raw Material" in the Chemistry of Wood, B.L. Browning ed., Interscience Publisher, Wiley and Sons, NY, (1963)
30. Springer, E.L. and L.L. Zoch, TAPPI 51 (1968)
31. Scaringelli, F.P., B.A. Strouse and I.S. Goldstein; NSF Grant No. PFR 77-12243A01, June (1979)
32. Trickett, R.C. and F.G. Neytzell-de Wilde; CHEMSA (South Africa) Mar (1982)
33. Root, D.F., J.F. Harris, J.F. Saeman, and W.K. Neil, For. Prod. Jr. 9 158 (1959)
34. Kwarteng, I.K.; Ph.D. Thesis, July, 1983, Thayer School of Engg., Hanover, NH

ONGOING ACTIVITIES IN THE EEC COMMUNITIES IN THE FIELD OF ACIDIC HYDROLYSIS

Dr. Theodor RIEHM
Formerly Technical Director of the Wood
Hydrolysis Centre in Rheinau

SUMMARY

This report discusses the potential uses of biomass worldwide and in individual regions as a basic raw material for the hydrolysis of wood celluloses.

An initial assessment of the potential of a process can be made on the basis of the ratio between the input (chemicals and energy materials) and the alcohol yield.

A review is given of the basic chemistry of the different old and new processes, i.e. using concentrated acids, using dilute acids and temperatures below the 200° C point, and the high-speed processes, in some of which the material is comminuted. These processes are also briefly described by means of simplified materials and process flow sheets.

The different processes are summarized and compared, and it is shown that to date no new process has reached the production stage.

It is estimated, in the long term, that no fuel will be developed before the 1990s in western Europe which, with the admixture of approximately 10 % alcohol, will be comparable in price and octane rating with premium-grade petrol. In other parts of the world, where conditions are more favourable however, it, might be possible to reacty economic viability sooner.

1. INTRODUCTION

To describe all the hydrolysis processes in 30 minutes calls for this report to read like a telegram. However, they are illustrated in a series of similar diagrams. These show the main economic parameters for the processes and describe them in the same operating conditions in order to compare the critical cost factors. The emphasis is on giving an overview rather than on details.

First we take a quick look at the availability of wood celluloses throughout the world, based on the comprehensive surveys carried out by the FAO (1). Fig. 1 shows the areas, the amount of wood felled, the amount of wood growing, and the yearly production of cereals and straw in the countries of Europe and other large areas of the world. Between 1963 and 1980 the total area of woodland increased by 7 %, but the amount of wood felled increased by 59 %. It is alarming to note that in the developing countries the amount of wood growing was 125 times the amount felled in 1963, but that this figure had dropped to 84 times by 1980.

Fig. 2 shows the hitherto untapped potential of wood celluloses on the basis of three cautious assumptions:

1. Increment in closed forests = 3 t/ha per annum
2. Increment in open forests = 1.5 t/ha per annum
3. 50 % of world production of straw is not used.

(Fig. 3) The pie chart shows the proportions of the earth's surface used for different purposes (in different colours and hatchings), and the bar chart uses the same colours and hatchings to give the annual Increment figures.

These should be compared with the production figures for cereals, for root crops and sugar and also with annual consumption of crude oil, first comparing t with t oven-dry, and then, on the right, with the energy contents. Even on the basis of energy content, potential production of wood and straw is greater than oil consumption for 1979.

Wood logging 1980, forest and country areas, wood consumption						
Country	Countryside*) 10^6ha	Forest*) Area 10^6ha	Forest/Countryside %	Wood logging**) 10^6 m^3	Wood logging**) m^3/ha	Quantity of wood 10^9 m^3
Germany	24,3	7,2	29,7	33,0	4,6	
Finland	30,5	23,6	77,4	47,9	2,0	
France	54,6	14,6	26,7	30,8	2,1	
Italy	29,4	6,3	21,4	8,3	1,3	
Yugoslavia	25,5	9,3	36,5	16,7	1,8	
Norway	30,8	8,3	27,0	8,8	1,1	
Poland	30,5	8,7	28,5	21,4	2,5	
Rumania	23,0	6,3	27,4	21,3	3,4	
Sweden	41,2	26,4	64,1	53,4	2,0	
Spain	50,0	15,3	30,6	11,7	0,8	
Europe (exclud. Turkey)***)	472,8	154,7	32,7	334,3	2,2	12
Africa	2964,7	636,6	21,5	433,9	0,66	3,8
North America } South America }	3889,1	1636,9	42,1	530,9 } 315,2 }	0,5	44 78
Asia (includ. Turkey)	2676,9	553,1	20,1	1017,1	1,84	17
Pacific Region	842,9	155,5	18,5	32,9	0,2	3,8
USSR	2227,2	920,0	41,3	356,0	0,28	79
World	13073,6	4056,7	31,0	3020,3	0,74	238
Data from FAO 1963	13034	3779	29	1900	0,46	
Variation 1963-1980	+0,3%	+7,4%		+59%		

Soft timber logging		1171,8 . 10^6 m^3	38,8%
Hard timber logging		1750,9 . 10^6 m^3	58,0%
Total:		2922,7 . 10^6 m^3	96,8%
Consumption:	Saw timber and veneer timber	841,5 . 10^6 m^3	27,9%
	Crown trees	34,6 . 10^6 m^3	1,1%
	Pulp wood	341,4 . 10^6 m^3	11,3%
	Other industrial wood	176,0 . 10^6 m^3	5,8%
	Industrial wood	1393,5 . 10^6 m^3	46,1%
	Fuel wood + charcoal	1626,8 . 10^6 m^3	53,9%
	Total consumption	3020,3 . 10^6 m^3	100,0%

*) From (152b);
**) From (152c);
***) This figure includes countries with less than 5000 ha forest area.

CROP-AREAS-YIELDS 1979

(FAO - production yearbook 1979 Vol. 33, Tables 9 - 17)

Country/Region	Area 1000 ha	Yield 1000 t/a	%	Yield t/ha	Straw *) 10^6 t/a
Germany	5207	22687	1,5	4,4	25,8 (1978)
France	9835	43768	2,8	4,5	48,1
Italy	5069	17753	1,1	3,5	19,5
Great Britain	3870	17302	1,1	4,5	19,0
Hungary	2921	12087	0,8	4,1	13,3
Poland	7872	17341	1,1	2,2	19,1
Rumania	6375	19275	1,2	3,0	21,2
Spain	7171	13830	0,9	1,9	15,2
Yugoslavia	4361	15650	1,0	3,6	17,2
Remaining countries	below	10000			
Europe (excluding USSR)	70442	239984	15,5	3,4	264,0
Africa	72389	66480	4,3	0,9	73,1
North America	98562	356703	23,0	3,6	392,4
South America	37743	63602	4,1	1,7	70.0
Asia	343972	629984	40,6	1,8	693,0
Oceania	16475	24312	1,6	1,5	26,7
USSR	121313	172011	11,1	1,4	189.2
World	760896	1553076	100	2,0	1708,4
By Cereal (World)					
Wheat	238723	425478	27,4	1,8	468,1
Rice	145268	379814	24,5	2,6	417,9
Maize	120540	394231	25,3	3,3	433,6
Barley, Rye, etc.	256365	353545	22,8	1,4	388,8
Grand Total	760896	1553076	100	2,0	1708,4

*) Calculated with 1,1 t Straw/t corn. Straw with 85% atro.

FIGURE 1 WORLDWIDE POTENTIAL ON LIGNOCCELOSES (10 ^{6}/A ATRO)

		FOREST AREA 1963 (FAO)					
		Managed		Not Managed		Together	
		2,240 10^6 ha		1,539 10^6 ha		3,779 10^6 ha	
		10^6t atro	t/ha	10^6t atro	t/ha	10^6t atro	t/ha
1	Growing stock 1963 (1 m^3 = 0.47 t atro)					111 860	29.6
2	Increment *	6 720	3.0	2 308	1.5	9 028	2.4
3	Removals 1980	1 420	0.63			1 420	0.38
4	Cereals 1979	1 553	2.0			1 553	2.0
5	Straw (calculated with 85% atro)	1 451	1.9			1 451	1.9
6	Straw + cereals	3 004	3.9			3 004	3.9
7	Sum of increment (2 + 6)	9 724		2 308		12 032	
8	Minus industrial wood + 50%	1 037				1 037	
9	Minus cereals + 50 % straw	2 278				2 278	
10	Potential on ligno-cellulose	6 409 = 66% of 7		2 308		8 717 = 72% of 7	

* from V.P. KARLIVAN at CHEMRAWN I Toronto July 78 3 - 8 m^3/ha
from H. LOFFLER, paper for discussion 8.12.80 for BMFT 3 - 3.2 m^3/ha comprising tree-crown and branches, but without herb and bush stratum, without roots.
It was set 3 t/ha for managed and 1.5 t/ha for not managed forest area.

FACTORS INFLUENCING THE UTILIZATION OF LIGNO-CELLULOSE

Negative Aspects	Positive Aspects
Land areas should not be increased	
Population growth, thus wood consumption in existing sectors about + 2.7%/a	Increment of yield because better forest management plant breeding (2 times yield per ha can be expected)
Raising of food consumption, thus larger cultivated areas and less forest areas	Further yield increment by plant breeding and manure however, the amount of straw increases with the increment of cereals production
Ecologic problems, especially in using tropic virgin forest	With fast growing plantation about 10 t/ha wood yield
At present, already important lack of energy in developing countries	In Table 5 are not included : Industrial wood residues : saw dust, slab Recycling of waste paper, pit wood, building and furniture wood etc. Cellulosic waste Other plant residues as : cotton stalks, fruits peel and shell etc. Regional available bagasse, because of its relevant carbohydrates content, not recommended for burning

FIGURE 2 WORLDWIDE POTENTIAL ON LIGNOCCELOSES (10 ^{6}/A ATRO)

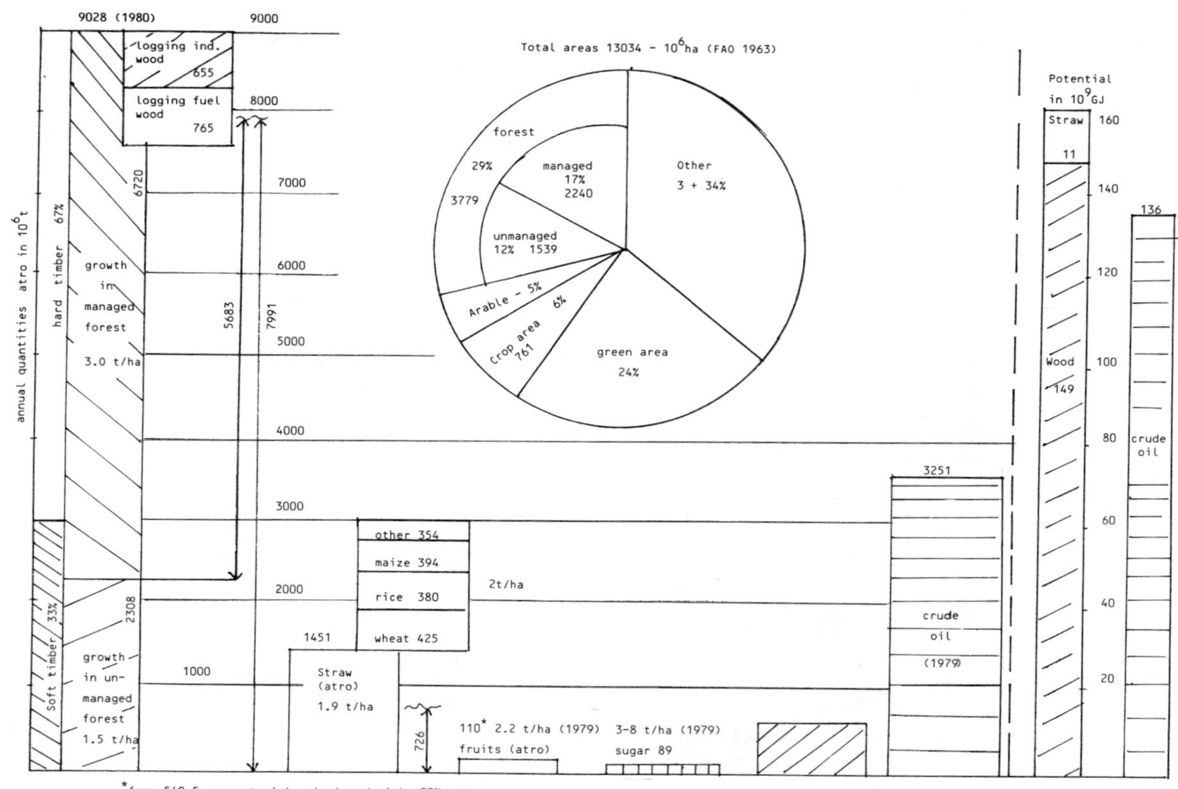

*from 549.5 raw material calculated with ~20% atro

FIGURE 3 WORLDWIDE GROUTH OF BIOMASS AND THE ANNUALY QUANTITIES TODAY OTHERWISE NOT USED

It is certain that the large quantities of ligno-celluloses which have not so far been used can be brought into use in new applications only slowly and only in part. Naturally, all general ecological requirements must be met when this is done. In the long term, however, further increases in world population will make it imperative for man to use this potential raw material.

Fig. 4 shows the chemical composition of the different types of lignocellulose. Note should be taken of the basic difference between the composition of the hemicelluloses which occur in coniferous woods, which have few pentosans, and those occurring in deciduous trees and all annuals, which have a large number of pentosans and few hexosans. This is of great importance in the later stages of sugar processing.

2. The main cost factors in the extraction of alcohol from wood celluloses

Fig. 5 gives a breakdown and comparison of production costs for three typical alcohol processes.

1. (right) alcohol from 'B' sugar beet (2)
2. (left) alcohol extracted from wood by the Scholler method, a proven process which has been used for decades in industry (3).
3. (centre) the HF process, as an example of a new development. So far this has been tested in the laboratory only. It would appear to offer the most favourable and certified production costs.

All processes are calculated for the same size of production unit, i.e. 60 000 m^3 pure alcohol/year using a factor $f^{0.65}$ for all costs linked to capital and a factor $f^{0.2}$ for labour costs.

Although this comparison produces a margin of uncertainty of approximately 10 % in the absolute figures of DM 1.13 - 1.86 per litre of pure alcohol, the percentage breakdown of costs certainly contain only minor errors.

≈2%
≈19% Lignin
≈29% Pentosane
≈50% Hexosane
13,6% Salts etc.
22,7% Lignin
≈1,4% Hexosane
21,9% Pentosane
≈40% Hexosane
10-12% org. acids salts, tannine
Lignin 20-25%
Hemicellulose 22-35%
Hexosane
Pentosane 20-30%
Cellulose 28-48%
pure cellulose
Holocellulose
extraktfreie Holztrockensubstanz
Holztrockensubstanz
4-5% org. acids usw.
Lignin 28-31%
Hexosane 12-15%
Pentosane 8-10%
Hemi-cellulose 20-25%
Cellulose 40-48%
BAGASSE
WHEAT STRAW
HARD TIMBER
SOFT TIMBER

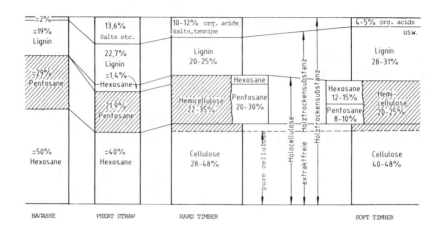

FIGURE 4 COMPOUND OF MOST IMPORTANT TYPES OF LIGNOCELLULOSES

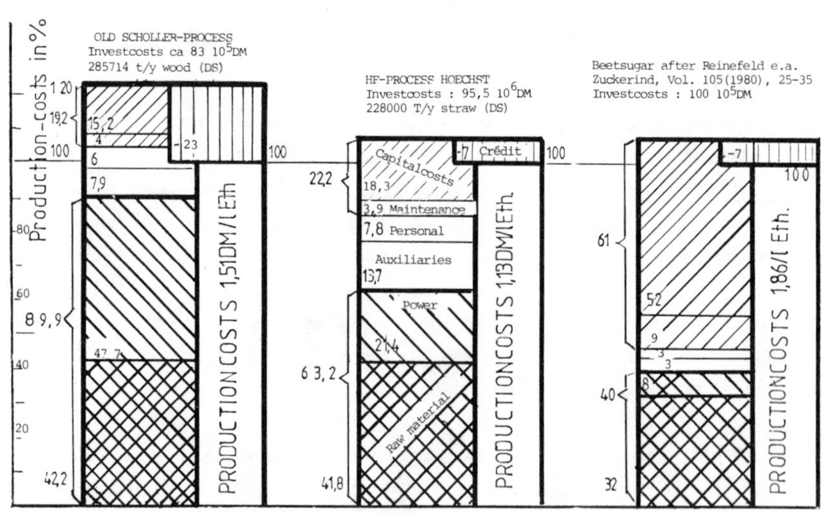

FIGURE 5 PRODUCTION COSTS OF ETHANOL : VARIOUS PROCESSES, ALL CALCULATED WITH 60 000 M3 ETHANOL/YEAR

Process	Scholler	HF process
Raw material costs	42,2 %	41,8 %
Energy costs	47,7 %	21,4 %
Raw materials and energy	89,9 %	63,2 %
Lignin (credit side)	23,0 %	6,9 %
Intermediate total	66,9 %	56,3 %
Auxiliary materials	7,9 %	13,7 %
Labour	6,0 %	7,8 %
Maintenance	4,0 %	3,9 %
Capital costs	15,2 %	18,3 %
Total production costs	100,0 %	100,0 %

Raw material and energy costs together account for either 90 % or two-thirds of the total production cost, depending on the process. The other cost components are small in comparison, and it may be assumed that this is also true of the other hydrolysis processes, which lie somewhere between the two processes illustrated above as far as their basic technology is concerned. Even when lignin as fuel is entered on the credit side, the proportion of total costs represented by raw materials and energy still ranges from well over 50 % to almost two-thirds of the production costs.

However, most of the energy costs (approximately 85 - 90 %) are the costs for raising steam, which depends on the amount of water involved in the processes. This is shown in blue and green or in diagonal hatching respectively in the materials flow sheet. The figures for these product flows are a preliminary indication of the efficiency of a process.

1. The raw materials costs (given the same price for raw materials) are expressed only in relation to the alcohol yield per tonne oven-dry of raw material or, expressed as a reciprocal value, i.e. raw material input in tonnes oven-dry per hl pure alcohol.

2. When a central steam energy unit is used, which is in any case imperative for plants on this scale, the primary energy vector, can be the same as the basic chemical material (wood cellulose). In remote districts where oil, natural gas or coal is not available there is in any case no other solution. Thus a figure for tonnes oven-dry of energy raw material per hl pure alcohol can also be calculated.

The sum of these two figures gives a process factor which represents between half and two-thirds of the production costs. This gives an initial comparison and assessment of the potential commercial viability of the different processes.

3. Concentrated acid processes

The quickest way to explain all the processes is by material flow sheets which are all to the same scale, starting top left with 1 000 kg oven-dry raw material in each case - represented as a black line (all organic substances) - leading up to the alcohol, lignin and the by-products. Water is represented by blue or diagonal hatching, and acid by green of again diagonal hatching. It is assumed that, in the case of deciduous wood, straw and other annuals that the pentoses can be fermented with 30 ml/100 g xylose.

On the right of each process a simplified production flow sheet is given, providing a general technical explanation of the process without going into long descriptions.

The fundamental chemical principle in the Bergius-Rheinau process (Fig. 6) is the hydrolysis of the cellulose and hemicellulose with concentrated hydrochloric acid at ambient temperature. It operates practically quantitatively and there are no losses, which is a great advantage of the process (4).

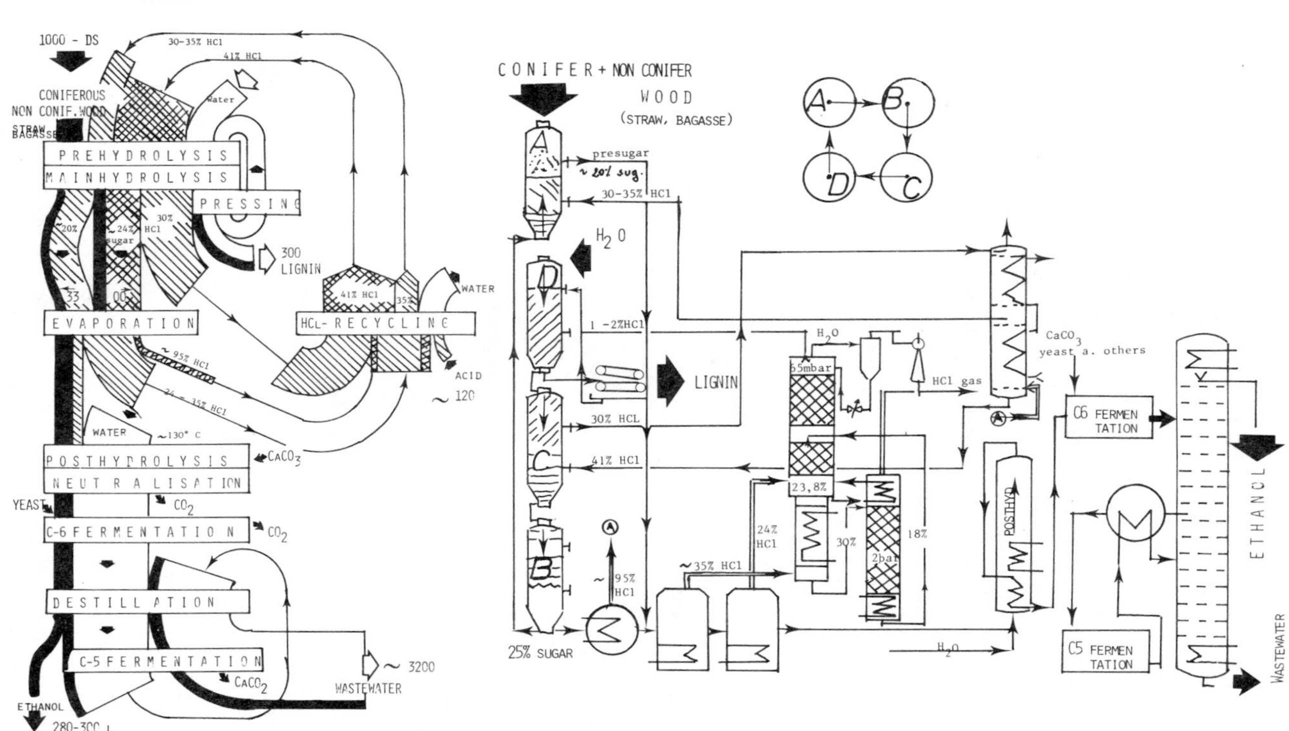

FIGURE 6 BERGIUS-RHEINAU-PROCESS MODIFIED SYSTEM RIEHM

All types of wood can be processed. At present straw, bagasse and other annuals can only be used under certain conditions. In addition to the 3 200 kg of water to be converted into steam (blue-below) large quantities of acid (green - centre) must be evaporated and concentrated. This is reflected in the energy costs and involves a comparatively complex technical plant (diagram on right). The wood loading, pre-hydrolysis, hydrolysis, sugar extraction and washing out of the lignin is carried out in four narrow towers (5). Through these towers flows a continual stream of a multi-phase liquid, beginning with water at D, and acquiring 41 % acid at C (middle), at B (below) a 25 % sugar solution is extracted and at A a 35 % acid is addedand pre-sugar solution is extracted. Only the lignin is discharged batchwise. The strong heat discoloration and the condensed water content of the raw material are extracted as a first run with the pre-sugar-solution extraction through the lateral outlets at A, so that there is no longer any need to dry the wood beforehand as was previously the case.

For details of the acid recovery process reference should made to more detailed literature (6).

The fermentation of the pentoses produces only very dilute solutions of alcohol (approximately 2 %), and this is the same for all processes. However, when the C6 and C5 fermentation processes are aligned as illustrated, with a total sugar concentration (before the C6 fermentation process) of approximately 15 %, the steam consumption for the alcohol concentration process increases by only a negligible extent compared with that required for the C6 fermentation process alone at the same sugar concentration.

Fig. 7 shows on the left the different hydrolysis processes for hemicellulose and cellulose with different hydrochloric acid concentrations, and on the right one of the many physico-chemical digrams needed for an understanding of the acid recovery process.

In Rheinau alcohol was produced only in the early 1930s, and latterly up till 1960 only crystalline glucose - the first xylito in the world to be produced on an industrial scale - and lignin-modified phenolic resins for muldings and foundry sands. The alcohol yield can be calculated accurately from the established sugar yield. The major parameters for all processes are summarized later on in this document.

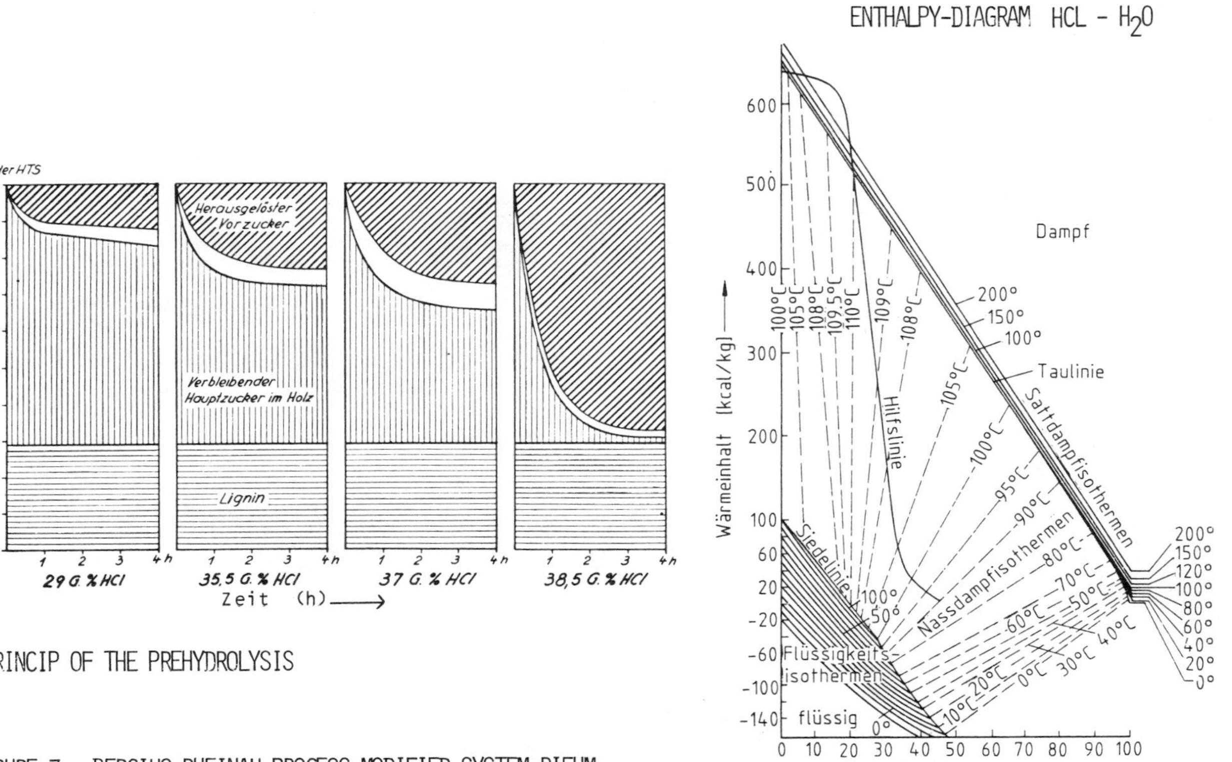

FIGURE 7 BERGIUS-RHEINAU-PROCESS MODIFIED SYSTEM RIEHM

According to FIAT reports, in the USA Goldstein (7) used a mainframe computer to optimize all the parameters in the Rheinau process, but even with a very large plant (350 000 t/a) he did not achieve any significant reductions in production costs.

In Japan Dr. Oshima of the Noguchi Institue hydrolysed wood in a research plant using gaseous HCl (8). Severe technical problems were encountered because material aggregated in a type of fluidized bed.

Recently, $CaCl_2$ + HCl have been proposed for hydrolysis (9). The capacity of the $CaCl_2$ to reduce the H_2O part-pressure allows to use smaller HCl-concentrations (219 g/l HCl and 159 g/l $CaCl_2$). In itself this is an advantage, but the drawback is that it is difficult to separate the sugar from the $CaCl_2$, both of which are readily soluble in water and neither of which can be distilled.

The basic chemical principle of the Phenol-HCl (Battelle Geneva) process is the preliminary separation of three components (10) at approximately 100° C. To one part ligno-cellulose are added 6 parts of a 40/60 Phenol/H_2 O solution, to which a little H_2SO_4 or HCl is added. After a short time the lignin and the hemicellulose go into solution (the prehydrolysis). The cellulose remains behind. This is washed out and pressed and a wuality, comparable with other celluloses. It can be hydrolyses (without drying) with gaseous HCl, and recently also with 41 % acid, and then acid recovery as in the Rheinau process.

When the prehydrolysis solution is cooled to approximately 30° C it separates into an organic phase containing lignin and phenol, and an aqueous phase with the semi-sugars and acid. The latter are then continues with the hexoses, as described for the Rheinau process, to produce alcohol. Phenol is distilled off from the lignin and a number of phenols are also recovered from the lignin by hydrocracking, which compensate to some extent for the losses. The residual phenols are sold. All types of wood, straw and other annuals can be processed. The process gives a yield practically the same as in the Rheinau process. At present plans are being drawn up for a pilot plant by an Italian alcohol-producing company.

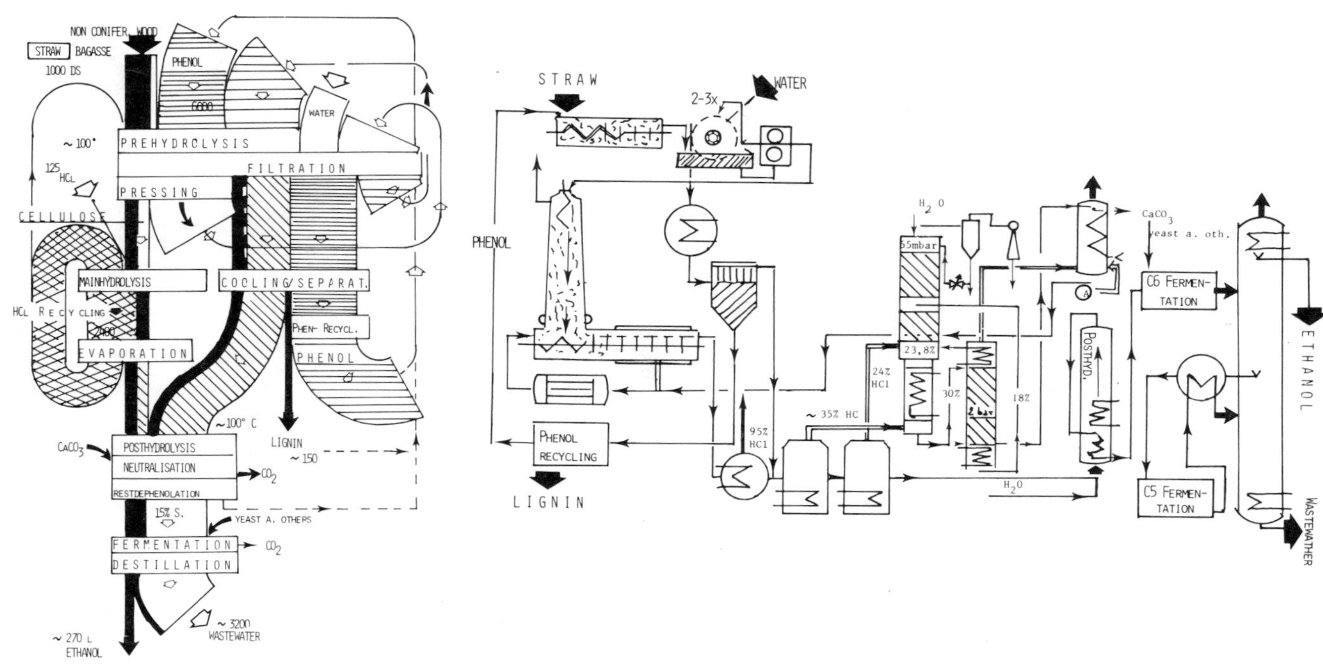

FIGURE 8 PHENOL- HCL - PROCESS

MATERIAL FLOW SHEET

1000 DS
STRAW BAGASSE
CONIFEROUS WOOD
NON CONIFER. WOOD
10
HF
DRYING
HF-ABSORBTION + HYDROLYSE ~40-50°
~1600 Σ
DESORPTION→80°C
H_2SO_4 H_2SO_4
POSTHYDROLYSIS 130°C
$CaCO_3$
NEUTRALISATION+FILTRATION
PRESSING
CO_2
14-15% SUGAR
YEAST
LIGNIN ~ 180
C6-FERMENTATION
CO_2
DESTILLATION
C5-FERMENTATION
$CaCO_2$
Wastewater
3200
ETHANOL 248 L

PROCESS FLOW SHEET

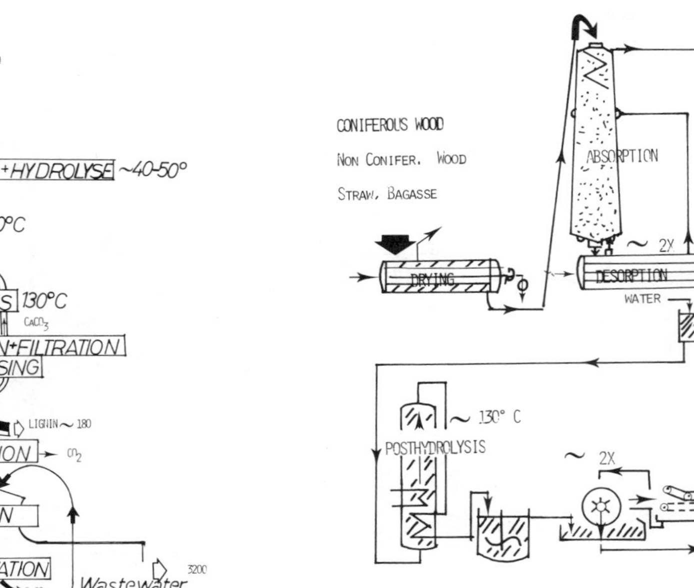

FIGURE 9 HOECHST - HF - PROCESS

A major criterion is whether the phenol recirculation can be achieved satisfactorily from the technical and economic points of view.

The basic chemical principle of the HF (Hoechst) process (11) is to hydrolyse cellulose and hemicellulose with HF gas at approx. 20-50° C, which is similar to the HCl-process. However, only approximately 0.6 parts of HF are required to one part of wood cellulose. This means that the energy input is far below that for the Rheinau process. The raw material must initially be dried to a residual moisture content of approximately 5 % before hydrolysis. Prehydrolysed material, and also all types of wood, straw and annual plants can also be processed. Absorbed HF is described again when the temperature of the hydrolised material is raised. The atmosphere kept out since the fresh material, which absorbs all the HF quantitatively, is introduced in the opposite direction to the HF flow, so that there are no traces of HF in the exhaust gases.

After the posthydrolysis of the sugar solution the lignin is filtered off and pressed. The rest of the process is the same as for the Rheinau process. Hoechst AG has now completed the laboratory development work. Plans had been drawn up for a pilot plant, but unfortunately work on constructing this has been stopped for reasons not connected with the process.

3.) The Scholler and similar processes

Fig. 10 shows the basic chemical principles of the processes using dilute acid which are still used today. In a simple autoclave test (bottom left) yields of only slightly more than 30 % of the potential value are obtained after 1 1/2 hours reaction time. The differences from the discontinuous upper curve represent the decomposition of the sugar. After that, there is a further drop in yield (upper left, continuous curve). Scholler solved the problem of the high degree of sugar decomposition by applying "percolation" (upper right), a feature of which is that the hydrolysis reaction time is independent of the residence time of the sugar in the reactor. The faster the percolation, for the same duration of the hydrolysis - i.e. the more water is pumped from above and solution drawn off below - the lower the degree of sugar decomposition, but the sugar

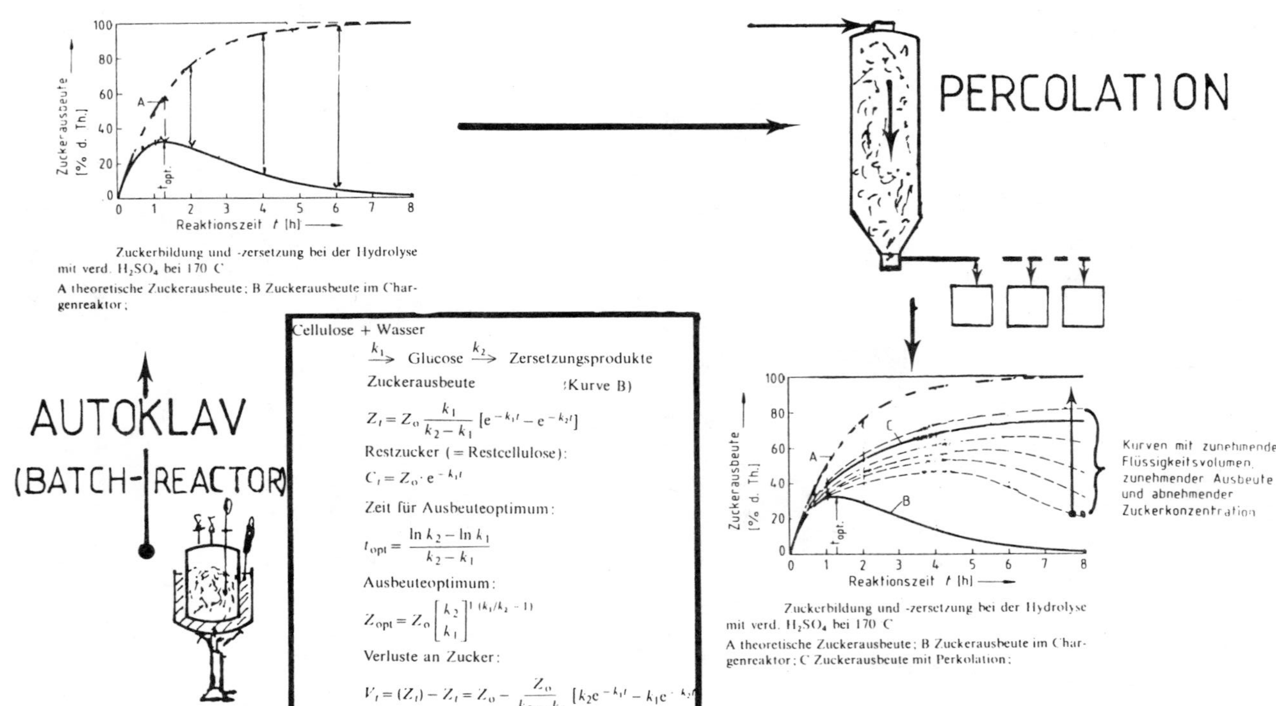

FIGURE 10 THE PRINCIPLES OF THE SCHOLLER - (AND LIKE) PROCESSES

concentration is also reduced, as shown by the series of curves (upper right). The rules and formulae (centre below) established by Scholler in 1923 were confirmed by Saemann (12) in 1945. Only slight correction was necessary, and they are also valid as they stand for all the modern processes in Chapter 4.

Following Scholler (13), a large number of hydrolysis plants were operated on the industrial scale for many years, and many plants are still operating today in Russia. A compromise between yield and dilution gave an approximately 4 % sugar concentration in practice. The corresponding flow of liquid (9 500 kg per t. oven-dry matter in Fig. 11 (left) is thus very broad. Normally the material from all the batches was fermented to produce alcohol. Later the initial batches (i.e. prehydrolysis) were also processed separately to produce furfural or yeast. The raw material was conifer wood. A limited amount of deciduous wood may be added, but this makes percolation more difficult. Straw and annuals cannot be processed.

The Madison process (14), shown in Fig. 11 right, differs very little from the Scholler process. As the liquid is eliminated continuously (and not batch-wise as in the Scholler process) the batch time is shorter, i.e. between 3.5 and 5 hours instead of 12 hours. Yields and concentrations are practically the same.

A comparable process is the T.V.A. process (15), which produces feed sugar instead of alcohol.

The most recent pilot plant in New Zealand (16) also operates in much the same way. The last dilute yields of hexoses are recovered for the initial percolation processes (i.e. prehydrolysis), which raises the average sugar concentration to 60 g/l.

4.) Short-time hydrolysis processes using dilute sulphuric acid

The short-time high-temperature processes developed after the war are based on the same physical and chemical principles as Scholler's process, but call for a totally different technology. As early as 1945 Saemann had discovered that, as conditions (temperature and acid concentration) became more critical, sugar formation increased at the expense of sugar

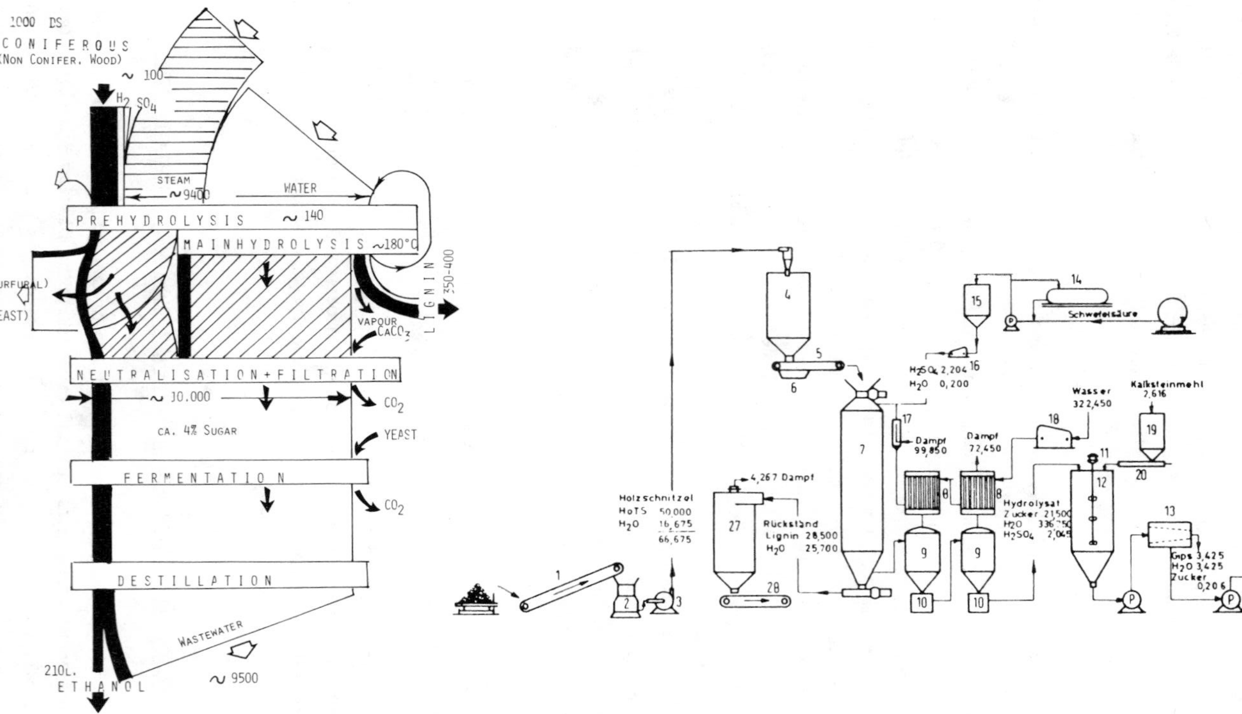

FIGURE 11 SCHOLLER - (AND LIKE) PROCESSES

decomposition. Fig. 12 left shows the k_1 and k_2 values on the logarithmic scale plotted against 1/T, as well as the k_1/k_2 values, which increase with higher temperatures and higher acid concentrations. The maximum yields from the autoclave test or from a continuous flow of homogeneous material also increase in the same way (above right). However, the optimum reaction time also falls progressively, and some sharp peaks of a few seconds' duration occur when the calculated yield is almost 70 % to the potential value. Percolation (to increase the yield still further) is not technically feasible. In this process it is of the greatest importance to keep exactly to the reaction time, with very short warming-up and cooling times.

The most detailed research into these physical and chemical basic principles was carried out by Grethlein (17) using laboratory apparatus (Fig. 13 right). He subjected a suspension of sawdust and dilute sulphuric acid to rapid heating with direct steam, pumped it along a pipe and then cooled it quickly. He achieved yields exceeding 60 % of the potential value (see table) and thus confirmed the measurements made by Seamann of the kinetics of the reaction.

Werner Pfleiderer USA (18) evaluated these data in a larger pilot plant over a period of several months. A double screw extruder (Fig. 14 right) was used for the reaction. The residence time varied between 6 and 20 seconds at 1 to 1.5 % H_2SO_4 and 220 to 240° C. Yields of 50 to approximately 55 % of the potential value were obtained with an 8-10 % sugar concentration. Accordingly, the amount of liquid accompanying the sugar, at 3 600 - 4 500 kg/t oven-dry matter, was considerably less than in Scholler's process (Fig. 14 left). All types of wood, straw and annuals could be processed, as well as waste paper. However, it is obvious that there were scale-up and corrosion problems with the complicated extruder. Plans to build an industrial plant did not go through and the pilot plant was closed down.

Reitter-Zellplan, Munich (19) divided the short-time hydrolysis process into three phases, the first two being carried out to the left of the optimum yield peak (Fig. 12). The sugar loss is relatively low in this process, so that higher total sugar yields could be achieved with the short-time hydrolysis process. Only the last phase operates at optimum reaction time.

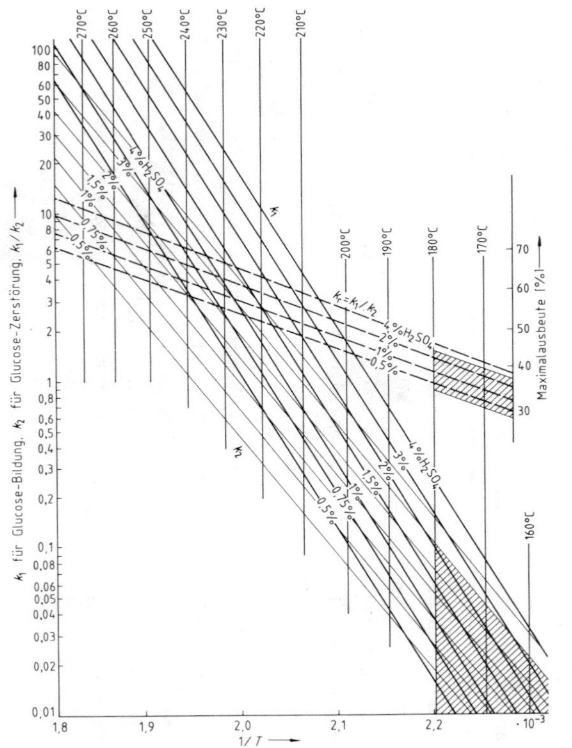

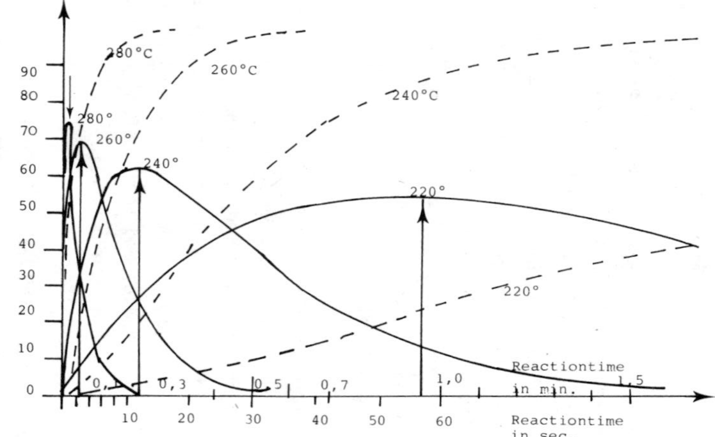

FIGURE 12 PRINCIPLE OF THE SHORTTIME HYDROLYSIS

Glucose and Xylose Yields from Mixed Hardwood at Selected Reaction Conditions

Temp. °C	Time min.	% Acid	% Glucose	% Residual Glucose	% Xylose	Residual Xylose
201	0.562	0.58	14.8	86.9	90.8	[illegible]
222	0.532	0.54	38.6	44.9	60.0	0.0
204	0.195	1.17	11.4	86.1	85.7	9.0
241	0.174	1.02	53.1	13.2	29.1	0.0
190	0.149	2.21	9.0	94.0	92.3	9.5
203	0.145	2.21	15.1	89.0	95.8	[illegible]
242	0.126	1.96	56.6	18.7	32.5	0.0

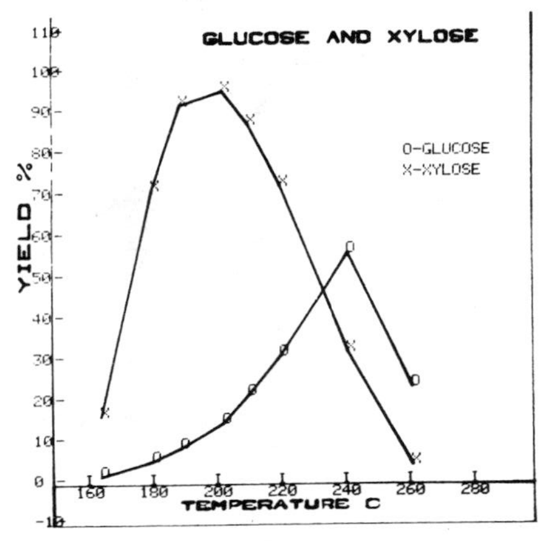

Glucose and xylose yields for acid hydrolysis of mixed hardwood in plug flow reactor 2% acid and 0.14 min.

FIGURE 13 GRETHLEIN-SYSTEM (USA)

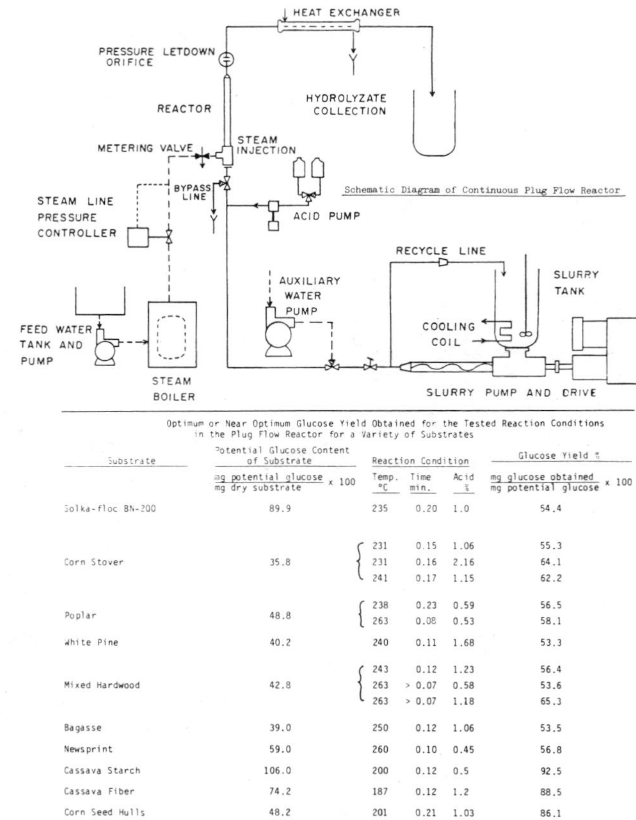

Schematic Diagram of Continuous Plug Flow Reactor

Optimum or Near Optimum Glucose Yield Obtained for the Tested Reaction Conditions in the Plug Flow Reactor for a Variety of Substrates

Substrate	Potential Glucose Content of Substrate: $\frac{\text{mg potential glucose}}{\text{mg dry substrate}} \times 100$	Reaction Condition: Temp. °C	Time min.	Acid %	Glucose Yield %: $\frac{\text{mg glucose obtained}}{\text{mg potential glucose}} \times 100$
Solka-floc BN-200	89.9	235	0.20	1.0	54.4
Corn Stover	35.8	231	0.15	1.06	55.3
		231	0.16	2.16	64.1
		241	0.17	1.15	62.2
Poplar	48.8	238	0.23	0.59	56.5
		263	0.08	0.53	58.1
White Pine	40.2	240	0.11	1.68	53.3
Mixed Hardwood	42.8	243	0.12	1.23	56.4
		263	> 0.07	0.58	53.6
		263	> 0.07	1.18	65.3
Bagasse	39.0	250	0.12	1.06	53.5
Newsprint	59.0	260	0.10	0.45	56.8
Cassava Starch	106.0	200	0.12	0.5	92.5
Cassava Fiber	74.2	187	0.12	1.2	88.5
Corn Seed Hulls	48.2	201	0.21	1.03	86.1

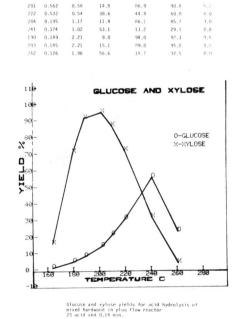

MATERIAL FLOW SHEET

CONIFEROUS WOOD
NON CONIF. WOOD
WASTE-PAPER
20 H_2SO_4
STEAM
WATER
HYDROLYSIS 220°C 0.5% H_2SO_4 20 SEC.
1:3 35 BAR
VAPOUR
$CaCO_3$
WATER
NEUTRALISATION + FILTRATION
8-10% SUGAR YIELD 60% TH
3600 | 4500
LIGNIN
CA 180 (?) L. ETHANOL

PROCESS FLOW SHEET

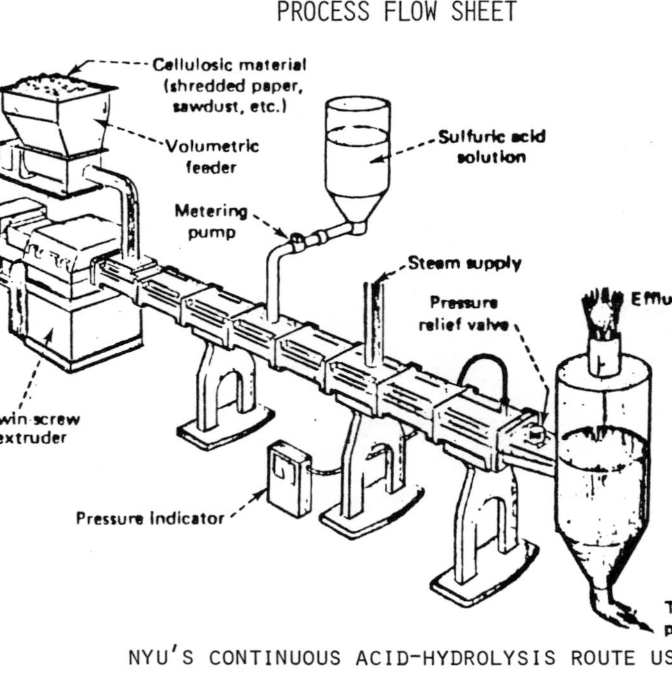

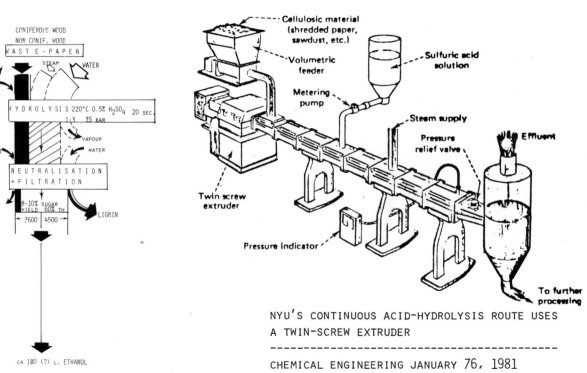

NYU'S CONTINUOUS ACID-HYDROLYSIS ROUTE USES A TWIN-SCREW EXTRUDER

CHEMICAL ENGINEERING JANUARY 76, 1981

FIGURE 14 WERNER-PLEIDERER (USA)

However, after each phase the hydrolysed material must be washed out in screen-belt presses or similar equipment, to ensure that the sugar formed is not further affected by acid or high temperatures (Fig. 15 right). However, in the large pilot plant in Vienna, which comprises equipment which is already on an industrial scale, problems were encountered with the filtering and washing out, as well as with the wide spectrum of reaction-times in the screw reactor, and these problems have still not been solved. Trials are now being carried out using only one main (cellulose) reactor following the initial pre-hydrolysis phase, the hexoses formed being fermented together with the lignin as a suspension.

All types of wood, straw and other annuals can be used as raw material. With regard to Fig. 15 left it should be noted that the materials flow sheet (original supplied by Zellplan) is not completely indentical in scale and representation with the other diagrams.

The summary tables gives the figures for a projected Zellplan plant in the USA. Objective and very thorough control trials to check the yield have so far failed to achieve these values.

Knauth (21) carries out the complete prehydrolysis, as does Scholler, in one and the same apparatus and using similar temperatures and acid concentrations, i.e. the steam treatment, addition of acid, hydrolysis and washing out (extraction). He differs from Scholler in that the process is fully continuous (Fig. 16 right). A low solids/liquid ratio gives a higher sugar concentration (approximately 80 g/l), and an extraction yield of approximately 85 % gives a sugar yield of approximately 20 % calculated on oven-dry weight of birch wood. Under more critical reaction conditions the pentoses can be converted into furfural. Straw and annuals cannot be used in this equipment.

The main (cellulose) hydrolysys requires a second reactor (autoclave). Further acid is added and the material is ground in refiner. It is then processed in the autoclave for 25 minutes at 185° C. Increased yields of hexoses can be obtained by processing the material twice in the autoclave (see Reitter). To date the filtering and washing out of the sugar solution can only be carried out in filter presses.

MATERIAL FLOW SHEET

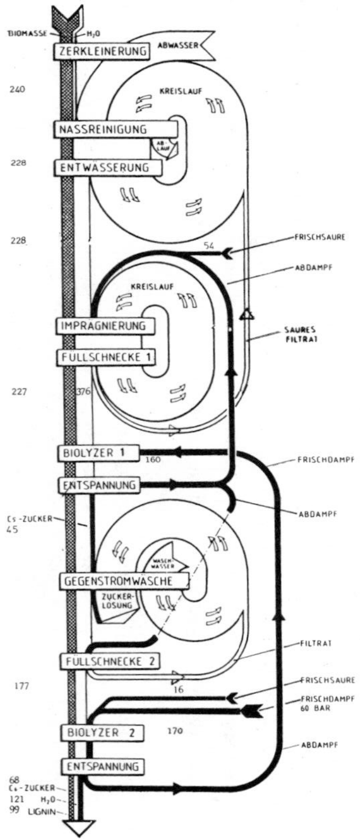

BIOMASSE
H_2O
ZERKLEINERUNG
ABWASSER
240
KREISLAUF
NASSREINIGUNG
ABLAUF
228
ENTWASSERUNG
228
54
FRISCHSAURE
ABDAMPF
KREISLAUF
SAURES FILTRAT
IMPRAGNIERUNG
FULLSCHNECKE 1
227
376
BIOLYZER 1
160
FRISCHDAMPF
ENTSPANNUNG
ABDAMPF
C_5-ZUCKER 45
WASCHWASSER
GEGENSTROMWASCHE
ZUCKERLOSUNG
FILTRAT
FULLSCHNECKE 2
177
16
FRISCHSAURE
FRISCHDAMPF 60 BAR
BIOLYZER 2
170
ABDAMPF
ENTSPANNUNG
68 C_6-ZUCKER
121 H_2O
99 LIGNIN

PROCESS FLOW SHEET

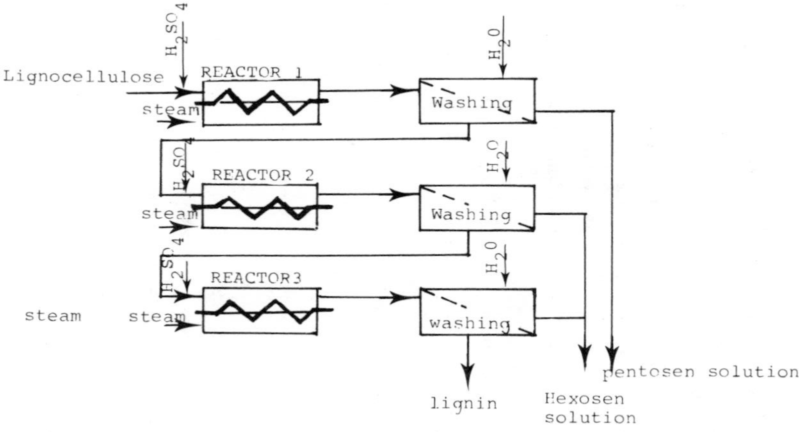

Lignocellulose
H_2SO_4
steam
REACTOR 1
H_2O
Washing
H_2SO_4
steam
REACTOR 2
H_2O
Washing
H_2SO_4
steam
steam
REACTOR3
H_2O
washing
lignin
Hexosen solution
pentosen solution

FIGURE 15 REITTER-PROCESS 1982

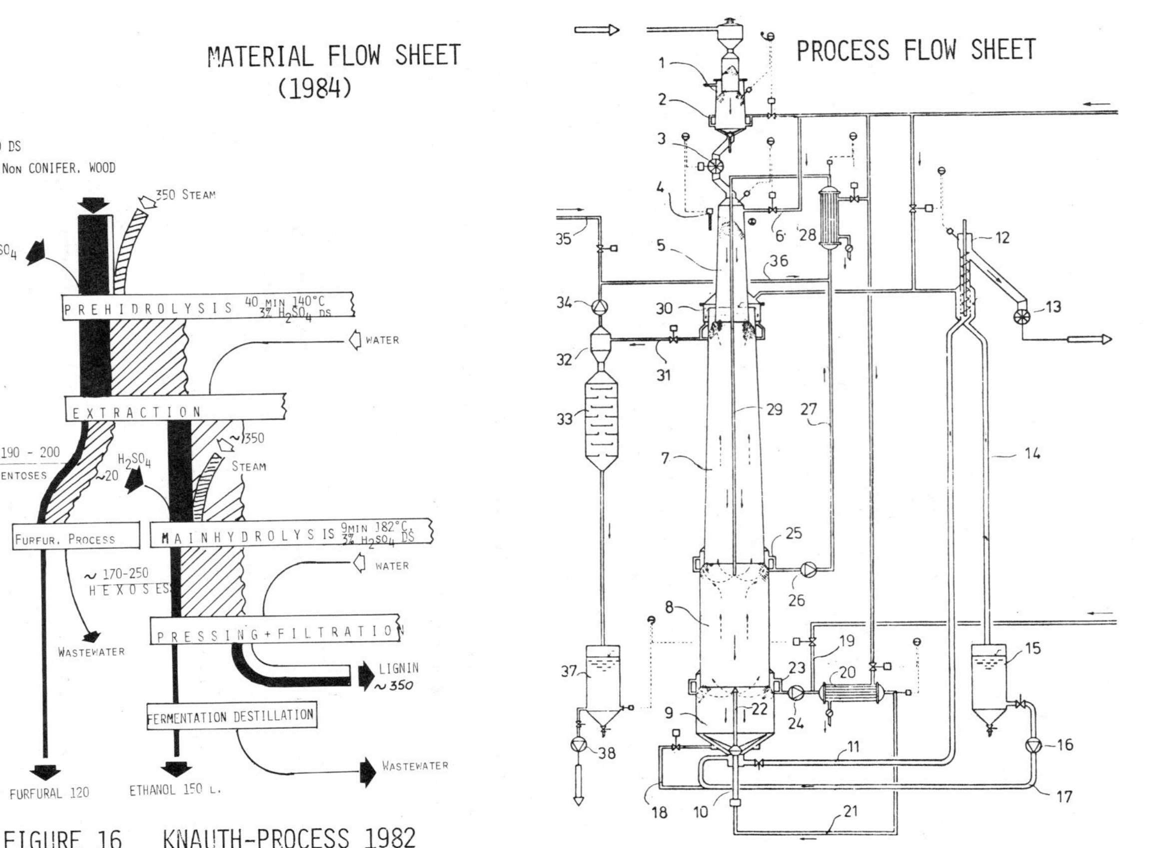

FIGURE 16 KNAUTH-PROCESS 1982

A new feature compared with the processes described before the comminution. In 1945 Saemann was unable to establish any significant differences in the kinetics of the hydrolysis process when particles of 0.075 - 0.75 mm were used. However, when very fine particles are used the reaction speed does appear to increase considerably. This appears to be the only explanation of the results obtained by Knauth with a reaction time of only 25 minutes at 185° C. The data available to me did not enable me to construct a material flow sheet (fig. 16 left) of adequate accuracy for the main hydrolysis process.

The figures given in the provisional summary tables are, as in the case of Reitter, theoretical. A control trial was recently carried out to produce firm figures, but it has not yet been evaluated.

For the salve of completeness, Fig. 17 shows the hydrothermolysis process of Bobleter, Innsbruck (23). Only water is used (plus the acetic acid produced). The laboratory equipment is apparently similar to that used by Grethlein, but with the major difference that, in Bobleter's process, percolation takes place three times at different temperatures, i.e. at 180° C, when the hemicelluloses go into solution, at 280° C, when the cellulose goes into solution, and lastly at 340° C, when the lignin also goes into solution. The solution remains in the reactor for approximately 45 seconds. The lignin obtained has mean molecular weight of approximate three phenyl propane units. The apparatus is still very small and the sugar concentrations very low.

The Buckau-Wolf design (23) can be described as extreme short-time hydrolysis. It is expected to operate in the same way as Grethlein's process, but the material will be comminuted beforehand. This process will use a 'Supraton' dispersing and reaction unit which has also proved useful for hydrolysing starch (Fig. 18). It is planned to construct a pilot plant in order to provide final confirmation of the effect of comminution.

The Tampella process (24) also uses the increased reaction speed offered by very small particles (Fig. 19 right). After the material has passed through a screw reactor, as in Reitter's process, the pressure suddenly falls and flash evaporation at the core of the material breaks down the particles, which then have a large inner surface area. The material passes through.

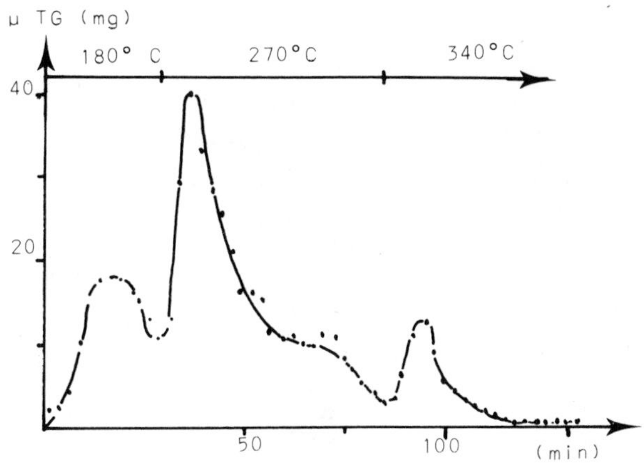

Hydrothermolysis in 3 steps with 180° (Hemicellulose-) 270° (Cellulose) and 340° C (Lignin-) - DEGRADATION only with WATER ! ca 1% Concentration. Reactiontime 45sec.

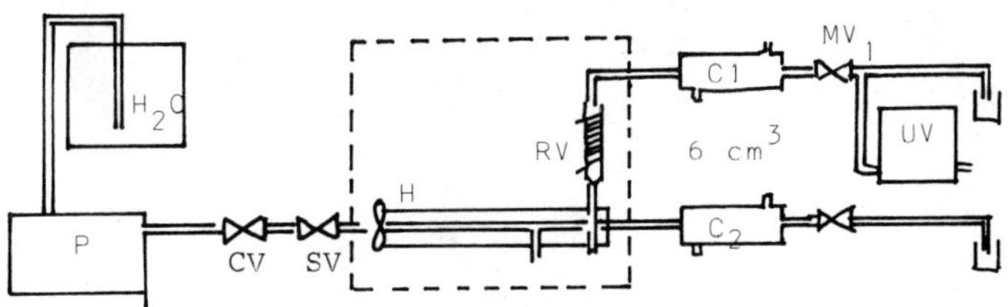

Scheme of the apparatus for hydrothermolysis p pump, CV check valve, T Thermocouples, RV reaction vessel, H preheating unit C cooler, MV metering valve, UV detector

FIGURE 17 BOBLETER-PROCESS

SUPRATON DISPERGIER-UND REAKTIONSMASCHINEN

FIGURE 18 BUCKAU-WOLF-SYSTEM (FRUPP)

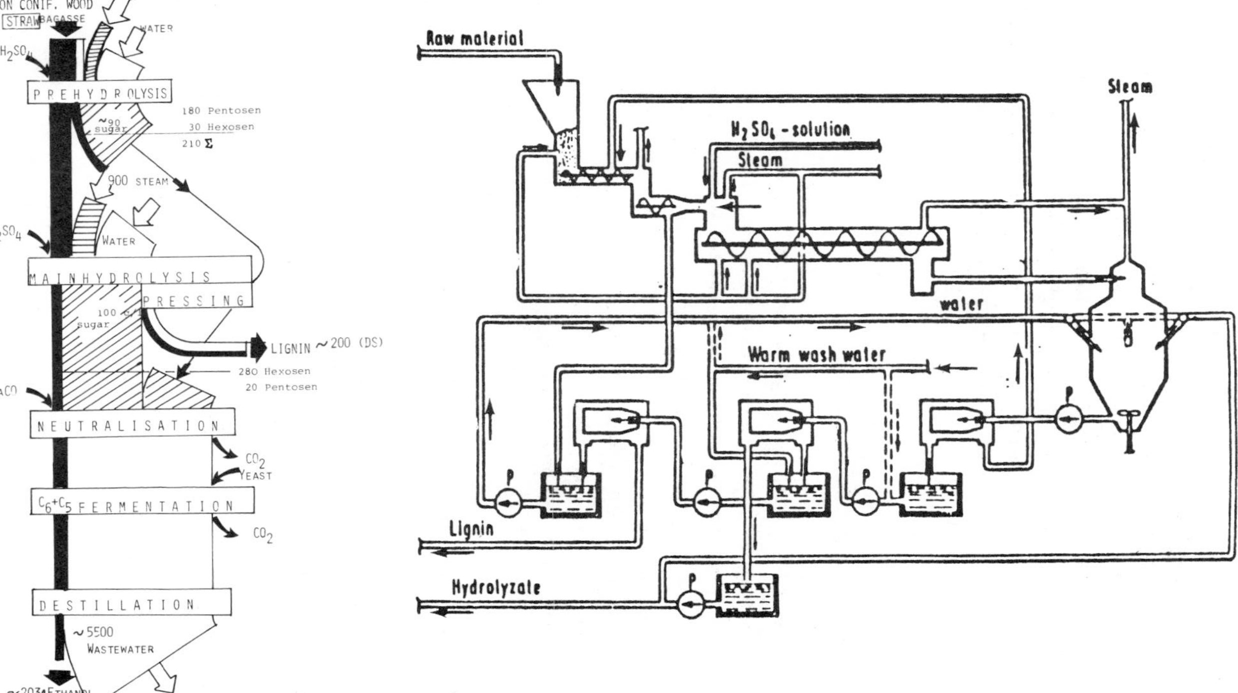

FIGURE 19 TAMPELLA AB. (FINLAND) SYSTEM

three decanters linked in series which separate it according to particle size. The largest particles are recirculated from the first decanter, the medium-sized particles are washed out in the second decanter, and the smallest particles are filtered out as lignin in the third decanter. The dilute sugar solution from the third decanter is used to moister fresh wood cellulose. In the absence of more accurate data the materials flow sheet has a few gaps (Fig. 19 left). All wood celluloses can be processed. The process is either 1- or 2-stage under the reaction conditions shown in Fig. 19.

5. Summary and future developments

The table below gives the major parameters determining the viability of the processes described. In each case when the alcohol yield is given, it is assumed that the pentoses can be fermented with 30 ml pure alcohol per 100 g pentoses, even though some writers wish to produce furfural or yeast from the pentoses. This makes the processes more comparable. In the 'raw material' line C = coniferous wood, D = deciduous wood and S = straw and other annuals. Underlining indicates the raw material on which the yields were based. Figures followed by (?) are not be confirmed.

PROCESS FIG. No	Bergius-Rheinau 6	Battelle-Geneva 8	HF-Hoechst 9	Scholler et al. 11	W.Pleiderer, USA 14	Reitter-Zellplan 15	Knauth 16	Tampella 19
Raw material	C/D	C/D/S	C/D/S	C/D	C/D/S + paper	C/D/S	C/D	C/D/S
Solution water kg/t oven-dry weight	3200	3200	3200	9500	3600-4500	-	-	
Recirculation acid " " "	3300	2400[x]	600 HF	-	-	-	-	-
Hexose alcohol l/t " " "	257/217	200	188	210	-	142 (?)	183(?)	150
Pentose alcohol l/t " " "	36/63	53	60	30	-	56 (?)	69(?)	53
Total alcohol l/t " " "	293/280	253	248	240	180	198 (?)	252(?)	203 (?)
Chemicals and energy input (see Section 2) kg/l pure alcohol	5,0/5,4	5,9	4,0	6,5	5,6 (?)	5,5 (?)	-	6,5 (?)

[x] = phenol-water

Any critical assessment will show that no plant could be built today using any new process, except in the case of processes 6 and 11, which are already being used on an industrial scale.

The most promising process should be process 9, for which more accurate production costs are given in Fig. 5. This has already been tested on the larger laboratory scale, using all modern analytical and monitoring methods, but even this process has not been operated on all circuits in a continuous pilot plant for a longer period.

With regard to process 8, the results of the planned trial runs in Italy will be of interest, in particular as regards the phenol section. No difficulties are anticipated with the main hydrolysis of the cellulose, since this had already been operated in Rheinau on the industrial scale. An alternative method without phenol is also to be tested in Italy with very hopeful results.

In none of the new developments, for which there were high hopes after the first oil crisis, have the yields of the Scholler process been achieved (4, 5, 19) or proven (16).

On the credit side, however, these processes can also use straw and other annuals as well as waste paper, which processes 6 an 11 cannot do. The difficulties which have been encountered, e.g. corrosion in almost all high-temperature processes, filtering and washing out problems in processes 15 and 16, and other problems, will probably be resolved. However, the question is what will be the long-term overall strategy for alcohol production (or, in more general terms, the production of sugar for biotechnological processes), if it is to be commercially viable.

The running battle which has been fought for half a century between the cold processes (6, 8 and 9) and the high-temperature processes (11) has not yet been decided. The former achieve high yields (6) approaching the potential value, but these are offset by the need for complex and capital-intensive plant, since large quantities of acid must be recirculated.

The latter processes use more simple plant and are less capital-intensive, but their yields are not so high. Both approaches, will remain important.

A higher degree of commercial viability can be expected - by developing countries from the simpler technology, and by countries with high raw materials costs from the higher yields.

On the basis of experience so far with the high-temperature processes there will have to be a more consistent approach to continuing development work on one of the two extremes, i.e. either the Scholler principle with reaction times measured in hours, or the short-time processes measured in seconds. In spite of the large percolators, in which the whole process from hydrolysis to washing out is carried out, the former are technically very simple an, in spite of their large dimensions, relatively economical. The other extreme, the short-time hydrolysis process, requires much more complex and expensive plant, and because of the considerable mechanical and chemical resistance which is required this plant must be made of very expensive and uncommon materials such as hasteloy, titanium or similar materials. However, such equipment can be designed, and the costs kept within acceptable limits, only if the residence time is measured in seconds, i.e. if the dimensions are small. When the reaction time is measured in minutes the problems of design and corrosion are hardly less, whereas the dimensions of the plant increase at least tenfold. This means that these compromise processes are probably not very promising.

The multi-stage processes achieve higher yields when reaction times are short. However, for the same reasons it is probable that the single-stage main hydrolysis process retains its edge when comparative calculations are made.

Almost all the newer processes try to use low liquid solids ratios, in order to save boiling-down costs. However, this trend should not go beyond the point at which the oligomer content after hydrolysis is noticeably higher. Otherwise, an post-hydrolysis process is required which more than cancels out the energy advantage.

Whether prehydrolysis is necessary before the main hydrolysis of cellulose is an individual decision in each case. The considerable increase in costs which this entails is often not proportional to the improvement in income, unless high-value products such as xylite can be produced and sold. The problems regarding furfural cannot be discussed here. New developments regarding pre-hydrolysis, whose reaction kinetics pose considerably, less problems - because of the flat maximum yield - than hydrolysis of cellulose, must still be compared with proven industrial processes used in the 1940s and in 1950s with sugar yields of 22 - 23 % from a total of approximately 30 % extracted from all types of ligno-celluloses, with sugar concentrations (after washing out) of approximately 100 g/l.

Fig. 20 represents an attempt to take a cautious look into the future, and compares the production costs for alcohol with those for premium-grade petrol. The upper section shows the variants on the Scholler process which are already technically possible and confirmed the more optimistic HF variants is compared with premium-grade petrol below (cf. also Fig. 5). New developments will come between the two extremes of this range of production costs.

It is assumed that the auxiliary, energy and labour costs for all processes will rise by 3 % per annum, and capital costs by 4 %. For the ligno-cellulose costs a lower limit of DM 100 tonne oven-dry weight is given for 1982, with a price rise of 3 % per annum, and an upper limit of DM 200 per tonne and a price rise of 4 % per annum. For premium-grade petrol the raw materials cost at 1982 prices represents 85 % of the production costs of DM 0.67 per litre (25). Between 1960 and 1980 crude oil prices rose by 23 % per annum in two large jumps. For future years an increase in the cost of crude oil of 8 - 10 % per annum, probably in stages, has been assumed (diagonally hatched field). In order to make the energy yield from alcohol comparable with that of premium-grade petrol a factor of 1.235 has been applied to the alcohol costs in Fig. 20 (25). The mean line for the HF process intersects that for the other processes in approximately 1998, and at the most optimistic estimate in 1993.

The process naturally becomes more economical if, instead of taking European conditions, the calculations are made for areas where ligno-cellulose is much cheaper. This is shown by the thin discontinuous line for a raw material input (1982) of DM 50/t oven-dry matter. This inter-

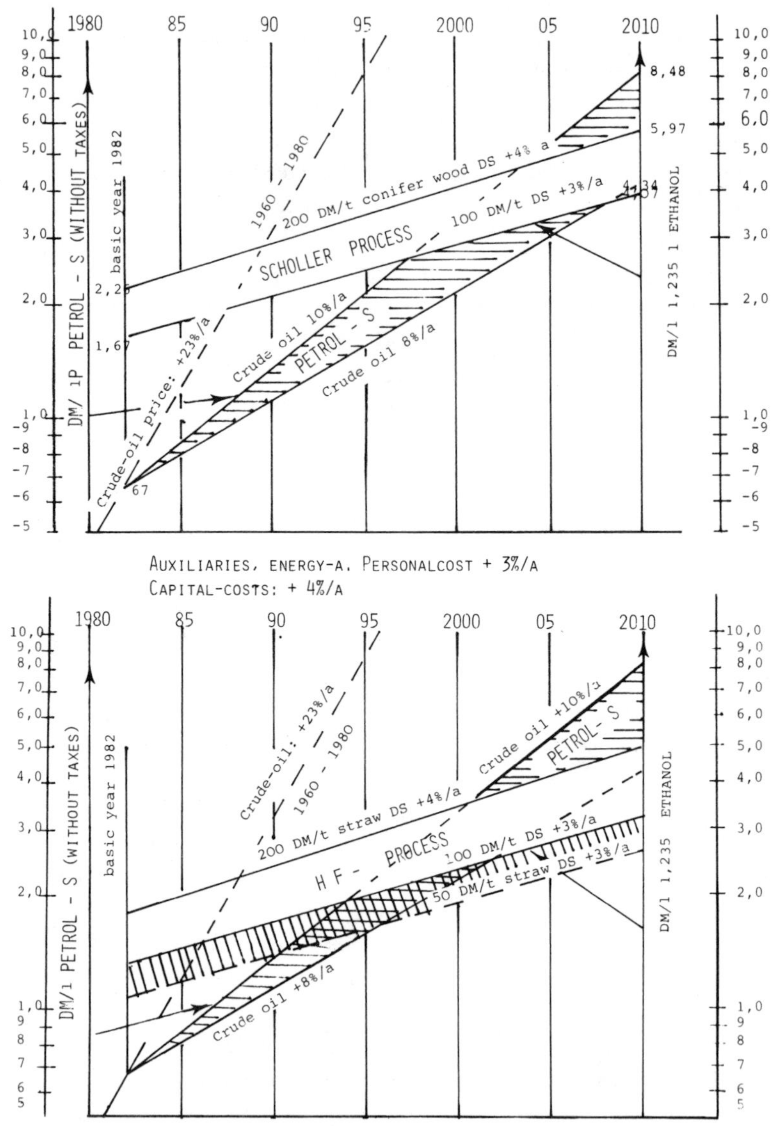

FIGURE 20 PROGNOSIS OF FABRICATION COST IN COMPARISON WITH COST OF PREMIUM GRADE PETROL

sects the graph for premium-grade petrol costs 3 - 4 years earlier. It is possible that raw material prices will be still lower, e.g. bagasse, which occurs almost free in some areas at a central point. Lastly, it is possible that the mixture of 10 % alcohol may not in practice give noticeably different consumption figures from those for pure premium-grade petrol. If the factor of 1.235 for ethanol could thus be eliminated from the marketing considerations, view, all the dates for achieving comparable costs are brought forward by 6 - 7 years.

No claims to absolute accuracy are made for these forecasts, particularly as regards the cost of premium-grade petrol. They will probably be attacked and criticized. However, long-term developments, such as the hydrolysis of biomass, require in each case a long-term viability forecast on which the author feels that development engineers and chemists must make a start in good time, so that the development work does not go astray. Management and economics experts, using other premises, might come to other conclusions, and Fig. 20 is designed to provoke this. In any event, a hasty glance is insufficient for making an assessment of the prospects for hydrolysis processes.

BIBLIOGRAPHY

1. FAO: World Forest Inventory, Rome 1963
   FAO-production yearbook 1079 V.33 table 9 - 17
   Statistisches Jahrbuch 1981 des Statistischen Bundesamtes Wiesbaden

2. Nach einem Gutachten ETH Zürich August 1955 über die Scholler-Anlage Ems (umgerechnet)

3. REINEFELD et al. Zuckerindustrie Bd 105 (1980) 25 - 35
   MISSELHORN, Die Branntweinwirtschaft Oct. 1981 5.326
   MEINHOLD, et al. IfBw-Arbeitsbericht 4/81

4. E. HÄGCLUND: TH. RIEHM, SVENSK PAPPERSTIDNING 61 (1958) 665

5. TH. RIEHM DT 927139 (1955), TH. RIEHM DT 1027150 (1958)

6. TH. RIEHM DT 1024063 (1958), US 2944123 (1957)

7. I.S. GOLDSTEIN, AiChe Symposium series Nr. 207, Vol 77, 85 - 92

8. I. KUSAMA, FAO Techn. Panel on Wood Chemistry Tokyo 1960, 1009

9. A.I.BEARDSMORE, ICI, Agricultural Div. Rand D. Department Contract No. B OS 084 UK (H)

10. BIOMASS Digest 4, Nr. 3 (1982)

11. R. ERKEL, Kollegium in Troisdorf 3.4.1981 (Verbindungsstelle Landwirtschaft - Industrie e.V. Essen)

12. I.F. SAEMAN, Ind. Eng. Chem. 37, 43 (1945)

13. SCHOLLER, Dissertation T.H. München 1923

14. E. HARRIS, E. BECKLINGER, Ind. Eng. Chem. 37, (1945) and Ind. Eng. Chem. 38 890, 896 (1946)

15. N. GILBERT et al. Ind. Eng. Chem. 44 1712 (1952)

16. D.A. WHITWORTH, 9. Australasian Conference on Chemical Engineering, August/September 1981

17. E. GRETHLEIN, Proceeding of the 2nd Joint US-USSR Enzyme Engineering Seminar, Febr. 1978, S. 441 ff

18. R. REMIREZ, Chem. Eng. 26, 51 (1981) and Chem. Ing. Techn. 52 A 318(1980)

19. F.J. REITTER, Kollogium Troisdorf, 3.4.1981 (Verbindungsstelle Landwirtschaft - Industrie e.V. Essen)

20. ZELL-PLAN The Biol-Process for the Conversion of Biomass (1983)

21. H. KNAUTH Bericht über Pilotanlage für Totalhydrolyse Sept. 1981, also DT 2744067 (1977) and DT 2827388 (1977)

22. O. BOLBLETER u.a. Energy from Biomass, Brighton 4.-7.Nov.1980 S 554 ff

23. Krupp Industrietechnik GmbH, zum Förderantrag an KFA Jülich PL Rohstofforschung 1982 and 1983

24. K. KESKINEN, et al. Helsinki University of Technology und OY Tampella Ab, Tampere 1982
V. TOHJOLA, 5th International Congress in Scandinavia on Chemical Engineering, Copenhagen 14.-16. April 1980
S. HÄMÄLÄ, u.a. Oy Tampella Ab, Tampere 10, Finnland

25. P. FAUL, 290 Dechema-Colloquium 13.1.1983

HYDROLYSIS OF CELLULOSE AND HEMICELLULOSE - VALORIZATION OF LIGNINE. A NEW PRETREATMENT FOR LIGNO-CELLULOSIC SUBSTRATES*

C. David, R. Fornasier
Université Libre de Bruxelles,
Faculté des Sciences, Campus Plaine, 206/1
Boulevard du Triomphe, 1050 Bruxelles.

ABSTRACT

A pretreatment of ligno-cellulosic materials with sodium hypochlorite-hypochlorous acid at controlled pH (between 7 and 9) considerably increases the accessibility of the cellulosic part of the substrate to chemical and biochemical reactants. As a consequence, the yield and rate of the enzymatic hydrolysis to glucose of the cellulosic part of the substrate is largely increased. Wheat straw, eucalyptus meal and white pine sawdust have been investigated. The increase in accessibility is assigned to degradation and (or) detachment of the lignin network. The loss in cellulose and hemicellulose is not important, lignin being preferentially degraded under carefully controlled pH conditions.
No change in the crystalline structure of cellulose (transformation of cellulose I to cellulose II) has been observed. The pretreatment when applied to pure cellulose decreases the yield of enzymatic hydrolysis : in the absence of lignin oxidation of the anhydroglucose units is important and results in the inhibition of the enzymatic hydrolysis.

* Research performed under contract with the E.E.C. and Solvay & Cie.

1. INTRODUCTION

The cellulosic fraction of biomass is a very important and renewable source of sugars which can be converted easily to ethanol and other basic chemicals. The yield of glucose obtained from cellulose is however low because crystallinity limits the accessibility of the substrate to enzymatic hydrolysis. In ligno-cellulosic compounds, both the lignine network and the crystallinity of the cellulosic fraction are responsible for the low yield in glucose obtained. This has been recognized many years ago and a large amount of research work has been devoted to optimize the yield of enzymatic hydrolysis (1). This process is very specific and by consequence very attractive when compared to acid hydrolysis in which competitive degradation of the sugars formed is a non negligible process.

A large number of physical and chemical pretreatments have been proposed to increase the accessibility of cellulose and ligno-cellulosic compounds to enzymatic hydrolysis. The most important are ball milling which reduces the crystallinity, alkali and ammonia swelling, explosive steam decompression which frees the cellulose fibers from lignine and solvents which, as Cadoxen, transform cellulose I into the more reactive cellulose II. All these processes have been reviewed recently (1).
Although these pretreatments are rather performant, none of them has attained industrial development because the increase in yield is not sufficient to justify the cost of the operation. Pretreatments thus remain a crucial problem for the economical utilization of the tremendously important quantity of waste ligno-cellulosic materials.

The present work is concerned with the pretreatment of cellulose (paper pulp) and ligno-cellulosic materials (straw, white pine sawdust and Eucalyptus Salinia meal) with NaClO-HClO at controlled pH (2). The rate and maximum yield (yield after 4 days) of the enzymatic hydrolysis by Trichoderma Viride cellulases in standard conditions are used as measurements of the accessibility of the substrate.

2. EXPERIMENTAL

2.1. Substrates

The cellulose used is paper pulp. The ligno-cellulosics are wheat straw, white pine sawdust and Eucalyptus Salinia meal.

2.2. Prehydrolysis

Hemicelluloses can be removed by refluxing the untreated substrate for two hours with either 2% H_2SO_4 or 2% HCl.

2.3. Pretreatment with NaClO-HClO (2)(3)

The pretreatment is performed in a thermostated bath with magnetic stirring, using NaClO 0.5M. The same results were obtained using "pro analysi", industrial and home-use NaClO solutions. 200 ml NaClO 0.5M are acidified to the chosen pH value with concentrated HCl. 6g of substrate are then introduced. The pH is maintained at constant value by addition of NaOH 5M. After 30 min, the solid is filtered on sintered glass, then washed by stirring in water, filtered and washed again on the filter. The pretreated substrate is then dried for 24 hours at 104°C.

2.4. Weight loss

It is measured by weighing the pretreated substrate at constant weight (24 hours in aerated oven at 105°C).

2.5. Loss in cellulose and hemicellulose

The quantitative saccharification of the substrate is performed before and after the pretreatment. C_5 and C_6 sugars are analyzed by HPLC on a 25 cm length and 0.46 cm internal diameter column, filled with the anion exchange resin Aminex A-28 in the borate form. A boric acid-potassium borate pH 8.8 buffer is used as mobile phase with flow rate of 0.4 ml min^{-1}. Spectrophotometric detection is made at 570 nm, using the in situ prepared complex of the sugars with copper 2,2' bicinchoninate

2.6. Quantitative saccharification

The substrate (250 mg) is dissolved in 3 ml H_2SO_4 72% (1h at 30°C). It is then diluted with 84 ml water and kept in the oven for 2h at 150°C. Neutralization of the acid is then performed with solid $Ba(OH)_2$.

2.7. Analysis of the lignine (K number)

The substrate is milled and the obtained particles (dimension $< 200\mu$) are suspended in a known amount of $KMnO_4$. After 10 min, the oxidation reaction is stopped by adding an excess KI which is titrated with $Na_2S_2O_3$. The K number is obtained from the results. This index multiplied by 0.13 gives the percent native lignine in the substrate (4).

2.8. Enzymatic hydrolysis

They were all performed in the same conditions (5). The substrate is first soaked for 24 hours at room temperature with 9 ml citrate buffer. 10 mg of cellulases from Trichoderma Viride (Onozuka R-10 from Kinki Yakult MGF Co. Ltd) are then dispersed in 1 ml citrate buffer and added to the substrate. The hydrolysis is performed at 45°C. A volume of 10 λ is taken after convenient time intervals and analysed for glucose by the glucosoxidase method (Boehringer GOD-perid glucose).

2.9. X-ray diagrams

They are obtained with a Philips PW 1010 generator equipped with a PW 1050/30 goniometer using K_{α}Cu. The samples are prepared by milling the substrates (particles < 400 μ) and compressing them between two glass plates.

3. RESULTS AND DISCUSSION

3.1. Untreated substrates

The quantity of glucose formed in the enzymatic hydrolysis as a function of time for the untreated substrates is given in Fig. 1 and 2, curves 1 and 2. The maximum yields obtained are respectively 41, 17, 14 and 0% for paper pulp, straw, pine wood and eucalyptus. Lignine is mainly responsible for the low yields obtained with pine and straw. Indeed, the same yield is obtained for untreated and for dilute acid treated pine and straw. Dilute acid treatment thus removes the hemicellulose without affecting the reactivity. Crystallinity limits the yield in glucose obtained for paper pulp. The non reactivity of eucalyptus is not explained. The kinetics of the enzymatic hydrolysis of cellulosic compounds will be discussed in details elsewhere. In all the experiments reported in the present paper, the hydrolysis reactions are performed in the same conditions for comparative purpose with the exception of those of paragraph 5.

3.2. Effect of the pH of the pretreatment

The weight loss, the loss in cellulose and the yield of glucose measured after 4 days for the enzymatic hydrolysis of cellulose have been measured for the different substrates pretreated at different pH between 2 and 11.5.

Fig. 3 shows that the weight loss is a function of the pH of the pretreatment. It presents a maximum in the range of pH 7-8 and is much less

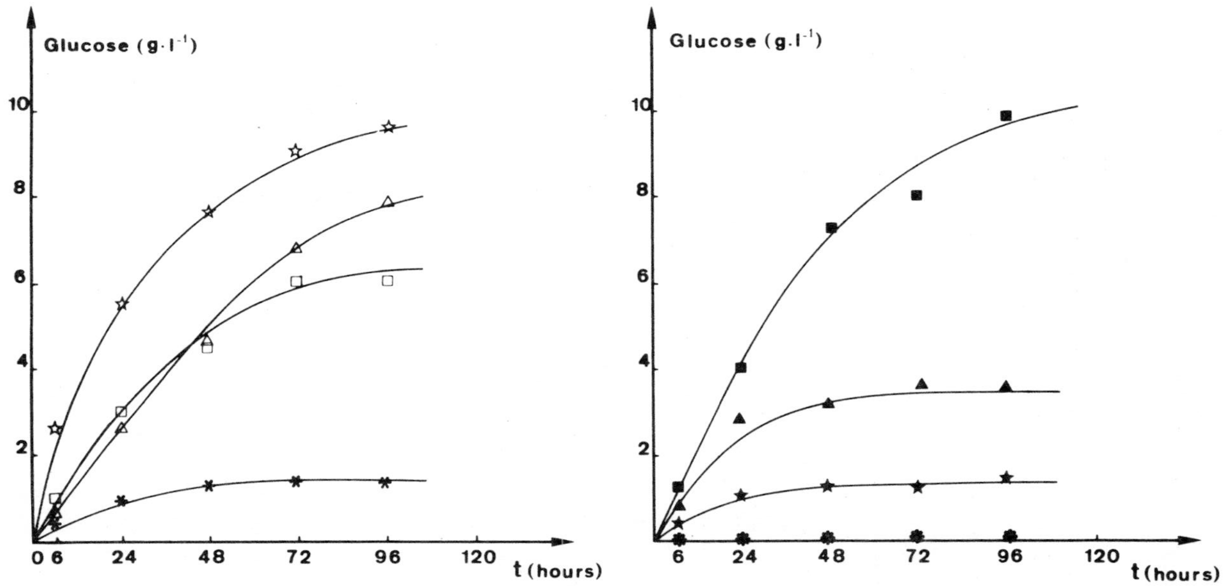

Fig. 1 : Glucose formed as a function of time in the enzymatic hydrolysis of :

☆ untreated paper pulp; ✱ untreated pine; □ paper pulp pretreated with NaClO-HClO at pH 9; △ pine pretreated with NaClO-HClO at pH 9.

Fig. 2 : Glucose formed as a function of time in the enzymatic hydrolysis of :

★ untreated straw; ✱ untreated eucalyptus; ■ straw pretreated with NaClO-HClO at pH 9; ▲ eucalyptus pretreated with NaClO-HClO at pH 9.

important for eucalyptus than for the other substrates.

The loss in cellulose presents a maximum in the same range of pH (Fig. 4) and is much more important for paper pulp than for the other substrates.

The yield in glucose obtained after 4 days for the enzymatic hydrolysis of the substrate (Fig. 5) also presents a maximum between pH 7 and 9 and decreases in the following order :

straw > pine wood > eucalyptus.

Paper pulp on the contrary presents a minimum in the range of pH 9-7. With the exception of paper pulp, the initial rate and maximum yield are much higher for the pretreated substrate than for the untreated one (Fig. 1 and 2) indicating an increase in the accessibility of the substrate and (or) a decrease of the inhibition by lignin.

The loss in C_5 sugars as determined for eucalyptus (Fig. 6) is not important.

The content in native lignin of the pretreated samples of the same substrate is also given in Fig. 6.

We shall now discuss the possible causes for the increase in accessibility of the ligno-cellulosic substrates.

When lignin increases, the yield in glucose after 4 days decreases. The relation between the two parameters is however not a simple proportionality (Fig. 7). Other factors could thus be important. Transformation of cellulose I into cellulose II is not important as indicated by Figures 8 and 9. These figures give the X-ray diagram of paper pulp (cellulose I), paper pulp treated with NaOH 18% (cellulose II), untreated pine and eucalyptus, pine and eucalyptus treated with NaClO-HClO and pine treated with NaOH 18%. The untreated woods differ from paper pulp by a larger part contribution of the amorphous part of the diagram due to lignin and hemicellulose. This amorphous part decreases on treating with NaClO-HClO. Some formation of cellulose II can be observed in the case of pine treated with NaOH 18%.

Oxidation of the anhydroglucose units of cellulose results in an inhibition of the hydrolysis reaction owing to the high selectivity of the cellulases as experimentally observed for paper pulp. Acid groups formation in the presence of NaClO in alkaline medium is well known from the literature (6-7).

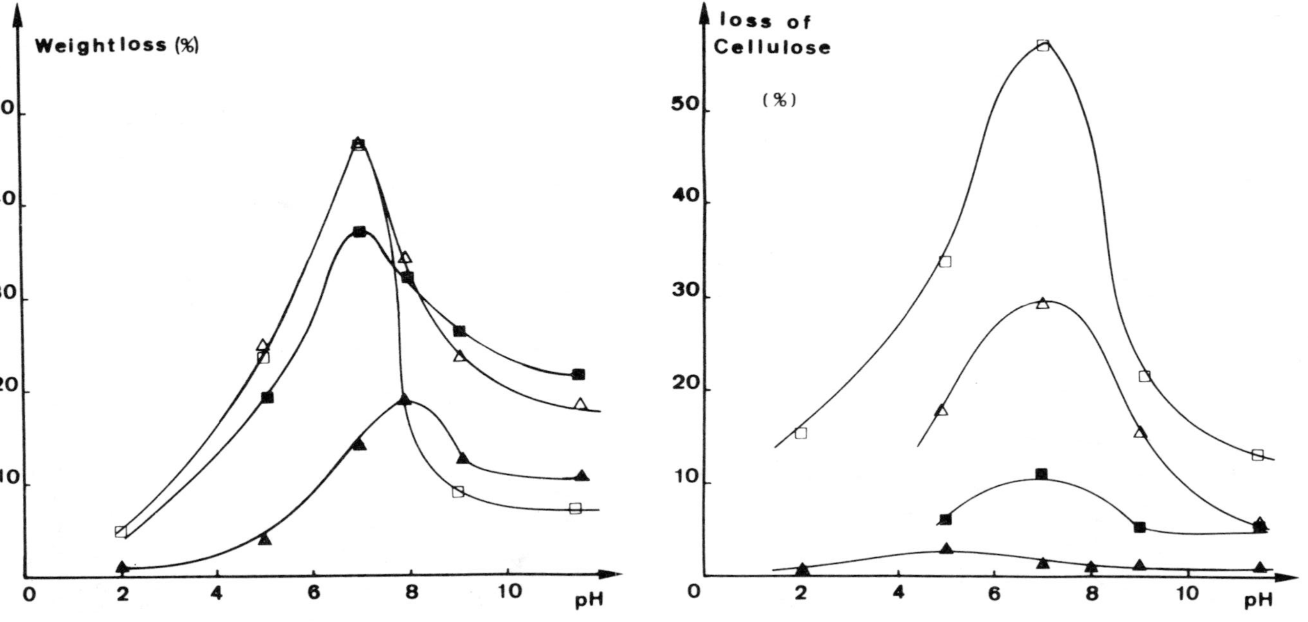

Fig. 3 : Weight loss as a function of the pH of pretreatment: □ cellulose; △ pine; ■ straw; ▲ eucalyptus.

Fig. 4 : Loss of cellulose as a function of the pH of the pretreatment : □ cellulose; △ pine; ■ straw; ▲ eucalyptus.

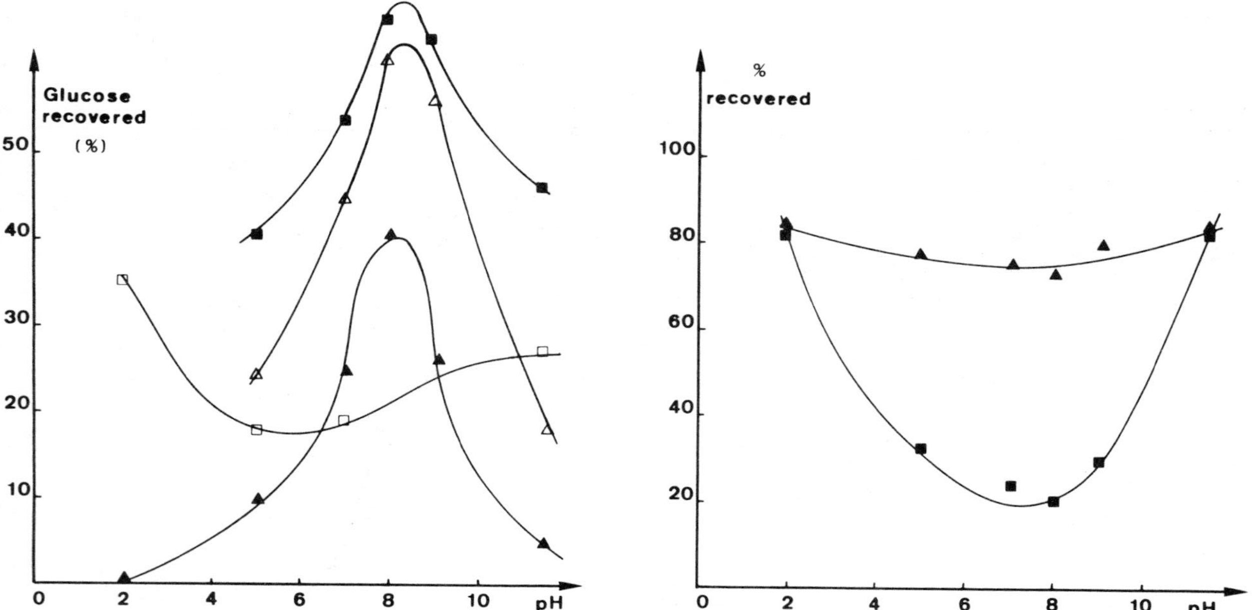

Fig. 5 : Glucose formed after 4 days in the enzymatic hydrolysis as a function of the pH of the pretreatment : □ cellulose; △ pine; ■ straw; ▲ eucalyptus.

Fig. 6 : Lignin (■) and C_5 sugars (▲) recovered as a function of the pH of the pretreatment for eucalyptus.

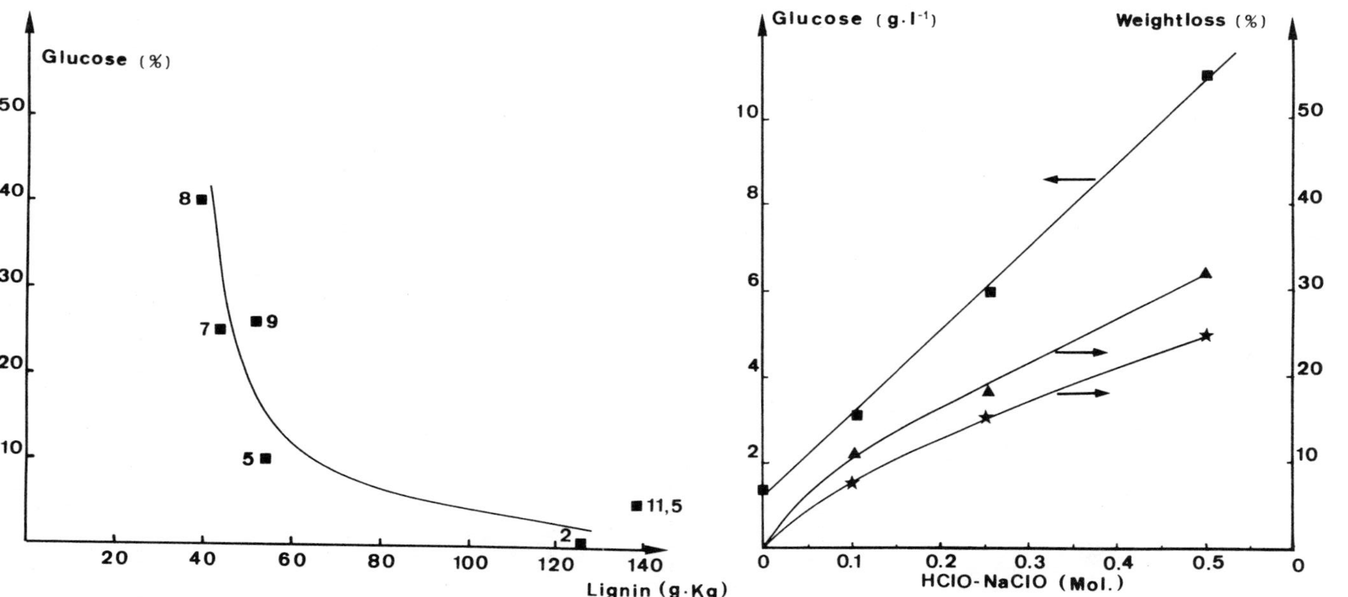

Fig. 7 : Glucose formed after 4 days in the enzymatic hydrolysis as a function of lignin content for eucalyptus pretreated at different pH (from 2 to 11.5).

Fig. 10 : Glucose formed after 4 days in the enzymatic hydrolysis and weight loss as a function of the initial molar concentration of NaClO-HClO. pH of pretreatment : 8.
Time of pretreatment : ■ and ▲ 30 min.; ★ 20 min.

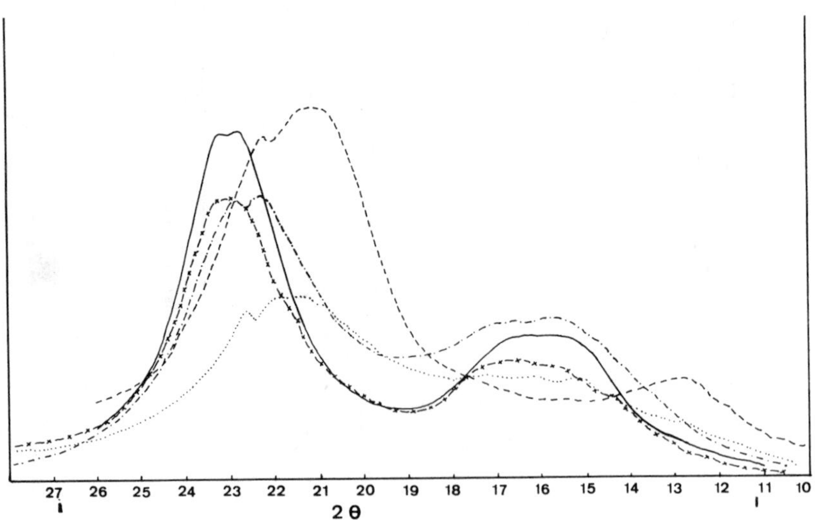

Fig. 8 : X-ray diagram of paper pulp ——————

paper pulp treated with NaOH 18% - - - - - - - - - - -

pine —·—·—·— ; pine treated with NaClO-HClO —x—x—x—

pine treated with NaOH 18% ....

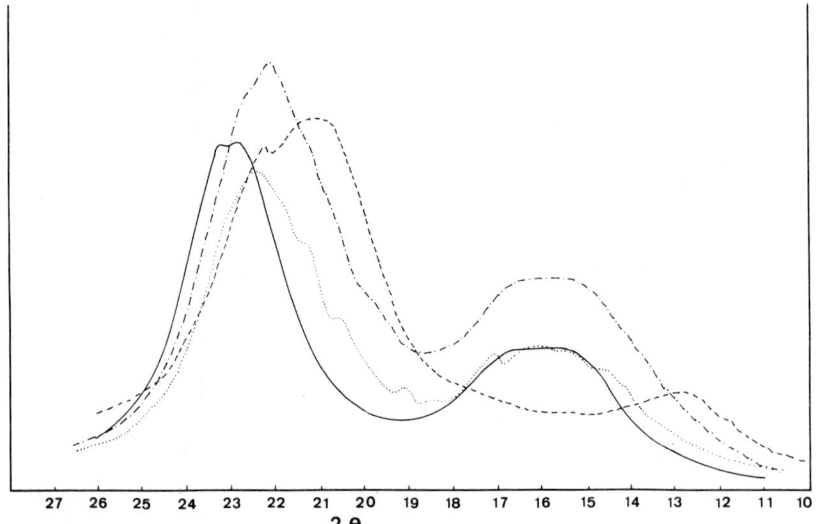

Fig. 9 : X-ray diagram of paper pulp ——————

paper pulp treated with NaOH 18% - - - - - - - - - -

eucalyptus —·—·—·—·; eucalyptus treated with NaClO-HClO ....

In neutral and acidic solution, reducing type celluloses are formed. Their relative rate of formation in different relative units are given in Table I. The rate of oxidation is maximum at pH 7 and could partly justify the minimum in the yield of glucose formed from paper pulp at pH 7 in the present work.

The main parameter responsible for the increase in accessibility seems thus to be the reaction of NaClO-HClO with lignine. The lignine is partly degraded into soluble products (8).
Detachment of the lignine from the cellulose fibers by breaking of the lignine-carbohydrate bonds also occurs. This detached lignine probably remains partly in the fiber walls protecting cellulose against NaClO attack but allowing accessibility to enzymes. Oxidation of lignine by NaClO is proposed by Sarkanen to result from an initial chlorination of the phenolate ions by HClO (8). The proportion of HClO and free phenolate ions present in the medium are given in Table I. Examination of this table shows that pH 7-9 corresponds to intermediate proportion of HClO and $\emptyset O^-$ and could justify the maximum yield in glucose and delignification observed in this range of pH.

3.3. Effect of the concentration of HClO-NaClO

The effects of the concentration of HClO-NaClO on the glucose formed and on the loss in weight in the case of straw are given in Fig. 10. Both parameters increase with increasing concentration. The loss in cellulose is unimportant. We therefore suppose that the loss in weight mainly represents the loss in lignine.

3.4. Effect of the time of pretreatment

The yield in glucose after 4 days and the loss in weight at different times of pretreatment are given in Fig. 11 for straw at two concentrations. These parameters first increase rapidly and then tend to a limiting value as a function of time. Similar results are obtained at pH 7 and 9.

3.5. Effect of the temperature

The experimental results are given in Fig. 12. They are not obtained at constant pH as in the preceding sections. The initial pH is about 11.5 and it changes as a function of time. The final pH after 15 minutes is respectively 11.5 and 11 for 20 and 30°C. At these temperatures, the loss in weight and the loss in cellulose are low in agreement with the results

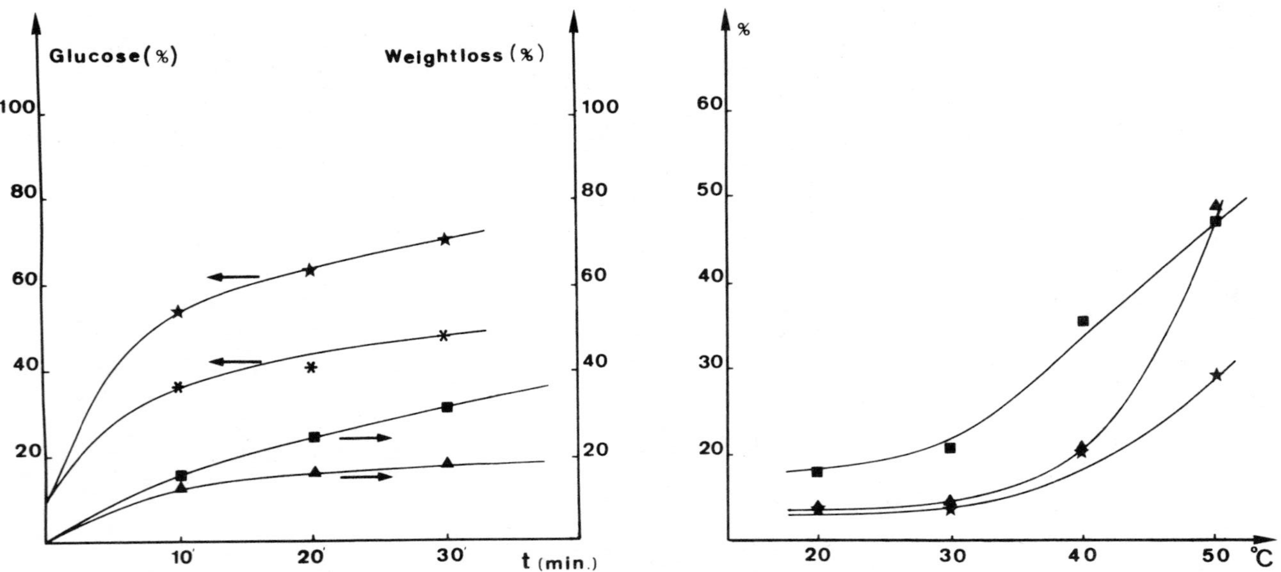

Fig. 11 : Yield in glucose after 4 days and weight loss as a function of the time of pretreatment at pH 8 for two initial concentrations of NaClO-HClO (★ , ■ : 0,5M - ✱ , ▲ : 0,25M).

Fig. 12 : Yield in glucose after 4 days (■), loss of cellulose (★) and weight loss (▲) as a function of the temperature of pretreatment.

obtained at a constant pH value of 11.5 reported in the preceding sections. At 40 and 50°C, the pH spontaneously decreases to 9 and 8 owing to the formation of acid groups by oxidation of cellulose and lignine (Fig. 12). In agreement with the experiments performed at a constant pH value of 8 and 9, the yield in glucose is higher but the loss in cellulose is also rather important.

4. CONCLUSION

Pretreatment of ligno-cellulosic materials with NaClO increases very effectively its accessibility to enzymes when performed near pH 8. The procedure seems to be of general application since it is valuable for straw, soft wood (white pine) and the very unreactive hard wood eucalyptus, but may not be applied to cellulose.

Optimization of the other conditions (time of pretreatment, temperature, ratio NaClO to substrate) needs however to be performed for every material and for every planned application taking the economic point of view into consideration. Indeed, the loss in weight and the loss in cellulose are different for different ligno-cellulosic materials in the same conditions owing to their particular morphology. In any case, pretreatments of half an hour at room temperature are good if not the best conditions. The increase in accessibility results mainly from competitive detachment and degradation of the lignine network and oxidation of the anhydroglucose units.

REFERENCES

(1) C.R. Wilke, B. Maiorella, A. Sciamanna, K. Tangnu, D. Wiley and H. Wong, Enzymatic Hydrolysis of Cellulose - Theory and Applications. Noyes Data Corporation, 1983.

(2) C. David, Ph. Thiry and R. Fornasier, Demande de Brevet Européen 83200306-5 (2-3-83) : Procédé de traitement de matériaux ligno-cellulosiques.

(3) The ratio $HClO/HClO + ClO^-$ as a function of pH is given in Table I.

(4) TAPPI method T236 05-76.

(5) C. David and Ph. Thiry, Eur. Pol. J. 17, 957 (1981).

(6) L.F. Mc Burney in Cellulose and Cellulose Derivatives, vol. V, part I, Interscience 1954.

(7) M. Lewin and J.A. Epstein, J. Pol. Sci. 58, 1023 (1962).

(8) Lignins, K.V. Sarkanen and C.H. Ludwig, ed. Wiley, 1971.

Table I

pH	Rate of formation of acid groups (relative units)	Rate of formation of reducing groups (relative units)	$\frac{\phi O^-}{\phi OH}$ [a]	$\frac{HClO}{HClO + ClO^-}$ [b]
5	1.5	0	3.10^{-6}	1
7	18	16	3.10^{-4}	0,77
8	3	1	3.10^{-3}	0,25
9	1	0,5	3.10^{-2}	0,03
11.5	-	-	10	$\sim 10^{-4}$

[a] pK_a substituted phenols $\sim$ 10.5

[b] pK_a HClO = 7.5

BACTERIAL HYDROLYSIS: A REVIEW

G. T. Tsao

Purdue University
West Lafayette, IN 47907 USA

Summary

Urban and industrial wastes contain large quantities of cellulose, other polysaccharides, proteins and other biological materials. Most biomass can be readily digested to generate methane; cellulose, however, is very resistant to biodegradation due to its strong crystalline structure and interference by lignin. Bacterial hydrolysis of cellulose is often the bottleneck of anaerobic digestion. Three types of bacteria are involved in methane formation from solid biomass. They are hydrolytic, acetogenic and methanogenic bacteria. The enzymes involved in hydrolysis of cellulose and other polysaccharides and proteins are fairly well known.

In order to enhance cellulose hydrolysis and thus also anaerobic digestion, physical and chemical pretreatments of various kinds have been investigated. Several common pretreatment techniques are described in this review. Lastly, a general discussion on potential alternative processes to bacterial hydrolysis in anaerobic digestion is also given.

Processes for conversion of biomass to methane may be classified into two categories, thermal and biological. Application of thermal processes is limited to feeds with low water content. Otherwise, the cost of heating and evaporation of water in order to reach high temperature is prohibitive. Biological gasification, commonly known as anaerobic digestion, is a low temperature process which is economic at a variety of scales. The product gas is composed primarily of methane and carbon dioxide. Some small amounts of sulfur compounds may also be present. The major limitations of the biological processes are two: 1) low conversion with often 50% of the organic matters unconverted, and 2) slow rate. To a large measure, these limitations are due to the strong resistance of cellulose towards hydrolysis by the bacterial cultures involved in the process. Bacterial hydrolysis is thus a subject of considerable importance to anaerobic digestion; any improvement that we may achieve in cellulose hydrolysis should lead to increase in process efficiency as measured by rate and/or yield.

Microbial methanogenesis is a natural process involving activities of often numerous unknown microbes, occurring in anaerobic environments such as lake sediments and animal digestive tracts. In fact, most of our knowledge of the microbiology of anaerobic digestion has been derived from the studies of rumen methanogenesis. In recent years, there has been a surge of research activities in the use of biomass for the production of not only methane as a fuel gas, but also other materials such as ethanol. Considerable progress has been made in studies of physical and chemical properties of cellulose, cellulolytic enzymes, techniques of pretreatment and so on.

I. Microbiology of Anaerobic Digestion

Methanogenesis has previously been considered a two-step process involving 1) formation of acids, alcohols, hydrogen, carbon dioxide, and so on by fermentation of the organic matters in the feed, and 2) conversion of these compounds to methane. As we understand the methanogenic bacteria better, we start to realize that true methanogens utilize primarily acetate, hydrogen and carbon dioxide to form methane; while another type of anaerobic microbes now referred to as hydrolytic bacteria are involved in conversion of proteins and polysaccharides such as cellulose, starch and hemicellulose to soluble intermediates and yet another group of microbes which are now commonly referred to as acetogenic bacteria are responsible for the formation of acetate, hydrogen and carbon dioxide. Therefore, the microbes in the process can be roughly classified in three categories: 1) hydrolytic; 2) acetogenic; and 3) methanogenic. The possibility of a high level of involvement of synergism and other inter-species interactions is obvious, through such mechanisms as substrate inhibition, product feedback inhibition, etc. A full understanding of these interactions may require years of future research.

II. Bacterial Hydrolysis

A. Cellulose. In average, municipal solid wastes contain 50% by weight of cellulosic materials including paper, cardboard boxes, sawdust, used furniture, dead trees, etc. Typically, hard wood contains 45% by weight of cellulose, 35% of hemicellulose of different kinds and also 20% or so of lignin and others. Cellulosic material is by far the most abundant biomass. It is available in a sufficiently large quantity that it may be considered a resource of some significance.

Bacteriodes succinogenes is the first important cellulose hydrolyzing organism that was isolated from rumen by Hungate (1). Its physiology has been extensively investigated. Other common rumen anaerobic bacteria include *Clostridium lochheadii* (2), *Clostridium cellobioporus* (3), *Ruminococcus flavefaciens* (4), *Ruminococcus albus* (2), and *Butyrivibrio fibrisolvens* (5).

Cellulose hydrolyzing enzymes from different microbial species have been isolated and extensively investigated. Even though there may be differences in fine details, properties of cellulases from different sources are amazingly similar. Generally, the cellulase complex is consisted of three major enzyme components: endoglucanase (β-1,4-glucan glucanohydrolase), exoglucanase (β-1,4-glucan cellobiohydrolase) and cellobiase (6).

As we mentioned before, the rate-limiting factor in anaerobic digestion is the slow rate of cellulose hydrolysis. What makes cellulose so resistant to bacterial hydrolysis can be traced to the physical structure of cellulose co-existing together with two other biomass constituents: hemicellulose and lignin. A few words about the structure of cellulose and cellulosic materials are thus in order (7).

Cellulose is a linear homopolymer of anhydroglucose units linked by beta-D-1,4 glucosidic bonds. Without side chains, parallel linear cellulose molecules are sterically compatible for close association and in forming intra- as well as inter-molecular hydrogen bonding. From a processing viewpoint, the beta-1,4 glucosidic linkage in cellulose may not be any more resistant to hydrolytic cleavage than the alpha-1,4 glucosidic linkage in starch if the cellulose molecules can be fully hydrated and exposed. In other words, the difficulty in obtaining fast and complete cellulose hydrolysis is due to not the primary linkage but rather the secondary and tertiary structures of cellulosic materials.

Individual linear polymeric cellulose molecules are linked together to form elementary fibrils. Depending upon its source, the degree of polymerization (DP) of cellulose ranges from 1000 to 10,000.

The crystalline structure of cellulose is still a subject of controversy. The fibrilized fringe unicellar model is probably the most popular, which envisions linear cellulose molecules to be parallel and closely associated with each other with some regions of disorder dispersed along the chain length. Meanwhile, models of cellulose fibrils involving folding chains have also been suggested. Shown in Figure 1 is the outline of a group of plant cells. The cellulosics are the cell wall materials of plants. Surrounding the cell is the middle lamella which is heavily lignified and contains lignin and hemicellulose in the proportion of approximately 70 to 30%. Cellulose is contained in the cell wall. Within the cell wall, four distinct concentric morphological layers can be distinguished. The outermost layer is called the primary wall which is about 0.1 mm thick and contains about one-third of cellulose. In the primary wall, the microfibrillar structure appears as a loose and random network.

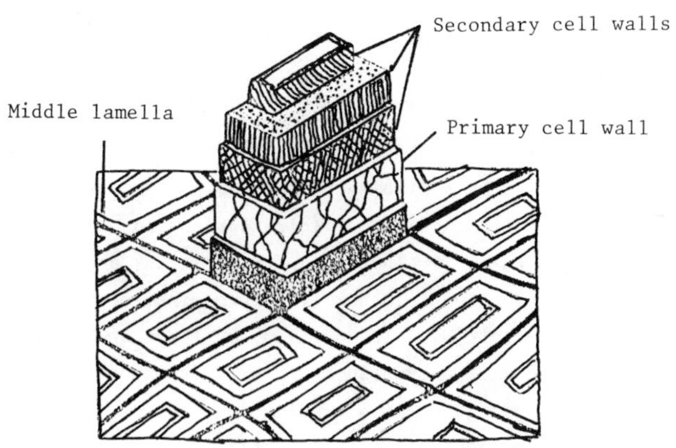

Figure 1. Architecture of plant cell wall

The secondary wall forms during growth and maturation of the cell and contains three layers, an outer layer, a middle layer, and an inner layer. The microfibrils of the outer layers are wound in flat helices. The direction of winding alternates to form overlapping spirals. In the middle layer, the cellulose fibrils are tightly packed in a steep parallel helix. The innermost layer consists of helical microfibrils. The structure of this layer is not clear. For a raw cotton fiber, the primary wall is about 0.1-0.2 μm thick, the secondary wall is 1-4 μm thick and contains almost pure cellulose. In general, cellulose occupies about 90% of the dry weight of the secondary wall in various fiber cells.

The above mentioned structural features are pertinent to bacterial hydrolysis of cellulosics. In enzymatic hydrolysis, the prerequisite for a reaction to take place is a direct association between the molecules of the enzyme and the substrate. In this case, the barrier represented by the lignin-rich middle-lamella makes the diffusion and the penetration by the macromolecular enzymes extremely difficult and slow, if not impossible. The phenolic groups in lignin might be even inhibitory towards enzymes.

A recent report by Lamed et al., (9) suggests the existence of a macromolecular complex with a distinct quarternary structure of a molecular weight of about 1.2 million. Such particles designated as "cellulosome" were isolated from cultures of *Clostridium thermocellum* and found to be able to dissociate into a multiple number of proteins, most of which showed cellulolytic activities. In addition, at least one of the protein fractions which showed weak enzymatic activity, might contribute to cellulose hydrolysis by serving as a binding factor. Besides binding to the surface of cellulose substrate, the quarternary structure of the cellulosome particles also enhance cellulose hydrolysis by bringing individual cellulolytic enzymes within close proximity of one another.

B. *Hemicellulose*. Unlike cellulose, hemicellulose is a family of heteropolymers with frequent side chains. Hexoses including glucose, mannose and galactose, pentoses including xylose and arabinose and uronic acids are the common monomeric components of hemicellulose. Again, unlike cellulose, hemicellulose is readily bio-degradable by many microbes. The predominant hemicellulose degraders found in rumen are *Bacteroides ruminicola*, *B. fibrisolvens*, *R. flavefaciens*, and *R. albus*.

Even though enzymatic hydrolysis of hemicellulose is a relatively easy process, the complexity of hemicellulolytic enzyme systems far exceeds that of cellulases, as hemicellulose is composed of far more varieties of monomers linked together by different types of bonds. The most common hemicellulose, xylan, has a D-xylose backbone linked by beta-1,4 bonds with extensive side chains often initiated by a L-anhydro-arabinose unit. Dekker and Richards (10) reviewed the hemicellulases of rumen microbes.

C. *Lignin* Lignin is the third major component of cellulosic materials. Under anaerobic conditions of typical methane generators, it is doubtful that lignin is at all biodegradable. Often, lignin is simply referred to as refractory, meaning non-biodegradable, and non-contributing to methane formation. While this may not be totally true; nevertheless, our knowledge of bacterial hydrolysis of lignin is far less than that of cellulose and hemicellulose. There have been reports on lignin biodegradation under aerobic conditions. As methane generators are often strictly anaerobic, oxygen-required lignin degradation is, therefore, of no direct relevance to the present review.

D. Pectin. Pectin comprises a group of complex polysaccharides containing mostly poly-D-galacturonic acid methylated in the carboxyl position. Pectinolytic bacteria found in rumen include *B. ruminicola*, *B. succinogenes*, *B. fibrisolvens*, and others. Three types of enzymes are involved in bacterial hydrolysis of pectinic materials. Pectinesterase demethylates pectin to form methanol and poly-D-galacturonic acid. Hydrolases produce oligomeric chains. The third type of enzymes, lyases, depolymerize the pectin chain forming galacturonic acid residues.

E. Starch. Biodegradation of starch is rarely a bottleneck in anaerobic digestion. The subject of enzymatic hydrolysis of starch is of great interest and has been very extensively investigated and understood. Many microbes found in methane generaters can digest starch. Some common ones include *Streptococcus bovis*, *Bacteriodes amylophilus*, *Selenomonas ruminantium*, *Succinomonas amylolytica*, *B. ruminicola*, and also a number of *Lactobacillus* species.

A complete hydrolysis of starch to glucose requires also a multiple number of enzymes functioning together in a synergistic manner (11). Two types of starch molecules exist in nature. Amylose is a linear polymer of anhydroglucose linked together by alpha-1,4-bonds. Amylopectin is a highly branched polymer of anhydroglucose. Besides the alpha-1,4-bond, amylopectin has alpha-1,6-bonds which initiate side chains. Three main types of enzymes are often found in starch hydrolysis. Alpha-amylase cleaves the alpha-1,4-bonds in a supposedly random fashion but it cannot hydrolyze the 1,6-bonds. Beta-amylase breaks of maltose (dimer of glucose) from the non-reducing end of amylose and amylopectin. It, however, cannot hydrolyze the 1,6-bonds. Glucoamylase has strong activity in cleaving individual glucose units from a amylolytic chain from its non-reducing end. It has some weak 1,6-bond cleaving activity. In addition to the above, there are also debranching amylolytic enzymes which cleave the alpha-1,6-glucosidic linkages.

F. Proteins. Beside polysaccharides and lignin, protein is also a family of macromolecules commonly found in solid wastes. Proteins are important in anaerobic digestion for providing not only carbon and energy sources, but also the nutritional value of the individual amino acids to the microbes. Proteolytic enzymes from two rumen bacteria, _Bacillus amylophilus_ (12), and _B. ruminicola_ (13) have been extensively studied.

III. Pretreatment of Substrates

Among various types of biomass found in solid wastes that can be used as carbon and energy sources for methane generation by anaerobic digestion, cellulosic material stands out for its great abundance in comparison with all other types of biomass, and also for the great difficulties we have encountered in achieving an efficient conversion. Pretreatment by some physical, chemical and biological means is of key importance in unlocking the huge amounts of cellulosic renewable resources to assure a high process efficiency in anaerobic digestion. Other raw materials from biological sources such as starch wastes can be readily fermented by many types of microorganisms to generate in good yields various products including methane gas. No pretreatment is necessary because starch can be readily metabolized by many living organisms. The current annual production of cellulosic materials of different types has been estimated to be about 50 billion tons; this appears to be the only natural resource that can meet our future needs of fuels in a significant proportion in a sustainable manner. Current research on cultivation of giant kelp in ocean water may eventually lead to development of new biomass supplies. Before some major breakthrough is made in such new cultivation technology, cellulosic material, is, indeed, the major available resource for biological methane generation.

Cellulosics are generally resistant to hydrolysis by enzymes due to two major hindrances represented by the cellulose structure and the lignin barrier. Other factors such as accessible surface, pore size distribution, degree of polymerization, moisture content, and so on also affect the rate and the extent of cellulose digestion. Pretreatment methods are applied to modify these physical and chemical characteristics in the hope of enhancing the cellulose hydrolysis. The effectiveness of a pretreatment should be evaluated in more ways than one. It may enhance the rate of cellulose hydrolysis and thus also the rate of methane generation. It may also affect the extent of cellulose hydrolysis and thus the ultimate yield of the final products.

Several recent reviews dealing with various pretreatment techniques have been authored by Millet, Baker, and Satler, (14); Dunlap, Thomson and Chiang (15); Tsao, et al., (7); Bungay (16); and Fan, Gharpuray, and Lee (17). Some selected pretreatment methods are summarized and discussed below.

1) Size Reduction. Cellulosic materials have been described as fibrous cellulose microfibrils wrapped in a continuous barrier of middle lamella consisting of lignin and hemicellulose. In the native state, the accessibility to cellulose is presumably limited to only the openings at both ends of a long and thin structure. Size reduction is viewed to increase the exposed ends by cutting the long structure into many short ones. If this is a correct piecture of the situation, size reduction should be expected to enhance more the rate rather than the yield of hydrolysis.

However, depending upon the techniques of milling, some of which may contribute more than just simple size reduction, ball milling can enhance not only the rate but also the ultimate glucose yield of cellulose hydrolysis.

Recently, Kelsey and Shafizadeh (18) applied simultaneous wet milling of various cellulosic materials during enzymatic hydrolysis and observed an extraordinary increase in the rate of glucose formation by increased accessibility. The mechanical force exerted on the cellulose substrate during simultaneous milling and hydrolysis may enhance the hydrolysis by two additional milling and mechanisms as speculated by Chang, Chou, and Tsao (19) and Tsao and Chiang (8). The force may "bend" or distort the structure such that macromolecular enzymes can readily penetrate into the interior of the crystalline cellulose microfibrils. The force may also help peel off the surface layers of oligomeric anhydroglucose chains from the shrinking solid substrate and thus expose fresh surfaces for hydrolytic attack.

2) Alkali. Alkali treatment of cellulosics is the best known method of enhancement of bio-digestion of cellulose. The application of alkali such as sodium hydroxide results in removal or rupture of lignin, an increase in surface area due to cellulose swelling, and also some degree of decrystallization of cellulose. The effectiveness of this pretreatment was demonstrated with an increased enzymatic digestibility by Guggolz, Kohler, and Klopfenstein (20); Han and Callihan (21); Dunlap, Thompson, and Chiang (15); and Detroy et al., (22). Jerger, Dolenc, and Chynoweth (23) reported their recent results in enhanced rate of methane

generation but not the ultimate yield after a pretreatment of woody materials with caustics.

Alkali removes lignin and also "swells" cellulose. What exactly takes place at a molecular level in swelling is still not clear. Under certain conditions, cellulose can be dissolved into a liquid form by alkaline solutions. The difference between swelling and dissolution could be only in the extent of the treatments. At a molecular level, it would seem to be difficult to draw a demarcation at where swelling ends and dissolution begins. Furthermore, one should also realize that re-crystallization of cellulose after swelling and/or dissolution is a spontaneous process according to thermodynamic considerations. Therefore, from the viewpoint of process design, the time span between caustic swelling and the subsequent neutralization by acid followed by biological digestion would be of importance in deciding the precise enhancement in both the rate and the extent of methane generation from the pretreated substrate.

3) Steam Explosion. It has been known (Chahal et al., 24) that pressuring wood with steam and then suddenly releasing it to atmospheric pressure would shatter the structure. The shattered cellulosic biomass is said to become susceptible to hydrolysis. Iotech Company in Canada has developed a process involving a steam pressure of 45 to 52 atmospheres and releasing the pressured biomass through dies to disintegrate the cellulosic material (Bungay 16). The resultant cellulose is 80% digestible by enzyme; however, the hemicellulose fraction suffers a large loss to degradation by the high steam temperature.

4) Freeze-Explosion. Steam explosion separates cellulose fibers in a popcorn-like manner due to fast evaporation of originally pressurized water in biomass. The 400 to 500°F of the high pressure steam degrades hemicellulose and causes liberation of organic acids from the plant material. The acids at the high temperature promote further degradation of the carbohydrates. Meanwhile liquid anhydrous ammonia has been known to be a swelling agent for cellulose. Dale and Moreira (25) reported the development of a freeze-explosion technique of pretreatment of cellulose with liquid ammonia. A quick release of pressure causes also fast evaporation of ammonia and thus shatters the solid substrate. The freezing temperature of the liquid ammonia eliminates the problem of heat and acid degradation of carbohydrates. The so pre-treated alfalfa and rice straw have produced greater than 90% of the theorectical maximum

yield of glucose by enzymatic hydrolysis. The residual ammonia in the pretreated biomass can presumably provide an additional nitrogen source supporting microbial growth.

5) Mild Acid Hydrolysis as a Pretreatment. Hemicellulose can be readily removed by a dilute acid hydrolysis under gentle conditions to yield a hydrolysate containing pentoses, hexoses and uronic acids. The hydrolysate should be a good liquid substrate for methane generation. A process involving the use of a minimum amount of liquid dilute acid to achieve a high hydrolysate concentration of some 200 grams of dissolved solids per liter has been developed by researchers of the Laboratory of Renewable Resources Engineering of Purdue University. The solid residue after removal of the solubilized hemicellulose contains mostly cellulose and lignin and is highly porous and weak in physical strength. The increased pore size due to the removal of hemicellulose can presumably increase the accessibility of cellulose to hydrolytic enzymes. The mild acid hydrolysis has therefore been looked upon as a pretreatment to enhance subsequent cellulose conversion.

While the Purdue Process involves a mild acid as a pretreatment at a moderate temperature of 80 to 100°C; a high temperature dilute acid pretreatment is also reported to remove hemicellulose by Knappert and Grethlein (26). With a 0.2% sulfuric acid, and a reaction time of 30 seconds at 180°C, cellulosic materials yield little glucose indicating a small extent of cellulose hydrolysis. However, the pretreatment does reduce the DP of the cellulose greatly and the residue can then be hydrolyzed by enzymes.

Organic acids, mostly acetic acid, are released from plant materials when they are heated in the presence of water. The released acids can then promote additional hydrolysis. This is a mild acid hydrolysis without the external addition of acid and thus it has been known as the autohydrolysis. Lora and Wayman (27) employed authohydrolysis to solubilize hemicellulose and depolymerize lignin. In a more recent report by Linden, Murphy and Moreira (28), autohydrolysis of wheat straw at 180°C for 45 minutes was found to enhance subsequent cellulose hydrolysis.

6. Dissolution of Cellulose by Solvents. A pretreatment by cellulose solvents can potentially disintegrate the biomass structure by dissolving cellulose and also rupturing the surrounding lignin barriers and thus remove the two major hindrances of cellulose hydrolysis in one operation (Tsao et al., 8). Ladisch, Ladisch, and Tsao (29), have reported almost

quantitative yields of glucose from enzymatic hydrolysis of cellulose following pretreatment with cellulose solvents. The key finding was that cellulose need not be dissolved but merely soaked or permeated with solvent to disintegrate the crystalline structure. This could prove important in process development since much less solvent is needed than would be required to dissolve cellulose to form a freely flowing liquid. Many cellulose solvents have been reported in the literature. Concentrated phosphoric acid was found to render cellulose highly reactive by Walseth in 1952 (30). Recently, solvents including Cadoxen, caustic ferric tartarate, concentrated sulfuric acid, concentrated aqueous zinc chloride have been applied in pretreatment to achieve nearly quantitative conversion of cellulose into glucose (Tsao et al., 8; Tanaka et al., 31).

Gaseous hydrogen chloride was also used in pre-treatment of cellulosics. HCl upon absorption by moisture normally existing in cellulosic materials will form in situ a concentrated hydrochloric acid which at a 40% plus concentration can also swell and dissolve cellulose. The exact function of HCl in the so-called Japanese Chisso process of cellulose hydrolysis is not fully understood. However, one can expect that the cellulose dissolving ability of the concentrated hydrochloric acid formed in situ by HCl absorption must have a major role in achieving the 95% plus cellulose conversion into glucose. The well known Bergins-Rheinau process uses 40% liquid hydrochloric acid to first disrupt the cellulose structure to achieve high glucose yield. The commercial success of the HCl processes will require a nearly complete recovery and recycle of the fairly expensive hydrogen chloride. This, so far, appears to be a task difficult to achieve. The corrosive nature of the hydrochloric acid poses another technical problem.

7. Irradiation. Gamma radiation causes in cellulosics depolymerization of biopolymers, decomposition of carbohydrates, reduction in cellulose crystallinity and increase in digestibility in subsequent hydrolysis by acid and enzymes (Han and Ciegler, 32, 33). It offers thus a potentially useful pretreatment method. Large quantities of gamma-ray emitting ^{137}Cs exist in fission wastes since the initiation of ^{239}Pu production. Pretreatment by irradiation could make use of the wastes and help reduce some of the disposal problems of radioactive wastes.

IV. Future Process Design and Future Pretreatment for Methane Formation

A. Process Design. Pretreatments often achieve cellulose decrystallization, rupture of lignin barriers, and/or fractionation of cellulose, hemicellulose and lignin into separate streams. A variety of process designs for biological methane generation are thus possible. It is not the purpose of this short review to describe details of potential process designs. We will simply state that a full-scale operation of methane generation from cellulosics may justify complicated process schemes that can include multiple processing steps and a number of co- and by-products. A simple process design may involve no more than a pretreatment of the biomass followed by feeding the whole pretreated biomass to an anaerobic digester. A somewhat more complicated version could include a separation of lignin, feeding the residual cellulose and hemicellulose into anaerobic digestion and processing the separated lignin into some marketable coproducts. As hemicellulose is easy to hydrolyze, another possible version could involve feeding the hemicellulose hydrolysate to methane generators but market the cellulose to pulp and paper industries and lignin to chemical industries for better economical returns. As cellulosic biomass becomes increasingly accepted in the process industries as a viable source of raw materials, we can expect that more sophisticated processing schemes with an integrated production of a multitude of fuels and chemicals will be the rule rather than the exception.

B. Other Possible Pretreatment Methods. Many of the current pretreatment methods briefly described here are aimed at improved cellulose hydrolysis to produce fermentable sugars with increased yield and/or enhanced rate. Fermentable sugars can be fed to methane generators, and they can also be used as the primary carbon and energy source for a host of fermentation processes to produce ethanol, organic acids, and other products. Sugars such as glucose are desirable target products of cellulose hydrolysis for this versatility in fermentation. On the other hand, cellulose hydrolysis to yield glucose is obviously not an easy task. With all the research and all the pretreatment methods developed in the past and in recent years, an unquentionably superior method of glucose production from cellulose has yet to emerge. Methane generation is done by microbial flora of mixed cultures of highly versatile biological capabilities; we may not have to feed them with fermentable sugars to achieve good methane production. We thus suggest a broadened view of processing of cellulosics to optimize specifically methane generation. We suggest searching for other potential pretreatment methods that may not necessarily enhance glucose production but some other intermediates so long as these intermediates can be made efficiently from cellulosics and can be converted efficiently into methane.

Furthermore, methane has been the target product for perhaps only historical reasons. If production of gaseous fuel is really the main objective, one wonders why it has to be methane. Methane from biodigestion is always rich in Co_2 and thus relatively poor in heating value. If we can go beyond the man-made boundaries mentioned here, perhaps we can start to examine many new potential pretreatment methods. For instance, by selective catalytic oxidation, and chemical modification, cellulose can be quantitatively converted into soluble, non-crystalline polymers of the following general structure.

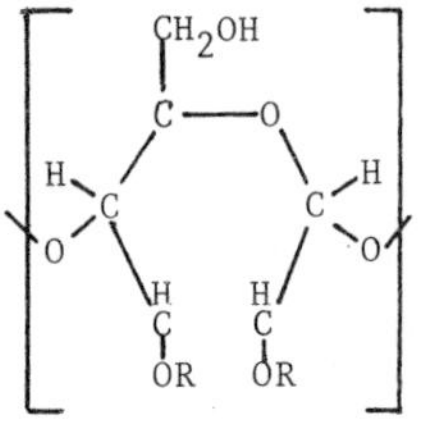

n = 20 to 1000

(I)

These polymers being polyols are bio-digestible and can be fed directly into anaerobic digestors. (I) can also be easily hydrolyzed into C_2 and C_4 fragments with or without simultaneous hydrogenation. The C_2 and C_4 fragments are again readily bio-digestible.

Carbohydrates with a general formula $-CH_2O-_{n}$ are really very rich in hydrogen. Carbohydrates do not serve as good gaseous fuel due to the excessive oxygen in the molecule. In order to produce hydrocarbons, we need selective removal of oxygen. If one heats up biomass, it loses oxygen in the form of H_2O, which does not help the generation of hydrocarbon gas. On the other hand, if we can remove oxygen from cellulose in the form of CO_2, we will be assured with a hydrogen enriched residue which should be an excellent feed for anaerobic digestors. Therefore we need to look for decarboxylation of carbohydrate as a general process and pretreatment scheme for generation of hydrocarbons. Several chemical pretreatments can promote decarboxylation from carbohydrates.

In brief, we are suggesting in this section, that if bacterial hydrolysis of cellulose to form soluble sugar is the main bottleneck of anaerobic digestion of solid wastes, other chemical reactions should also be examined for possible alternatives to hydrolysis for generation of soluble intermediates to be fed to acetogenic and methanogenic bacteria.

REFERENCES

1. Hungate, R.E., "Studies on Cellulose Fermentation. III. The Culture Isolation of Cellulose-Decomposing Bacteria from the Rumen of the Cattle," J. Of Bacteriology 53:631-45 (1949).

2. Hungate, R.E., "Microorganisms in the Rumen of Cattle Fed a Constant Ration," Can. J.. Microbiology 3:289-311 (1957).

3. Hungate, R.E., "Studies on Cellulose Fermentation. I. The Culture and Physiology of an Anaerobic Cellulose Digesting Bacterium," J Bacteriology, 48:499-513 (1944).

4. Hungate, R.E., "The Anaerobic Mesophilic Cellulolytic Bacteria," Bact Review, 14:1-49 (1950).

5. Shane, B.S., Gouws, L. and Kistner, A., "Cellulolytic Bacteria Occurring in the Rumen of Sheep Conditioned to Low Protein Teff Hay," J. Gen. Microbiol., 55:445-51 (1969).

6. Gong, C.S., Ladisch, M.R., and Tsao, G.T., "Biosynthesis, Purification, and Mode of Action of Cellulases of *Trichoderma reesei*," Adv. Chem. Ser. 181, 261-287 (1979).

7. Tsao, G.T., Ladisch, M., Ladisch, C., Hsu, T.A., Dale, B., and Chou, T., "Fermentation Substrates from Cellulosic Materials: Production of Fermentable Sugars," Ann. Reports Ferm. Processes, 2:1-22 (1978).

8. Tsao, G.T. and Chiang, L.C., "Cellulose and Hemicellulose Technology," The Filamentous Fungi, 4:296-326 (1983).

9. Lamed, R., Setter, E. Kenig, R., and Bayer, E.A., "The Cellulosome - a Discrete Cell Surface Organelle of *Clostridium thermocellum* which exhibits separate antigenic, cellulose-binding and Various Cellulolytic Activities," Paper presented at 5th Conf. on Biotech. for Fuels & Chemicals 1983. Gatlinburg, TN.

10. Dekker, R.F.H. and Richards, G.R., "Hemicellulases: Their Occurrence, Purification, Properties, and Mode of Action," Adv. Carbohydrate Chem. Biochem., 32:277-352 (1976).

11. Fogarty, W.M. and Kelly, C.T., "Starch Degrading Enzymes of Microbial Origin," Progress in Industrial Microbiol, 15:87-150 (1979).

12. Blackburn, T.H., "The Protease Liberated From Bacteroides Amylophilus Strain H18 by Mechanical Disintegration," J Gen Micro., 53:37-51 (1968).

13. Hazelwood, G.P. and Edwards, R., "Proteolytic Activities of a Rumen Bacterium, Bacteriodes ruminicola R8/4," J. Gen Micro. 125:11-15 (1981).

14. Millet, M.A., Baker, A.J., and Satter, L.D., Physical and Chemical Pretreatments for Enhancing Cellulose Saccharification. Biotech Bioeng. Symp., 6:125-153 (1976).

15. Dunlop, C.E., Thompson, J., and Chiang, L.C., Treatment Processes to Increase Cellulose Microbial Digestibility. Am. Inst Chem. Eng. Symp. Series, 72:58-63 (1976).

16. Bungay, H.R., Energy, The Biomass Option, New York, Wiley. (1981).

17. Fan, L.T., Gharpuray, M.M. and Lee, Y.H., Evaluation of Pretreatments for Enzymatic Conversion of Agricultural Residues. Biotech. Bioeng. Symp., 11:29-45 (1981).

18. Kelsey, R.G. and Shafizadeh, F., Enhancement of Cellulose Accessibility and Enzyme Hydrolysis by Simultaneous Wet Milling. Biotech. Bioeng., 22:1025-1036 (1980).

19. Chang, M.M., Chou, T.Y.C., and Tsao, G.T., Structure, Pretreatment, and Hydrolysis of Cellulose, Adv. Biochem. Engr., 20:15-42 (1981).

20. Guggolz, J., Kohler, G.O. and Klopfenstein, J.J., Composition and Improvement of Grass Straw for Ruminant Nutrition. J. Animal Sci., 33:151-156 (1971).

21. Han, Y.W. and Callihan, C.D., Cellulose Fermentation: Effect of Substrate Pretreatment on Microbial Growth. Appl. Microbiol., 27:159-165 (1974).

22. Detroy, R.W., Lindenfelser, L.A., Sommer, S. and Orton, W.L., Bioconversion of Wheat Straw to Ethanol: Chemical Modification, Enzymatic Hydrolysis and Fermentation, Biotech. Bioeng., 25:1527-1535 (1981).

23. Jerger, D.E., Dolenc, D.A., and Chynoweth, D.P., Biogasification of Woody Biomass Following Physical and Chemical Pretreatment, Paper presented at 7th Symp. on Energy from Biomass and Wastes, Florida (1983).

24. Chahal, D.S., McGuire, S., Pikor, H. and Noble, G., Bioconversion of Cellulose and Hemicellulose of Wood into Fungal Food and Feed with Chaetomium cellulolyticum. Paper presented at the 2nd World Congress of Chemical Engineering, Montreal. (1981).

25. Dale, B.E. and Moreira, M.J., A Freeze-Explosion Technique for Increasing Cellulose Hydrolysis., Biotech Bioeng. Symp. 12:31-43 (1982).

26. Knappert, D., Grethlein, H. and Converse, A., Partial and Hydrolysis of Cellulosic Materials as a Pretreatment for Enzymatic Hydrolysis, Paper presented at Am. Inst. Chem. Eng. Natl. Meeting, San Francisco, (1979).

27. Lora, J.H. and Wayman, M., Delignification of Hardwood by Autohydrolysis and Extraction, TAPPI, 61:47-50 (1978).

28. Linden, J.C., Murphy, V.G. and Moreira, A.R., Wheat Straw Autohydrolysis, Adv. in Biotech., 2:41-46 (1980).

29. Ladisch, M.R., Ladisch, C., and Tsao, G.T., Cellulose to Sugars: New Path Gives Quantitative Yield, Science, 201:743-745 (1978).

30. Walseth, C.S., The Influence of the Fine Structure of Cellulose on the Action of Cellulases, TAPPI, 35:233-238 (1952).

31. Tanaks, M., Taniguchi, M., Morita, T., Motsuno, R., and Kamikubo, T., Effect of Chemical Treatment on Solubilization of Crystalline Cellulose and Cellulosic Wastes with *Pellicularia filamentosa* Cellulase. J. Ferm. Tech., 53:186-190 (1979).

32. Han, Y.W. and Ciegler, A., Effect of Gamma-ray Irradiation of Sugar Production from Plant Biomass. Biotech Bioeng. Symp., 12:73-77 (1982).

33. Han, Y.W. and Ciegler, A., Effect of High Energy Radiation on Lignocellulose Conversion, Annual Reports on Fermentation Processes, 6:299-322 (1983).

ONGOING ACTIVITIES IN EUROPE IN THE FIELD OF ENZYMATIC HYDROLYSIS

M. LINKO

VTT Biotechnical Laboratory
Espoo, Finland

Summary

So far no processes based on the enzymatic hydrolysis of cellulosics have progressed beyond pilot scale despite intensive research efforts. However, this has not discouraged people responsible for research strategies. Research on various aspects within this field is going on in most European countries.

The areas involved include production of the enzymes capable of breaking down cellulose and hemicellulose, pretreatment of cellulosic materials, hydrolysis of cellulose and hemicellulose (primarily xylan) as well as utilization of the hexose and pentose sugars formed. The utilization of the sugars may be simultaneous with the hydrolysis. Direct conversions of cellulosics to ethanol have also been studied.

Efficient development of the hydrolysis processes is possible only provided that detailed knowledge of the biochemistry involved is available. Some European institutes are carrying out important research work in this field.

Much of the pioneer work has been carried out outside Europe, primarily in Japan and the U.S.A. However, Europe has been catching up. In this connection it should be mentioned that Europe has a strong position in the industrial production of enzymes, including cellulases. Current European research on bioconversion of cellulosic materials covers all the areas involved.

1. INTRODUCTION

Research on enzymatic hydrolysis includes a number of different features such as production of the enzymes capable of breaking down cellulose and hemicellulose, pretreatment of cellulosic materials, hydrolysis of cellulose and hemicellulose, utilization of the hexose and pentose sugars formed, possibly simultaneously with the hydrolysis, and direct conversion of cellulosics to ethanol, SCP or other products. Thorough scientific knowledge is necessary for development of process applications. Biochemistry, chemistry, microbiology and genetics are all involved. A review describing the present status of enzymatic hydrolysis has been recently presented (1).

There has been considerable research activity in all these fields in Europe during recent years. However, it is not possible to review the work of all European groups in this presentation. Moreover, it is impossible to know what activities are ongoing. This paper aims at reviewing all the various aspects of enzymatic hydrolysis by pointing out examples of recent results considered especially interesting and valuable, rather than trying to present a complete list, which would lead to repeating more or less similar descriptions. There are, no doubt, several important groups in Europe not mentioned in this presentation, and they could have been picked as examples with equal motivation.

In order to keep this presentation up-to-date, all publications referred to are from the years 1983 and 1984, and in order to keep it European, no American, Canadian, Australian, Japanese, Indian etc. research results have been included, although it would have been very tempting to do so, because very important research work in this field is being and has been carried out outside Europe.

2. ENZYME PRODUCTION

Enzymatic hydrolysis starts with production of enzymes. Some groups in France, Finland and Germany are working with *Trichoderma reesei* (2,3,4, 5,6). This organism is generally considered to be the most efficient producer of extracellular cellulases. Various strains have been compared (2,5). Some new European strains compare favourably with the efficient American strains described in the literature. A special advantage of some strains is their ability to produce efficiently cellulases also with non-cellulosic carbon sources such as lactose or even glucose (2,3,4). This also facilitates the use of fed batch (3) or continuous processes (6).

In addition to *Trichoderma reesei* some other organisms are also being studied for cellulase production. *Trichoderma pseudokoningii* has been studied in Austria (7). The addition of carboxymethyl cellulose to a medium containing crystalline cellulose (avicel) as a carbon source resulted in increased production of several enzymes, especially β-glucosidase and exo-β-1,3-glucanase.

Talaromyces emersonii has been extensively studied in Ireland (8,9, 10). Mutants with enhanced cellulase activity have been isolated and different inducing substrates have been tested.

Alternaria alternata has been studied in Greece (11). This organism is known to produce a thermostable lactase in a whey medium, but it is also an interesting producer of cellulase and β-glucosidase.

Aspergillus species are known as good producers of β-glucosidase, xylanase and β-xylosidase, but for example *Aspergillus fumigatus* mutants also capable of producing high cellulase activities have recently been developed in Great Britain (12). A mixture of *Trichoderma reesei* and *Aspergillus niger* has been studied in Italy for hydrolysis of cellulose (13).

Sporotrichum pulverulentum, anamorph of *Phanerochaete chrysosporium*, has been studied in Sweden for many years. These investigations are being continued. Mutants of this white-rot fungus with increased cellulase and β-glucosidase production have been isolated (14). The cellulase activity obtained is comparable to that reported for *Trichoderma reesei* QM 9414.

Several *Penicillium* species are known to be cellulase producers. *Penicillium funiculosum* is being studied in Great Britain (15).

Some cellulase producing organisms such as *Trichoderma reesei* also produce β-glucosidase, xylanase and β-xylosidase. In addition, several other microbes have been tested specifically for the production of these enzymes. This work is probably being continued in many European laboratories. In addition to conventional submerged fermentation processes solid state fermentation has also been used for production of enzymes. β-glucosidase production by *Aspergillus phoenicis* in solid state fermentation has been studied in France (16).

3. RECOMBINANT DNA TECHNIQUE

Recombinant DNA technique provides new tools for enzyme technology. In the field of cellulases practical applications cannot be expected in the near future, but some interesting results have recently been reported. At least four groups in Europe are active in this field in Denmark (17), Finland (18,19), France (20), and Ireland (21).

Application of recombinant DNA technology to any given area requires extensive knowledge of the biochemistry and molecular biology of the system to be exploited. It is only recently that this type of information regarding cellulases and cellulolytic organisms became available.

A French group (20) has reported the cloning of two different genes from *Clostridium thermocellum*, both coding for an enzyme capable of hydrolyzing CMC. These genes were isolated by screening a cosmid bank for the expression of cellulase genes in *E. coli*. One gene coded for an enzyme able to hydrolyse trinitrophenylated CMC, whereas a second gene coded for an enzyme able to hydrolyse CMC but not trinitrophenyl CMC. These genes have been subsequently analysed in detail (51).

A similar approach to cloning cellulases has been used in Ireland (21). A *Bacillus subtilis* gene coding for an endo-β-1,3-1,4-glucanase has been transferred to *E. coli* using bacteriophage λ and plasmid cloning vectors. The *Bacillus* β-glucanase produced in *E. coli* digested barley glucan and lichenan but not CMC or laminarin as did the original enzyme. A similar enzyme has been cloned in Denmark from *Bacillus circulans* (17).

In Finland, partly because of earlier work on the development of more efficient cellulase producing strains of the fungus *Trichoderma reesei*, much effort has been focused on cloning the cellulase genes from this organism. Unfortunately eukaryote genes are rarely expressed in bacterial hosts. This has meant that more sophisticated cloning techniques were required to identify the genes of interest. However, the cloning and characterization of the gene for the major cellobiohydrolase, CBH I, from *Trichoderma reesei* has recently been reported (18). All the major cellulase genes have now been isolated and are being characterized in detail. The expression of the β-glucosidase from *Aspergillus niger* in the yeast *Saccharomyces*, leading to the production of a yeast capable of growing on cellobiose has also been reported (19).

4. BIOCHEMISTRY OF CELLULASES

The biochemistry of the enzymatic hydrolysis of cellulose is extremely complicated. A review on microbial cellulases has recently been published (22). After years of extensive studies there are still many open questions. The application of recombinant DNA methods to improve the production of cellulases has created a new demand for detailed information concerning these enzymes. Several European countries are active in this field of research.

Electron microscopy has been used in France to study the action of cellobiohydrolase from *Trichoderma reesei* on microcrystalline cellulose (23). It has been shown that cellobiohydrolase is capable of breaking down cellulose without endo-glucanases. This is in accordance with recent Finnish studies according to which a cellobiohydrolase from *Trichoderma reesei* acting alone is capable of hydrolysing native cotton (24). Enzyme-substrate affinity and immunoaffinity methods were used for purification of the enzymes. Further work is in progress.

The protein sequences of cellulolytic enzymes of *Trichoderma reesei* have been analyzed in Sweden (25). The sequence of about 490 amino acid residues of the main cellobiohydrolase has been established by automatic liquid phase Edman degradation. A partial sequence of endoglucanase has been determined with the same technique in combination with solid phase technique (26). These investigations are being continued.

In addition to *Trichoderma reesei* and *Sporotrichum pulverulentum* (*Phanerocaete chrysosporium*), *Clostridium thermocellum* has also been studied in Sweden. An interesting yellow affinity substance involved in the cellulolytic system of *C. thermocellum* has recently been found (27).

The biochemistry of *Trichoderma* cellulases has also been studied in the Federal Republic of Germany (28), United Kingdom (29), Czechoslovakia (30) and the USSR (31).

The present situation in biochemical studies of cellulases can be summarized as follows. New purification and analysis techniques adapted for cellulolytic enzymes have confirmed their complex nature. Enzymes survive the new purification procedures in an active form. Enzyme components not previously described are being discovered. The theories of the mode of action of cellulolytic enzymes should be rechecked once again.

5. PRETREATMENT OF CELLULOSIC MATERIALS

One of the major technical and economical problems in the enzymatic hydrolysis of cellulosic materials is the development of an efficient inexpensive pretreatment method. Efficient pretreatments tend to be costly and, furthermore, destructive at least to the hemicellulose components, which are easily decomposed into non-utilizable reaction products. The same problem is encountered in acid hydrolysis: loss of sugar and formation of inhibitory compounds.

The main alternatives of pretreatments for practical applications seem to be steaming, organosolv treatments and mild acid hydrolysis. Mild acid hydrolysis can also be combined with steaming.

Steaming treatments are being studied in a German-Finnish cooperation project (32,33). During steaming the hemicelluloses become water soluble and can subsequently be extracted. The accessibility of the fibrous material to cellulases is markedly increased. However, the optimal conditions of pretreatment for enzymatic hydrolysis of cellulose and for fermentation of pentoses do not coincide.

A method called MD-Organosolv has been developed in FRG (34). This method is based on the use of alcohol as solvent. A demonstration plant has been constructed for further development of this process with a capacity of 3 t/d.

In the Battelle-Geneva process (35) phenols are used as delignification solvents but resinification is avoided. The delignification is carried out at 100°C and atmospheric pressure. Under these conditions the phenols are totally miscible with the water phase, forming a homogenous liquid phase into which lignin is dissolved. The hemicellulosic fraction is hydrolyzed simultaneously. The remaining fibrous cellulose fraction is separated by filtration and the liquid phase is allowed to cool. During cooling the phenolic portion with the dissolved lignin separates spontaneously from the aqueous phase containing the pentoses.

A method called hydrothermolysis has recently been developed in Austria (36). A percolater type reaction vessel is used. Biomasses such as aspen wood or wheat straw are hydrolyzed in two stages using temperatures of 180-200°C for the easily hydrolyzable polysaccharides and about 260°C for the polysaccharides which are difficult to hydrolyze. The first stage could perhaps serve as a pretreatment for enzymatic hydrolysis.

6. HYDROLYSIS

Some details of the hydrolysis of cellulosic materials depend on the final product. If the final product is ethanol, hydrolysis is often combined with fermentation in order to minimize end product inhibition.

A possible way of decreasing enzyme costs is recirculation of the enzymes. Recirculation in an ultrafiltration membrane reactor has been studied in Sweden (37,38,39). Ultrafiltration also removes sugar from the reaction mixture, thus decreasing end product inhibition. In the Swedish investigation the degree of conversion of pretreated sallow was in this way improved from 40 per cent in a batch hydrolysis to 95 per cent (within 20 hours), and the initial hydrolysis rate was increased up to seven times (38). Aqueous two-phase systems for enzyme recirculation are also being studied in Sweden (40).

Significant losses of cellulases are caused by strong adsorption of the enzymes onto undigested material. In a semi-continuous hydrolysis in which fresh substrate was added intermittently, the amount of reducing sugars produced was 26 g/g enzyme compared to less than 5 g/g enzyme in a conventional batch hydrolysis (38).

Enzymatic hydrolysis of steam-pretreated lignocellulosic materials is being studied in German-Finnish cooperation (33). Both separate hydrolysis and combined hydrolysis and ethanol fermentation are used for different pretreated materials. Fed-batch hydrolysis for increasing the total substrate concentration is also being investigated. The inhibition of cellulases by glucose and ethanol has been studied. Consecutive hydrolyses are used to elucidate the possibilities of enzyme reuse.

A pilot reactor for enzymatic hydrolysis of cellulosic waste materials has been developed in FRG (41). The volume of this reactor is 45 litres and its length is 1.80 m. The diameter of the spiral is 170 mm and its turning speed can be regulated to give retention times from 0.5 hours to 350 hours. The materials tested have been potato pulp, sugar beet cuttings and wheat straw.

Acid and enzymatic hydrolysis of cellulose have been compared in Belgium (42,43). The substrate studied was papermill sludge. A problem in the acid hydrolysis was the high mineral load. This problem is not encountered in enzymatic hydrolysis.

The utilization of hemicellulose, in addition to cellulose, is also coming into focus. Many European countries are certainly involved in this field of research, including FRG (44) and Finland (32). A key problem in the utilization of xylan seems to be finding suitable final products, other than SCP or ethanol. Ideally, it should be a bulk product for which a pentose sugar would be a more favourable raw material than hexoses, but such products are not easy to come by.

7. SINGLE-STAGE PROCESSES

Cellulosic materials can also be converted into final products in a single stage, without any separate enzyme production process. Most of these direct processes are based on *Clostridia*. These is considerable research activity in this field in France. *Clostridium thermocellum* has been studied for conversion of sugar beet pulp (45). Without any pretreatment (or with short milling) 60% of the dry matter of beet pulp was consumed in 60 hours in a ten litre reactor. The conversion products are acetic acid and ethanol. *Clostridium acetobutylicum* has been used in coculture with a mesophilic cellulolytic *Clostridium* sp. (46).

One interesting approach to the utilization of cellulosic materials is their protein enrichment in solid state fermentation. This may be more feasible than SCP processes at present. Solid state fermentation processes for protein enrichment have been studied particularly in France (47) and also in some Eastern European countries such as GDR (48) and the Soviet Union (49).

A *Trichoderma album* mutant strain has been used in France. This organism has a high protein content, a favourable amino acid composition and a low nucleic acid content. It does not produce toxins or an unpleasant smell. Consequently, it seems to be a suitable organism for feed production. An estimate for a plant producing 200000 t/a of protein-enriched product (18% protein, 90% dry matter) gave a price of 0.14 USD/kg (47). Present research aims at further reduction of this price. Results from nutritional assays are promising.

Conversion of cellulose into fungal cell mass in a stirred tank reactor is being studied in FRG (50). The organism used is *Chaetomium cellulolyticum*. Steam-pretreated straw has been found more suitable than NH_3-pretreated straw. A microprocessor-controlled fed batch process has been studied. The control is based on alkaline consumption.

8. CONCLUSIONS

There is significant research activity in many European countries in the field of enzymatic hydrolysis. All the sub-areas are included. However, it can hardly be stated that Europe is at present in a key position in general. On the other hand, in some areas Europe is quite advanced, for example in the biochemistry of cellulases and in the production of these enzymes. Also, the general level of knowledge in the field of enzymatic hydrolysis is in Europe up-to-date, enabling technical applications as soon as they become feasible. This readiness must be maintained and further technical improvements must be created. It is not enough to be just a quick follower.

ACKNOWLEDGEMENT

Several people from many European countries have kindly sent information necessary in writing this paper.

REFERENCES

1. Linko, M. Enzymatic hydrolysis - present status and future development. Statusseminar "Bereitstellung und Verwertung von Lignincellulosen". KFA Jülich, 21.-22. November 1983.
2. Bailey, M. & Oksanen, J. Cellulase production by mutant strains of *Trichoderma reesei* on non-cellulosic media. Third European Congress on Biotechnology, München, 10-14 September 1984 (Accepted).
3. Pourquie, J. Fed batch production of cellulolytic enzymes by *Trichoderma reesei* on a lactose based culture medium. VII International Biotechnology Symposium, New Delhi, February 19-25, 1984, Abstracts Vol. 2, p. 617.
4. Warzywoda, M., Ferre, V. & Pourquie, J. Development of a culture medium for large scale production of cellulolytic enzymes by *Trichoderma reesei*. Biotechnology and Bioengineering 25 (1983) 3005-3010.
5. Warzywoda, M., Vandecasteele, J.P. & Pourquie, J. A comparison of genetically improved strains of the cellulolytic fungus *Trichoderma reesei*. Biotechnology Letters 5 (1983) Nr. 4, 243-246.
6. Ross, A., Schügerl, K. & Scheiding, W. Cellulose production by *Trichoderma reesei*. European Journal of Applied Microbiology and Biotechnology 18 (1983) 29-37.
7. Harrer, W., Kubicek, C.P., Röhr, M., Wurth, H. & Marihart, J. The effect of carboxymethyl cellulose addition on extracellular enzyme formation in *Trichoderma pseudokoningii*. European Journal of Applied Microbiology and Biotechnology 17 (1983) 339-343.

8. Moloney, A.P. & Coughlan, M.P. Sorption of *Talaromyces emersonii* cellulase on cellulosic substrates. Biotechnology and Bioengineering 25 (1983) 271-280.

9. Moloney, A.P., Considine, P.J. & Coughlan, M.P. Cellulose hydrolysis by the cellulases produced by *Talaromyces emersonii* when grown on different inducing substrates. Biotechnology and Bioengineering 25 (1983) 1169-1173.

10. Moloney, A.P., Hackett, T.J., Considine, P.J. & Coughlan, M.P. Isolation of mutants of *Talaromyces emersonii* CBS 814.70 with enhanced cellulase activity. Enzyme and Microbial Technology 5 (1983) 260-264.

11. Macris, B.J. Enhanced cellulase and β-glucosidase production by a mutant of *Alternaria alternata*. Biotechnology and Bioengineering 26 (1984) 194-196.

12. Wase, D.A.J. & Vaid, A.K. Isolation and mutation of a highly cellulolytic strain of *Aspergillus fumigatus*. Process Biochemistry 18 (1983) Dec., 35.

13. Cantarella, M., Pezzullo, L., Scardi, V. & Alfani, F. Enhancement of cellulose hydrolysis using cellulase mixtures. VII International Biotechnology Symposium, New Delhi, February 19-25, 1984, Abstracts Vol. 2, p. 432.

14. Eriksson, K-E. & Johnsrud, S.C. Mutants of the white-rot fungus *Sporotrichum pulverulentum* with increased cellulase and β-D-glucosidase production. Enzyme and Microbial Technology 5 (1983) 425-429.

15. Wood, T. This symposium.

16. Deschamps, F. & Huet, M.C. β-glucosidase production by *Aspergillus phoenicis* in solid state fermentation. Biotechnology Letters 6 (1984), Nr. 1, p. 55-60.

17. Jørgensen, P.L. Personal communication.

18. Teeri, T., Salovuori, I. & Knowles, J. The molecular cloning of the major cellulase gene from *Trichoderma reesei*. Biotechnology 1 (1983) 696-699.

19. Penttilä, M.E., Nevalainen, K.M.H., Raynal, A. & Knowles, J.K.C. Cloning of *Aspergillus niger* genes in yeast. Expression of the gene coding *Aspergillus* β-glucosidase. Molecular and General Genetics 1984 in press.

20. Cornet, P., Tronik, D., Millet, J. & Aubert, J-P. Cloning and expression in *Escherichia coli* of *Clostridium thermocellum* genes coding for amino acid synthesis and cellulose hydrolysis. FEMS Microbiology Letters 16 (1983) 137-141.

21. Cantwell, B.A. & McConnell, D.J. Molecular cloning and expression of a *Bacillus subtilis* β-glucanase gene in *Escherichia coli*. Gene 23 (1983) 211-217.

22. Enari, T-M. Microbial cellulases. In: Microbial enzymes and biotechnology, Fogarty, W.M., ed. Applied Science Publishers, London 1983, pp. 183-223.

23. Chanzy, H., Henrissat, B., Vuong, R. & Schuelein, M. The action of 1,4-β-D-glucan cellobiohydrolase on *Valonia* cellulose microcrystals. An electron microscopic study. FEBS Letters 153 (1983) 113-118.

24. Nummi, M., Niku-Paavola, M-L., Lappalainen, A., Enari, T-M. & Raunio, V. Cellobiohydrolase from *Trichoderma reesei*. Biochemical Journal 215 (1983) 677-683.

25. Fägerstam, L.G., Pettersson, L.G. & Engström, J.Å. The primary structure of a 1,4-glucan cellobiohydrolase from the fungus *Trichoderma reesei* QM 9414. FEBS Letters 167 (1984) 309-315.

26. Bhikhabhai, R. & Pettersson, L.G. The cellulolytic enzymes of *Trichoderma reesei* as a system of homologous proteins. FEBC Letters 167 (1984) 301-308.

27. Ljungdahl, L.G., Pettersson, B., Eriksson, K-E. & Wiegel, J. A yellow affinity substance involved in the cellulolytic system of *Clostridium thermocellum*. Current Microbiology 8 (1983).

28. Sprey, B. & Lambert, C. Titration curves of cellulases from *Trichoderma reesei:* demonstration of a cellulase-xylanase-β-glucosidase-containing complex. FEMS Microbiol. Letters 18 (1983) 217-222.

29. Thomas, D.A., Stark, J.R. & Palmer, G.H. Purification of glucan hydrolases from commercial preparation of *Trichoderma viride* by chromatofocusing. Carbohydrate Research 110 (1983) 343-345.

30. Labudová, I. & Farkaš, V. Multiple enzyme forms in the cellulase system of *Trichoderma reesei* during its growth on cellulose. Biochimica and Biophysica Acta 744 (1983) 133-140.

31. Rabinovich, M.L., Chernoglazov, V.M. & Klyosov, A.A. Isoenzymes of endoglucanase in cellulase complexes: various affinity for cellulose and different role in the hydrolysis of the insoluble substrate. Biokhimiya 48 (1983) 369-378.

32. Puls, J., Poutanen, K. & Viikari, L. The effect of steaming pre-treatment on the biotechnical utilization of wood components. BioEnergy '84, Goethenburg, 18-21 June 1984.

33. Poutanen, K. & Puls, J. Enzymatic hydrolysis of steam-pretreated lignocellulosic materials. Third European Congress on Biotechnology, München 10-14 September 1984 (Accepted).

34. Edel, E. & Feckl, J. Das MD-Organosolv-Zellstoffverfahren. Zellstoffherstellung mittels Alkohol als Extraktionsmittel. Statusseminar "Bereitstellung und Verwertung von Lignincellulosen". KFA Jülich, 21.-22. November 1983.

35. Anonymous. The development of a new process for the fractionation and hydrolysis of biomass. Rintekno Oy & Battelle Geneva Research Centres, Geneva, December 6., 1983.

36. Bonn, G., Concin, R. & Bobleter, O. Hydrothermolysis - a new process for the utilization of biomass. Wood Science and Technology 117 (1983) 195-202.

37. Ohlson, I., Trädgårdh, G. & Hahn-Hägerdal, B. Recirculation of cellulolytic enzymes in an ultrafiltration membrane reactor. Acta Chemica Scandinavica B37 (1983) 737-738.

38. Ohlson, I., Trädgårdh, G. & Hahn-Hägerdal, B. Enzymatic hydrolysis of sodium hydroxide pretreated sallow in an ultrafiltration membrane reactor. Biotechnology and Bioengineering 26 (1984) in press.

39. Ohlson, I., Trädgårdh, G. & Hahn-Hägerdal, B. Evaluation of UF and RO in a cellulose saccharification process. Proceedings of the 4th Tübingen Symposium on: Synthetic Membranes in Science and Industry, September 6-9, 1983. Accepted for publication in Desalination.

40. Hahn-Hägerdal, B. Biokonvertering av cellulose till flytande bränsle (etanol). Report 30.6.1983.

41. Borchert, A. Entwicklung und Erprobung eines Pilotreaktors für die enzymatische Hydrolyse cellulosehaltiger Abfallstoffe. Zwischenbericht 3.2.1984.

42. Paquot, M. & Hermans, L. Alternatives possibles à la mise en décharge des boues de papeteries. Un example: Wiggins Teape (Belgium) S.A. Tribune Cebedeau 36 (1983) No. 473, p. 147-155.

43. Thonart, P., Marcoen, J.M., Desmons, P., Foucart, M. & Paquot, M. Etude Comparative de l'Hydrolyse Enzymatique et de l'Hydrolyse par Voie Acide de la Cellulose. Holzforschung 37 (1983) 173-178.

44. Puls, J. Verwentung von Hemicellulosen. Statusseminar "Bereitstellung und Verwentung von Lignocellulosen". KFA Jülich, 21.-22. November 1983.

45. Spinnler, E., Rossero, A. & Blachere, H. Bioconversion of cellulose by *Clostridium thermocellum*, salt concentration sensitivity and amino acid metabolism. VII International Biotechnology Sympsium, New Delhi, February 19-25, 1984, Abstracts Vol. 2, p. 437-438.

46. Fond, O. & Petitdemange, H. Cellulolysis by a coculture of a mesophilic cellulolytic *Clostridium* and *Clostridium acetobutylicum* and influence of the sugar production on the acetone-butanol fermentation. VII International Biotechnology Symposium, New Delhi, February 19-25, 1984, Abstracts Vol. 2, p. 428-429.

47. Durand, A., Teilhard de Chardin, O., Chereau, D., Larios de Anda, G. & Blachere, H. Protein enrichment of sugar beet pulps by solid state fermentation. VII International Biotechnology Symposium, New Delhi, February 19-25, 1984, Abstracts Vol. 2., p. 464-465.

48. Klappach, G., Weichert, D. & Meyer, D. Protein enrichment in lignocellulosic material. VII International Biotechnology Symposium, New Delhi, February 19-25, 1984, Abstracts Vol. 2., p. 461.

49. Golovleva, L.A., Golovlev, E.L., Chermensky, D.N., Okunev, O.N., Ganbarov, Kh. & Brustovetskaya, T.A. Solid state fermentation of plant raw materials. Symposium "Bioconversion of plant raw material by microorganisms". Reports from Department of Microbiology, University of Helsinki, 26/1983, pp. 22-43.

50. Hecht, V., Schügerl, K. & Scheiding, W. Optimization of cellulose conversion into fungal cell mass. Journal of Chemical Technology and Biotechnology 33B (1983) 231-240.

51. Cornet, P., Millet, J., Béguin, P. & Aubert, J-P. Characterization of two *CEL* (cellulose degradation) genes of *Clostridium thermocellum* coding for endoglucanases. Biotechnology 1 (1983) 589-593.

IMPROVEMENT OF PRETREATMENTS AND TECHNOLOGIES FOR ENZYMATIC HYDROLYSIS OF CELLULOSE FROM INDUSTRIAL AND AGRICULTURAL REFUSE AND COMPARISON WITH ACIDIC HYDROLYSIS

M. Paquot, P. Thonart, M. Foucart, P. Desmons and A. Mottet
Département de Technologie
Faculté des Sciences Agronomiques
5800 Gembloux (Belgium)

SUMMARY

Hydrolysis of cellulose may be effected by chemical or enzymatic methods. Each of these two technologies has its specific advantages and disadvantages, which were compared in the course of the research.
Enzymatic hydrolysis of cellulose depends on the interaction between the enzyme complex and the substance to be hydrolysed. The basic research carried out on paper mill celluloses shows that it is mainly the endo- and exocellulase activities which are absorbed on the substrate. In the case of the Trichoderma reesei complex enriched with Aspergillus niger β-glucosidase, the exocellulase activity has proved to be the limiting factor in the design of a reactor which recycles the enzyme complex.
Tests were carried out of various pretreatments for substrates of more complex chemical composition: straw, maize stalks, beet pulp, etc.
The rate and overall degree of hydrolysis, especially in the case of straw, were improved by chemical treatments such as cooking with lime-soda in combination with mechanical processes.
A first pilot plant for straw hydrolysis was then set up with the aim of attaining higher sugar concentrations in the fermenter.

The hydrolysis residues were also characterized and compared with the residues from acidic attack.

1. HYDROLYSIS OF CELLULOSE: BASIC RESEARCH

1.1 Comparison of acidic hydrolysis and enzymatic hydrolysis

The difficulties arising with regard to acidic hydrolysis and enzymatic hydrolysis are fundamentally different. The main problems posed by the two techniques are set out in Table I.

Problems of acidic hydrolysis
- cellulose crystallinity
- attainable percentage hydrolysis
- rapid furfural formation
- formation of hydroxymethlyfurfural
- non-specificity of reaction
Problems of enzymatic hydrolysis
- enzyme production
- concentration in substrate
- lignin-carbohydrate relationships
- kinetics of hydrolysis
- risks of microbial contamination
- inhibition of the enzyme complex by the products of hydrolysis

Table I : Problems of acidic and enzymatic hydrolysis

It is therefore not surprising that acidic hydrolysis may be used as a preparation for enzymatic hydrolysis or as an effective method of treating enzymatic hydrolysis residues. These two possibilities are illustrated by Figures 1 and 2. Enzymatic hydrolysis is carried out with the Trichoderma reesei enzyme complex (0.3 IU/ml) enriched with Aspergillus niger β-glucosidase (5 IU/ml).

1.2 The enzyme complex and its adsorption on the substrate

The need for the Trichoderma reesei enzyme complex to be enriched with β-glucosidase is the first factor which makes it possible to increase the efficiency of hydrolysis, as cellobiose has a powerful inhibiting effect on the cellulases.

The study of the composition of the enzyme complex described in the article by Desmons (Desmons et al, 1983) made it possible to isolate, though not totally, various exocellulase and endocellulase activities, β-glucosidase, β-xylosidase and xylanase activities. Adsorption of the enzyme complex on various cellulosic substrates was observed, in

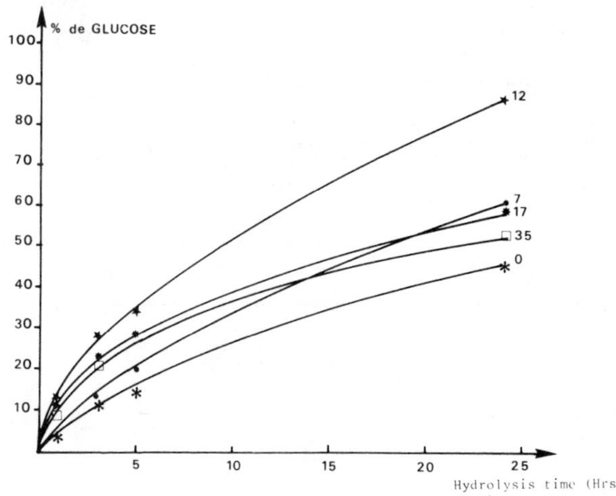

Figure 1: Behaviour of residues from acidic hydrolysis of cellulose when subjected to enzymatic hydrolysis

0 = control - non-hydrolysed Ardennes cellulose
7 = after 7% acidic hydrolysis
12 = after 12% acidic hydrolysis
17 = after 17% acidic hydrolysis
35 = after 35% acidic hydrolysis

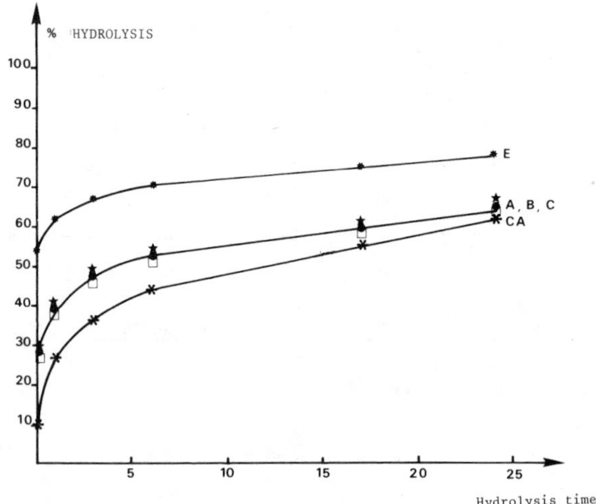

Figure 2: Behaviour of residues from enzymatic hydrolysis of cellulose when subjected to acidic hydrolysis

CA : defibred pulp (control)
A : residue after 18.1% hydrolysis
B : residue after 38.2% hydrolysis
C : residue after 56.3% hydrolysis
E : residue after 94.4% hydrolysis

particular by HPLC. The results shown in Tables II and III, for example, indicate that exo-β-1,4 cellobiohydrolase and endo-β-1,4 glucanase activities are sharply reduced when the enzyme complex is recycled. These two enzymes are very rapidly adsorbed on the cellulosic substrate in the first minutes of the hydrolytic process.

	ENZYMATIC ACTIVITIES				
Sample	FPase (%)	CMCase (%)	β-glucosidase (%)	xylanase (%)	β-xylosidase (%)
Commencement	100	100	100	100	100
1st recycling	90	70	93	102	107
2nd recycling	80	70	88	104	85
3rd recycling	45	63	96	65	105
4th recycling	49	48	84	81	105
5th recycling	28	42	86	62	107

Table II: Recycling of the enzyme complex after 30 minutes of action on a cellulosic hardwood pulp (sulphate pulp)

	ENZYMATIC ACTIVITIES				
Sample	FPase (%)	CMCase (%)	β-glucosidase (%)	xylanase (%)	β-xylosidase (%)
Commencement	100	100	100	100	100
1st recycling	81	74	96	97	100
2nd recycling	46	48	94	79	98
3rd recycling	19	41	91	69	96
4th recycling	8	25	95	60	96
5th recycling	1	26	86	52	91

Table III: Recycling of the enzyme complex after 30 minutes of action on a cellulosic softwood pulp (sulphate pulp)

The xylanase, β-glucosidase and β-xylosidase activities show little change. The results for adsorption of the cellulolytic enzymes clearly demonstrate that exo-β-1,4-cellobiohydrolase activity is the limiting factor in the design of a reactor which recycles the enzyme complex.

In addition, the results for endo-β-1,4-cellobiohydrolase (the CMCase test) seem to show a very rapid loss of activity in the first cycles and then a relative stabilization. It would appear that some of the endo-β-1,4-cellobiohydrolase can no longer be adsorbed on cellulose or that the endocellulase activity has various components which do not have the same adsorption/desorption capacities with respect to cellulose.

1.3 Characteristics of the cellulosic substrate and the hydrolysis residues

Hydrolysis of cellulose by the acidic or enzymatic methods depends primarily on the accessibility of the substrates to the hydrolysing agents. Such accessibility is largely determined by the lignin/carbohydrate relationships and the greater or lesser degree of compactness of the cellulose itself.

The crystalline structure is of great importance in acidic hydrolysis but is less important in enzymatic hydrolysis, in which enzymatic accessibility correlates with properties such as porosity, specific surface area and degree of hydration, which must be altered to improve the yields from hydrolysis.

These properties also provide a basis for characterizing the hydrolysis residues. Table IV shows some results obtained during hydrolysis of a hardwood sulphate pulp.

At the same time, it was demonstrated that the residue from acidic hydrolysis lost part of its susceptibility to acidic attack (Fig. 3). The residues from enzymatic attack, on the other hand, showed equal or even increased susceptibility to further hydrolysis provided the yields of the first hydrolysis did not exceed a limit value - 60% in our experiment (Fig. 4).

As already mentioned, endocellulases and exocellulases had been shown to be adsorbed on cellulose. This could explain the above phenomenon despite the fact that the residue of the first hydrolysis was washed.

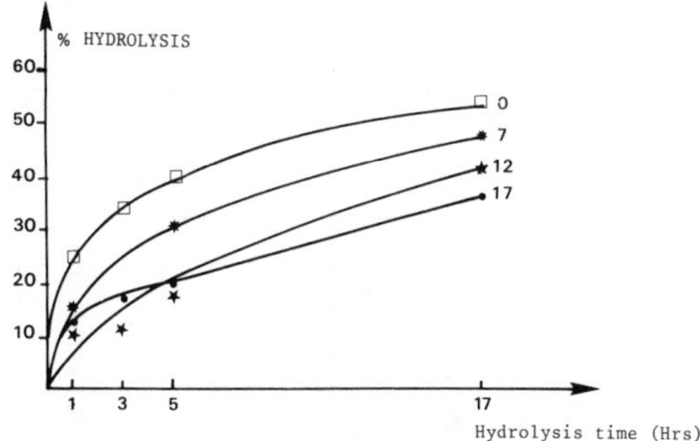

Figure 3: Behaviour of residues from acidic hydrolysis of cellulose when subjected to a second acidic hydrolysis

0 : untreated
7 : residue after 7% acidic hydrolysis
12 : residue after 12% acidic hydrolysis
17 : residue after 17% acidic hydrolysis

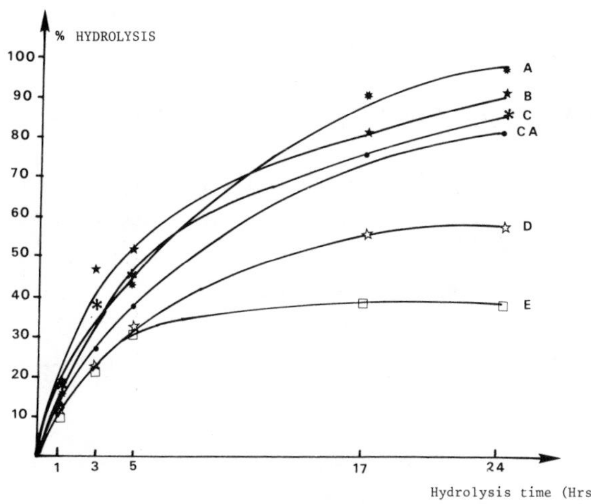

Figure 4: Behaviour of residues from enzymatic hydrolysis of cellulose when subjected to a second enzymatic hydrolysis

CA : defibred pulp (control)
A : residue after 18.1% hydrolysis
B : residue after 38.2% hydrolysis
C : residue after 56.3% hydrolysis
D : residue after 72.1% hydrolysis
E : residue after 94.4% hydrolysis

PROPERTIES	Enzymatic hydrolysis		Acidic hydrolysis	
	Before ±60% hydrolysis	After ±60% hydrolysis	Before ±10% hydrolysis	After ±10% hydrolysis
Reducing power	↗	≃	↗	↗↗
Microporosity	↗	↘		
Macroporosity	↗	↗	↗	↗
Solubility NaOH 18%	↗↗	↗	↗	≃
Degree of polymerization	≃↘	↗	↘↘	≃ LODP
Index of crystallinity	≃	≃	↗	↗≃
Ash content	↗	↗	↗	↗

Table IV : Change in the characteristics of the residue from acidic or enzymatic hydrolysis of a cellulosic hardwood sulphate pulp

LODP: level off degree of polymerization

2. ENZYMATIC HYDROLYSIS OF VARIOUS AGRICULTURAL WASTES

In addition to the problems of accessibility with regard to cellulose, enzymatic hydrolysis of various agricultural substrates is impeded by the presence of lignin. Studies were carried out of various substrates (straw, maize stalks, bran, brewery draff, beet pulp, jute, sisal, etc.) and various pretreatments, both chemical (cooking in the presence of lime, neutral sulphite, etc.) or mechanical (wet grinding, refining, etc.).

These treatments cause the fibres to swell in water and increase their accessibility. Complete delignification is not required: all that is necessary is to alter the structure of the lignin and its relationship with the cellulosic compounds.

Cooking with lime-soda proved most effective for maize stalks, straw, sisal and jute. Neutral sulphite treatment was particularly beneficial for maize and beet pulps. Both treatments permitted hydrolysis yields in the region of 90% after 48 or 72 hours' saccharification but potentially useful results were obtained after as little as 8 hours' saccharification (Table V). The results for maize, bran, draff and pulps are not improved by mechnical treatment in the form of wet grinding. This process does, however, result in an improvement of about 50% for straw treated with lime.

Type of pulp	Maize	Straw	Bran	Draff	Beet pulp	Jute	Sisal
Mechanical	16	15	21	17	9	-	-
Semi-chemical (lime-soda)	79	61	45	46	5	67	63
Semi-chemical (neutral sulphite)	84⁻	60	67⁻	66⁻	87	-	-

⁻ not ground

Table V: Degree of hydrolysis obtained for the various substrates after eight hours' saccharification (0.3 IU cellulase/ml; 5 IU cellobiase/ml)

The results shown in Table V are in the form of a percentage hydrolysis related to the α-cellulose and pentosan content. They must be weighted if it is proposed to take account of the residual lignin or the pre-treatment losses. In this respect, the most efficient pretreatment is not always the most cost-effective.

Generally speaking, if cooking yields are taken into account, the best overall degree of hydrolysis is obtained by cooking with lime-soda. The yield obtained by lime-sode cooking of straw, for example, is 82% (straw/liquor ratio = 1/12; liquor composition: 10% $Ca(OH)_2$ + 1% NaOH, t^o : 80^oC, time: 5 hours).

When cooking yields are taken into account, it appears that this type of pretreatment is not suitable for brewery draff, beet pulps and bran since it is too exhaustive. Mechnical treatment in the form of refining has proved to be a very useful aid to improving the rate and overall degree of hydrolysis. Figure 5 shows that refining of straw with powdered lime improves the degree of hydrolysis by at least 10% and that the treatment is effective after the first few minutes of the refining process.

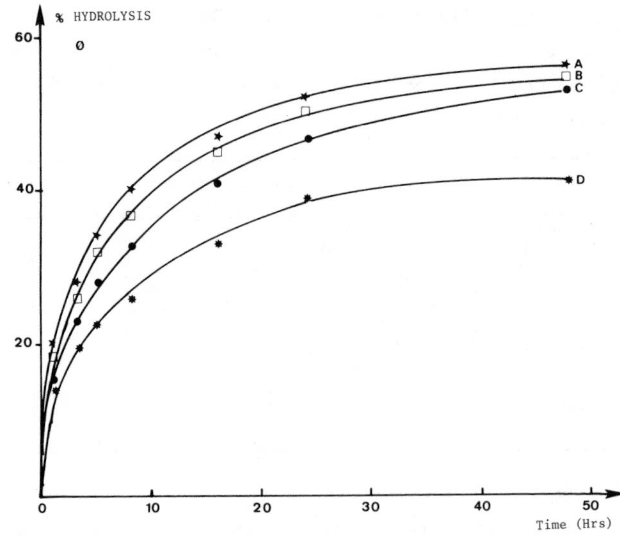

Figure 5: Hydrolysis of straw/powdered lime pulp as a function of time and of refining treatment in the presence of 0.3 IU cellulase and 5 IU cellobiase. A : 60' refining B : 15' refining C : 5' refining D : unrefined

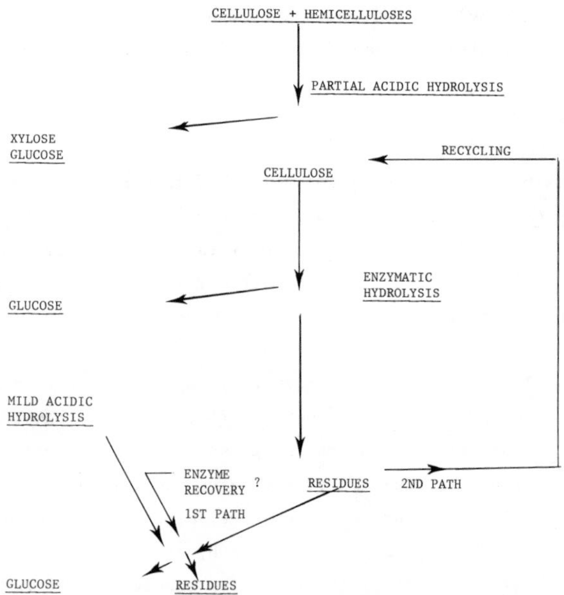

Figure 6: Flowchart for production of fermentable sugars from delignified cellulosic substrates

3. REACTOR FOR ENZYMATIC HYDROLYSIS OF STRAW/LIME PULP

For the purposes of post-hydrolysis treatment, it is essential in practice to obtain as high a sugar concentration as possible.

Most work on enzymatic hydrolysis is carried out with a substrate concentration of about 4% in the fermenter.

Trials were therefore carried out with increased proportions of dry matter in the fermenter: 5% and then 10% dry matter.

These experiments indicated

- that the substrate concentration in the fermenter had to be increased gradually to ensure satisfactory rheological properties;
- that extra cellulose had to be provided because of the problems connected with enzyme adsorption;
- that it would be useful to recycle the hydrolysis residues;
- that it would useful to develop techniques for recovery and reuse of enzymes.

When the pulp concentration is increased, the hydrolysis yields do not increase proportionately.

For a dry-matter substrate concentration of 10%, 60% conversion of the cellulosic matter was attained for straw. In the last analysis, this represents a sugar concentration of 4.6% in the fermenter when allowance is made for the presence of lignin.

The sugars obtained consist of approximately 25% xylose, 75% glucose and no cellobiose.

4. CHARACTERIZATION OF RESIDUES FROM STRAW HYDROLYSIS

Both acidic and enzymatic hydrolysis leave a residue whose utilization could help to make the saccharification process more viable economically. Before the question of viability can be investigated, the residues must be characterized and distinguished where possible.

It is immediately clear that such residues, whatever the type of hydrolysis practised, always contain substantial proportions of cellulosic and/or hemicellulosic substances.

These residues do not have the same recycling potential (see § 1.3).

Residues from straw have a fairly high lignin content. Table VI shows some of the results obtained.

Composition	Residue 1	Residue 2	Residue 3
Alcohol-benzene extract (%)	8	9	37.5
Ash (%)	5	9	8.5
Pentosans (%)	12	10	0
Klason lignin (%)	26	34	28
O CH_3	4.3	6	5
G/S/P ratio[1)]	39/60/1	53/44/2	49/46/4

Table VI: Chemical composition of hydrolysis residues

Residue 1: residue from 45% enzymatic hydrolysis
Residue 2: residue from 60% enzymatic hydrolysis
Residue 3: residue from 45% acidic hydrolysis

1) G = guaiacyl group; S = syringyl group;
P = p-substituted phenoxy group

The chemical composition of the residues from acidic hydrolysis is such that they are very susceptible to alcohol-benzene extraction. The essential difference between the residues from acidic hydrolysis and those from enzymatic hydrolysis indeed lies in the different degrees of degradation they have undergone. The greater degradation which occurs during acidic attack is apparent from the molecular weight distribution curves. The residual lignin after enzymatic attack has higher molecular weights.

5. PROSPECTS

Enzymatic hydrolysis of agricultural and industrial wastes depends to a large extent on the accessibility of the fibres to the enzymes. Such accessibility is affected by:

- the presence of lignin in the substrates;
- the structure of the cellulose chains.

Pretreatments are necessary to improve the hydrolysis yields. Low-temperature cooking with lime-soda proved effective for certain substrates including straw, especially since it can be combined with a mechnical treatment (grinding, refining) of the pulps obtained.

Initially, it appeared that complete delignification during pretreatment was not absolutely necessary. However, when 10% of dry matter is used in the reactor to increase the sugar concentration, the sugar content is only 4.5%. The role of the residual lignin should be reconsidered in this light, at least for purposes of reactor design.

- Firstly, recycling of the hydrolysis residues is possible but introduces a very large quantity of lignin to the reactor.
- Secondly, the design of an enzyme-recycling reactor is made harder by the irreversible adsorption of certain components of the enzyme complex.

In the case of straw, it became clear that there was a risk of adsorption of certain components (β-xylosidase) which were not adsorbed on pure cellulose.

The lignin or the pretreatment may thus play a part in the adsorption of the enzyme complex.

In view of the foregoing, it is important to consider the possible uses of the residual lignin. Unless these uses are based on the high molecular weight of the lignin no benefit is likely to be derived from using the mildest pretreatments of the lignocellulosic substrate.

The flowchart for production of fermentable sugars from delignified substrates could encompass the various aspects investigated in our research. Such a production flowchart might be as shown in Figure 6.

ARTICLES AND PAPERS PUBLISHED IN THE COURSE OF THE RESEARCH PROGRAMME

- PAQUOT, M.; THONART, P., 1982
  Hydrolyse enzymatique de la cellulose régénérée
  Holzforshung, 36, 177-181.

- PAQUOT, M.; HERMANS, L., 1983
  Alternative possible de la mise en décharge des boues de papeterie
  Un exemple : WIGGINS TEAPE S.A. (Belgium)
  Tribune CEBEDEAU, 473, 36, 147-155.

- THONART, P.; MARCOEN, J.M.; DESMONS, P.; FOUCART, M.; PAQUOT, M., 1983
  Etude comparative de l'hydrolyse enzymatique et de l'hydrolyse par voie acide de la cellulose.
  Partie I : Morphologie du substrat en cours d'hydrolyse enzymatique
  Holzforschung, 37, 173-178.

- PAQUOT, M.; FOUCART, M.; DESMONS, P. et THONART, P.
  Etude comparative de l'hydrolyse enzymatique et de l'hydrolyse par voie acide de la cellulose.
  Partie II : Morphologie du substrat au cours de l'hydrolyse acide
  Holzforschung (à paraître).

- PAQUOT, M.; FOUCART, M.; DESMONS, P.; THONART, P., 1983
  Conversion des déchets agricoles et industriels pour l'hydrolyse de la cellulose
  Proceeding of Cost Worshop, Avril, ZURICH, 115-117
  CEE Applied Sci. Publishers, Ed. FERRANTI, M.P.; FIECHTER, A.

- PAQUOT, M.; HERMANS, L., 1983
  Cellulose Hydrolysis of Paper mill Sludges
  Proceeding of Cost Worshop, Avril, ZURICH
  CEE Applied Sci. Publishers, Ed. FERRANTI, M.P.; FIECHTER, A.

- DESMONS, P.; PAQUOT, M.; THONART, P., 1983
  Utilisation de l'HPLC pour séparer les activités cellulolytiques de *Trichoderma reesei*
  Acte du Colloque Pasteur Biosciences, 6-9 sept., PARIS.

ENZYMATIC CONVERSION OF THE CARBOHYDRATES OF STRAW INTO SOLUBLE SUGARS

T.M. WOOD, R.M. HOFFMAN and J. ANNE BROWN
Rowett Research Institute, Bucksburn, Aberdeen, Scotland, U.K.

Summary

An agar plate screening method has been devised for the selection of mutant strains of the fungus *Penicillium pinophilum* which hyperproduce cellulase, β-glucosidase and xylanase enzymes in the presence of a catabolite repressor and/or end product inhibitor. Several mutants have been obtained by mutagenesis using UV-light or nitrosoguanidine. When cultured in a fermenter one mutant yielded eight times more filter-paper-hydrolysing activity than the wild-type, with a productivity of 20-25 $Ul^{-1}h^{-1}$. In one of the mutants, xylanase and β-glucosidase was substantially derepressed in the presence of the catabolite repressor glycerol. Reference is made to the different enzymes comprising the cellulase system and to some factors affecting the regulation of the enzymes in the fungal cell.

INTRODUCTION

The enzymatic conversion of the cellulose and hemicellulose components of straw into soluble sugars offers an attractive alternative for the utilization of this agricultural waste which amounts to some 85 million tonnes per annum in the EEC. The soluble sugars which are generated can be used as chemical feedstocks or as growth substrates for a number of microorganisms in the production of fuel, single cell protein or high cost microbial products such as antibiotics.

Soluble sugars, of course, can also be generated from cellulosic materials by acid hydrolysis. The advantage of the enzymatic route to the industrialist, however, is that it is low in energy consumption, it is non-polluting and it produces sugars free from undesirable side products. One disadvantage of the enzymatic method is that the cost of the microbial enzymes is high. Clearly, to improve the economics of the conversion considerable effort must be focused on improving the yield of the enzymes, their properties, or ideally, both.

One approach being actively pursued in these laboratories and elsewhere (1,2,3) is to isolate, from a cellulolytic fungus, mutant strains that have improved properties; namely (a) the controls regulating the synthesis of the enzymes are modified (derepressed), and (b) the enzymes synthesised are not affected or inhibited by the end products of their activity.

Only a few microorganisms secrete cellulase complexes (so-called, 'true' or 'complete' cellulases) capable of degrading all forms of cellulose (3,4). The work carried out to date on improving 'true' cellulase production has mostly concentrated on the cellulase complex of the fungus *Trichoderma reesei* (1-3). Mutants with more than 3 times the cellulase (measured as activity to filter paper) activity of the wild type have been produced in fermenters (5). Synthesis of the enzymes in some of the mutants appears to be less subject to catabolite repression than in the wild type fungus (1). Despite these achievements, the specific activity of the cellulase has not been increased (5). It is clear from the many published reports that further work is still required on improving the cellulase of *T. reesei* before this organism can be used in an economic fermentation system for the saccharification of cellulose.

Cellulases from other fungi have received far less attention than that of *T. reesei*. However, some show considerable promise, with several fungi producing 'complete' cellulases with activities comparable to, or better than, that of wild type of *T. reesei* (4). Notable in this respect are the cellulases of *Fusarium solani* (6) and *Penicillium pinophilum*(4).

During the tenure of the present contract with the EEC, the fungus *Penicillium pinophilum* (*P. funiculosum*) IMI 87160iii has been studied with a view to producing, by physical and chemical methods, mutant strains hyperproducing, with modified properties, the enzymes necessary to generate soluble sugars from straw.

Enzymes involved in the breakdown of the polysaccharides in straw.

Straw consists, on a dry matter basis, of about 33-44% cellulose and 16-36% hemicellulose, together with lignin and other minor components. Xylan is the principal polysaccharide present in the hemicellulose fraction. Consequently both cellulases and xylanases are required for the hydrolysis of straw. Extracellular enzyme preparations obtained from P. pinophilum contain both types of enzyme when the fungus is cultured with straw or pure cellulose as the sole carbon source. It appears that xylanase is constitutively produced by this fungus.

Xylanase of P. pinophilum appears in multiple forms in all cultures but little is known, as yet, about their properties or their modes of action. The cellulase, which is equally complex, has been studied in more detail in the hope that the strain improvement can eventually be explained in terms of changes in the enzyme constituents.

Basically, three types of enzyme are involved in the solubilization of highly ordered celluloses by P. pinophilum; namely, exo-1,4-β-glucanases, endo-1,4-β-glucanases and cellobiases or β-glucosidases. These enzymes by acting sequentially and in concert can solubilize crystalline cellulose. Each of the enzymes exists in several forms. The exoglucanases are of particular interest in that this type of activity is shown by two immunologically unrelated cellobiohydrolases and a glucohydrolase (8).

A range of substrates is required for assaying the different enzymes. Carboxymethyl cellulose (CM-cellulose), which is a soluble derivative of cellulose, is attacked by endoglucanases only, but cellulose that has been swollen in H_3PO_4 is solubilized by either endo-glucanase or cellobiohydrolase. Cotton fibre, filter paper, Solka Floc (a hemicellulose-free wood pulp) and Avicel (a microcrystalline α-cellulose) contain high proportions of crystalline cellulose, and these are hydrolysed extensively only by the combined action of endoglucanase and cellobiohydrolase.

In the present and other strain improvement studies,filter paper-hydrolysing activity is the activity most widely measured to indicate and evaluate the presence of 'true' cellulase activity (i.e.combined endo- and exo-1,4-β-glucanase activity)(9).

Plate technique for screening of mutants P. pinophilum.

For the efficient screening of large numbers of isolates of P. pinophilum produced by mutation, an agar plate technique based on that developed by Montenecourt and Eveleigh (10) has been used. Cellulase secretion by these mutants was examined by incorporating a source of cellulose into the agar. A growth inhibitor, such as oxgall, was included in the medium, thus allowing about 100 colonies to be screened per plate. When clearing zones developed around the fungal colonies growing on the agar, cellulase was being secreted.

Clearing and growth of P. pinophilum IMI 87160iii on agar plates containing a number of different cellulose sources and at different concentrations of oxgall was examined. There was very little clearing of agar plates containing Avicel or Solka Floc. Clearing zones developed, readily around colonies grown on H_3PO_4-swollen cellulose or delignified, ball-milled straw (Fig.1a), but not on a straw preparation that had not been delignified.

0.2% oxgall was optimum on all cellulose sources for producing discrete colonies. With no oxgall the colonies grew in a diffuse manner: high concentrations of oxgall (0.5% or 1%) increasingly inhibited growth. Phosfon D (10) (0.005% v/v) seemed to enhance the clearing round the colonies, but the mechanism of this action is not understood.

5% glycerol has been used as a repressor of cellulase synthesis in studies with T. reesei (1). P. pinophilum IMI 87160 iii grew better on ball-milled delignified straw plates containing 5% glycerol and oxgall (0.2%) than on similar plates not containing glycerol. Large, spreading colonies were produced in the presence of glycerol, but by increasing the oxgall concentration to 1% discrete colonies were obtained.

There were no clearing zones around colonies of P. pinophilum IMI 87160 iii grown on delignified straw plus glycerol plates (Fig.1b) although good clearing was produced in the absence of glycerol (Fig.1a). It was concluded that the cellulase enzymes were being repressed in the presence of glycerol.

Mutagenesis of P. pinophilum with UV-light.

Attempts to enhance cellulase secretion by P. pinophilum IMI 87160 iii were undertaken by mutation. Spores were irradiated with ultraviolet light for 165 seconds, but which time 99.9% of the spores were killed. Appropriate dilutions of the spores were streaked-out on plates containing delignified straw plus 5% glycerol. This plating regime allows for the direct screening of mutants with the ability to degrade straw under repressed conditions: other cellulose substrates would be less precise in this respect.

The plates were examined after incubating them for 5-6 days at 27°C. The colonies which produced good clearing were sub-cultured on to fresh straw plus glycerol paltes not containing oxgall. This allowed rapid growth of the colonies. The most promising of these isolates were sub-cultured several times in order to ensure that the mutants were stable.

Several mutants of IMI 87160 iii produced by UV irradiation were able to effect clearing of delignified straw medium containing glycerol. However, the extent of clearing by even the best of these (mutant C22C) was somewhat less (Fig. 1c) than that seen on the straw medium containing no glycerol (Fig. 1d). Mutant C22c produced a greater area of clearing on straw medium containing no glycerol (Fig. 1d) than was shown by the wild type (Fig. 1a).

Mutant C22c and wild type IMI 87160 iii were compared for their abilities to produce extracellular enzyme in shake flasks in the presence and absence of 5% glycerol. The results showed that enzyme activities of the wild-type strain IMI 87160 iii were greatly repressed in the presence of glycerol (Fig. 2). In contrast,the synthesis of xylanase and β-glucosidase were largely derepressed in the mutant when glycerol was present (Fig. 2): filter paper-hydrolysing activity, and CM-cellulase, however, were only slightly derepressed. The differential effect of glycerol on the repression of β-glucosidase and CM-cellulase suggests that these two enzymes are under separate control.

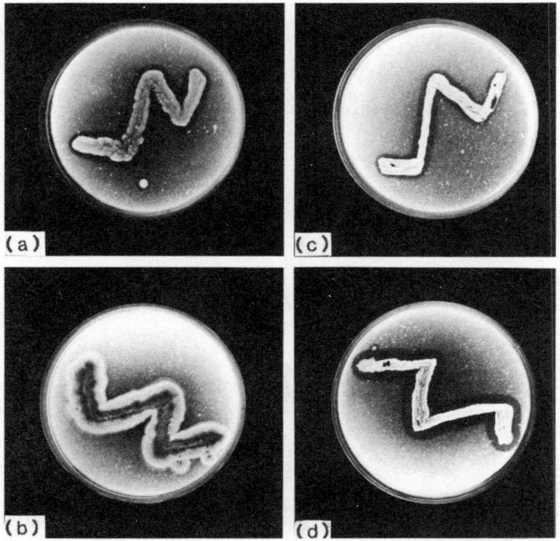

Fig. 1. Penicillium pinophilum spores were streaked across the agar plates containing Vogel medium (10) and ball-milled, delignified straw (0.5% w/v) plus 5 µg of Phosfon D per ml and 1.0% w/v oxgall. Incubation was at 27°C for 6 days. (a) wild-type IMI 87160 iii with no glycerol, (b) wild-type IMI 87160 iii + 5% (v/v) glycerol, (c) mutant C22c + 5% (v/v) glycerol, (d) mutant C22c with no glycerol.

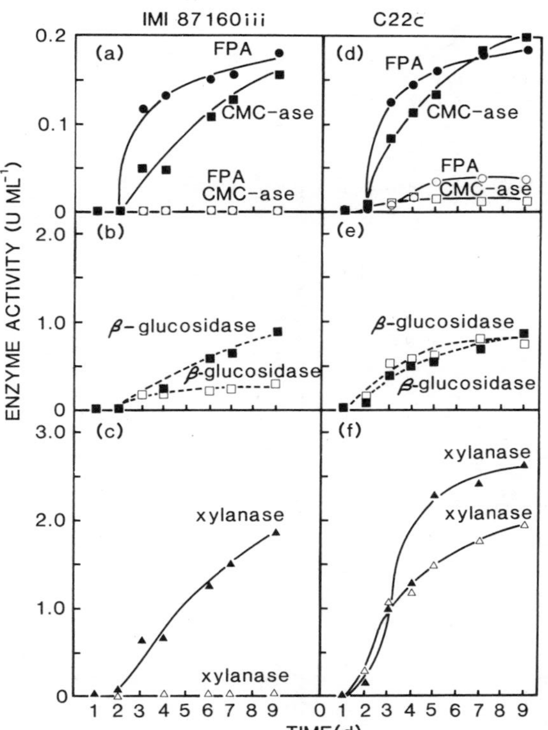

Fig. 2. Enzyme production by Penicillium pinophilum IMI 87160 iii (wild type) (a-c) and mutant strain C22c in milled-straw medium in the absence (▲ , ■ , ●) and presence (△ , □ , ○) of the catabolite repressor, glycerol (5%).

Strain improvement studies were extended using mutant C22c as the parent strain, UV light as the mutagen and with glucose replacing glycerol as the catabolite repressor in the selection medium. Ten mutants with enhanced abilities to produce extracellular cellulase, β-glucosidase and xylanase activities were isolated from approximately 8000 colonies. The best of these (I/10 Table 1) produced, when cultured in shake flasks, 2.6 times the CM-cellulase, 4.0 times the β-glucosidase and approximately twice the activity to filter paper synthesised by the wild type strain. Specific activities were increased by approximately 50% in the case of CM-cellulase and β-glucosidase activities, but there was little change in the specific activity of the enzyme showing activity to filter paper. Since activity to filter paper is a measure of the presence of both cellobiohydrolase and CM-cellulase enzymes, the above observation could indicate that only one of the two types of enzyme has been 'improved'.

Gluconolactone is a powerful inhibitor of microbial β-glucosidase (cellobiase). Gluconolactone was therefore added to the agar-plate screening medium along with cellobiose in an attempt to select a mutant that could synthesise a β-glucosidase less affected by end product inhibition. At a gluconolactone concentration of 10 μM a few small colonies were detected which showed a remarkably wide area of clearing of the straw. The correlation between the ability to produce clearing on the plates and to produce extracellular enzyme in shake flasks, was not, however, particularly good, but one of the isolates (III/16F Table 1) synthesised enzymes in improved yield and with higher specific activities than mutant C22c.

On the basis of their abilities to produce enzymes in shake flasks mutant strains C22c, I/1, I/10 and III/16 are the best of those isolated so far from about 30,000 colonies examined: I/1, I/10 and III/16 are significantly better than mutant C22c. A comprehensive comparison of the enzyme synthesising abilities of these mutant strains is presented in Table 1. This shows that in all cases, straw, either untreated or delignified, is a better carbon source for induction of the required enzymes than the purified cellulose, Solka Floc.

Mutagenesis with nitrosoguanidine (NTG)

Dose response experiments carried out with the complex medium developed by Anne *et al.*, (11) indicated that spores of *P. pinophilum* could be suitably mutated when exposed to NTG for 60 mins. at a concentration of 200 μg/ml. Under these conditions, 90% mortality of the spores was effected. The surviving spores formed colonies 3-4 mm in 4-6 days.

A series of six mutation experiments was carried out with mutant C22c involving approximately 100 agar-straw-plates and 12000 surviving colonies. Those colonies which produced good clearing of the straw in the presence of the catabolite repressor glucose (0.5% w/v) were subcultured on potato-dextrose-agar plates and tested for their abilities to produce extracellular enzyme in shake flasks.

Table 1. Comparison of enzyme activities produced in shake flasks by wild-type and mutant strains of _Penicillium pinophilum_ when grown on different cellulosic materials for 14 days.

Strain	Substrate	β-glucosidase ($U\ ml^{-1}$)	Activity to filter paper* ($U\ ml^{-1}$)	CM-cellulase ($U\ ml^{-1}$)	Xylanase ($U\ ml^{-1}$)	Soluble protein ($mgml^{-1}$)
87160 iii (wild-type)	Straw	0.82	0.30	1.08	0.45	0.55
	Delignified straw	1.33	0.15	0.58	-	0.51
UV C22c	Solka Floc	2.00	0.37	1.5	1.76	0.31
	Straw	2.13	0.32	1.81	3.06	0.56
	Delignified straw	1.78	0.28	1.51	2.32	0.72
UV I/1	Solka Floc	2.35	0.49	2.27	1.80	0.55
	Straw	2.87	0.70	2.75	4.05	0.88
	Delignified straw	2.55	0.37	2.43	3.27	0.93
UV I/10	Solka Floc	3.30	0.64	2.83	2.91	1.16
	Straw	2.94	0.54	2.85	4.14	1.11
	Delignified straw	3.14	0.70	3.73	4.64	1.43
UV III/16	Solka Floc	1.93	0.17	1.23	1.71	ND
	Straw	3.22	0.67	2.99	4.53	0.91
	Delignified straw	3.41	0.53	3.38	3.60	1.36
NTG III/6	Straw	7.7	1.57	13.3	6.5	2.09
	Straw	16.4*	1.57	22.9*	8.9*	

* Assayed at 50°C; all other assays done at 37°C.

From this study, NTG III/6 was outstanding (Table 1). It yielded an enzyme solution which, in comparison with that obtained from wild-type 87160 iii, contained twelve times the endo-1,4-β-glucanase activity five times the activity to filter paper, nine times the β-glucosidase and fourteen times the xylanase activity. The production of extracellular protein was nearly four times that of the wild-type and there was a significant increase in specific activity in each of the enzyme activities (CM-cellulase, 3.2; activity to filter paper, 1.5; β-glucosidase, 2.6). Some of the other mutants isolated liberated enzyme showing a very large increase in specific activity in comparison with the wild-type enzyme, but the concentration of extracellular enzyme produced was very low.

Mutation studies will continue. Three selection lines are being followed using UV- and NTG- mutagenesis. By alternating these physical and chemical mutagens it is hoped that suitable strains can be obtained. Protoplasts of these mutant strains will then be prepared and fused in an attempt to obtain new strains carrying as many as possible of the characteristics desired. This new initiative will help to circumvent the problems of reversion that may be encountered in strain improvement studies involving many steps.

Auxotrophic mutants have been obtained requiring added asparagine or methionine for growth. These labelled mutants will be used as models for the fusion studies.

Studies on cellulase and xylanase production by a mutant strain of *P. pinophilum* in an instrumented fermenter.

The fermentations were carried out in a New Brunswick Microgen fermenter 10 l working capacity with controls for stirrer speed, pH, temperature, foam and dissolved oxygen. DO was not allowed to fall below 30% saturation by varying the stirrer speed between 250 and 600 rev/min^{-1} and temperature was maintained at 27°. pH was adjusted by the automatic addition of NH_4OH or HCl.

A typical fermentation profile obtained with mutant NTG III/6 using 6% ball-milled barley straw and a 5% inoculum from a 4-day shake flask culture, is shown in Fig.3. Growth was normally very rapid indeed, and in approximately 48 hours the broth became very viscous: this fell after 80-90 hours. The pH was maintained at pH 4.0 when it was found that β-glucosidase,xylanase and CM-cellulase all exhibited maximum stability at this pH.

Maximum filter paper-hydrolysing activity of 2.3-2.4 Uml^{-1} was reached in 120-144 hours, giving a productivity of approximately 20 Ul^{-1} h^{-1}. Using a 10% inoculum productivity was increased to 23-24 Ul^{-1}h^{-1}.

Slightly higher filter paper hydrolysing activities (2.5 Uml^{-1}) were obtained from *P. pinophilum* cultures using 6% Solka Floc in the medium. However, this was not obtained until 144 hours of fermentation.

These activities, although encouraging, at this early stage in the strain improvement programme with *P. pinophilum*, fall short of the activities quoted for *T. reesei* cultures (5). QM 9414 cultures, for example, have been reported to reach 5.7 Uml^{-1} filter paper hydrolysing activity in batch cultures in cellulose medium (5). The best *T. reesei* mutant (C30) obtained after many years of world-wide activity in strain improvement studies, has been reported to yield a remarkable 15 Uml^{-1} of filter paper hydrolysing activity when cultured using a cellulose substrate with a very high surface area (5). Clearly, *P. pinophilum* has some way to go before it can compete successfully with *T. reesei* as a source of enzyme for generating glucose from cellulosic materials. However, productivity values of 20-24 Ul^{-1}h^{-1} for *P. pinophilum* compare

Fig. 3. Production of enzyme and protein by P. pinophilum mutant strain NTG III/6 in an instrumented fermenter.

Ball-milled straw 6% w/v in salts medium (13). The pH was not allowed to fall below 4.

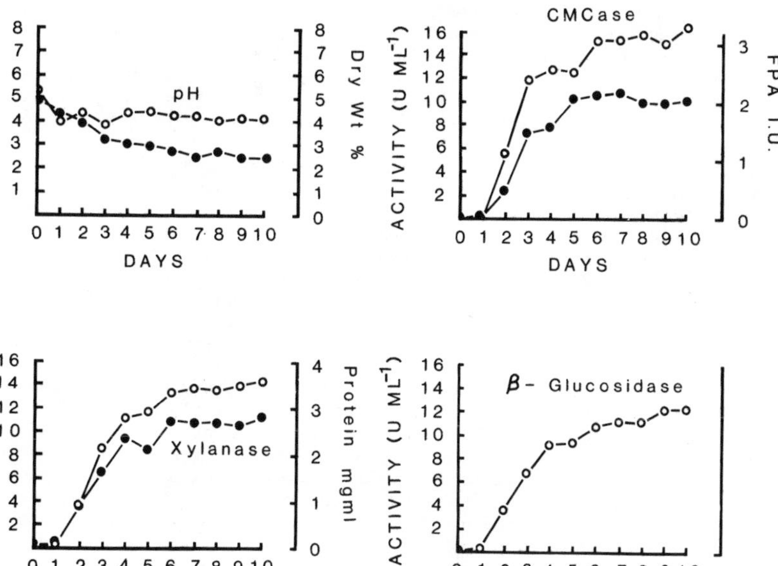

more favourably with the values of 30-40 $Ul^{-1}h^{-1}$ quoted for batch cultures of T. reesei.

P. pinophilum, however, has one advantage over T. reesei. Whereas wild type T. reesei and mutant strains are poor producers of extracellular β-glucosidase (5), wild type P. pinophilum is relatively good and mutant NTG III/6 is excellent. NTG III/6, for example, yielded 16 Uml^{-1} of β-glucosidase in the fermenter when assayed at 37°C and 34 Uml^{-1} when assayed at 50°C. This compares with reported values of <1U ml^{-1} for T. reesei QM6a and mutant QM 9414 (5). For efficient saccharification T. reesei cellulase must be supplemented with β-glucosidase from another fungus (12).

Further improvement in 'true' cellulase activity of P. pinophilum can probably be obtained by concentrating on isolating a mutant with an ability to produce more CM-cellulase activity. The values of 16 U ml^{-1} of CM-cellulase obtained in the 6% straw medium (Fig. 3) with the NTG III/6 mutant in the fermenter were estimated at 37°C. When assayed at 50°C the CM-cellulase activity was found to be 33 U ml^{-1}; but this still compares unfavourably with reported activities of 150 U ml^{-1} for T. reesei C30 (5).

However, to imply that the desired enzyme preparation will be obtained merely by improving the yields of the various enzymes, is, judging by the experience with T. reesei, not correct. It seems likely that enzymes with properties so altered that they are not affected by end product inhibition is the desired goal. This may ultimately be achieved by the empirical approach presently followed, but the breakthrough may only be obtained, in the end, by careful manipulation of the fungal genes after the development of a full understanding of the mode of action of the enzymes and the factors regulating their production in the cell.

Preliminary studies carried out in this laboratory on regulation in P. pinophilum would seem to indicate that endo-1,4-β-glucanases and β-glucosidases are co-ordinately induced by cellobiose and that β-xylosidase and xylanase are co-ordinately induced by xylose. However, the relatively low activities of enzyme induced by cellobiose and xylose suggest that these sugars were not the true inducers. Sophorose, which is an extremely potent inducer of cellulase in T. reesei is not an inducer of the same enzyme in P. pinophilum. It may be that transglycosylation products produced by the fungus may be required to induce high enzyme activities, and this is an aspect that needs concerted study.

As far as the mode of action is concerned, several problems have emerged during the present investigation, and these must be addressed. These are: (a) the apparent synergistic activity observed between the two immunologically unrelated cellobiohydrolases in solubilizing highly ordered cellulose, (b) the apparent existence of a complex between one of the cellobiohydrolases and a small peptide (?) component that renders the enzyme unable to solubilize highly ordered cellulose, and (c) the fact that only one of the two cellobiohydrolases can act in synergism with endoglucanase from another fungus.

A solution to these and other problems concerned with the interaction of endo and exo-1,4-β-glucanases on the face of the cellulose crystallite may make it possible to proceed with strain improvement on a more scientific basis.

REFERENCES

1. MONTENECOURT, B.S. and EVELEIGH, D.E. (1977). Preparation of mutants of Trichoderma reesei with enhanced cellulase production. Appl. Environ. Microbiol. 34, 777-782.
2. MANDELS, M., WEBER, J. and PARIZEK, R. (1971). Enhanced cellulase production by a mutant of Trichoderma viride. Appl. Microbiol. 21, 152-154.
3. GALLO, B.J., ANDREOTTI, R., ROCHE, C., RUY, D., and MANDELS, M.(1979). Cellulase production by a new mutant strain of Trichoderma reesei MCG 77. Biotech. Bioeng. Symp. No. 8, 890191.
4. WOOD, T.M. and McCRAE, S.I. (1979). Synergism between enzymes involved in the solubilization of native cellulose. Adv. in Chem. Series, 181, 186-209.
5. RUY, D.D.Y. and MANDELS, M. (1980). Cellulases: biosynthesis and applications. Enzyme Microb. Technol., 2, 91-102.
6. WOOD, T.M. and McCRAE, S.I. (1977). The mechanism of cellulase action with particular reference to the C_1 component. Proc. Bioconversion Symp., IIT, Delhi, 111-141.
7. WOOD, T.M. McCRAE, S.I. and McFARLANE, C.M. (1980). The isolation, purification and properties of the cellobiohydrolase component of Penicillium funiculosum cellulase. Biochem. J. 189, 51-65.
8. WOOD, T.M., McCRAE, S.I. and WILSON, C.A. (1984). Observations on the complex interactions involved in the enzymatic hydrolysis of cellulose. Appl. Biochem. & Biotechnol. (in press).
9. MANDELS, M., ANDREOTTI, R. and ROCHE, C. (1976). Measurement of saccharifying cellulose. Biotechnol. and Bioeng. Symp. 6, 21-33. Interscience.
10. MONTENECOURT, B.S. and EVELEIGH, D.E. (1977). Semiquantitative plate assay for determination of cellulase production by Trichoderma viride. Appl. and Environ. Microbiol., 178-183.
11. ANNE, J., EYSSEN, H. and De SOMER, P. (1974). Formation & Degeneration of Penicillium chrysogenum protoplasts. Arch. Microbiol. 09, 159-166.
12. STERNBERG, D., VIJAYAKUMAR, P. and REESE, E.T. (1977). β-Glucosidase: microbial production and effect on enzymatic hydrolysis of cellulose. Canad. of Microbiol. 23, 139-147.
13. MANDELS, M. and WEBER, J. (1969). The production of cellulase. Adv. in Chem. Series, 95, 391-414.

BIODEGRADATION OF LIGNOCELLULOSES IN AGRICULTURAL WASTES

J.M. LYNCH
Glasshouse Crops Research Institute, Worthing Road,
Littlehampton, West Sussex, UK

S.H.T. HARPER, S.J. CHAPMAN and D.A. VEAL
Agricultural & Food Research Council, Letcombe Laboratory, Wantage,
Oxon, UK

Summary

Straw represents one of the major lignocellulosic wastes from agriculture. The basis for a microbiological treatment process which hastens decomposition while turning the negative or zero value resource to a positive value is outlined. Cellulolytic fungi convert the cellulose of straw to cellobiose and glucose and these products are channelled to a N_2-fixing anaerobe (*Clostridium butyricum*) to upgrade the plant nutrient value of the straw. A third organism, *Enterobacter cloacae*, which produces extracellular polysaccharide is co-inoculated to support the co-operative microbial community. The treated material has the potential to stabilize soil structure and control soil-borne diseases. These laboratory studies demonstrate the potential to produce novel agricultural composts with fertilizer, soil conditioning and plant protection value.

1. INTRODUCTION

Straw represents one of the major lignocellulosic materials produced in modern agriculture. In mixed farming systems, some of the straw produced is used as animal feed and bedding and there are a host of minor uses, such as in mushroom production. Traditionally the straw surplus has been ploughed into the soil. Some farmers believe this to be beneficial to soil conditions and plant growth but until recently there has been relatively little scientific evaluation of this effect.

In modern agriculture there has been a reduction in mixed farming resulting in large straw surpluses in some localities. The large bulk of the material makes transport uneconomic and disposal, therefore, becomes a problem. In Britain, direct drilling of crops provides timeliness for cultivation, especially in wet autumns, and can make the difference between getting a winter crop sown and having to resort to the lower-yielding spring crops. It is also economic in energy terms. The presence of straw in direct-drilling impedes crop establishment, this effect having been clearly demonstrated over many years, both in Britain and in the Pacific Northwest of the United States (1). Microbiological degradation of straw, including the immobilization of nutrients, production of phyto-toxins and build-up of pathogens and other antagonistic micro-organisms, is in part responsible, but mechanical interference of drilling can also be involved. Commonly yield reductions as high as 20% occur. Thus for successful direct-drilling, it is recommended that straw should be burnt. Smaller (about 5-10%) yield reductions commonly occur even if straw is ploughed into the soil.

In Britain, there is presently a strong environmentalist lobby against straw burning. In other European countries, there are fewer reported problems from surplus straw, which in southern climates may be due to earlier harvests allowing longer periods for straw decompositon prior to drilling succeeding crops. In any event, burning is actively discouraged. Burning is also restricted in the United States. However in some countries, such as Germany, the problems are only overcome by the energy-intensive cultivation methods of deep-burial of the residues.

There is, therefore, considerable incentive to find convenient and cheap methods of overcoming the negative effects of straw, while perhaps providing a positive function, thus harnessing some of the photosynthetic energy which the plant has expended in producing this plant biomass. Our strategy has been to control the biodegradation of straw, for which there is an analagy in the composting of straw to provide a substrate for the cultivated mushroom (2).

Straw is about 80% w/w cellulose and hemicellulose of which about one half is readily available to micro-organisms and serves as a potential substrate for phytotoxin (acetic acid) production under fermentative conditions and the growth of pathogens, particularly those with powerful cellulase enzymes. By colonizing the straw with other non-harmful cellulolytic fungi, the availability of the substrate to harmful organisms might be minimized and decomposition might be hastened. The C:N ratio of straw is about 120:1 and decomposition is usually limited by the availability of N which has to be met ultimately by fertilizer N. Saprophytic N_2-fixation by bacteria in soil is usually limited by available carbohydrate energy because the enzyme, nitrogenase, is very expensive in terms of its ATP requirements (3). We have, therefore, further explored if the carbohydrate energy of straw could be channelled to this enzyme, thereby eliminating the need to provide a source of fixed N to hasten straw decomposition and to make available any surplus N to the plant. The second point in our strategy is, therefore, to inoculate with a microbial consortium. Simultaneously we have searched for other useful agricultural attributes of the inocula. The studies have ranged from pure cultures on pure substrates to non-axenic cultures on natural substrates. The former enables the principles for study to be established under controlled conditions whereas the latter is the first step to application.

2. CO-OPERATIVE DEGRADATION OF CELLULOSE BY CELLULOLYTIC FUNGI WITH *Clostridium butyricum*

When a range of cellulolytic fungi (*Fusarium* spp., *Penicillium* spp., *Sordaria* sp. and *Trichoderma* spp.) were grown on CF 11 cellulose, they released soluble sugars into the growth medium after oxygen had been removed, concentrations being maximal 2 days after oxygen removal. The amounts of sugar were far greater under anaerobic than aerobic conditions. Amongst the fungi with the ability to release large amounts of sugar was *Trichoderma harzianum*. This fungus was grown asociatively with *Clostridium butyricum*, an anaerobic N_2-fixing bacterium which does not possess a cellulase. Thus when cellulose was supplied as the sole carbon and energy substrate, N_2 was only fixed and the fermentative end-products of the bacterium, acetate and butyrate were only formed when a source of cellulase was provided. This could either be the pure enzyme or the fungal culture. The associative growth took place in the presence of air, indicating that the fungus uses available O_2 and thereby provides respiratory protection to the anaerobe. Small amounts (up to 0.1 g l^{-1}) of N added to the growth medium as $(NH_4)_2SO_4$ hastened the decomposition rate constant of both the pure fungal culture and the mixed culture by a factor of about 4. In the mixed culture the added N increased the N gain by fixation by a factor of c. 9. The function of the added N appears to be a primer action in stimulating fungal growth and concomitant cellulase production.

Once the cellulase is functional and the cellobiose and glucose are provided to the bacterium, studies with ^{15}N demonstrated that the fixed bacterial N becomes available to the fungus to complete the mutualistic partnership.

3. STRAW AS A SUBSTRATE FOR CO-OPERATIVE NITROGEN FIXATION

Initial studies were made by inoculating straw with the cellulolytic fungus *Penicillium corylophilum* in association with *C. butyricum* (4). The studies were made using wheat straw aseptically in a solution of mineral salts contained in conical flasks or non-aseptically with the straw contained in glass columns and moistened by continuous recirculation of the mineral salts solution. In the flask study, *P. corylophilum* in pure culture decomposed straw polysaccharides (cellulose and hemicellulose) only when N was added to the straw. *C. butyricum* did not grow or fix N_2 in pure culture, but grew and supported cellulolysis by *P. corylophilum* in co-culture, presumably by the provision of a source of fixed N.

In the column study the combined inoculants increased the decomposition rate constant from 0.0096 d^{-1} to 0.0139 d^{-1} compared with non-inoculated straw. Over a period of 8 weeks at 25^oC, the N gain was 11.5 mg (g straw lost)$^{-1}$, although the major nitrogenase activity was measured in the first 4 weeks. Extrapolating to a field scale for straw produced at 7 t ha^{-1}, the N gain figures would equate to 35 Kg N ha^{-1}. As the st aw already contained 3.1 mg N g^{-1}, the total N content of straw available for subsequent crops would be 57 Kg N ha^{-1} or about half of the annual fertilizer applied to arable crops.

In a further study (5), the cellulolytic communities occurring at aerobic-anaerobic interfaces of non-inoculated straw were studied by packing the straw into glass jars and partially submerging it. Acetylene reduction (nitrogenase) reached a peak after 2 weeks when the atmosphere was maintained anaerobic but when the jars were aerated, reduction was much greater and continued to increase up to 3-4 weeks. However, in the latter situation, the greatest activity was in zones where there were aerobic/anaerobic interfaces. Decomposition was also most intense in these regions.

Table 1. Decomposition rate constant (k day^{-1}) of polysaccharides in straw decomposed, partially waterlogged at 25°C for 4 weeks (5)

Position of sample	Aerated	Anaerobic
Non-waterlogged	0.0011	0.0006
Interface	0.0145	-
Waterlogged	0.0101	0.0057

The anaerobic bacteria which appear responsible for the nitrogenase activity were C. butyricum. The aerobic bacterial population was dominated by Pseudomonas spp. (>90% of isolates) but it included Enterobacteriacae. Cellulolytic bacteria were not isolated under aerobic or anaerobic conditions.

A diverse fungal population was isolated and these were usually restricted to the aerobic and anaerobic interfaces of jars; fungi were not isolated from jars maintained fully anaerobic. The isolated fungi were assayed for their cellulolytic activity using pure cellulose and straw. One such set of tests is shown in Table 2.

Table 2. Weight loss from wheat straw in shaken flask culture and clearing of cellulose from agar beneath colonies after 3 weeks at 20°C

Fungus	Weight loss from straw (mg/g)	Clearing of cellulose agar (mm)
Acremonium persicinum	305+	4
Fusarium culmorum	210+	5
Fusarium oxysporum	210+	0
Penicillium chrysogenum	99	6
Penicillium corylophilum	69	0
Penicillium hirsutum	125	5
Penicillium janthinellum	240+	6
Penicillium purpurascens	61	8
Sordaria alcina	267+	5
Unidentified	351+	6
Trichoderma hamatum	250+	11
Trichoderma harzianum	238+	10
Trichoderma viride	236+	15

+ indicates significant decomposition of straw celluloses.

Other tests based on measurement of hyphal growth on sterile wheat straw internodes and on agar plates containing a hot water extract from straw showed that the powerfully cellulolytic species could colonise straw rapidly by using water-soluble components initially. The effective cellulolytic species also had the ability to rapidly penetrate sterile straw internodes.

When *F. culmorum*, *P. purpurascens*, *S. alcina* and *T. harzianum* were inoculated onto straw separately with *C. butyricum*, there was growth of the bacterium and the fungi.

It appears, therefore, that the cellulolytic fungi and *C. butyricum* have the potential to co-operate in the degradation of straw, particularly where the oxygen supply is restricted but not excluded. A short-list of fungi with potential as straw inoculants to hasten straw decomposition and co-operatively fix N_2 can thus be produced by excluding the plant pathogenic species (*Fusarium* spp.) and those which are unable to produce a diffusible enzyme, based on the cellulose agar clearing test.

4. SOIL STABILIZING EFFECTS OF MICRO-ORGANISMS DEGRADING STRAW

Enterobacter cloacae is a bacterium isolated as a colonist of straw. It produces copious extracellular polysaccharide on nutrient agar. When the cells and associated polysaccharide were added to the volcanic ash from Mount St. Helens, Washington (equivalent to the silt loam soil of the Pacific Northwest but with no organic matter), an aggregating effect was produced which made aggregates more stable to shaking in water (6). The bacterial cells appear to interact with clay particles to produce a cementing action. Such stabilization was also achieved by the fungi *Penicillium purpurascens* and *T. harzianum* and, to a lesser extent, *Mucor plumbeus*. In these initial experiments, wheat straw could act as a substrate for micro-organisms producing the soil stabilizing agents but there was no advantage of inoculation over and above the effect produced by the native population of the straw.

In subsequent experiments, it was found that the lower the N content of straw, the greater the amount of aggregating materials produced (7). The increase in viscosity of solutions produced during straw decomposition and the aggregating effect is largely the result of microbial polysaccharide production, N-limitation resulting in maximal production rates. The polysaccharide extracted by hot water from the decomposed straw was composed mainly of galactose, glucose and mannose with smaller quantities of arabinose, xylose, rhamnose, fucose and ribose (8). The presence of these sugars indicates a mainly microbial origin for the polysaccharide but with some hemicellulose breakdown products. The polysaccharide precipitable with 70% (v/v) ethanol accounted for 0.5% (w/w) of the degraded straw.

The polysaccharides of a range of bacteria and yeasts which had been isolated from straw were characterised and compared. They were then grown as co-cultures with each of the fungi *Sordaria alcina*, *T. harzianum* and *T. hamatum* with straw as the carbon substrate. This time inoculation of straw with the fungus alone always improved the aggregate stabilizing potential of the straw and the addition of the bacteria and yeasts increased the effect still further. Amongst the bacteria effective in this action were strains of E. cloacae. Maximal polysaccharide production in the straw was observed within 2-3 weeks following inoculation. The amount of polysaccharide produced and the specific viscosity it generates in solution was not directly related to the stabilizing effect. From this it appears that both quality and quantity are the determining factors in the effectiveness of the microbial polysaccharides. There were only small differences in the relative effect of the polysaccharides on volcanic ash and a clay soil; this is perhaps surprising because not only is organic matter present in the soil but also the particle size distribution and mineralogy of the two were quite different.

5. THE BIOCONTROL PROPERTIES OF STRAW COLONISTS

It has been recognized for some time that *Trichoderma* spp. have the potential to control a range of soil-borne diseases (9). In testing this potential a variety of pathogens (*Botrytis cinerea*, *Fusarium avenaceum*, *F. culmorum*, *F. oxysporum*, *F. solani*, *Phomopsis sclerotioides*, *Pyrenochaeta lycopersici*, *Sclerotinia sclerotiorum*) were inoculated onto one side of a Petri dish containing malt agar and the fungal isolates on the other side. Always, *T. harzianum* and *T. hamatum* overgrew the pathogens but other *Trichoderma* isolates, including *T. viride*, were not as broad-spectrum in their activities. The other genera of fungi isolated from straw were seldom as effective as the *Trichoderma* spp. Of the limited tests so far, we have also found the *Trichoderma* spp. to be active against some pathogens on the root itself. It is also of interest that *E. cloacae* appears to have the potential to control damping-off diseases caused by *Pythium* spp. (10).

6. CONCLUSION: TOWARDS A STRAW INOCULATION PROCESS FOR AGRICULTURE

A major potential economic advantage of treatment of straw with micro-organisms appears to be that the readily available components of the straw which are the substrates for micro-organisms responsible for the adverse effects on plant growth are rapidly exhausted (11). The acceleration of straw decomposition appears to be most effective with cellulolytic fungi in co-culture with *C. butyricum*. The added advantage is that N is gained by fixation of dinitrogen from the atmosphere. Part of this N fixed seems to be used by the fungus. We have recently found that the introduction of *E. cloacae* to make a three-membered series is even more effective in generating the association. The third bacterium may provide a growth factor to the association but most likely its metabolism and production of polysaccharide provides respiratory protection to the anaerobic bacterium and to the 0_2-sensitive nitrogenase, thereby optimizing its efficiency. Although relatively low numbers (10^4 g^{-1} straw) of each bacterium are needed to establish them on the straw, it is vital that E. cloacae is co-inoculated to etablish C. butyricum on the straw. Good establishment (up to 2×10^9 cells g^{-1} straw) can occur within six days.

The soil stabilizing and biocontrol potentials of the treated straw appear to be a bonus to us but in some agricultural or horticultural systems this may be the major properties sought. For example, in some North American soils subject to water erosion, microbial soil conditioners are already being evaluated. Ideally for agricultural use a spray system would be developed to treat straw in situ if the promise of the present laboratory observations are sustained in pot and field trials. Probably some form of incorporation into the soil would be necessary to conserve moisture. For horticulture, stacks of straw sheeted with polyethylene might be inoculated. The exothermic reaction might be a problem in the stack system, we have recorded temperatures up to 38^oC in straw bales with an ambient temperature of 13^oC and up to 52^oC when the ambient temperature was 24^oC. However, this and any other operational problems can be tackled when the scale-up of the process from the laboratory is fully justified.

7. ACKNOWLEDGEMENTS

We acknowledge the Commission of the European Communities for financial aid under contract RUW-033-UK and the Agricultural & Food Research Council for a research studentship to D.M.V. We thank Mrs Lynda Gussin, Mrs Susan Parsons, Mrs Anne Uglow and Miss Nina Jenkins for excellent technical assistance.

REFERENCES

1. LYNCH, J.M. and ELLIOTT, L.F. (1984). Crop residues. In: Crop Establishment: Biological Requirements and Engineering Solutions, ed. M.K.V. Carr, in press. London: Pitmans.

2. LYNCH, J.M. and WOOD, D.A. (1984). Controlled microbial degradation of lignocellulose: the basis for existing and novel approaches to composting. In: Composting of Agricultural and Other Wastes, ed. J.K.R. Gasser, in press. Luxembourg: CEC.
3. POSTGATE, J.R. and HILL, S.. (1979). Nitrogen fixation. In: Microbial Ecology. A Conceptual Approach, pp. 191-213, ed. J.M. Lynch and N.J. Poole. Oxford: Blackwell Scientific Publications.
4. LYNCH, J.M. and HARPER, S.H.T. (1983). Straw as a substrate for co-operative nitrogen fixation. Journal of General Microbiology 129, 251-253.
5. HARPER, S.H.T. and LYNCH, J.M. (1984). Nitrogen fixation by cellulolytic communities at aerobic-anaerobic interfaces in straw. Journal of Applied Bacteriology, in press.
6. LYNCH, J.M. and ELLIOTT, L.F. (1983). Aggregate stabilization of volcanic ash and soil during microbial degradation of straw. Applied Environmental Microbiology 45, 1348-1401.
7. ELLIOTT, L.F. and LYNCH, J.M. (1984). The effect of available carbon and nitrogen in straw on soil aggregation and acetic acid production. Plant & Soil, in press.
8. CHAPMAN, S.J. and LYNCH, J.M. (1984). A note on the formation of microbial polysaccharide from wheat straw decomposed in the absence of soil. Journal of Applied Bacteriology, 56, 337-342.
9. CHET, I., HADAR, Y., ELAD, Y., KATAN, J. and HENIS, Y. (1979). Biological control of soil-borne plant pathogens by Trichoderma harzianum. In: Soil-borne Plant Pathogens, eds. B. Schippers and W. Gams, pp. 585-91. Academic Press, London.
10. HADAR, Y., HARMAN, G.E., TAYLOR, A.G. and NORTON, J.M. (1983). Effects of pregermination of pea and cucumber seeds and of seed treatment with Enterobacter cloacae on rots caused by Pythium spp. Phytopathology 73, 1322-1325.
11. LYNCH, J.M. and ELLIOTT, L.F. (1983). Minimizing the potential phytotoxicity of wheat straw by microbial degradation. Soil Biology & Biochemisty 15, 221-222.

PRODUCTION OF FERMENTABLE SUBSTRATES BY ENZYMIC HYDROLYSIS OF BEET-PROCESSING WASTE

M.P. COUGHLAN[1], A.P. MOLONEY[1], P.J. CONSIDINE[2], A. O'RORKE[2], T.J. HACKETT[2] and M. THOMPSON[3]

[1]Department of Biochemistry and [2]Institute of Industrial Research and Standards (Biochemical Unit), University College, Galway and [3]School of Physical Sciences, Regional Technical College, Galway, Ireland.

Summary

Culture filtrates of *Talaromyces emersonii* UCG 208, or of *Trichoderma reesei* MCG 77, when grown on beet pulp synthesize a range of enzymes that catalyze the conversion of the polysaccharide components of this agricultural residue to soluble sugars. The enzymic saccharification process is facilitated if the pulp is pretreated by milling or by incubation with alkali or peracetic acid. However, treatment of unmilled pulp with commercial pectinase prior to addition of *Talaromyces* or *Trichoderma* culture filtrates is even more effective. Almost complete hydrolysis of the polysaccharide content has been achieved under suitable conditions.

1. INTRODUCTION

Implicit in the title of this symposium is the realization that environmental protection is no longer the sole reason for the treatment of wastes. Rather, it is hoped that anaerobic digestion or hydrolysis of carbohydrate-containing wastes of urban, industrial and agricultural origin may be economically justifiable per se. We have chosen beet processing waste, i.e. beet pulp, as the raw material for several reasons. On the one hand it is relatively cheap and available in large quantities at central locations. Thus, purchase and transport costs are minimized. Secondly, the combined contents of cellulose, hemicellulose and pectin, all of which may be hydrolyzed to fermentable monomers, make up about 85% of the dry weight of the pulp. Moreover, the lignin content is low, c. 4% of dry weight, relative to other wastes. This together with the fact that the pulp has been subjected to considerable "pretreatment" during processing should, one would hope, facilitate hydrolysis of the cellulose fraction. Beet pulp may therefore be used as a model for other such substrates with low lignin and high pectin contents. Two possible uses of this material were considered. Solid state cultivation of a suitable cellulolytic microorganism on pulp could improve its nutritional value as an animal food by adding cellular protein and by beginning the degradation of crystalline cellulose. Alternatively, the polymeric carbohydrates in this waste could be hydrolyzed to their constituent monomers for subsequent fermentation to the desired endproduct. We have concentrated on the latter approach.

For some time we have had an academic interest in the enzyme systems produced by *Talaromyces emersonii*. This thermophilic fungus, when grown on appropriate media produces a complete extracellular cellulase system comprised of 3 β-glucosidases (EC 3.2.1.21), 4 endoglucanases (EC 3.2.1.4) and at least 4 exoglucanases (EC 3.2.1.91) and a number of other enzymes relevant to the carbohydrates found in wastes (1-5; and see below). We therefore chose to investigate the hydrolysis of beet pulp carbohydrates by this organism, and, for comparative purposes, that by the mesophilic fungus *Trichoderma reesei*.

2. RESULTS

Since enzyme production may account for much of the cost of saccharification of cellulosic wastes we have endeavoured to enhance cellulase productivity by Talaromyces emersonii. Of the various strains of this organism examined for cellulase (FPase) activity, CBS 814.70 was best, giving about 0.5 IU/ml of culture filtrate (Table I).

Table I. Cellulase activity of Talaromyces emersonii

Strain	Relative cellulase activity A	B
CBS 814.70	100	86.7
CBS 393.64	11.2	4.4
CMI 146.499	55.6	44.4
CMI 154.254	22.2	11.2
CMI 155.697	44.4	40.0
CMI 158.743	46.7	44.4

The various strains were grown on cellulose-peptone medium for 84 h at 45°C (A) or at 50°C (B). The cellulase (FPase) activities in culture filtrates are expressed as a percentage of that (c. 0.4 IU/ml) of CBS 814.70 grown at 45°C (from ref. 2).

By medium optimization and genetic manipulation of this strain we obtained a number of mutants with enhanced cellulase activity, viz. up to 2 IU/ml and productivities of 20-25 IU/l/h (Table II). During these studies we found the extent of clarification of cellulose-containing agar by "complete" cellulase to be a linear function of the logarithm of enzyme activity (7). This allowed convenient and quantitative estimation of cellulase activity in large numbers of culture filtrates to be made without having to carry out tedious dilution or remove inhibitors such as cellobiose (7). A number of "pure" (viz. Avicel, Solka floc, filter paper, cotton) and "waste" lignocellulosic substrates (viz. newspaper, straw, beet pulp) were found to induce the synthesis of the complete cellulase system by T. emersonii (8). Moreover, the above substrates are all digested to different extents by this extracellular system. However, saccharification is most readily achieved when the substrate to be hydrolyzed is also used as the inducer (8). Clearly, the more heterogeneous the composition of the waste the more complex the battery of enzymes needed for digestion. In the project under investigation hemicellulase and pectinase activities

Table II. Cellulolytic activities of wild type and mutant strains of Talaromyces emersonii (from ref. 6)

Strain	Total cellulase (U ml^{-1})	Endo-glucanase (U ml^{-1})	β-D-Gluco-sidase (U ml^{-1})	Cotton activity (mg eq ml^{-1})
CBS 814.70[a]	0.40	70	0.6	-
CBS 814.70[b]	0.77	56	1.15	2.23
UCG 42[b]	1.00	85	0.80	-
UCG 73[b]	1.50	149	1.80	4.02
UCG 146[b]	1.62	140	1.49	4.18
UCG 164[b]	1.63	150	2.15	4.65
UCG 181[b]	1.4	111	0.58	3.96
UCG 182[b]	1.30	112	0.55	4.0
UCG 183[b]	1.76	96	0.93	2.97
UCG 184[b]	1.24	101	1.10	3.22
UCG 208[c]	0.98	-	-	-

[a]The wild type, CBS 814.70, was grown at 45°C in 0.5% (w/v) corn steep liquor/1.0% (w/v) ammonium nitrate/mineral salts/2% (w/v) Solka Floc.
[b]The wild type and mutants were grown in a modification of the above medium. Ammonium sulphate replaced ammonium nitrate and 0.1% (v/v). Tween 80 was included. The ability of filtrates (generally harvested between 72 and 96 h) to hydrolyse filter paper (cellulase activity) was measured at 60°C as were endoglucanase and β-glucosidase activities. Cotton activity was measured by incubating an aliquot of culture filtrate, diluted five-fold with 0.05 M sodium acetate buffer, pH 5, with 50 mg cotton linters (final volume 1 ml) at 60°C for 24 h in sealed tubes without shaking. After 24 h the concentration of reducing sugar in each supernatant was measured.
[c]As for medium b, above, except that 2% (w/v) Solka Floc was replaced by 4% (w/v) beet pulp.

are as important as cellulase activity. Indeed, the mutant, i.e. *T. emersonii* UCG 208, which most effectively brings about digestion of beet pulp, did not exhibit greatest cellulase activity but did produce more polygalacturonase activity than did the others (Table III).

Table III. Cellulase and polygalacturonase activities of mutant strains of _Talaromyces emersonii_ CBS 814.70

Strain	Cellulase activity (IU/ml)	Polygalacturonase activity (IU/ml)
CBS 814.70	0.38	4.5
UCG 208	0.98	33.9
UCG 204	1.19	22.6
UCG 185	0.82	14.7
UCG 203	0.98	14.1
UCG 219	0.78	12.4
UCG 229	0.71	12.4
UCG 186	0.97	7.9

Strains were grown at 45°C for 96 h on the medium containing 4% (w/v) beet pulp described in Table II[c].

From an economic point of view it is important that enzyme use and reuse be maximized. Thus, one should like to be able to recover enzyme from incompletely digested substrates. We find that all of the components of the cellulase system of _T. emersonii_ are rapidly adsorbed by cellulose and gradually returned to the liquid phase as hydrolysis of the substrate proceeds (9). The desorbed enzyme is then available for digestion of newly added substrate. The extent of adsorption is influenced by pH, temperature, the nature of the substrate and its concentration (9).

Operational stability of the enzyme systems in use is also an important economic consideration. Digestion of beet pulp by _Talaromyces emersonii_ UCG 208 culture filtrates is maximal at 60°C, pH 4.2 (though for convenience hydrolyses are carried out at pH 5 at which pH activity is about 10% lower than at 4.2; ref. 13). At 60°C, pH 5, in static reaction mixtures only 19% of the original cellulase activity is lost in 24 h and only 21% when mixtures are shaken. In the presence of substrate, no further loss of cellulase activity is observed over a further period of 5 days at 60°C whether static or shaken. Even at 70°C, pH 5, 36% of the original activity is retained after 4 h compared with 30% after 4 h on shaking. The xylanase, endocellulase (carboxymethyl cellulase) and β-glucosidase activities in culture filtrates of _T. emersonii_ UCG 208 are even more thermostable, no loss in activity being observed after 96 h at 60°C, pH 5: Indeed, the latter had a half-life of 6 h at 70°C, pH 5 (5) while under these conditions the half-lives of the four endocellulases ranged from 4.5 to 9.5 h (10). By contrast, the extracellular polygalacturonase activity of this organism was surprisingly unstable with a half-life at 60°C, pH 5.0 of only 30-40 min or of 200 min at 60°C, pH 3.8 (12). All of the Trichoderma reesei MCG 77 activities examined were much less thermostable than those of Talaromyces emersonii under similar conditions (6,11,12).

Saccharification of beet pulp was monitored by measuring the release of reducing sugars or of glucose or by analysis of hydrolyzates by high pressure liquid chromatography (2,6,8,11-13). When culture filtrates of T. emersonii UCG 208 were used as the source of enzyme little digestion took place unless the incubation temperature exceeded 40°C (Fig. 1), the optimum, as stated above, being at 60°-65°C.

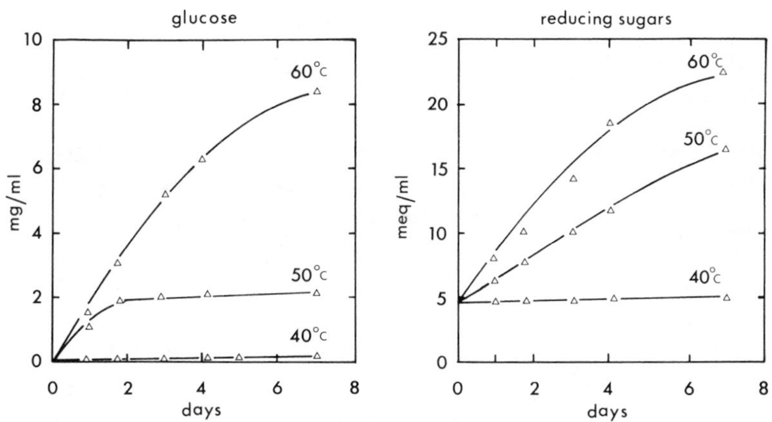

Fig. 1. Effects of temperature on the hydrolysis of 8% (w/v) beet pulp in 0.1 M sodium acetate buffer, pH 5, by T. emersonii UCG 208 culture filtrate.

Investigation of the effects of enzyme and substrate concentration on the saccharification process showed that the greater the concentration of pulp the greater was the amount of enzyme required to effect equivalent digestion in a given time (Fig. 2).

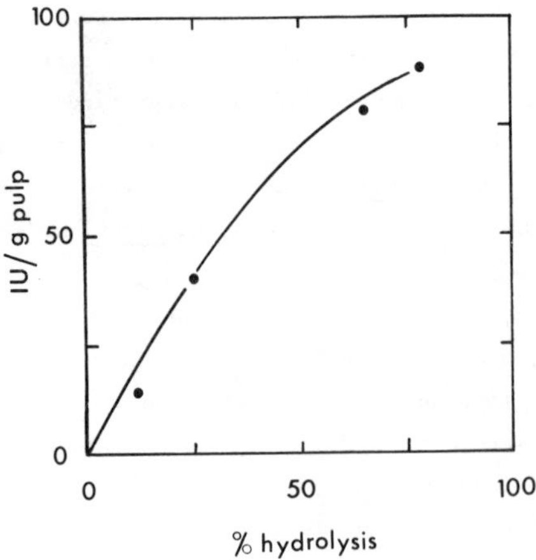

Fig. 2. Hydrolysis of beet pulp at 60°C in 0.1 M sodium acetate buffer, pH 5, by culture filtrates of *T. emersonii* UCG 208 as a function of the ratio of cellulase activity to pulp concentration.

In this context, we find that culture filtrates of *Trichoderma reesei* MCG 77 or QM 9414 were usually, but not always, more effective than those of *Talaromyces emersonii* UCG 208 at least in the earlier stages of the reaction (Fig. 3). Mixtures (equal volumes) of *Talaromyces* and *Trichoderma* culture filtrates were somewhat better than either filtrate alone indicating that each organism provided some specific enzyme activities lacking in the other, i.e. they were complementary (6).

Processes designed to saccharify cellulosic wastes must contend with the crystalline nature of cellulose itself and with its protective covering of lignin. The lignin content of beet pulp is, as we have stated before, low relative to most wastes. Nevertheless, the above results indicated that without some physical or chemical pretreatment extensive enzymic digestion would be difficult to achieve. Accordingly, the ability of culture filtrates to digest pulp which had been subjected to various pretreatments was investigated (Table IV). Hydrolysis of total polysaccharides was improved by some 30% by ball-milling or grinding the pulp from an average size of 1.54 mm to an average size of 10 μm.

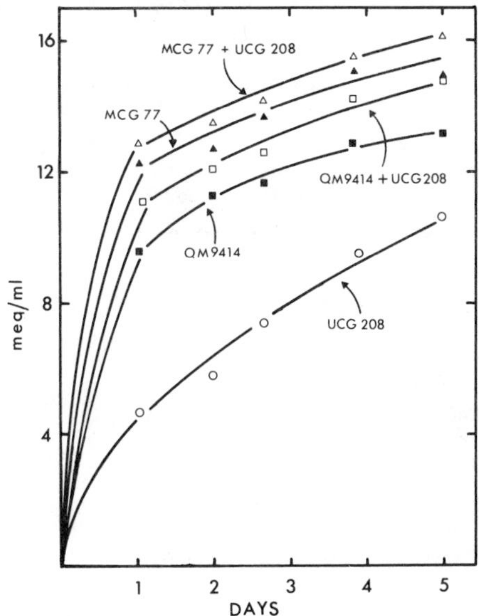

Fig. 3. Digestion of beet pulp by Talaromyces emersonii and Trichoderma reesei. Ten ml of each culture filtrate adjusted to pH 5 with sodium acetate buffer (final concentration 0.1 M) and mixtures (equal volumes) of the T. emersonii and T. reesei pH adjusted culture fluids were incubated at 60°C (T. emersonii) or 50°C (T. reesei and mixtures) with 400 mg dry homogenized beet pulp with shaking in sealed universal bottles. Samples were taken at intervals for measurement of reducing sugars. The initial cellulase activities (U ml^{-1}) were: T. emersonii UCG 208, 1.1; T. reesei QM 9414, 1.0; T. reesei MCG 77, 1.74; QM9414/UCG 208 assayed at 50°C, 0.87; MCG 77/UCG 208 assayed at 50°C, 1.25.

Such treatment benefited hydrolysis of the cellulose fraction rather more than it did that of the pectin and hemicellulose fractions. Alkali pretreatment of the ball-milled pulp gave a further slight improvement in saccharification but only in the earlier stages of reaction. By contrast, peracetic acid treatment per se not only effected the release of reducing sugars but also facilitated the enzymic hydrolysis of the hemicellulose and pectin fractions by about 100%. Comparison of the values for reducing sugars and glucose released shows that the effects of this acid pretreatment on hydrolysis of the cellulose fraction was less marked. Nevertheless,

Table IV. Effects of pretreatment on saccharification of beet pulp polysaccharides by *T. emersonii* UCG 208 (from ref. 11).

Pretreatment	% hydrolysis of total polysaccharide content		% hydrolysis cellulose content	
	excluding control	including control	excluding control	including control
1 Homogenized by Waring blender (av. diam. 1.54 mm)	22.0	25.2	33.7	35.5
2 Ball-milled or Cross-beaten (av. diam. 10 μM)	27.3	34.5	50.6	54.2
3 As in 2 plus 4 N NaOH for 24 h @ 30°C, or 0.25 N NaOH, 1 h, 121°C	26.9	37.0	-	-
4 As 1 plus 20% (v/v) peracetic acid, 1 h, 100°C	35.9	62.6	51.1	54.0
5 As 2 plus 20% (v/v) peracetic acid, 1 h, 100°C	50.2	76.7	58.4	63.1

Samples of treated beet pulp (4%, w/v) were incubated with culture filtrate (cellulase activity 0.83 IU/ml) for 5 d @ pH 5, 60°C in sealed vessels. Aliquots of reaction mixtures were taken at intervals for measurement of reducing sugar and glucose concentrations. The percentage hydrolysis of total polysaccharide content (cellulose, hemicellulose and pectin) was calculated on the basis that it comprised 85% of dry wt. of pulp and that complete hydrolysis would yield 34.4 meq reducing sugar/ml. The extent of cellulose hydrolysis was calculated on the basis that it comprised 22% of dry wt. and that its complete conversion would yield 8.9 mg glucose/ml. The control values refer to reducing sugars or glucose made soluble by the pretreatment alone.

....

as much as 77% digestion of the total polysaccharide content of pulp so treated was effected in 5 days at 60°C, pH 5 by *T. emersonii* UCG 208 exhibiting a cellulase activity of only 0.85 IU/ml of culture filtrate.

Filtrates of *T. emersonii* UCG 208 and of *T. reesei* MCG 77 were assayed for their ability to hydrolyze pectin (methoxyl content 9.7%). In neither case was linearity maintained for more than a few minutes. By contrast, no such difficulty was observed in assays of several different commercial pectinase preparations even though the extent of substrate hydrolysis was much greater than that achieved by the Talaromyces or Trichoderma filtrates (11). This would suggest that neither filtrate contained the full range of enzymes necessary for pectin degradation or that the relevant enzymes are unusually susceptible to product inhibition. They are, as we have stated above, markedly less stable than the corresponding cellulases and hemicellulases. We have also noted that the mutant of T. emersonii, i.e. UCG 208, that most effectively catalyzed hydrolysis of pulp polysaccharides exhibited the greatest polygalacturonase activity (Table III). For these reasons and because the pectin content of pulp is high (c. 24-32% of dry weight) we examined the effects of pretreatment with commercial pectinase on the saccharification process. Such pretreatment markedly enhanced the ability of culture filtrates of T. emersonii UCG 208 and of T. reesei MCG 77 to effect hydrolysis of pulp polysaccharides (Figs. 4,5). In fact, digestion was calculated to be almost quantitative under appropriate conditions and could be speeded up by increasing the amounts of enzyme used.

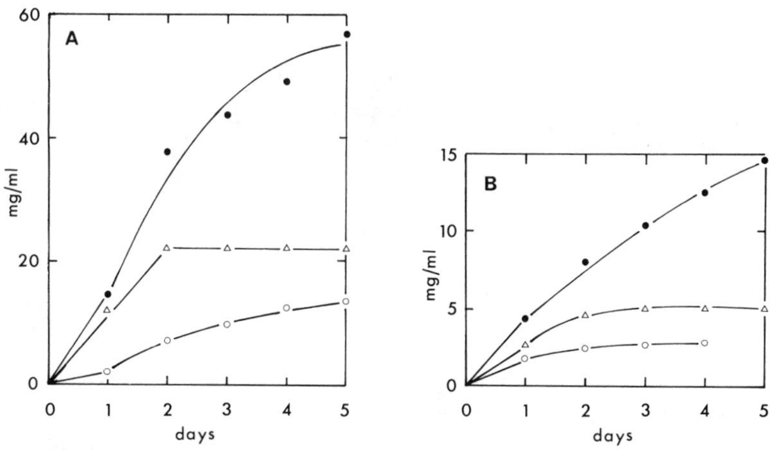

Fig. 4. Effects of pectinase pretreatment on the hydrolysis of beet pulp by culture filtrate of T. reesei MCG 77 (see legend to Fig. 5 for details)

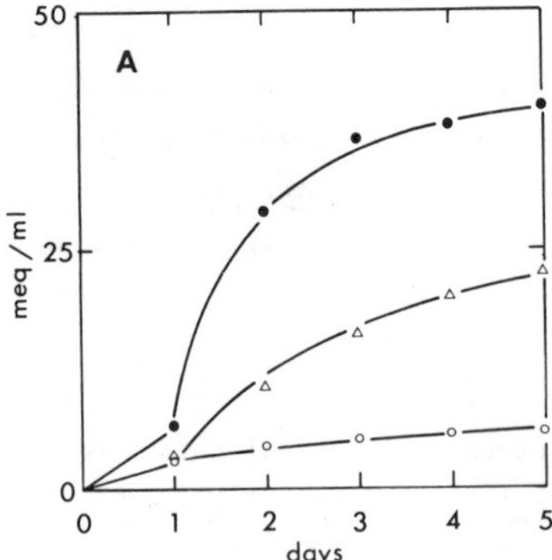

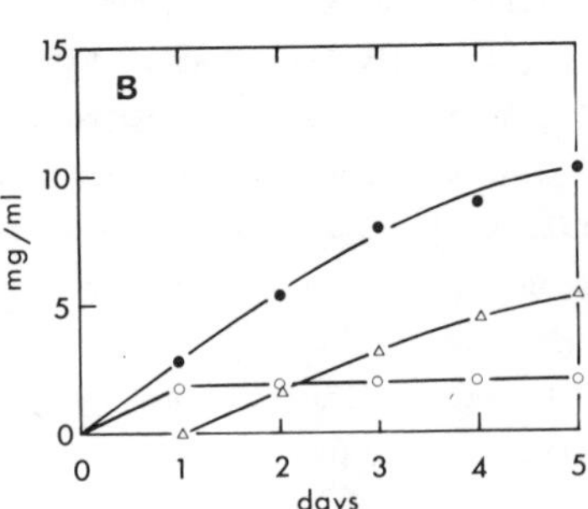

Fig. 5. Effects of pectinase pretreatment on the hydrolysis of beet pulp by culture filtrate of _T. emersonii_ UCG 208. 8 g of pulp and 50 μl of commercial pectinase (Sigma P5146) in 50 ml of 0.1 M sodium acetate buffer, pH 5, were incubated at 25°C. At 24 h 50 ml of _T. reesei_ MCG 77 (Fig. 4) or of _T. emersonii_ (Fig. 5) culture filtrate were added (●) and the incubation temperature was increased to 50°C (_T. reesei_) or 60°C (_T. emersonii_). The concentrations of reducing sugars (A) and of glucose (B) in hydrolyzates were measured at intervals. The following controls were also run: (Δ) Reaction mixtures containing buffered pulp but without added pectinase were incubated at 25°C for 24 h. Filtrates were then added and the temperature increased to the appropriate value. (○) Pulp and pectinase in buffer were incubated at 25°C for 24 h. At this time 50 ml of buffer was added and the temperature was stepped up as before. The activities in the commercial preparation and in the culture filtrates are given in Table V.

The commercial pectinase preparations used exhibited a number of enzyme activities (Table V) any one or more of which in conjunction with the _Trichoderma_ or _Talaromyces_ filtrates may have expedited the hydrolysis of pulp. In this context, pectinase, pectin lyase and hemicellulase are obviously important. So also may β-glucosidase. While most of the enzymes

Table V. Enzyme activities in *T. emersonii* UCG 208 and *T. reesei* MCG 77 culture filtrates and in Sigma (P5146) pectinase preparation (from ref. 13).

Activity	UCG 208 ($U.ml^{-1}$)	MCG 77 ($U.ml^{-1}$)	Sigma P5146 ($U.ml^{-1}$)
Cellulase (FPase)	3.5	4.0	0.26
CMCase	454.7	197.2	132.0
β-glucosidase	2.1	4.5	30.0
Xylanase	18.5	20.7	66.7
β-xylosidase	0.9	2.6	+
Pectinase	-	-	135.0
Polygalacturonase	70.8	294.8	1076.0
Pectin lyase	0	0	14.6
Pectinesterase	0.11	0.11	0.03
Arabinogalactanase	0	0	0

The *T. emersonii* and *T. reesei* culture filtrates which had been concentrated some 3-fold before use were assayed at 60°C and 50°C respectively. The Sigma enzyme activities were assayed at 25°C. The + sign indicates that the activity known to be present was not quantitated, 0 means that activity was not detected.

in the commercial preparations are totally inactivated within 5 min at 60°C the latter was found to have half-life at 60°C, pH 5 of 10 h (11). Thus, it may have stimulated cellulose hydrolysis by cleaving the inhibitory cellobiose even after the step-up to the higher temperature (Fig. 4,5). Indeed, supplementation of *Trichoderma* culture filtrates with β-glucosidase from other sources is common practice in cellulose hydrolyses. In this context we have demonstrated the conversion of cellulose to ethanol in a model system comprised of cellulose, fungal cellulase and calcium alginate gel containing yeast and immobilized β-glucosidase (14,15).

We note that Pilnik's group have also reported (16) that a combination of cellulolytic (*T. viride*, Maxazyme Cl 2000) and pectolytic (*A. niger*, Rapidase C 80) activities gives almost complete liquefaction of beet pulp and potato fibre. The Rapidase C 80 used by these workers also contained a number of enzymes other than those involved in pectin degradation.

Nevertheless, we consider the latter to be the key to the enhanced hydrolysis of pulp by culture filtrates of *T. reesei* or *T. emersonii*. By solubilizing the pectin they not only contribute directly to saccharification but in so doing render the cellulose and hemicellulose fractions more accessible to the appropriate hydrolytic enzymes. In view of this an organism producing the complete range of thermostable pectin-, hemicellulose- and cellulose-degrading enzymes would be most useful.

ACKNOWLEDGEMENT

Much of the work reported here was supported by EEC Contract RUW-035-81 to M.P.C. A.P.M. was in receipt of a Dept. of Education, Ireland, graduate student maintenance grant.

REFERENCES

1. M.A. Folan & M.P. Coughlan Int. J. Biochem. (1978) 9, 717-722.
2. M.A. Folan & M.P. Coughlan, Int. J. Biochem. (1979) 10, 505-510.
3. A. McHale & M.P. Coughlan (1980) FEBS Lett. 117, 319-322.
4. A. McHale & M.P. Coughlan (1981) Biochim. Biophys. Acta. 662, 145-151.
5. A. McHale & M.P. Coughlan (1981) Biochim. Biophys. Acta. 662, 152-159.
6. A.P. Moloney, T.J. Hackett, P.J. Considine & M.P. Coughlan (1983) Enzyme Microb. Technol. 5, 260-264.
7. P.J. Considine & M.P. Coughlan (1982) Bioscience Reports, 2, 299-302.
8. A.P. Moloney, P.J. Considine & M.P. Coughlan (1983) Biotechnol. Bioeng. 25, 1169-1173.
9. A.P. Moloney & M.P. Coughlan (1983) Biotechnol. Bioeng. 25, 271-280.
10. A.P. Moloney, M.P. Coughlan, S.I. McCrae & T.M. Wood (1984) submitted to Biochem. J.
11. A.P. Moloney, A. O'Rorke, P.J. Considine & M.P. Coughlan (1984) Biotech Bioeng. in press.
12. M.P. Coughlan, M.A. Folan, A. McHale, P.J. Considine & A.P. Moloney (1984). In proceedings Lignocellulose Biodegradation Conference (eds. D.A. Wood & J.M. Lynch) G.C.R.I., Littlehampton, Sussex, U.K. (1983). Appl. Biochem. Biotechnol. in press.
13. M.P. Coughlan et al. (1984) in preparation.
14. M.P. Coughlan, A. McHale, A.P. Moloney, A. O'Rorke, P.J. Considine & M.P.J. Kierstan (1982) Biochem. Soc. Trans. 10, 173. 24, 1461-1463.
15. M.P.J. Kierstan, A. McHale & M.P. Coughlan (1982) Biotechnol. Bioeng. 24, 1461-1463.
16. G. Beldman, A.G.J. Voragen, F.M. Rombouts & W. Pilnik, in Proc. E.C. Contractors' Meeting on "Energy from Biomass", Capri (1983)

BIODEGRADATION OF CELLULOSE BY MICROBIAL PROCESSES

*J.B. PARRY and J.H. SLATER
*Dept of Environmental Sciences, University of Warwick, Coventry CV4 7AL, U.K.
*Dept of Applied Biology, UWIST, Cardiff, CF1 3NU, U.K.

Summary

Microbial communities were isolated from soil and compost and their ability to degrade pure cellulose and complex lignocellulosic materials examined in pure and mixed culture. The compost community consisted of five fungi, one of which predominated in all growth studies. This organism was identified as *Aspergillus fumigatus* and was further examined in pure culture to optimize conditions for extracellular enzyme production and cellulolysis. A buffered medium was introduced which increased rates of cellulolysis and production of extracellular endoglucanase and β-glucosidase. Growth of *A. fumigatus* on waste celluloses revealed that complex ligno-cellulosic materials provided the best substrates for free enzyme production. The soil community consisting of eight fungi and two bacteria was examined in pure and mixed culture under batch and continuous flow conditions. Cellulose breakdown and extracellular enzyme production achieved by the mixed culture grown on pure cellulose was never significantly better than that achieved by two of the isolates in pure culture. Similar results were recorded using hay and straw substrates. The two cellulolytic isolates identified as *A. fumigatus* and *Penicillium simplicissimum* were shown to produce all of the components of the cellulase complex. Conditions for extracellular enzyme production and cellulolysis by *P. simplicissimum* were optimized and compared with enzyme levels and breakdown rates achieved by *Trichoderma reesei* QM9414 and *A. fumigatus* IMI 246651.

1. INTRODUCTION

Lignocellulosic material is abundant in municipal, agricultural, industrial and forestry wastes. In recent years the interest in the enzymes degrading such renewable resources of biomass has resulted in the study of lignocellulolytic organisms and the mode of action of their enzymes. Cellulosic substrates can be used directly for the production of single cell protein or by hydrolysis to produce fermentable sugars which can further yield liquid fuels and chemicals. Lignocellulase enzymes could be used to increase the digestibility of crop fodder, to increase the rate of breakdown in crop stubble fields or to accelerate the disposal of lignocellulosic waste materials. In order to achieve such goals, the unit cost of the enzyme must be reduced as this represents a considerable proportion of the process cost (1). This has been achieved to some extent by the optimization of culture conditions and by use of hyperproducing cellulase mutants (2), mutants which are less sensitive to end-product inhibition (3, 4, 5) and constitutive cellulase producing mutants (6).

Fungi are generally considered to be the best producers of extracellular enzymes capable of degrading cellulose but in many cases this only takes place in the presence of the organism. There are relatively few organisms which produce a cell-free culture filtrate capable of extensive saccharification of cellulose and the majority of purification and separation work has been carried out using *Trichoderma viride* (and the *T. reesei* mutants derived from *T. viride*) (2), *Sporotrichum pulverulentum* (7), *T. koningii* (8, 9, 10), *Fusarium solani* (11) and *Penicillium funiculosum* (12, 13) all of which produce culture filtrates which can totally hydrolyse crystalline cellulose.

Interest in cellulolytic bacteria continues particularly with respect to recent advances in the fermentation of cellulosic materials to ethanol by mixed cultures of *Clostridium* spp (14). In general, bacteria produce low yields of extracellular cellulases due to one of more of the cellulase components existing as cell-bound or intracellular enzymes which necessitates substrate-cell association. Cellulolytic bacteria located in the rumen appear to be directly associated with the substrate. In *Bacteroides*, cell-wall bound enzymes come into close contact with substrate due to adhesion to the surface whilst for *Ruminococcus* spp extracellular cellulases bind strongly to the substrate surface (14). Both bacterial and fungal enzymes achieve considerable breakdown of cellulose although the systems by which this is achieved and the subsequent utilization of products may be different. The transfer of bacterial cellulase genes into other organisms e.g. *Zymomonas mobilis*, an ethanol producer, is an area of research which is becoming increasingly important. Cornet *et al*., (16) report the cloning of genes for amino acid synthesis and cellulose hydrolysis into *E. coli*.

In many cases work carried out on the optimization of enzyme yield and cellulose breakdown has used pure cultures grown on pure cellulose. The initial aim of work crried out under contract no. RUW-032 UK was the isolation of microbial communities growing on cellulose or cellulose-containing wastes. It was proposed that such a community be established under continuous-flow conditions which would mimic the situation in the natural environment. Conditions for cellulolysis would be optimized and the rates achieved in the continuous flow system compared with the mixed culture and pure isolates under batch conditions. Once conditions for cellulolysis and enzyme production were established using a pure cellulosic substrate further work would examine the effects of using a variety of lignocellulosics as substrates.

2. BIODEGRADATION OF CELLULOSE AND CELLULOSE-CONTAINING MATERIALS: A STUDY IN THE MAXIMIZASTION OF THE RATES OF CONVERSION BY MICROBIAL PROCESSES.

2.1 Isolation of cellulolytic microbial communities.

The work commenced with two sources of microorganisms (a) soil, from a wheat straw stubble field and (b) compost from a mature garden compost heap. Isolation of the organisms was carried out using several techniques including (a) batch (closed) enrichment, (b) continuous flow enrichment using a chemostat, (c) Warcup poured plates (17) and (d) column enrichment in which cellulose was mixed with a corrugated support matrix within a glass column and supplied with air and medium. The enrichments were carried out at temperatures of 25 and 30°C for the soil samples and 37°C and 50°C for the compost samples.

The organisms isolated by the above techniques were tested for cellulolytic activity by the cellulose-azure method (18), by growth on cellulose agar plates containing ball-milled cellulose and by growth on filter paper agar plates. Under batch conditions the organisms isolated were predominantly fungi although bacteria were present during the early stages of the enrichment as indicated by growth on nutrient agar and malt extract agar plates. It is likely that few of the organisms within the community would be truly cellulolytic and a large number of bacteria would exist upon the by-products of cellulolysis but would not exhibit cellulolytic activity in pure culture.

The continuous flow enrichment of soil microorganisms was established using a one litre chemostat supplied with 0.5% (w/v) cellulose/mineral salts medium. The community isolated from the soil inoculum consisted of seven fungi and two baceria whilst the compost organisms consisted of five spore-forming fungi. The soil isolated community could be maintained in mixed culture in a chemostat whilst in the compost community a single green spored isolate predominated in all growth studies. As a result, the compost isolate was examined in pure culture and work was carried out to determine the enzyme components and to optimize conditions for extracellular enzyme production and cellulose breakdown. The soil isolates were examined as a community and as pure cultures.

2.2 Optimization of culture conditions for the compost isolate *Aspergillus fumigatus*

The green spored isolate was identified as *Aspergillus fumigatus* by the Imperial Mycological Institute, Kew. During growth on 1% (w/v) Avicel in basic salts medium *A. fumigatus* produced all the enzymes of the cellulase complex although β-glucosidase activity decreased corresponding to the decrease in the pH of the culture medium. A variety of different buffered media were examined in an attempt to retain the acid-labile β-glucosidase activity throughout the course of the incubation. It was found that citrate and acetate buffers did not support growth whilst a 0.1M sodium phosphate buffered medium allowed β-glucosidase to be retained and to increase over the incubation period.

It was found that a large amount of flocculant growth occurred when sodium phosphate buffer was used which would be difficult to deal with in a continuous system and an attempt was made to reduce growth and retain cellulase activity by reducing levels of ammonium sulphate. It was found that higher specific activity was obtained together with less growth when the concentration of ammonium sulphate in the medium was reduced from 2.5 gl^{-1} to 0.5 gl^{-1}.

When substrate concentrations were varied between 1 and 5% (w/v) Avicel in sodium phosphate buffer it was found that extracellular enzyme production was dependent on incubation time and substrate concentration. A substrate concentration of 1% (w/v) was found to be optimum for extracellular enzyme production as free enzyme decreased with increasing substrate concentration. It is possible that at cellulose concentrations over 1% enough binding sites are available on the cellulose to 'mop up' any free enzyme. A comparison of the sodium phosphate buffered medium with potassium phosphate buffered medium showed that extracellular enzyme production was highest when phosphate buffer was used at a concentration of 0.1 - 0.2M.

The possibility that enzyme substrate binding was responsible for the decrease in extracellular enzyme levels was investigated by incubating cell free culture filtrate with 1% (w/v) Avicel at various pH values over a period of several days. No detectable binding of endoglucanase occurred. As a result, the cellulose-mycelium cellulose mycelial solids from a five week culture of A. fumigatus grown on 1% (w/v) Avicel/basic salts medium were examined. The solids were washed with (a) 0.1M acetate buffer pH 4.8 (b) 0.1% (v/v) Triton X-100 (c) 0.1M acetate buffer pH 4.8 and finally with (d) 0.5M NaCl. After washing the solids were resuspended in buffer containing 0.5% (w/v) azide to prevent any further growth of the organism and re-incubated at 37°C. Plated samples of the solids gave no sign of growth at any time during the azide incubation. There was no significant release of enzymes from the cellulose-mycelial solids by any of the washing procedures but reducing sugars were detected in the incubation buffer suggesting that enzymes remained bound to the surface of the substrate and/or organism. Buffer was removed and replaced with fresh buffer at regular intervals and it was found that reducing sugars continued to be released over the eighty day period examined. Paper chromatography carried out on the incubation buffer showed that glucose alone was liberated from the culture solids. Dialysed incubation buffer was assayed and traces of endoglucanase activity were detected possibly as a result of enzyme release due to substrate degradation. These results suggested that enzyme-substrate binding was responsible for the decrease in free enzyme activity observed when substrate concentrations were increased from 1.0 - 5.0% (w/v).

Continuous and fed-batch culture of A. fumigatus

Optimization of culture conditions for extracellular enzyme production and cellulose breakdown were initially carried out under batch conditions with a view to setting up a continuous system. The problem of delivering a solid substrate at very low feed rates was overcome by the use of a metred dosage membrane pump and time switch. The two litre chemostat was maintained under batch conditions for ten days to allow the organism time to establish and it was then fed at a rate of 3 ml h^{-1} with 0.5% (w/v) sigmacell. A one litre batch-grown culture was set up under the same conditions as the chemostat to compare enzyme levels achieved. Despite the low nitrogen medium which reduced mycelial growth, pellet formation and cellulose clumping occurred causing outlet valves in the chemostat to block. When aeration and agitation were increased to prevent these problems, spore formation occurred which was particuarly undesirable due to the pathogenic nature of the organism. In addition, enzyme levels which increased while the chemostat was maintained under batch conditions and also in the comparative batch culture, rapidly decreased when the fresh medium was supplied probably as a result of adsorption of free enzyme to fresh cellulose and of loss of bound enzyme via the outlet system.

In view of the problems encountered using a pathogenic organism in continuous culture a further investigation was carried out using a fed-batch system in which 50% of the flask contents were removed and replaced with fresh media at 30 day intervals. In non-buffered media extracellular enzyme levels decreased with each subsequent addition of medium but in the buffered system enzyme levels increased after each addition of substrate although a plateau was reached after which no further increase occurred. The possibility that this plateau was due to substrate depletion was investigated by carrying out a dry weight estimation on the solids removed from the flask and the material remaining when the experiment was terminated. Results showed that over 90% of the substrate had been utilised and as a proportion of the solid material would be cell mass actual values were probably nearer 100%. The rate of cellulolysis increased at each substrate feed stage.

Time course for substrate degradation by **A. fumigatus**

The time interval between each addition of substrate in the fed-batch system was approximately 30 days and in order to determine the time required for substrate depletion to occur, an activity/cellulolysis time course was commenced. Flasks containing 1% (w/v) Avicel in 0.2M phosphate buffered medium were inoculated, incubated in an orbital shaker at 37°C and harvested at time intervals, in triplicate. Culture filtrates from all three flasks were assayed for exoglucanase, endoglucanase, β-glucosidase and xylanase activities and their pH values measured. The dry weight of solids from two of the flasks were estimated and a cellulose determination carried out on this material. The solids from the third flask were resuspended in buffer, sonicated and a protein estimation carried out. The percentage weight loss figures did not correspond to the disappearance of cellulose as measured by the anthrone method (19) suggesting that a large proportion of the harvested material was cell mass. Virtually all the cellulose had been broken down by the twelfth day after inoculation confirming that substrate depletion was probably a factor influencing the plateau in enzyme levels found in the fed-batch system. The feed additions should be made at approximately ten day intervals. Thus, the use of the phosphate buffered medium improved the rate of cellulolysis 5-6 fold over previously reported values (20).

Effect of buffered and unbuffered media at different substrate concentrations

Having developed a buffered medium in which extracellular enzyme production and the rate of cellulolysis were improved, the optimum cellulose concentration was re-examined over the range 0.25-10.0% (w/v) Avicel and compared with the unbuffered system over the same concentration range. The most immediate difference between the two media was the presence of β-glucosidase only in the buffered medium. Levels of β-glucosidase increased during the course of the incubation and with increasing substrate concentration up to 5% (w/v) Avicel. Endoglucanase levels in the buffered medium were approximately eight-fold higher than the unbuffered system. Endoglucanase levels increased during the course of the incubation and with increasing substrate concentration up to 5% (w/v) whereas in the unbuffered system endoglucanase levels either plateaued or decreased with time and decreased with increasing substrate concentration. Extracellular exoglucanase levels in the unbuffered medium also decreased with increasing substrate concentration but in this case extracellular enzyme levels were highest in the unbuffered medium. In spite of the low levels of free exoglucanase activity in the buffered medium over 95% of the cellulose had been degraded at substrate concentrations from 0.25-5.0% (w/v) whereas the percentage cellulose utilised in the unbuffered system decreased from 96% at 0.25% (w/v) cellulose to 22% at 5% (w/v) cellulose. Thus, using the buffered medium an increase in the rate of cellulolysis has been achieved and extracellular enzyme levels can be improved by increasing substrate concentration. In both media the slurry formed at 10% (w/v) substrate concentration was too thick to allow aeration and little growth was recorded.

Enzyme production resulting from growth of **A. fumigatus** on simple and complex substrates

Having optimized cellulolysis and extracellular enzyme production using pure crystalline cellulose, the same buffered medium was used to investigate enzyme production by A. fumigatus grown on substrates varying from simple sugars to complex lignocellulosics, including materials found in both municipal and agricultural wastes. Pure cellulose materials have been most widely used in characterizing cellulases and cellulolytic microorganisms. Such materials have little or no hemicellulose and no lignin unlike the majority of substrates of commercial interest. In addition there is no compartmentalization of pure cellulose such as is found in natural substrates which will have all three components in close association with one another. It is important, therefore that natural lignocellulosic materials be studied as substrates for enzyme production. Pre-treatment of lignocellulosic materials to increase their accessibility has been found to be desirable where untreated pure cellulose gave better results than biomass (14). However, in the case of A. fumigatus the lignocellulosic substrates including newspaper, glossy paper, cardboard, hay, straw and sawdust produced higher levels fo extracellular enzyme than did the pure celluloses. Hay, in particular, enhanced extracellular enzyme production in agreement with findings of Stewart et al (21) but, interestingly, similarly high values were obtained from growth on straw and sawdust.

Under buffered conditions, soluble sugars did not completely repress cellulase production whereas the low pH values attained in non-buffered systems resulted in enzyme inactivation. These results agree with the observation by Mandels & Sternberg (22) that the so called 'glucose effect' resulting in feedback inhibition/repression of the enzymes is in fact likely to be a pH effect.

Preliminary work on the separation and purification of culture filtrates resulting from growth of A. fumigatus on buffered and unbuffered media and on the different substrates has commenced.

3.2 Soil isolated community

The soil community consisting of eight fungi and two bacteria was isolated by a continuous enrichment followed by plating onto Luria agar and malt agar containing streptomycin and crystal violet. The fungi were identified by the Imperial Mycological Institute, Kew as (a) Penicillium nigricans (b) Penicillium simplicissium (c) Aspergillus fumigatus (d, e and f) three strains of Fusarium oxysporum (g) Paecilomyces lilacinus and (h) Gliocladium roseum. All eight fungi have been grown in mixed and pure culture on cellulose media and their enzyme activities determined. The organisms fell into three groups, those which produced the full complement of cellulase enzymes, those which produced endoglucanase and β-glucosidase enzymes and one which produced only β-glucosidase. The bacteria grew on carboxymethyl cellulose agar plates and on filter paper agar plates but only very small colonies were produced on ball-milled cellulose agar plates. No colonies were produced under anaerobic conditions. In batch culture the bacteria did not produce detectable quantities of extracellular enzymes although the cells remained viable and increased in number. This is probably due to the fact that many bacterial cellulases are cell bound or substrate-cell associated enzymes (14).

Extracellular enzyme production and cellulose breakdown by the soil community was examined under continuous flow and batch conditions. In the case of the soil community pellet formation and cellulose clumping did not occur and it was possible to study the community under continuous flow conditions. Initial extracellular enzyme levels increased while the chemostat was maintained under batch conditions but declined when fresh substrate was supplied although all members of the community remained in the chemostat. It was thought that the result could be due to pH inactivation of the enzymes but the addition of a pH control unit did little to rectify matters and it seems likely that free enzymes were lost as a result of binding to fresh substrate and through the overflow mechanism.

Cellulose breakdown and extracellular enzyme production by the mixed culture was compared with that achieved by the individual isolates in pure culture. Results consistently showed that the activity of the mixed culture was no better than that of two of the individual isolates i.e. A. fumigatus and P. simplicissimum both of which produced the full complement of cellulase enzymes. Thus, the mixed culture was not expressing its full potential when growing on pure cellulose and as a result the mixed culture and the two cellulolytic organisms were grown on hay and straw to investigate whether the mixed community would be more successful using natural substrates. Extracellular enzyme levels were higher in all three cases during growth on the lignocellulosic materials compared with growth on pure cellulose probably due to the higher pH of the culture filtrates during growth on these substrates. However, as in the case of pure cellulose, substrate breakdown and free enzyme levels achieved by the mixed culture were not significantly better than the two pure cultures although different enzyme profiles were produced by the three systems. The composition of the lignocellulosic materials was examined before and after growth of the community and the two cellulolytic isolates to determine whether there were any differences in the mode of attack of the substrate. All three systems degraded the hemicellulose component of the lignocellulose faster than the cellulosic component and no significant difference in the mode of attack could be detected by this method. It appeared to be more important that the complete range of enzymes was produced rather than high levels of any one component. This finding is in agreement with earlier work on A. fumigatus (20) and relates to the observation by Mandels (2) that any one component of the enzyme complex may be limiting but as it is increased other components will, in turn, become limiting.

Optimization of culture conditions for the soil isolate P. simplicissimum

The success of P. simplicissimum IMI 267966 with respect to cellulolysis and extracellular enzyme production during the mixed culture work has led to a more detailed investigation of this organism in pure culture. No information on the cellulolytic activity of this organism has been reported in the last 35 years. During the mixed culture work P. simplicissimum was shown to produce the full complement of cellulase enzymes although β-glucosidase activity was lost as pH values of culture filtrates fell. It was found that optimum extracellular enzyme production and cellulolysis were achieved at 25°C in 0.1M phosphate buffered medium with nitrogen supplied as peptone. The optimum assay temperatures for the individual enzymes were found to be 60°C for endoglucanase and β-glucosidase, 40°C for the exoglucanase whilst the xylanase exhibited broad range temperature tolerance between 35 and 70°C.

Enzyme activity of *P. simplicissimum* was compared with that of *Trichoderma reesei* QM9414, one of the early hyperproducing cellulase mutants, and *Aspergillus fumigatus* IMI 246651. *Trichoderma* spp are reputed to produce the most active cellulases (2) although they have the disadvantage of low levels of β-glucosidase as it is predominantly an intracellular enzyme. *Aspergillus* cellulases are generally found to be high in endoglucanase and β-glucosidase production but low in exoglucanase. Under laboratory conditions extracellular exoglucanase and endoglucanase production by *P. simplicissimum* was equivalent to that by *T. reesei* and β-glucosidase was equivalent with that produced by *A. fumigatus*. *P. simplicissimum* has the advantage over *T. reesei* that it produces a predominantly extracellular β-glucosidase and is safer to use than *A. fumigatus* in that it is not a known pathogen. Pure cellulose was broken down at a similar rate and to a similar extent by all three organisms. *P. simplicissimum* appears to be an ideal organism for further optimization and mutation work.

Preliminary work on the separation and purification of the cellulase and xylanase components of the *P. simplicissimum* enzyme complex has commenced.

P. simplicissimum has also been grown on a variety of substrates ranging from simple sugars to complex lignocellulosics using the buffered medium. Results indicate that municipal and agricultural wastes again provide the best substrate for free enzyme production.

Conclusions

To conclude, it would appear that under the conditions investigated within the scope of this work, a community of organisms does not degrade either pure cellulose or lignocellulosic materials at a significantly better rate than individual organisms from the community in pure culture. It seems probable that the cellulolysis achieved by the mixed culture is dependant on the cellulolytic ability of the isolates within the community which produce the full complement of cellulase enzymes.

Conditions for extracellular enzyme production and cellulose breakdown have been optimized for the compost isolate, *A. fumigatus*, and a six-fold increase in the rate of cellulolysis has been achieved. Higher levels of extracellular enzyme were achieved when substrate concentrations were increased under buffered conditions and by use of lignocellulosic substrates.

Extracellular enzyme production and cellulose breakdown by the soil isolate, *P. simplicissimum*, has been shown to compete favourably with *T. reesei* QM9414 and *A. fumigatus* 246651 under laboratory conditions. It produces an extracellular β-glucosidase. An increase in extracellular enzyme levels was achieved by growth on lignocellulosic substrates.

References

1. Wilke, C.R., Cysewski, G.R., Yang, R.D. and van Stockar, U. (1976). Biotechnology and Bioengineering, 18 1315.
2. Mandels, M. (1982) Annual Reports on Fermentation Processes 5, 35.
3. Montenecourt, B. and Eveleigh, D. (1979). Tappi. Proceedings Annual Meeting, New York, March 1979, 101.
4. Montenecourt, B. and Eveleigh, D. In "Hydrolysis of Cellulose: Mechanisms of Enzymatic and Acid Catalysis" (R.D. Brown and L. Jurasek, eds.). Advances in Chemistry Series 181, 289.
5. Montenecourt, B., Kelleher, T.J., Eveleigh, D.e. and Pettersson, L.G. (1980). In "Biotechnology in Energy Production and Conservation" (C.D. Scott, ed.) Biotechnology and Bioengineering Symposium no. 10, 25.
6. Shoemaker, S.P., Raymond, J.C. and Bruner, R. (1981). In "Trends in the Biology of Fermentations for Fuels and Chemicals" (A. Hollander, ed.) Plenum Press, New York.
7. Eriksson, K.E. (1978). Biotechnology and Bioengineering, 20, 317.
8. Wood, T.M. and McCrae, S.I. (1972). Biochemical Journal, 128, 1183.
9. Wood, T.M. and McCrae, S.I. (1975). Symposium on Enzymic Hydrolysis of Cellulose. (Bailey, Enari and Linko eds.). Aulanko, Finland, 231.
10. Wood, T.M. and McCrae, S.I. (1978). Biochemical Journal, 171, 61.
11. Wood, T.M. and McCrae, S.I. (1979). In "Hydrolysis of Cellulose: Mechanisms of Enzymatic and Acid Catalysis" (R.D. Brown and L. Jurasek, eds.) Advances in Chemistry Series, 181, p. 181.
12. Wood, T.M. and McCrae, S.I. and Farlane, C.c. (1980). Biochemical Journal, 189, 51.
13. Wood, T.M. and McCrae, S.I. (1982). Carbohydrate Research, 110, 291.
14. Ladisch, MR., Lin, K.W., Voloch, M. and Tsao, G.T. (1983). Enzyme and Microbial Technology, 5, 82.
15. Ng, T.K., Weimer, P.J. and Zeikus, J.g. (1977). Archives Microbiologica, 144, 7,
16. Cornet, P., Tronik, D., Millet, J. and Aubert, J.-P. (1983). FEMS Microbiology Letters, 16, 137.
17. Warcup, J.H. (1950). Nature (London) 166, 117.
18. Smith, R.E. (1977). Applied and Environmental Microbiology, 33, 980.
19. Updegraff, D.M. (1969). Analytical Biochemistry, 32, 420.
20. Stewart, J.C. and Parry, J.B. (1981). Journal of General Microbiology, 125, 33.
21. Stewart, J.C., Lester, A., Milburn, B. and Parry, J.B. (1983). Biotechnology Letters, 5, 543.
22. Mandels, M. and Sternberg, D. (1976). Journal of Fermentation Technology, 54, 267.

CHARACTERIZATION OF CELLULASE IN SOILS AND SEDIMENTS AND THE EVALUATION OF SYNTHETIC HUMIC-CELLULASE COMPLEXES

J.M. SARKAR and R.G. BURNS

Biological Laboratory, University of Kent, Canterbury,

Kent CT2 7NJ, U.K.

Summary

Cellulase and β-D-glucosidase were copolymerised with several phenolic compounds. These phenolic-enzyme copolymers were similar to the stable humic-enzyme complexes commonly reported in soil in many ways: precipitation with acid; colour; E4/E6 ratios; C, H, N and S content and IR spectra. The enzyme activity of the copolymers showed varying degrees of resistance to proteolysis, organic solvents, and storage at high temperatures. All immobilized enzymes had increased K_m values and decreased V_{max} values in comparison with soluble β-D-glucosidase (9.2 m$\underline{M}$, 190 μmol $\underline{p}$-nitrophenol mg^{-1} h^{-1}); the β-D-glucosidase-resorcinol copolymer was the most active (10.5 m$\underline{M}$; 104 μmol $\underline{p}$-nitrophenol mg^{-1} h^{-1}). β-D- Glucosidase, cellulase and the enzyme-resorcinol copolymers were further immobilized on PM10 ultrafiltration membranes, κ-carrageenan, DEAE cellulose and calcium phosphate gel. The best support for immobilization (in terms of high V_{max} values) was the ultrafiltration membrane. An ultrafiltration cell, containing the membrane-immobilized β-D-glucosidase- resorcinol copolymer, was operated as a continuous reactor with substrate flow rates from 0.1 ml to 0.7 ml min^{-1} without decrease in product formation.

1. INTRODUCTION

Cellulose is the most abundant polymeric constituent of plant material and thus is regarded by fuel, food and chemical industries as an important renewable resource. The net world wide production of cellulose is in the region of 1 x 10^{11} tonne per annum; 5% of which occurs as agricultural waste (19).

The production of glucose from cellulose is a depolymerisation process that may be achieved by mineral acid or enzymic hydrolysis. A comparison of the activities of acid and enzyme catalysts on three cellulosic substrates at 50°C (14) showed that 100,000 times by weight more acid than enzyme was required to bring about the same degree of hydrolysis. Thus there are strong arguments in favour of enzyme hydrolysis of cellulose especially if the cost is reduced by prolonging the life of the enzyme and using a semi-continuous reactor.

Cellulose is hydrolysed by the combined activities of at least three enzymes: an exoglucanase (EC 3.2.1.91), an endoglucanase (EC 3.2.1.4) and β-D-glucosidase (E C 3.2.1.21) (3). Endo- and exoglucanase are inhibited by cellobiose and thus the presence of high levels of β-D-glucosidase is essential to maintain the momentum of hydrolysis (6). Currently attention has been drawn to the possible use of immobilized β-D-glucosidase in cellulolysis and β-D-glucosidase has been immobilized on chitosan and on cellulose carbonate (2, 20). Based on the knowledge that many extra-cellular enzymes are immobilized in soil and sediments through association with humic colloids and that this association confers on enzymes a remarkable stability in comparison with soluble enzymes (4, 12), we have attempted to copolymerise soluble β-D-glucosidase and a mixture of exo-and endo-glucanases with various synthetic 'humic-like' phenolic compounds.

This paper describes the preparation and properties of a variety of stable β-D-glucosidase-phenolic copolymers and presents evidence to suggest that they are comparable to the naturally-occurring humic-enzyme complexes of soil. As a first step to discovering the applicability of a β-D-glucosidase-resorcinol copolymer in the process of cellulolysis, we have further immobilized the copolymer on an ultrafiltration membrane, κ-carrageenan and DEAE cellulose. Cellulase from *Aspergillus niger* and the cellulase-resorcinol copolymer were also immobilized on κ-carrageenan, an ultrafiltration membrane and on calcium phosphate gel. Much of this work has been reported in two recent publications (15,16).

2. MATERIALS AND METHODS

Preparation of β-D-glucosidase-L-tyrosine copolymer. L-tyrosine (2g) and mushroom tyrosinase (EC 1.14.18.1) (35 mg) were dissolved separately in 50 ml of 50 mM-phosphate buffer (pH7.2). The enzyme solution was added to the L-tyrosine solution and the mixture agitated at 25°C. At the time of quinone formation (which occurs immediately or after a few min - depending upon the concentration and the activity of the tyrosinase - and can be identified by the appearance of a purple/yellow colour accompanied by high absorption at 305 nm) sweet almond β-D-glucosidase (40 mg) dissolved in 10 mM-acetate buffer (pH5.4) was added. The mixture was shaken (50 rpm) at 25°C for a further 12 h and then centrifuged (20,000 g). The pellet was washed several times with double distilled water and recentrifuged. The final pellet was resuspended in 75 ml of 15 mM-acetate buffer (pH5.4), dialysed against 2 mM acetate buffer (pH5.4) for 48 h and freeze-dried. The resulting black powder, containing immobilized β-D-glucosidase, was stored at -25°C.

Preparation of β-D-glucosidase- and cellulase- resorcinol and pyrogallol copolymers. A number of phenolic compounds commonly found in soil humic matter were chosen: particularly resorcinol and pyrogallol but also phloroglucinol, orcinol, catechol, protocatechuic acid, caffeic acid, ferulic acid, 2,4-dihydroxybenzoic acid, 3,5-dihydroxybenzoic acid and p-hydroxybenzoic acid. Phenolics were used separately for making homopolymers except for one heteropolymer - a combination of resorcinol and pyrogallol (1:1). Each phenolic (250 mg) was dissolved in 100 ml of 100 mM-acetate buffer (pH5.4) and the solution added to the cylinder of a gradient mixer. A second cylinder contained H_2O_2 (1% w/v). The phenolic solution and the H_2O_2 were mixed (1:1) and passed into a ice-cooled peroxidase solution. The horseradish perioxide (EC 1.11.1.7) (10 mg) was dissolved in 10 ml of 100 mM-acetate buffer (pH5.4). At the time of quinone formation (see above) addition of either β-D-glucosidase or cellulose (40 mg) was begun and continued dropwise for 10 h. At the end of this period the β-D-glucosidase-phenolic copolymers were concentrated *in vacuo*, dialysed against distilled water, freeze-dried and stored at -25°C. All phenolics and enzymes were supplied by Sigma, Dorset, England.

Stability of copolymerised β-D-glucosidase and cellulase preparations. The stability was examined aginst protease, organic solvents, and at elevated temperatures. The details of the procedures have been published elsewhere (16).

Determination of carbon, hydrogen, nitrogen and sulphur. β-D-Glucosidase-phenolic copolymers (2 mg) were analysed for C, H and N using an elemental analyzer (Carlo Elba 1106). Sulphur was determined by burning each preparation (5 mg) in oxygen and titrating against barium acetate using a mixture of thorin and methylene blue (2:1) as indicator (8). Oxygen was determined by difference.

Immobilization procedures. β-D-Glucosidase and β-D-glucosidase-resorcinol copolymer (40 mg) were immobilized on κ-carrageenan (21), DEAE (DE52) cellulose (17) and on ultrafiltration membrane (15).

Cellulase and cellulase-resorcinol copolymer (100 mg) were immobilized on κ-carrageenan (21) and on an ultrafiltration membrane (15). A calcium phosphate gel was suspended in 50 ml of 100 mM-acetate buffer (pH5.0) and centrifuged at 10,000g prior to its use as a support. Cellulase or cellulase-resorcinol copolymer were solubilized in 10 ml of 100 mM acetate buffer (pH5.0) and 10 g of calcium phosphate slurry was added while the solution was stirred at 4°C. After 1 h the suspension was centrifuged at 20,000g and the supernatant fluid was decanted. The calcium phosphate was subsequently washed at least five times by resuspension in acetate buffer and centriguation. The cellulase activity was measured in supernatant fluid and pellets.

Measurement of β-D-glucosidase activity. Enzyme activity was measured at 37°C using p-nitrophenyl-β-D-glucopyranoside (PNPG) in 100 mM-acetate buffer (pH5.4) (concentration in reaction mixture, 15 mM). K_m and V_{max} values were determined using six substrate concentrations ranging from 1.87 mM to 60 mM. Continuous activity was monitored using an ultra-filtration cell and a high level of substrate (final concentration 24 mM) added by pipette or peristaltic pump. The details of the procedure have been reported (16).

Measurement of cellulase activity. Enzyme activity was measured at 40°C using carboxymethyl cellulose (CMC - type 9M8F; Hercules, Wilmington, U.S.A.) in 100 mM-acetate buffer (pH5.0) (final concentration 0.3%). The total reducing sugars (RS) released as glucose equivalents was estimated by the DNS method (18).

3. RESULTS AND DISCUSSION

Activity of copolymers. Retention of β-D-glucosidase activity following copolymer synthesis varied according to the phenolic compound use. The β-D-glucosidase-resorcinol copolymer retained 55% of the activity of the free enzyme; resorcinol/pyrogallol, 36%; 2, 4-dihydroxybenzoic acid, 32%; 3, 5-dihydroxybenzoic acid, 29%; phloroglucinol, 27%; ferulic acid and L-tyrosine, 25%; catechol, 21%; caffeic acid and pyrogallol, 19%; orcinol, 15%; and protocatechuic acid, 14%. Based on the nitrogen contents of four copolymers (Table I) the specific activities retained were resorcinol (22%), pyrogallol (10.5%), resorcinol/pyrogallol (12.4%) and L-tyrosine (6.7%). Probably this variation is due to either the quinones or any residual aromatic monomers causing varying degrees of inactivation during the formation of the copolymer, or that the degree of incorporation of the protein varied. Furthermore, the chemical structure and degree of polymerisation of the final product will affect the expression of enzyme activity.

Although we have not investigated the exact mechanism of enzyme attachment to the polyphenolics, it is clear that the enzyme becomes associated *after* the peroxidase- or tyrosinase-induced oxidative polymerisation of phenolic units. Significantly, β-D-glucosidase added to the reaction before the quinone stage was not immobilized to any great extent and β-D-glucosidase adsorbed to the polyphenolics once formed, or entrapped by vigorous shaking with the polymer was comparatively inactive and unstable. Recent investigations (9,10) have shown that an invertase-humate complex was formed due to adsorption of the enzyme but that the enzyme-humic complex was unstable.

Analysis of β-D-glucosidase-phenolic copolymers. All the synthetic copolymers were light brown/black in colour, had a pH of 6.2 and like humic acids, flocculated upon addition of 1 M-HCl. Elemental and spectral (UV, visible and IR) analyses have confirmed that these synthetic polymers are analogous to native soil-humic-enzyme-complexes. The details have been published (16).

Table I Elemental analysis of C, H, N, S and O in β-D-glucosidase-phenolic copolymers and a soil humic-enzyme extract. Average values for fulvic acid and humic acid from various soils are included for comparison.

Element %	Glucosidase-resorcinol	Glucosidase-pyrogallol	Glucosidase-resorcinol/pyrogallol	Glucosidase-L-tyrosine	Humic-enzyme complex[1]	Fulvic acid[2]	Humic acid[2]
C	34.4	33.5	38.5	47.2	43.7	43.4	58.1
H	5.8	4.4	4.2	5.9	5.3	4.1	4.5
N	2.5	1.8	2.9	3.7	3.5	2.8	4.6
S	1.0	0.8	0.7	0.9	-	-	-
O	56.3	59.5	53.7	42.3	46.0	47.2	33.6

(1) Extracted from a silt loam soil (11)
(2) Average of three soils; % sulphur not reported (5)

Stability of β-D-glucosidase- and cellulase-phenolic copolymers.

Effect of protease. Fifty three percent of the activity of soluble β-D-glucosidase was lost after 1 h exposure to protease and only 30% of the activity remained after 24 h (Fig.1). In contrast, after 24 h the β-D-glucosidase-L-tyrosine copolymer showed only a 12% loss in activity and the β-D-glucosidase- resorcinol copolymer only a 25% loss of activity. Soluble cellulase remained stable for 5 h in the presence of protease; after 5 h the activity of soluble cellulase began to decline and 17% loss in activity was recorded after a 20 h period. The cellulase-resorcinol copolymer showed no loss in activity after 20 h exposure to protease. In general, immobilized enzymes in both soil and on synthetic supports are protected from proteolytic attack but to varying degrees depending their mode of attachment (24).

Effect of organic solvents. The copolymerised β-D-glucosidase preparations were less sensitive to acetone and butanol than was the soluble β-D-glucosidase (Fig.2). However, at high solvent concentrations (20% v/v) the β-D-glucosidase-L-tyrosine copolymer was most resistant. Cellulase copolymers were unaffected by butanol and more stable in the presence of acetone than soluble cellulase.

Immobilization of enzymes often produces enhanced stability towards organic solvents (22). The resistance of cellulolytic enzymes towards organic solvents is particularly important where lignin materials must be digested in solvents prior to cellulolysis.

Comparison of the properties of free and immobilized enzymes.

pH Optima. The pH optimum for both preparations of free β-D-glucosidase, and those immobilized on DEAE cellulose, was 5.5 and is in agreement with earlier observations (1). In contrast, when β-D-glucosidase and β-D-glucosidase-resorcinol copolymer were immobilized on κ-carrageenan or an ultrafiltration membrane there was a broader peak of maximal activity between pH5.5 and 6.5. When cellulase or cellulase-resorcinol copolymer was immobilized on κ-carrageenan and calcium phosphate gel the pH optimum shifted from pH5.0 to 5.6.

Thermostability. The thermostability of the immobilized forms of β-D-glucosidase and β-D-glucosidase-resorcinol copolymer was compared by maintaining preparations at elevated temperatures for 60 min (Fig.3). The soluble β-D-glucosidase rapidly lost activity above 45°C and no activity was recorded at 65°C. In contrast, β-D-glucosidase-resorcinol copolymer retained 60% of activity at 65°C. The thermostability was increased when both preparations of β-D-glucosidase were immobilized on κ-carrageenan and

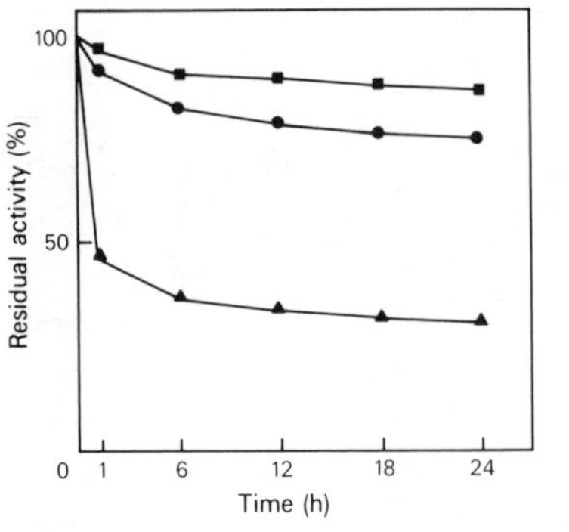

Figure 1.

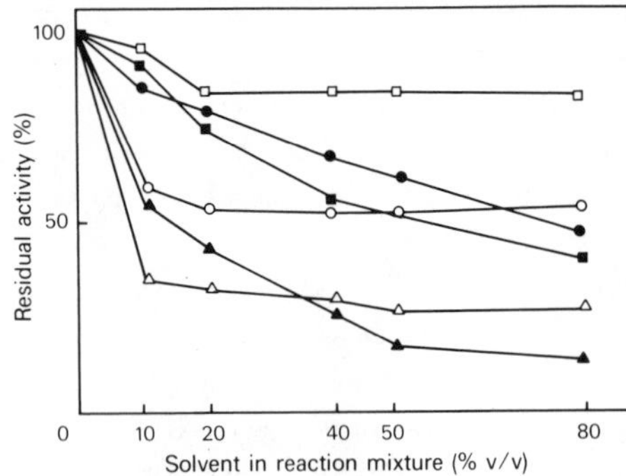

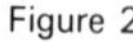

Figure 2.

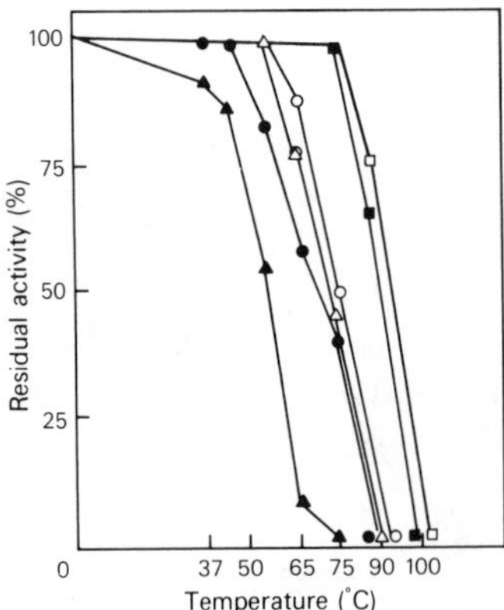

Figure 3.

Figure 1 Effect of protease on: β-D-glucosidase (▲); β-D-glucosidase-L-tyrosine copolymer (■); and β-D-glucosidase-resorcinol copolymer (●). Corrected for loss of activity with time from soluble β-D-glucosidase without protease.

Figure 2 Effect of organic solvents on: β-D-glucosidase + acetone (▲); β-D-glucosidase + butanol (△); β-D-glucosidase-L-tyrosine copolymer + acetone (■); β-D-glucosidase-L-tyrosine copolymer + butanol (□); β-D-glucosidase-resorcinol copolymer + acetone (●); and β-D-glucosidase-resorcinol copolymer + butanol (○).

Figure 3 Inactivation in 60 min at stated temperature of: β-D-glucosidase (▲); β-D-glucosidase-resorcinol copolymer (●); β-D-glucosidase-κ-carrageenan (△); β-D-glucosidase-resorcinol copolymer-κ-carrageenan (○); β-D-glucosidase-ultrafiltration membrane (■); and β-D-glucosidase-resorcinol copolymer-ultrafiltration membrane (□).

ultrafiltration membranes.

The soluble and immobilized cellulase preparations remained stable up to 52°C, but above this temperature the activity of both preparations declined (e.g. 48% loss to soluble enzyme at 78°C and 35% loss to immobilized enzyme). The increased thermostability of the immobilized enzymes is likely to be of importance since we intend to use thermophilic microorganisms (7) in our cellulolysis process.

Re-use of copolymerised β-D-glucosidase and cellulase. The possibility of reusing β-D-glucosidase and β-D-glucosidase-resorcinol copolymer after immobilization on an ultrafiltration membrane was examined. The results (Fig.4) showed that enzyme activity did not decrease by more than 20% during seven cycles of adding substrate, removing products, and washing the cell and membrane.

Ultrafiltration cells containing membrane-immobilized β-D-glucosidase and β-D-glucosidase-resorcinol copolymer were operated as continuous stirred reactors in the hydrolysis of PNPG. PNPG solutions (24 mM) were pumped into the cell continuously and the effect of flow rate on product formation measured (Fig.5). Product formation did not decrease upon increasing the flow rate from 0.1 ml min^{-1} to 0.7 ml min^{-1}. When trypsin was immobilized on ultrafiltration membranes, the rate of casein hydrolysis remained almost constant at a range of substrate flow rates (13).

The possibility of reusing soluble and copolymerised cellulase preparations was also examined. The results (Fig.6) showed that the enzyme was strongly attached to the calcium phosphate gel and ultrafiltration membrane. Several washings, either with 100 mM-acetate buffer or 2 M-NaCl, could not desorb appreciable quantity of enzyme. Furthermore, in the process of adsorption the active site of the cellulase remain accessible to the substrate. The enzyme was not strongly adsorbed on κ-carrageenan as it desorbed on washing with 2 M-NaCl. The enzyme adsorbed on ultrafiltration membrane and calcium-phosphate gel was reused in batch reactors and the enzyme activity did not decrease by more than 25% during seven cycles of adding products, and washing the pellets and membrane.

Kinetic values. K_m and V_{max} values of the immobilized preparations of β-D-glucosidase are compared to those of the free enzyme (Table II). The K_m values of free β-D-glucosidase and β-D-glucosidase-resorcinol copolymer when immobilized on κ-carrageenan were significantly higher compared with those associated with other supports. However, these preparations had considerably higher maximal rates of activity than DEAE cellulase immobilized preparations. A decline in activity of immobilized

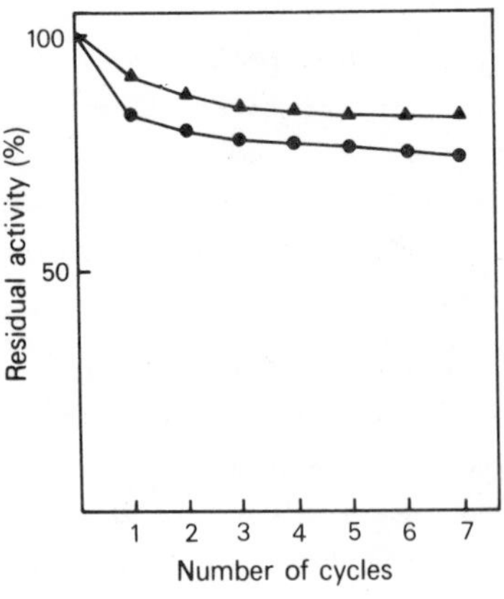

Figure 4.

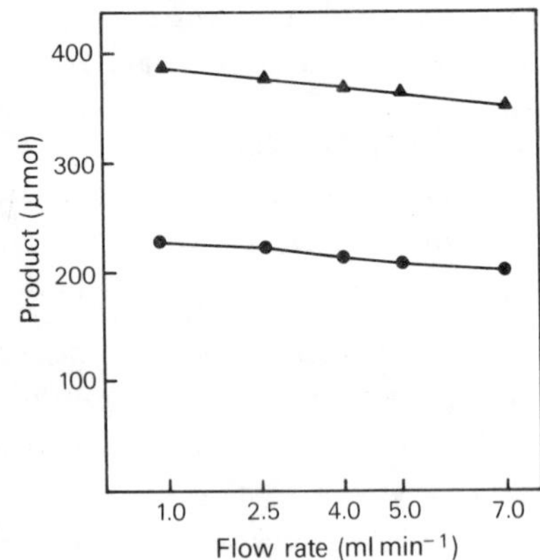

Figure 5.

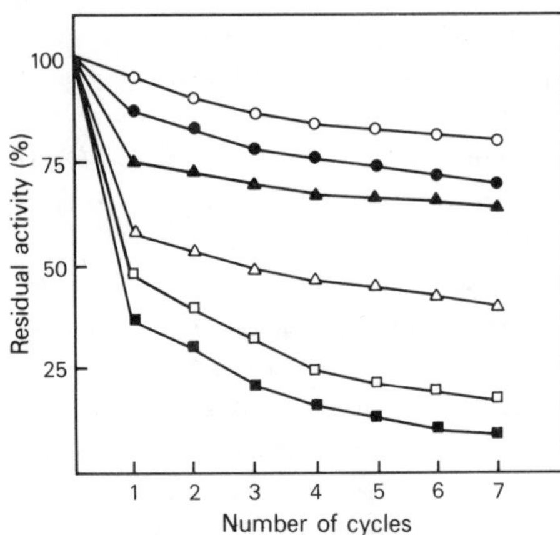

Figure 6.

Figure 4 Effect of reuse of β-D-glucosidase (▲) and β-D-glucosidase-resorcinol copolymer (●) attached to ultrafiltration membranes.

Figure 5 Effect of flow rate on product formation of: β-D-glucosidase (▲) and β-D-glucosidase-resorcinol copolymer (●) attached to ultrafiltration membranes.

Figure 6 Effect of reuse of cellulase attached to various supports: Cellulase-ultrafiltration membrane (○); cellulase-resorcinol copolymer ultrafiltration membrane (●); cellulase-calcium phosphate gel (△); cellulase-resorcinol copolymer-calcium phosphate gel (▲); cellulase-κ-carrageenan (□); and cellulase-resorcinol copolymer-κ-carrageenan (■).

Table II K_m and V_{max} values of β-D-glucosidase and β-D-glucosidase-resorcinol polymer before and after their attachment to various supports.

Enzyme support	K_m (mM)		V_{max} (μmol mg^{-1} h^{-1})	
	β-D-glucosidase	β-D-glucosidase resorcinol polymer	β-D-glucosidase	β-D-glucosidase-resorcinol polymer
Ultrafiltration membrane	15.3	16.8	83.5	42.4
κ-Carrageenan	97.0	46.2	17.3	6.0
DEAE Cellulose	17.6	15.6	1.5	1.0
Soluble preparation	9.3	10.5	190.0	104.0

β-D-glucosidase on cellulose was also reported (20). The kinetic contents of the free β-D-glucosidase-resorcinol polymer (compared with commercial soluble β-D-glucosidase) were K_m 10.5 mM (9.3) and V_{max} 104 μmol mg^{-1} h^{-1} (190). The immobilization of β-D-glucosidase and β-D-glucosidase-resorcinol copolymer on ultrafiltration membranes gave a combination of high V_{max} and low K_m values in comparison with enzyme immobilized on κ-carrageenan and DEAE cellulase. Recently ultrafiltration membranes have been used to immobilize enzymes for the continuous transformation of benzaldehyde to benzylalcohol (23).

4. CONCLUSION

The results have shown that β-D-glucosidase - and cellulase - copolymers immobilized on ultrafiltration membranes retain high levels of enzyme activity. We are now evaluating the advantages of these enzyme-copolymers adsorbed to ultrafiltration membranes in comparison to non-adsorbed enzyme-copolymers, soluble enzymes and absorbed enzymes. It is already known that enzyme copolymers are resistant to protease, solvents and elevated temperatures and experiments are in progress to discover whether these stresses affect the stability of the membrane. We are also developing single or two stage semi-continuous reactors (using various combinations of stabilized β-D-glucosidase and endo- and exocellulase) for the continuous catalysis of cellulosic substances.

Acknowledgements. We acknowledge the support of the European Economic Community who financed many of the studies reported in this paper. The facilities of the CSIRO Division of Soils, Adelaide, Australia were essential to the preparation of this paper.

REFERENCES

(1) Barker, S.A., Doss, S.H., Gray, C.J., Kennedy, J.F., Stacey, M. and Yeo, T.H. (1971). *Carbohyd. Res.* 20, 1.

(2) Bissett, F. and Sternberg, D. (1978). *Appl. Envir. Microbiol.* 35, 750.

(3) Burns, R.G. (1982). Carbon mineralization by mixed cultures. In: *Microbial Interactions and Communities*, A.T. Bull and J.H. Slater, eds. Vol.1, pp.475-543, London: Academic Press.

(4) Burns, R.G., Pukite, A.M. and McLaren, A.D. (1972). *Soil Sci. Soc. Am. Proc.* 36, 308.

(5) Hayes, M.H.B. and Swift, R.S. (1978). The chemistry of soil organic colloids. In: *The Chemistry of Soil Constituents*, D.J. Greenland and M.H.B. Hayes. eds. pp.179-320, Chichester: John Wiley.

(6) Hsu, T.A., Gong, C.S. and Tso, G.T. (1980). *Biotechnol. Bioeng.* 22, 2305.

(7) Johnson, E.A., Sakajoh, M., Halliwell, G., Madia, A. and Demain, A. (1982). *Appl. Envir. Microbiol.* 43, 1125.

(8) MacDonald, A.M.G. (1961). *Analyst* 83, 3.

(9) Maignan, C. (1982). *Soil Biol. Biochem.* 14, 439.

(10) Maignan, C. (1983). *Soil Biol. Biochem.* 15, 651.

(11) Mayaudon, J., El Halfawi, M. and Bellinck, C. (1973). *Soil Biol. Biochem.* 5, 355.

(12) Mayaudon, J., Batistic, L. and Sarkar, J.M. (1975). *Soil Biol. Biochem.* 7, 281.

(13) Mukherjea, L.N., Bhattacharya, P., Ghosh, B.K. and Taraphdar, B.K. (1977). *Biotechnol. Bioeng.* 19, 1259.

(14) Reese, E.T. (1956). *Appl. Microbiol.* 4, 39.

(15) Sarkar, J.M. and Burns, R.G. (1983). *Biotechnol. Letts.* 5, 619.

(16) Sarkar, J.M. and Burns, R.G. (1984). *Soil Biol. Biochem.* 16, (in press).

(17) Sarkar, J.M. and Mayaudon, J.M. (1983). *Biotechnol. Letts.* 5, 201.

(18) Sinclair, P.M. (1965). *Chem. Eng.* 72, 90.

(19) Spano, L.A. (1976). Enzymatic hydrolysis of cellulosic wastes to fermentable sugars and the production of alcohol. In: *Clean Fuels, Biomass Sewage, Urban Refuse and Agricultural Wastes Symposium*, pp.324-348. U.S. Army Natick Res. and Dev. Massachussets, U.S.A.

(20) Srinivasan, V.R. and Bumm, M.W. (1974). *Biotechnol. Bioeng.* 16, 1413.

(21) Takata, I., Kayashima, K., Tosa, T. and Chibata, I. (1982). *J. Ferment. Technol.* 60, 431.

(22) Tosa, T., Mori, T. and Chibata, I. (1969). *Agr. Biol. Biochem.* 33, 1053.

(23) Wiśniewski, J., Winnieki, T. and Majewska, K. (1983). *Biotechnol. Bioeng.* 25, 1441.

(24) Zaborsky, O.R. (1973). *Immobilized Enzymes.* CRC Press, Cleveland, Ohio.

Summary of the discussion of Session I :

CARBOHYDRATE HYDROLYSIS

Chairman : J.P. BELAICH

Rapporteur : J.M. LYNCH
T.M. WOOD

The following points were raised in discussion:

1. Pretreatment. Dilute acid leaches hemicellulose and thus creates larger pores than those in the native substrate. It can improve the rate of subsequent enzymatic hydrolysis.

2. Cellulose form. Cellulose II is more reactive than Cellulose I because there are more beta-linkages in II than I. However, after the removal of the $beta_L$-rich regions the residual crystalline short cellulose fractions are still very resistant to enzymatic hydrolysis.

3. Enzymatic hydrolysis of straw and other refractory materials. It has been observed that percolation of straw and other refractory materials with enzyme improves the rate of hydrolysis. It was suggested that the acceleration of hydrolysis may be a result of the removal of soluble sugars and the consequent relieving of the so-called feed back inhibition of the enzymatic reactions. Some of the enzymes will be adsorbed on the lignin; renewing the enzyme will therefore help the hydrolysis process.

4. Models. There are many models to describe the kinetics of cellulose degradation. Often, however, they describe only homogenous kinetics ignoring the fact that cellulosic substrates are often not pure cellulose and are not homogenous. In a complete model, factors such as the quaternary structure of the cellulase enzymes and the so-called "cellulose binding factors" must be considered.

5. Reactors. There do not seem to be any special problems in the development of reactors for enzymatic hydrolysis. In principle the reactors can be very simple, compared with acid hydrolysis.

6. Thermostability. The need for thermostable cellulases depends on the final product. If the product is ethanol, combined hydrolysis and fermentation is probably used. This means that the thermal stability of the ethanol-producing organism may be a bottleneck for some other applications. Thermal stability of the enzymes is an advantage. For example, *Talaromyces emersonii* produces rather thermostable cellulases.

7. Recombinant DNA technology. This really offers new tools, at least theoretically. A major problem at present, however, is the low specific activity of cellulase compared with, for example, amylase. The solution to this problem could be achieved by finding an organism which produces a cellulase with a high specific activity and then cloning the gene coding for this enzyme in another organism; for example a *Bacillus* species which could then rapidly produce this powerful enzyme in large quantities. The rapid development of recombinant DNA technology makes this goal almost realistic, provided that the appropriate type of enzyme is found. Therefore it may be recommended that a number of cellulase-producing organisms, other than *Trichoderma*, should be screened for efficient production of cellulase.

8. The pH and temperature optima of micro-organisms producing cellulase. All the fungi investigated in the laboratory of Dr T. Wood elaborate cellulases that have a pH optimum in the region of pH 4.5-5.0. The temperature optimum of the cellulase of *Penicillium funiculosum* is 50-55°C. Some of the other cellulases have temperature optima of 45-50°C.

9. Environmental problems. There are no known hazards in the environment of any cellulolytic organisms which have been introduced. The bacterial species such as *Clostridium* which may be associated with the cellulolytic fungi are not known to be hazardous but it is clear that if any of these are introduced into farming systems any potential hazards will need to be checked by suitable regulatory agencies.

SESSION II

ANAEROBIC DIGESTION

ANAEROBIC DIGESTION IN THE EEC

H. NAVEAU

Bioengineering Unit,
Catholic University of Louvain,
Place Croix du Sud, 1/9,
1348 Louvain-la-Neuve, Belgium

Summary

A review of the research, development and demonstration activities in anaerobic digestion, financed by the Commission of the European Communities (CEC) from 1979 to 1983, is done with special emphasis on the "Research and Development Programme in the Field of Recycling of Urban and Industrial Wastes" (DG XII).

Within this programme, research has been led through 21 research contracts dealing with various aspects of biomethanation and a number of residues from agriculture, industry and household. They include aspects from fundamental research (microbiology) to building and running large-scale digesters, through research and development of new processes. The main results are highlighted.

The coordination activities of this programme are described; they consist of contractor's meetings and exchange of information and propositions on various topics such as ways of reporting results or starting of digesters.

Research done in the frame of the Solar Energy Programme, Project E "Energy from Biomass" (DG XII) is summarized with special attention paid to the study "Biogas plants in Europe"; it is a survey of biogas plants in the EEC and Switzerland and their problems, together with an assessment of bottlenecks and propositions for the future development of biomethanation in Europe.

Finally, an outline of the activities within the Demonstration Programme of DG XII (Directorate-General for Energy) is presented as well as the first results obtained.

1. INTRODUCTION

Anaerobic digestion in the EEC covers a large spectrum of activities in research, development, demonstration and full-scale field digesters, financed by community, state, industrial and private money. The aim of this paper is to make a review of those activities financed by the Commission of the European Communities, with special emphasis on the results obtained within the research and development programme in the field of "Recycling of Urban and Industrial Wastes" managed by the Directorate-General for Science, Research and Development (DG XII). This seminar on "Anaerobic Digestion and Carbohydrate Hydrolysis of Wastes" is indeed the place where the results from this programme 1979-1983 are to be presented. Other programmes dealing with anaerobic digestion have taken place in DG XII Solar Energy Programme, Project E "Energy from Biomass" and DG XVII (Directorate-General for Energy), Demonstration Programme.

2. THE PROGRAMME ON RECYCLING OF URBAN AND INDUSTRIAL WASTES

In this programme, the research area III "Fermentation and Hydrolysis" is divided into three topics : Anaerobic Digestion, Carbohydrate Hydrolysis and Composting.

The research conducted into anaerobic digestion falls into two distinct categories, most of the work being covered by cost-sharing contracts, but some by concerted action.

2.1. Research under the cost-sharing contracts

The 21 research contracts concluded can be classified either by type of waste - i.e. urban waste (household refuse), agricultural waste, slaughterhouse waste, effluent and fundamental research - or by the scientific discipline or objective, e.g. bacteriology, specification of operating conditions, effect of temperature (thermophily), development of digesters for substrates with a high dry matter content (solids), construction of full-size digesters or large-scale pilot units, development of monitoring and process-control methods and so forth. As can be seen, the research spans not only the main categories of waste from which biogas (methane) can be produced, but also the scientific and technical problems encountered, ranging from the bacteriological aspects to operation of large-scale digesters in practice.

A list of the contracts together with a brief description of their objectives has been published[1] and can be obtained from the DG XII of the CEC.

This section sums up the main findings obtained, classified by substrate. Note that most of the contracts have covered the years 1980 or 1981 to 1983.

2.1.1. Urban waste

a. Pretreatment :

Crushing urban wastes with a ball-mill at elevated temperature clearly increases the biogas yield and seems to be the most effective pretreatment amongst mechanical, chemical (acid or base) and thermal methods and their combinations. However a simple and fast composting step has a very strong positive influence on biogas yield of urban solid wastes biomethanised in solid state, when the moisture content is high enough ($>$ 35 %). Up to 80 m^3 of methane per ton of urban solid wastes have been obtained in 4 weeks. Pretreatment with enzymes is also effective.

b. Operating parameters :

The stable biomethanation of urban solid waste in one step completely mixed digesters is not straightforward. Volatile fatty acids build up very easily and acidification of the mixture is a real problem. It has been shown that a sufficient alkalinity, or a sufficient high concentration of bicarbonate ions (HCO_3^-) together with the corresponding cations (M^+), is crucial for stable digestion. The composition of the mixture also plays on important role, paper having a stabilizing action through its cellulosic and hemicellulosic fraction.

The mixture of urban solid wastes with other substrates such as liquid manure or sewage sludges may allow to overcome these problems. It also helps to equilibrate the C:N:P ratio around 100:5:1 since the urban residues are poor in nitrogen and phosphorus. A concentration of the feed within 45-125 kg dry matter/m^3 does not modify the methane yield.

The sludges from urban solid wastes digestion are difficult to decant. Addition of flocculants allows filtration resulting in a dry matter content of up to 35 %.

c. Methane productivity with different processes :

In completely mixed digesters, a methane yield of around 0.20 m^3 methane per kg organic matter added is generally found. In batch fermenters, a production of 80 m^3 methane per ton of fresh urban refuse has been reached, corresponding to about 75 % of the theoretical yield. With a two-step process consisting of a percolation and a biomethanation step, a yield of 0,32 m^3 methane per kg organic matter added has been reached.

d. Full-scale digesters :

A 2000 m^3 completely mixed digester has been built to digest mixtures of urban solid wastes, sewage sludges and agricultural wastes such as manure, straw or yeast. Starting the digestion with a mixture of bovine and pig manure and brewery yeast at 12 kg volatile solids/m^3 has been slow.

f. Sanitary landfills :

Experimentations have been made in six experimental landfills to evaluate the influence on gas production rate of various parameters, including waste composition, particle size, density and moisture content. The methane concentration in biogas reaches faster 50 % with more compacted residues. When biogas is not pumped but only collected, great variations in production appear in correlation to atmospheric pressure. However, methane concentration is stable. Collection of biogas using horizontal piping is difficult because pipes are easily smashed and/or filled with water. Only drains made of tyres would work.

2.1.2. Agricultural wastes

A comparison of various manures - bovine, pig, poultry and rabbit - and their mixture with pig manure shows that poultry and pig manures have a synergistic effect; digestion of their mixture at 10 or 15 days mean retention time gives the same methane yield but mixtures of pig and bovine manure on the contrary yield less biogas at 10 than at 15 days mean retention time. The bovine manure needs a longer retention time, probably because of its fibre content. When maximum energy production is sought, the optimum is 15 days whilst from a waste treatment (depollution) point of view, it is 20 days. Also, concentration should not be too high to avoid a kind of inhibition.

Poultry and pig manures can be digested at retention times below 10 days, but, depending upon a given situation, a compromise has to be found between net energy production and depollution.

A digestion system for pig manure composed of 12 digesters of 18 m^3 each (216 m^3 total), has been built and operated. It runs with a single pump and a valve system. Temperature ranges between 15-18 °C in winter and 25-28 °C in summer and conversion in DCO is 70 % in winther and 90 % in summer. Volumetric productivity rate ($r_{V.CH4}$) over 1 m^3 biogas/m^3 digester.day has been obtained in summer. Modular system of this type should cut down installation costs of the digester, although this digestion system has been built in a relatively warm region of Europe so that savings on heating system were possible.

A plug-flow digester for solid agricultural residues has been designed and used at laboratory level for substrate with relatively high (18-20 %) dry matter content. A liquid phase helps to get a rapid substrate inoculation. Straw digestion at 26 days retention times gives a yield of 0.18 l methane per g volatile solids provided daily additions of ammonium carbonate are made. Hydrolysing activity increases fast and keeps constant up to the end of the digestion time.

Thermophilic digestion of solids (a mixture of a fine fraction from urban solid wastes and pig manure) at 25 % dry matter content and 10.5 % volatile solids shows that retention times of 10 to 20 days give a methane volumetric production rate ($r_{V.CH4}$) of over 2 l methane per liter digester and per day but a methane yield increasing from 0.2 to 0.4 l methane per kg volatile solids introduced ($Y_{CH4/VSo}$ = 0.2... 0.4 l CH_4/g VS_o) when increasing the retention time from 10 to 20 days. The level of volatile fatty acids (over 3 g/l) indicate that the biology in this digestion process is not running smoothly. Is this related to the diffuculty of the intermediate metabolites to move between the bacteria so that interspecies exchange is slow ?

2.1.3. Slaughterhouse residues

Studies using mixtures of sewage sludges with bovine or pig abattoir wastes indicate, as with the manure (see § 2.1.2) a problem in including bovine residues at retention time below 10 days.

Conversion in COD ($Y_{CODr/CODo}$) for bovine and pig abattoir wastes reached respectively 37 and 50-58 % at mesophilic temperature. In thermophilic conditions, instability was observed at retention time below 8 days but could be overcome using continuous in place of semi-continuous feeding.

2.1.4. Industrial liquid effluents

The processes studied in this section are based on the principle of bacteria (active biomass) accumulation.

Two contracts have worked on wastewater from olive oil production. One approach uses a contact process (sedimentation tank and recycle of the sedimented sludge in the digester); a volumetric loading rate (B_V) of 6 g COD per liter digester and per day has been maintained with a conversion ($Y_{CODr/CODo}$) of 85 % when nitrogen and alkalinity are provided. An other research uses 5 anaerobic filter in serie. With a volumetric loading rate (B_V) of 2,7 g COD per liter digester per day and a retention time of 12.5 days, a conversion ($Y_{CODr/CODo}$) of 70 % is reached.

Industrial effluents, with varying compositions (concentration) and flows often present difficult problems. One study is trying to find ways of running a digester on such a difficult industrial effluent. Nutrients balance may be a key factor. Elsewhere, two separate pilot-scale fixed-bed anaerobic digesters have been installed on process waters from a sugar refinery. Conversions of 70 % in COD and 80 % in BOD_5 have been obtained at one day retention time with a biogas volumetric production rate ($r_{V.gas}$) of 11.5 m^3/m^3.d. A full-scale plant (twice 1500 m^3) has then been built and operated for one year with success.

Biogas purification may be a necessity, especially as it concerns hydrogen sulfide. It may also be valuable when large quantities are involved such as with biogas from landfills. Various techniques to eliminate CO_2 and H_2S have been compared and a process based on adsorption followed by desorption by heating and a heat exchanger has been found stable. A net yield of 81 % of the methane introduced has been obtained when using methane for heating and running the compressor.

2.1.5. Fundamental research

Cellulose fermentation using co-cultures has been studied in view of producing solvents or acids. Cellulose degradation is much enhanced using in co-culture a mesophilic cellulolytic clostridium and a Clostridium acetobutylicum. The mutual influences of both culture in this case as with cellulolytic and methanogenic cultures have been investigated.

Enrichments of wild population in batch digesters with cellulolytic or methanogenic bacteria show that only addition of methanogenic bacteria increases the biogas production.

Production of single metabolites (acids) by pure culture anaerobic fermentation of agricultural wastes has been the subject of a contract. After classification of 135 cellulolytic bacteria, some of them have been studied. Production of mixtures of acids and of extracellular compounds are the common problem. Trials are made to modify this pattern by using adapted fermentors, but it remains difficult to combine specificity and yield.

Being able to run digesters automatically in a stable way would certainly do a lot for the success of anaerobic digestion. This is difficult because a high number of parameters need to be known and that methods to measure them do not exist or are not applicable continuously. A method has been devised to determine by a simple measure (titration) and calculations the values of the main parameters which are necessary for the description and the control of the digestion process. An automatic titrating instrument has been designed, built and tested. It should help to maintain a digester within the limits of safe running.

2.2. Concerted action

The concerted action to date has taken five directions : coordination of the research conducted under cost-sharing contracts, dissemination of information on all the research being carried out into anaerobic digestion in the Member States, publication of a reference work defining the parameters and assessing the methods of analysis to apply to anaerobic digestion processes, the compilation of a reference work defining guidelines for the starting of digesters and this seminar on anaerobic digestion and carbohydrate hydrolysis. This concerted action is carried out and supervised by the contact/coordination group for research topic III (fermentation and hydrolysis), which consists of national delegations plus CEC representatives.

2.2.1. Coordination of the research conducted under cost-sharing contracts

The contractors met twice a year to coordinate their research, with gatherings at Brussels on October 8, 1981, Wageningen (Netherlands) on March 3, 1982, Nancy (France) on September 29 and 30, 1982, Broni (Italy) on May 17-19, 1983 and this final meeting in Luxembourg. Some of these meetings have taken place in connection with contractor's meeting in the Carbohydrate Hydrolysis topic, since hydrolysis is often a first necessary step in anaerobic digestion. Each contractor outlines the results of his research, which are then discussed by all the participants. En fact, these meetings are like seminars for circulating and discussing scientific findings to the benefit of all contractors and Contact-Coordination Group members.

2.2.2. Dissemination of information on the research being carried out in EEC countries

The national delegates contribute towards a survey of the research in hand. The survey consists of data sheets giving basic details of the research, including the title of the projects, location, programmme, resources, scheduled duration and findings. This databank is put on computer and its organization is continuing.

2.2.3. Reference document on the parameters and methods of analysis

The scientific literature available is chaotic in its description of experiments, and application of the variables for description of the state of the substances and the course of the process respectively. A wide range of parameters - many of them completely unsuited - are applied. Very often the characterization of the starting material and of the effluents, the definition of the process under consideration and the description of the digester used are incomplete or badly expressed. A wide variety of parameters are applied with ill-defined units. All this makes it harder to understand and compare the results and, consequently, severely hampers exchanges of information.

To improve this situation, the contact/coordination group has compiled a compendium of recommendations which could serve as reference standards for defining parameters and analyses. This proposal is published in an international scientific journal for use and comments by the scientific community[2]. Comparison of analytical methods is under way so as to have at least a set a reliable and well defined methods to be used when determining the most common and necessary data.

2.2.4. Reference work for the starting of digesters

A survey of data available on the starting of digesters is being launched; the aim is to compare the various methods used so as to produce guidelines on how to start digesters successfully and what to avoid to reduce problems.

2.2.5. Seminar on Anaerobic Digestion and Carbohydrate Hydrolysis

Finally, this seminar is organised in Luxembourg where all contractors in Anaerobic Digestion and Carbohydrate Hydrolysis will present the final results of the 1st phase of their work together with some presentations by internationally remowned experts and contractors in DG XII Solar Energy Programme (Energy from Biomass).

2.3. Conclusions

The conclusions all point to the benefit of the research programme. Although the most immediate result is to produce encouraging scientific results in the various fields dealt with - generating certain industrial spin-offs already including patents - the emphasis must also be placed on its success in forging links between scientists, industrialists and those responsible for the scientific and industrial policy relating to anaerobic digestion of wastes.

It has brought a visible improvement in relations between research scientists plus far faster, more pertinent exchanges of information, all of which is bound to improve the quality of the R & D work and, hence, the return on the investment put into it.

3. RESULTS FROM THE SOLAR ENERGY PROGRAMME, PROJECT E "ENERGY FROM BIOMASS"

Within this programme which ended in December 1983, research under cost-sharing contracts has been done on strating up and inhibition in thermophilic processes and anaerobic digestion of agricultural wastes and energy crops (including algae) with various digester's design, the most interesting being two-phase digestion of solid substrate or liquid from farm manure. In addition, a survey on "Biogas plants in Europe" has been done.

Results show interesting results on inhibition by ammonium salts and volatile fatty acids in thermophilic conditions, and developpment of two phases processes at laboratory or pilot scale. Treatment of solid residues has been achieved with good results. For liquid manure, treatment with an anaerobic filter has been proven feasible, reliable and probably cheap.

3.1. The study "Biogas plants in Europe"

The study "Biogas plants in Europe"(3) has shown that the 550 biogas plants existing in 1983 in the European Community member states and in Switzerland, represent about 95 000 m^3 of methane digester working volume on farm and 174 000 m^3 of methane digester working volume in industry, all together a methane digestion capacity of 269 000 m^3.

Assuming average running conditions of the methane digestion and, hence, an average methane productivity of 1 m^3 biogas per m^3 digester working volume and per day, a daily output of 269 000 m^3 of biogas may be expected. Assuming that, on the average, 30 % of the produced biogas is used to run the biogas plant itself and that the 70 % remaining are effectively used, this represent already in 1983 a yearly output of 33 kilotonnes oil equivalent (ktoe), worth 7 million ECU.

From the evolution of the number of methane digesters as a function of time, it appears that the learning curve for biogas plants on farms started in 1979-1980, followed by the learning curve for biogas plants in industry which started in 1980-1981.

3.1.1. Potentiality of biomethanation in the European Community

The potential contribution of biomethanation to the energy requirements of the European Community has been evaluated. Biomethanation of animal wastes can provide yearly 14.1 Mtoe, crop residues 8.5 Mtoe(4), and of municipal wastes 8.9 Mtoe[5]. These 31.1 Mtoe represent only 2 % of the estimated european energy consumption in 1983[5], but a market value of 7 000 million ECU. This means an important but local potential. When it will become customary to grow energy crops as catch crops, algae or aquatic plants, these values may increase three-fold[5].

3.1.2. Economics of biogas plants in 1983

Limited information is as yet available concerning the economics of the biogas plants existing in the countries covered by the survey by the end of 1982. Not many methane digesters have already been in routine operation for more than one year and, if so, their economics has not always been properly evaluated. Nevertheless, valuable data have been obtained for 32 methane digesters on farm.

Clearly, a biogas plant can already be economic, clearly, only few biogas plants are presently economic. Why then do most methane digesters on stream in 1983, appear uneconomic ? Three major reasons have been found. The first reason concerns the poorly reliable performance of the methane digestion engineering. The second reason concerns the poor uses of the biogas produced by the methane digestion. An average daily productivity of 1 m^3 of biogas per m^3 of methane digester is the lower limit for economy and can be obtained, even for full-scale methane digesters. However, few biogas plants in the practice meet this average daily productivity of 1 m^3 per m^3 methane digester and the reasons for this are analysed. Besides, all produced biogas needs to be properly used. This is as yet not often the case for a number of reasons which are evaluated.

To be economic, the investment costs for the overal biogas plant must not exceed 300 to 400 ECU per m^3 digester working volume. These cost limits can generally be met in the states surveyed by the study.

3.1.3. Bottlenecks in the implementation of biomethanation

Bottlenecks in the implementation of biogas plants are of two orders, first the general context in which biomethanation is developing and second, the particular problems encountered in biogas plants. These bottlenecks presently deter as well farmers as industries from investing in a biogas plant.

More particular to the biomethanation technology itself, there does still not exist a reliable methane digestion system for a number of feedstocks potentially available for biomethanation. This is still especially true for bedding manure, although some of the early biogas plants were already dealing with this biomass substrate. As far as the engineering, the operation and the monitoring of biogas plants is concerned, a number of problems were highlighted by the survey. Many methane digesters still exhibit a lack of reliability especially during the first year of operation.

3.1.4. Scenario for the future

Recommendations are made to overcome the bottlenecks in biomethanation and a scenario for implementation of biogas plants is proposed, would the bottleneck be overcome.

4. RESULTS FROM THE DEMONSTRATION PROGRAMME, ALTERNATIVES ENERGY SOURCES

The demonstration programme is the logical extention of the Community "Research and Development" activities as it provides the opportunity for exploiting projects from national "R and D" programmes, private research programmes and activities of the Commission itself. Demonstration differs from "R and D" and from pilot stage due to the industrial scale of the projects. There is a requirement for prospects of economic viability because in the normal investment stage, the inherent risks are still considered by the entrepreneurs to be too high. The demonstration also open up a european scale market for the successful technology.

Demonstration projects covers biogas plant utilizing animal manures and industrial residues. Animal manure will be used in 2 collective systems for two villages (mixed manures), pig manure, for 5 digesters, including a project using solar energy to heat the digester, cattle manure, for 1 digester, poultry manure, for 1. Various design are tested such as completely mixed or plug-flow, steel or plastic material, numerous heating and mixing devices, etc. Size is from 20 up to 300 m^3 for individual digesters but 2000 m^3 for the village digesters.

The following industrial residues are also considered for treatment in demonstration projects : solid and liquid distillery waste, brewery wastewater, slaughterhouse and meat packing industry wastes. There is also a landfill operation.

Various ways of utilizing the biogas produced are also investigated such as cogeneration of heat and electricity, electricity generation only, house heating. Some new projects have been recently awarded but are not considered in this short list.

The main results of this demonstration programme to-day are an improved knowledge of the design and engineering problems related to building and operating a biogas plant and the way to solve them. It has also led the industry, the builders, to begin a reflexion based on their experience on the socio-economics aspects, the motivation of the owners and various factors related to the development perspective of biogas plants. This is a very good experience and to convey it to the biogas community will be very usefull.

REFERENCES

1. Naveau H. La recherche en "Digestion Anaérobie" dans le Programme R & D de la CEE "Recyclage des déchets urbains et industriels". In Bio + 82, lères Rencontres Internationales Bio-Industries, Nancy, 28 sept. - ler oct. 1982, Proceedings (1983). Doc. CEC XII/1193/82-EN/IFB (in english)

2. Colin F., Ferrero G.L., Gerletti M., Hobson P., L'Hermite P., Naveau H.P. and Nyns E.-J. Proposal for the definition of parameters and analytical measurements applicable to the anaerobic digestion processes. Agricultural Wastes, 7, 183-193 (1983).

3. Demuynck M., Nyns E.-J. and Palz W. Biogas plants in Europe, A practical handbook. Solar Energy R & D in the European Community, Series E Energy from Biomass, vol. 6, D. Reidel Publ. Co., Dordrecht, NL, (1984)

4. Palz W., and P. Chartier. Energy and economical analysis. In Energy from biomass, Applied Science, London. p. 215, (1980).

5. Hall D.O. Food versus fuel, a world problem. In A. Strub, P. Chartier and G. Schleser (Eds.), Energy from biomass, 2nd E.C. Conference. Applied Science, London. pp. 43-62, (1983).

ACKNOWLEDGMENTS

This work has been supported by CEC contract n° RUW-066-B (N).

ANAEROBIC FERMENTATION OF SEMI-SOLID AND SOLID SUBSTRATES

L. DE BAERE* and W. VERSTRAETE**

*ACEC-NOORD, Dok Noord 5, 9000 Gent, Belgium
** Laboratory of Microbial Ecology, State University of Gent, Coupure L 653, 9000 Gent, Belgium

Summary

This paper gives an overview of the anaerobic digestion of semi-solid and solid substrates. For the EEC, two important potentials are identified i.e. energy crops and municipal solid wastes (MSW).

At the current gas prices, energy crops do not offer economical perspectives as such. However, if the 540 ECU per hectare, which are now used to deal with dairy surpluses, would be granted as a subsidy, energy crop farming would become a viable alternative in the EEC. Subsidies could then be reduced as energy prices rise. In contrast with reforestation, a switch from conventional to energy crops would not seriously affect current social structures or employment levels, and would have a positive impact on the EEC import-export balance. Digestion of energy crops and crop residues could produce about 20 million tons of oil equivalents per year, or 1.5 % of the total EEC-energy demands estimated for 1985.

MSW can be digested in a variety of ways. Recent developments with regard to handling, fermentation and posttreatment of MSW are reviewed. Particular attention is given to dry digestion. In the process developed at Gent, loading rates of 11 to 13 kg VS/m^3 reactor.day are possible, corresponding to retention times of 2 to 3 weeks, with methane production amounting to 60 to 95 m^3 CH_4/ton of solid waste. The economics of dry fermentation are very attractive compared to aerobic stabilization of MSW. About 200 ECU of methane and compost can be produced annually per m^3 of fermentor, with an investment cost of 1000 ECU per m^3 of fermentor.
In case 20 % of the MSW in the EEC would be treated by anaerobic fermentation, an annual methane production of 1.1×10^9 m^3 would be obtained with a value of 200×10^6 ECU.

1. THE POTENTIALS

Anaerobic digestion has long been considered as an inefficient and unreliable treatment process, applicable to a limited number of substrates, such as sewage sludge and animal manures. The energy crisis, together with a growing understanding of the microbiological, biochemical and thermodynamic aspects of anaerobic digestion, has stimulated a number of important developments in the field of anaerobic digestion in the past decade. The development of the upflow anaerobic sludge blanket (UASB) reactors or other flow-through reactors has made anaerobic purification of dilute wastewaters feasible. Two-phase anaerobic treatment processes, consisting of an acidification and methanogenesis phase, allow high-rate anaerobic digestion of high-strength industrial wastewaters. Research indicates that anaerobic digestion of raw sewage at ambient temperatures (5 to 20°C) and at short hydraulic retention times (less than 6 hours) may be an economical means of pretreating municipal wastewaters before aerobic purification. Attention is now also being drawn towards the anaerobic digestion of semi-solid and solid substrates. Recent advances in this area are making methane production from semi-solid and solid substrates more and more feasible. It is the purpose of this paper to give an overview of the work done and being done in the field of anaerobic fermentation of semi-solid and solid substrates.

Semi-solid substrates can be defined as substrates with a total solids (TS) content between 12 to 20 % or substrates which have to be handled and manipulated as solids.

First of all, there are the semi-solid wastes produced by the sugar beet industry, canning factories, breweries, greenhouses etc. The purpose of anaerobic digestion of these substrates is primarily waste stabilization and volume reduction. The quantities of these wastes available for fermentation are rather limited because the wastes can often be recycled on the land or sold as a feed additive.

Energy crops are a second group of semi-solid substrates and represent a much larger potential than the semi-solid wastes. Energy crops are crops, rich in sugar or starch, such as sugar beets, potatoes, cassava, etc., which are grown specifically as substrates for energy production. The potential for growing energy crops in the EEC is increasing every year. In the past two decades, agricultural production has risen 1.5 to 2 % annually, while consumption increased only at a rate of 0.5 %. The discrepancy between production and consumption is particularly disturbing for the European dairy industry : production has risen at an annual rate of 2.5 % in the past 10 years, while consumption has been increasing at a rate of only 0.5 %. The difference is even greater in the last two years, as production has increased 3.5 to 4.0 % per year while consumption has had a tendency to stagnate. The intervention policy for the dairy industry alone is costing the member countries 4 700 million ECU per year or for the tax-payer 45 ECU for every ton of milk produced. This corresponds with about 540 ECU for every hectare of land cultivated for milk production (1).

It is clear that the problem of the EEC is overproduction. Some measures have been proposed to impose penalties on farmers producing more than 15 tons of milk per hectare cultivated. Such counterproductive measures would, however, undermine agricultural research and development on a long term basis. An alternative option is to reduce the surface area used for milk production, and to utilize the available acreage for growing energy crops. Economic analyses, based on the current state-of-the-art, indicate that methane production from energy crops cannot compete with current gas prices. However, if the subsidies used for milk-production were diverted to cultivating energy crops, at 540 ECU per hectare as mentioned above or up to 1620 ECU per hectare of overproduction capacity for Belgium, then energy crop farming is certainly an attractive possibility. Employment opportunities and social structures would not be affected by a switch to energy crops and instead of hampering agricultural research, it would stimulate both agricultural and biotechnological research. There is certainly a large genetic potential yet to be discovered in the field of selecting crops that are highly efficient in their photosynthetic process and are easily converted to methane. Hall postulates that 7 % of the total energy could be provided through biomass in the EEC by the year 2000 with 0.7 % coming from energy crops (15). Verstraete estimates that 1 % of the total energy needs in Belgium could be obtained from methane digestion of energy crops and crop residues (22). The development of highly active methane digestion processes for these mostly semi-solid substrates can play an important role in solving the agricultural dilemma of the EEC and its high dependency on foreign crude oil.

The main source of solid substrates actually being considered for methane fermentation, is municipal solid waste (MSW). MSW is produced in very large quantities : each person produces about 0.7 to 0.8 kg of MSW per day, amounting to an annual production of 70 x 10^6 tons in the EEC. MSW consists for 55 % of solids, of which 60 - 70 % are biodegradable. As will be discussed further, about 100 m³ of CH_4 can be produced per ton of MSW. In case 20 % of the MSW in the EEC would be digested and used for biogas recovery, a yearly methane production of 1.1×10^9 m³ would be obtained. This corresponds with a value of 200 million ECU.

Obviously the potentials for producing biogas in the EEC from semi-solid and solid substrates warrant careful further examination of the available technology.

2. DIGESTION TECHNOLOGIES

Mainly 3 digestion concepts have been investigated and applied for the anaerobic fermentation of semi-solid and solid substrates :

1. the conventional completely mixed digestion;
2. the two-phase digestion;
3. dry anaerobic digestion.

2.1. Conventional completely mixed digestion

The completely mixed digestion process is the type of digestion process which is normally applied to sewage sludge and animal manure. The feed is introduced into a completely mixed vessel and an equal amount of effluent is withdrawn from the reactor. The substrate remains on the average 2 to 4 weeks in the reactor under anaerobic conditions at a temperature of 35°C. It is obvious that this process can be used for semi-solid and solid wastes, if the substrates are first diluted. In the case of MSW however, this means that, for example, 10 m³ of water needs to be added per ton MSW to reduce the TS-content from 55 % to 5 %.

Some semi-technical and full-scale plants have already been constructed in order to demonstrate the feasibility of this process. A facility, capable of digesting an input stream of 100 tons of waste per day, was started up in 1978 in Pompano Beach, Florida. Mixing and scum formation problems allowed the digester to be operated at only 3.5 % TS, instead of the 6 to 8 % TS as planned, at a hydraulic retention time of 20 days. The digesters have been operated principally on recycled filtrate without any buildup of inhibitors. The filtercake obtained after dewatering contains 30 % TS and will be dewatered further in order to produce a burnable fuel. In a following phase, it was planned to operate the installation under thermophilic conditions (60°C) at a 5-day retention time and a 5 % TS concentration in the reactor (23).

An alternative process for the digestion of MSW has been developed by Ghosh and Klass (13). The MSW is carefully prepared before adding to the digester. The raw MSW is coarse shredded and air-separated, to provide a highly organic wastestream. Grits, stones and glass particles are removed and the organic fraction is finally fiberized to a product with a median size of 0.6 mm and a moisture content of only 6 %. The low moisture content allows long-term storage. The digester feed is finally prepared by blending 80 parts-by-weight of the fiberized organic fraction and 20 parts-by-weight of sewage sludge solids on a dry total solids basis. The sewage sludge had a TS-content of 2 to 3%, so that 8 to 12 m^3 of sewage sludge are required per ton of fiberized refuse, resulting in a concentration of $\pm$ 10 % of total solids in the feed. The optimal loading rate for this substrate was determined to be 2.24 kg of VS/m^3r.d at a hydraulic retention time of 12 days. Gas production of 400-l reactors reached 0.56 m^3 CH_4/m^3r.d. Plans for a 100-ton-per-day facility provide for a dewatering of the digested residue with a vacuumfilter. The water would subsequently be aerobically purified during a period of 3 to 5 days while the solids could be landfilled or used as soil conditioner. Intensive refuse separation causes the fraction of MSW available to anaerobic digestion to be rather limited. Diaz and Trezek reported that they obtained only 10 % of the biodegradable fraction of MSW after shredding, air classification and sieving (9). This organic substrate was 0.25 to 1.25 cm long and was digested in a completely mixed reactor (1.6 m^3). Gas production rates of 0.44 m^3 CH_4/m^3r.d were obtained at a hydraulic retention time of 30 days and a loading rate of 1.1 kg VS/m^3r.day.

The carbon content of the organic fraction of MSW is relatively high. An Italian researcher (11) found an average C:N:P ratio of 100:1.3:0.2 while a ratio of 100:5:1 is considered to be optimal. He investigated various solid and semi-solid substrates, which can be added to the digesters in order to satisfy the N and P requirements. Sewage sludge and liquid swine manure were found to be the most likely wastes that can be utilized for correcting the nutritional imbalance. Laboratory and pilot scale experiments were carried out with the organic fraction in combination with sewage sludge in a ratio of 1:6.3 of MSW to sewage sludge (w/w). This mixture gives a feed with a TS-content of 10.7 %, corresponding to a concentration of approximately 7 to 8% TS in the completely mixed reactor, and a C:N:P ratio of 100:3.3:0.8. This substrate was digested at a loading rate of 5 kg VS/m^3r.d and a hydraulic retention time of 10 days (35°C). The biogas production was reported to be 0.43 to 0.53 m^3/kg of VS or 1.3 to 1.6 m^3 of CH_4/m^3r.d. A demonstration plant with a digester of 2 000 m³ has been constructed at Broni with the support of the European Community. This is the largest anaerobic plant handeling solid urban refuse in Europe (11).

2.2. Two-phase digestion

Methanogenesis of complex organic matter, as found in semi-solid and solid substrates, is a process which consists out of three steps : the complex insoluble polymers must first be degraded to soluble polymers (liquefaction), which in turn are decomposed to form mainly short-chain volatile fatty acids (acidification). These components then form the substrate for the methanogenic associations (methanogenesis). In a two-phase process, the liquefaction and acidification of the substrate is accomplished or stimulated in a first reactor, sometimes called the liquefaction-acidogenesis reactor (LA-reactor), while methanogenesis takes place in the second reactor, usually a flow-through reactor.

The two-phase digestion process has first been promoted by Ghosh et al. (12). They suggest a 2-phase process with 2 completely mixed reactors in series for the combined digestion of sewage sludge and fiberized MSW, as described above. The conversion of refuse to volatile acids and volatile acids to methane can best be obtained by operating a 1st reactor at a detention time of 3 days and the 2nd methanogenic reactor at a detention time of 5 days. The total digestion time is then 50 % less than the 12 days deemed optimal for conventional completely mixed digestion (13).

Ishida et al. added a thermochemical pretreatment of the mixture of sewage sludge and MSW prior to a 2-phase digestion. A pretreatment of the feed slurry (13 to 15 % solids) at a pH of 9.8 and 60°C during 3 hours markedly increased volatile fatty acids production. After 24 hours, the pretreated substrate had produced 3 times as much volatile fatty acids per kg VS as the untreated. After pretreatment, a two-phase digestion, consisting of a liquefaction phase at 60°C and at a pH of about 6.0 during 36 to 48 hours, and a gasification phase at 60°C and a pH of 7.5 to 8.2 during a time of 108 to 144 hours. The treatment time for the total process amounts to 6 to 8 days. The results of a pilot plant with a 1 m³ pretreatment reactor, a 0.5 m³ liquefaction fermentor and a 1.5 m³ gasification reactor were very satisfactory. The feed contained a VS-content of 8 to 16 %. Gas production rates of 3.2 m³ of CH_4/m³r.d at a loading rate of 9.4 kg VS/m³r.d were obtained for the methane reactor (16).

Augenstein proposed a packed-bed reactor for the 1st phase and a non-biological activated carbon system as the 2nd phase. Water was pumped continuously through the solids bed and a sieve separated the solids from the liquid. The polluted wastewater was pumped to the activated carbon, where the toxic substances were removed, and recycled to the 1st phase. Shredded MSW was digested at a retention time of 20 days. The total solids content in the reactor reached 12 to 15 % and a methane production rate of 0.5 m^3/m^3r.d was observed (2).

With the development of the UASB, the activated carbon system was soon replaced by an upflow reactor (20). This process concept has been tested on a pilot-scale, with 5 first-phase batch reactors (60 m^3 each) and a UASB reactor (10 m^3). Tomato plants, sugar beet wastes and MSW have been treated. The retention times for the various wastes in the first phase reactors were in the range of 2 to 4 weeks. When sugar beet waste was digested, acidification was so rapid that the UASB-reactor had to be loaded at 40 kg of COD/m^3r.d. Because of gas leakage in the 1st phase, it is up until now not possible to specify actual methane production (14).

The liquefaction-acidification reactor can also be a partially or completely mixed reactor, which is then followed by an upflow reactor. Cohen proposes a partially mixed LA-reactor with a solids-liquids separator on the outside of the reactor. Pieces of leaves, sugar beets and roots have been used as a substrate. An LA-reactor (800 l) and a UASB reactor (250 l) were fed continuously and loading rates of 13.8 kg TS/m^3r.d were obtained. Mean solubilization at a load of 3.6 kg TS/m^3r.d was 86 % of the solids added, while volume reduction was 93 %. The recycled effluent from the UASB-reactor had a strong buffering effect on the LA-reactor, while short hydraulic retention times of the recycled water increased the rate of the liquefaction-acidogenesis stage. The experiments are now being carried out on a 30 m^3 LA-reactor and a 30 m^3 UASB-reactor. The TS-content of the feed (sugar beet waste) is 12 % resulting in an 8 % TS concentration in the LA-reactor. A full-scale reactor (200 m^3) treating sugar beet wastes is planned for 1984 (5).

A two-stage methanisation process for semi-solid vegetable canning wastes has been investigated by Roy (21). The process includes a completely mixed thermophilic LA-reactor in connection with a mesophilic anaerobic filter. A methane productivity of 2.4 m^3 of CH_4/m^3r.d was reached using this system. Simultaneous experiments with completely mixed mesophilic and thermophilic digesters for comparative purposes, yielded methane production rates of 1.4 and 2.3 m^3 of CH_4/m^3r.d respectively.

2.3. Dry fermentation

Dry fermentation can be defined as the digestion which takes place at a TS-concentration of 20 % or higher. An uncontrolled spontaneous dry fermentation occurs in landfills at solids contents of 40 to 70 %. The biogas produced in these landfills can be recovered via a gas collection system in the landfill. The first recovery plant was on line at the landfill Palos Verdes, California, in 1974, where purified natural gas is supplied to customers at a daily rate of 35,000 m^3 of CH_4. Many installations have been constructed in Europe since then, especially in Switzerland and Germany. The quantities of gas that can be recovered vary greatly because the efficiency of the recovery depends on many factors, such as the site itself, the age and composition of the landfilled material, etc. Gas yield per kg of waste can be estimated at a maximum of about 200 l, while about 20 to 25 % of the gas is actually recoverable. The rate of gas production is estimated at 6 to 40 l per kg per year. Although decomposition in a landfill site may continue for more than 50 years, the period of economical gas recovery is currently thought to be 10 to 20 years after the landfill has been closed (24).

Legislature in Germany has caused the planning of only large landfill sites with a holding capacity of 1 to 10 million m^3 of MSW. These large sites make gas recovery very attractive. For example, the landfill site "Am Lemberg" has a surface area of 16 hectares and will have a final depth of 40 m. The site with a total capacity for 3.4 million m^3 of MSW, has been filled for 2/3 and the biogas is collected through a number of vertical pipes in the fill. Gas recovery amounts to 1200 m^3 of biogas per hour ($\pm$ 50 % CH_4). The gas can be either utilized to produce electricity in 2 generators of 144 kW each, or can be used to heat greenhouses during the winter that are in the proximity of the site (18). The rate of recovery from the site can be calculated to be in the range of 12 l biogas/m^3r.d.

Biogas has been recovered at the Pforzheim landfill since 1981. The landfill site has a surface area of 10 ha and a filling capacity of 3 million m^3 of municipal refuse. The gas recovered has been of a remarkably constant and high quality, containing 60 % ($\pm$ 2 %) of methane. This is thought to be due to a close control of gas pressure in the landfill. As soon as the pressure reaches 1 mbar, the gas compressor withdraws the gas out of the landfill until the gas pressure has dropped to -0.5 mbar. Methane, CO_2 and oxygen are constantly monitored, with the latter parameter being the most sensitive and useful for control purposes. The biogas is converted to electricity in two motors of 130 kW, while two more will be installed shortly. The price paid for the electricity has been high because the landfill management has been able to assure a constant production level to the electrical company (10).

Anaerobic digestion in landfills is slow but can be enhanced considerably. Buivid et al. investigated various parameters that could increase the rate of gas production in a landfill. They were able to produce 142 l CH_4/dry kg MSW within a period of 3 months under simulated landfill conditions. Digested sewage effluent, recycle of leachate and high compaction of the MSW were all shown to stimulate gas production (3).

The dry fermentation process can not only be carried out in landfills, but if the process is fast enough, it can be carried out in controlled anaerobic fermentors. Some of the advantages of dry anaerobic fermentation of solid wastes are the following :

- The addition of water is limited, decreasing posttreatment costs and reactor volume.
- Heating requirements during the process are minimized. The exothermic energy released during anaerobic degradation of 1 mole of glucose (401 kJ) amounts to only 15 % of the energy released through aerobic decomposition. However, under high substrate concentrations and sufficient degradation rates, a well-insulated dry fermentor may require little or no additional heat input to keep the fermentation process at optimum temperature conditions.
- Chemical additions are less than the amount required for liquid fermentation.

Dry anaerobic fermentation, however, necessitates the development of new reactor designs, capable of handling solid substrates under anaerobic condition, at a price that remains economically justifiable. Ducellier and Isman used silos filled with manure and bedding as methane digesters (18). Jewell proposed low-cost batch fermentors to treat all organic wastes and crop residues on the farm by means of a dry anaerobic digestion process. His research indicates that significant methane production rates can be obtained at TS-contents of up to 40 %. The rate of digestion is influenced by the TS-content at solids concentrations of 32.5 % and higher (17).

Currently, some new semi-continuous and continuous dry fermentation processes are being developed. A semi-continuous process for the dry fermentation of a precompost (shredded MSW after a short aerobic decomposition) in under investigation in France. The solid substrate is diluted to a TS-content of 30 % before feeding and fermented under mesophilic conditions in a continuous dry fermentor. Gas production rates amount to 3 m^3 biogas/m^3r.d. The digested residue is dewatered and processed further and a commercial high quality compost is obtained. An advanced dry fermentor was designed in order to receive an incoming waste of 30 % solids. After the reactor has been fed, the reactor is completely closed except the outlet for the digested residue. The gas production causes the pressure to rise and the digested residue is pushed out of the reactor. When the pressure reaches 500 mbar, the gas is suddenly released, the reactor contents rise rapidly and this has a mixing effect. Per ton of MSW entering the plant, about 90 m^3 of biogas and 390 kg of compost (75 % TS) is produced. A plant, handling 42 tons of MSW per day, would require 2 reactors of 500 m^3 each (4).

An alternative dry fermentation process (Figure 1) has also been investigated in Gent, Belgium (7,8). The anaerobic process treats solid wastes coming in at a TS-content of 30 to 40 % and is applicable to precompost as well as shredded MSW (< 5 cm). The process operates at 35°C and a gas production of 60 m^3 CH_4/ton of MSW and 90 m^3 CH_4/ton of precompost was produced at retention times of 2 and 3 weeks respectively. Gas production rates for laboratory scale reactors (40 l) vary from 4 to 5 m^3 biogas/m^3r.d. The methane concentration in the biogas is 50 to 55 %, resulting in methane production rates of 2.1 - 2.8 m^3/m^3.reactor.day. Exothermic energy produced at these elevated rates, becomes an important factor in the energy requirements for the dry fermentation process. Calculations point out that this energy can meet the heating requirements of a well-insulated digestor, not taking into account the energy needed to raise the temperature of the substrate. For MSW, a short aerobic treatment of the waste could raise the temperature of the substrate to 60°C, so that a maximum amount of methane produced is available for other purposes.

The economic feasibility of the anaerobic dry fermentation process appears to be quite attractive. One m^3 of reactor produces netto about 620 m^3 of CH_4 and 8 tons of dry commercial compost per year with a total value of 200 ECU (see flow-sheet). Investment costs for the dry digestion process are of the order of 1 000 ECU, not taking into consideration the investment for final gas utilization. This compares very favorably with aerobic composting processes, where the production cost of 1 ton of compost is on the average 15-20 ECU, while there is a small net benefit possible (5 to 10 ECU/ton of compost) through dry anaerobic fermentation. The eco-

Table I. Process parameters for the digestion of semi-solid and solid substrates

	Substrate	Loading rate	Retention time	T	Total Solids Feed	Gas productivity
		(kg $VS/m^3r.d$)	(days)	(°C)	(%)	(m^3 $CH_4/m^3r.d$)
COMPLETELY MIXED						
Walter (100 tpd)	MSW	1.9	20	35	3.5*	0.6 - 0.7
Ghosh & Klass (400 l)	fiberized MSW + sewage sludge	2.24	12	35	10	0.56
Diaz & Trezek (1.6 m^3)	MSW	1.1	30	35		0.44
Gerletti	MSW + sewage sludge	5	10	35	10.7	1.3 - 1.6
TWO-PHASE						
Ghosh & Klass	fiberized MSW + sewage sludge	3.36	8	35		0.8 (2°)
Ishida (0.5 + 1.5 m^3)	pretreated MSW	9.4	8	60	8 - 16	3.2 (2°)
Augenstein (208 l)	MSW			35	12 - 15*	0.5 (1°+2°)
Cohen (250 + 800 l)	sugar beet wastes	3.6(TS)		35		1.1 (1°+2°)
Roy	vegetable canning wastes			60 - 35		2.4 (1°+2°)
DRY FERMENTATION						
Rettenberger (landfill)	MSW		∞	Ambient		0.012
Buivid (210 l)	MSW		90	35		0.3
Carantino	precompost MSW			35	30	1.8
De Baere & Verstraete	precompost MSW	11	21	35	32 - 36	2.8
(40 l)	MSW	12.9	14	35	34 - 40	2.1

* Total solids concentration in the reactor

FIGURE 1

FLOW SHEET FOR HIGH-RATE ANAEROBIC COMPOSTING (GENT-PROCESS)

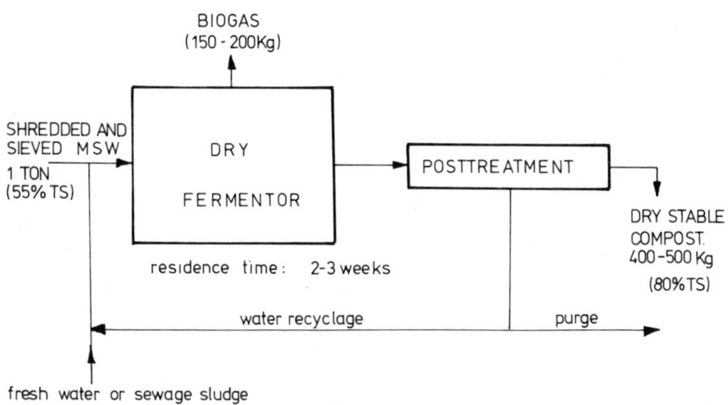

nomics of the process can be made even more attractive by co-digesting other substrates, such as sewage sludge or seasonal semi-solid wastes, which will increase methane production and/or the quality of the compost. The process has been investigated on a scale of 7 m³ under batch-conditions. Two continuous dry fermentors will be installed in 1984, with a volume of 30 and 100 m³ and a handling capacity of 1.5 and 5 tons of substrate respectively.

3. DISCUSSION

The anaerobic fermentation of semi-solid and solid substrates is rapidly gaining momentum due to new advances especially in the area of 2-phase digestion and dry anaerobic fermentation. Two-phase digestion seems primarily applicable to semi-solid substrates, such as sugar beet and canning wastes. The primary purpose for treating these wastes is stabilization and volume reduction, while methane production is only an additional benefit. The two-phase process appears also applicable to the fermentation of energy crops, which might become economically attractive in the near future. The primary goal for the digestion of energy crops is to produce the methane gas, with the extra benefit that the stabilized substrate can be recycled to the land as an organic fertilizer.

The use of completely mixed reactors for the digestion of semi-solid and solid substrates appears to be only warranted if the substrate can be combined with sewage sludge or animal manure. The semi-solid or solid substrate functions then more as an auxiliary substrate in the digestion process, increasing the methane productivity of the anaerobic reactor. Co-digestion is especially attractive for wastes that are deficient in certain nutrients, such as some petrochemical and vegetable wastes (6, 11).

Gas recovery is practiced more and more on large landfills. Certain measures could be taken to increase productivity in the landfill and to assure a constant biogas quality.

Continuous dry fermentation of MSW or a precompost from MSW offers good perspectives for the near future. Dry fermentation of solid waste minimizes heating requirements and post-treatment costs, while allowing high gas production rates at retention times similar to alternative anaerobic processes (Table I). As has been the case in the treatment of liquids, anaerobic fermentation of semi-solid and solid substrates is becoming an attractive alternative as compared to the aerobic treatment methods. Anaerobic fermentation of solids produces large quantities of gas and could make a small but non-negligible contribution to the natural gas consumption in the EEC. It remains to be tried out in practice to prove that the biomethanation of MSW is technically and economically possible.

LITERATURE

1. ANONYMOUS. (1983). Adapting EC-agricultural policies (in Dutch). Bulletin van de Europese Gemeenschappen. Supplement 4/83.
2. AUGENSTEIN, D.C.; WISE, D.L.; COONEY, C.L. (1977). Packed bed digestion of solid wastes. Resources Recovery and Conservation. Elsevier Scientific Publishing Company, A'dam, 2, 257-262.
3. BUIVID, M.G.; WISE, D.L. (1981). Fuel gas enhancement by controlled landfilling of municipal solid waste. Resources and Conservation, 6, 3-20.
4. CARANTINO, S. (1983). Instantanés. Une nouvelle source de méthane : les ordures ménagères. Biofutur, Juin, 54-56.
5. COHEN, A.; KOEVOETS, W.A.A.; ZOETEMEYER, R.J. (1983). Fast anaerobic digestion of solid vegetable wastes on semi-technical scale. Proc. European Symposium Anaerobic Waste Water Treatment, Noordwijkerhout, Netherlands, p. 171.
6. DE BAERE, L.; VAES, H.; VERSTRAETE, W. (1982). Proc. Int. Symp. on Advances in Anaerobic Digestion 25-27 Oct., Mexico City, Ed. G. Viniegra, UAM, Mexico City, Mexico.
7. DE BAERE L. (1983). Biogas from MSW through dry digestion (in Dutch). Flanders Technology. Water, May, 115-116.
8. DE BAERE L.; VERSTRAETE, W. (1984). High-rate anaerobic composting with recovery of biogas. Biocycle. In press.
9. DIAZ, L.F.; TREZEK, G.J. (1977). Biogasification of a selected fraction of municipal solid wastes. Compost Science, 18(2), 8-12.
10. DUPONT, R. (1982). Kriterien bei der Verstromung und Netzeinspeisung, Abrechnungsfragen. VDI Berichte Nr 459, 37-42.
11. GERLETTI, M. (1981). Anaerobic digestion of MSW improved by other organic wastes in large pilot plants. E.E.C. Research contract nr. RUW-015-I. Progress Report.
12. GHOSH, S.; CONRAD, J.R.; KLASS, D.L. (1975). Anaerobic acidogenesis of wastewater sludge. Jour. Water Poll. Control Fed., 47, 30-45.
13. GHOSH, S.; KLASS, D. (1979). The production of substitute natural gas and recyclables from municipal solid waste. Recycling Berlin '79. E. Freitag-Verlag für Umwelttechnik, Berlin, K.J.Thomé-Kozmiensky, 789-796.

14. HOFENK, G. (1983). Anaerobic digestion of solid organic wastes (process selection and pilot plant research). Proc. European Symp. Anaerobic Waste Water Treatment, Noordwijkerhout, Netherlands, 475-491.

15. HALL, D.O. (1983). Biomass for energy - Fuels now and in the future. In "Biogass Utilization", ed. W.A. Cote, NATO Advance Study Institute, Plenum Press, N.Y.

16. ISHIDA, M.; ODAWARA, Y.; GEYO, T.; OKUMURA, H. (1979). Biogasification of municipal waste. Recycling Berlin '79. E. Freitag-Verlag für Umwelttechnik, Berlin, K.J. Thomé-Kozmiensky, 797-802.

17. JEWELL, W.J. (1979). Future trends in digester design. The First International Symposium on Anaerobic Digestion. Cardiff, Sept. 17-21.

18. LESAGE, E.; ABERT, P. (1952). Gas de fumier : La diffusion nouvelle du livre. (Biogas : Latest techniques of production and utilization). In French. Published by Marcel Armand, 27 Luneville Street, Saint Quentin France. April. 132 pp.

19. RETTENBERGER, G. (1981). Recent developments in recovery and utilization of landfill gas. Universität Stuttgart, Bandtäle 1, 7000 Stuttgart 80.

20. RIJKENS, B.A.; VOETBERG, J.W. (1981). Two-step process for anaerobic digestion of solid wastes. The second International Symposium on Anaerobic Digestion. Travemunde, Sept. 6-11.

21. ROY, F.; VERRIER, D.; FLORENTZ, M. (1983). Importance of the liquefaction step in a two-stage methanisation process of solid cannery vegetable wastes. Proc. European Symp. Anaerobic Waste Water Treatment, Noordwijkerhout, Netherlands, 175.

22. VERSTRAETE, W. (1981). Biogas (in Dutch). Leefmilieu dossier nr. 6 Stichting Leefmilieu, v.z.w. Keizerstraat 8, 2000 Antwerpen.

23. WALTER, D.K. (1982). Anaerobic digestion of municipal solid waste to produce methane. Recycling International : Recovery of energy and material from residues and wastes. Ed. K.J. Thomé-Kozmiensky, Berlin, 206-212.

24. WILKEY, M.; ZIMMERMAN, E. (1982). Landfill gas recovery in the U.S.A. Recycling International : Recovery of energy and material from residue and wastes. Ed. K.J. Thomé-Kozmiensky, Berlin, 213-219.

PRODUCTION OF METHANE BY ANAEROBIC DIGESTION OF DOMESTIC REFUSE

A. PAUSS, E.-J. NYNS and H. NAVEAU

Unit of Bioengineering, Catholic University of Louvain,

B-1348 Louvain-la-Neuve, Belgium

Summary

The non-reliable character of the biomethanation of domestic refuse in completely mixed process is essentially due to the alkaline content and to the cellulose plus hemicellulose content of the refuse. The best results are obtained with a high alkaline content in the influent (coming either from the refuse or from external addition) and with a proportion of cellulose plus hemicellulose turning around 40 per cent of COD introduced. So as to resolve the problem of unreliability, we have tried two types of two-steps processes in which the refuse is little diluted; higher results are obtained compared to the CSTR.
Finally, the biomethanation of domestic refuse combined with septic tank residues has been shown feasible, more reliable and more productive than biomethanation of refuse alone.

1. INTRODUCTION

Biomethanation of domestic refuse is a well known process; some industrial installations are working at this time, using in most cases classic completely stirred tank reactors (CSTR). Almost all installations produce about 100 Nm^3 of CH_4 per ton of refuse introduced and degrade about 70 per cent of the organic matter introduced. However, almost all installations observe great variations in the production and a low reliability at long time (a).

Therefore it seems important to identify the factors responsible for the unreliability of the process, to find solutions and, if necessary, to develop a new process more adapted to biomethanize domestic refuse.

For this reasons, the research has been focused on four topics : (1) to examine the reliability of biomethanation of domestic refuse in CSTR and the factors that cause it, (2) to correct the unreliability by use of physico-chemicals additives, (3) to develop new process more efficient and (4) to study the feasability of the biomethanation of domestic refuse combined with septic tank residues.

2. SUBSTRATE SPECIFICATIONS

The organic fraction from domestic refuse is separated manually and the papers discarded. A typical analysis of the refuse is given in Table 1; it can be observed that the organic matter content (80 to 90 % of total solids) is the only constant in the refuse.

Table 1. Typical analysis of the organic fraction of domestic refuse paper discorded

TS	($g\ kg^{-1}$)	230-300	Lignin	($g\ kg^{-1}$)	5-19
VS	(% TS)	82-87	COD	"	300-400
Ashes	"	10-14	$COD_{soluble}$	"	70-100
Alkalinity	"	6-16	N_{TK}	"	5-10
Cellulose	"	5-28	N_{NH4^+}	"	0.3-1
Hemicellulose	"	6-22	$PO_4^=$	"	2-3

Table 2 gives the alkalinity, cellulose and hemicellulose content of the various samples used in this study. In spite of the setting-up of a logical analytical procedure to prepare samples through grinding and homogenizing, it appears that, at laboratory scale, domestic refuse of reproductible composition is very difficult to obtain.

Table 2. Alkalinity, cellulose and hemicellulose contents of samples of refuse used.

Samples of domestic refuse	4	5	6	7	8	9	10	11	12
Alkalinity (% TS)	6.1	6	16.4	10	9	7	8.4	9.4	9.7
Cellulose "	22.5	25.6	25	27	25		28		5.4
Hemicellulose "	24.1	22	11	10	6		8.5		10.3

At last, to improve the quantification and the parametrization of the process, the F_{420} measuring method has been set up for the mixed liquor of the digester of domestic refuse. As it can be seen in Figure 1, the F_{420} content is correlated with the state of digestion.

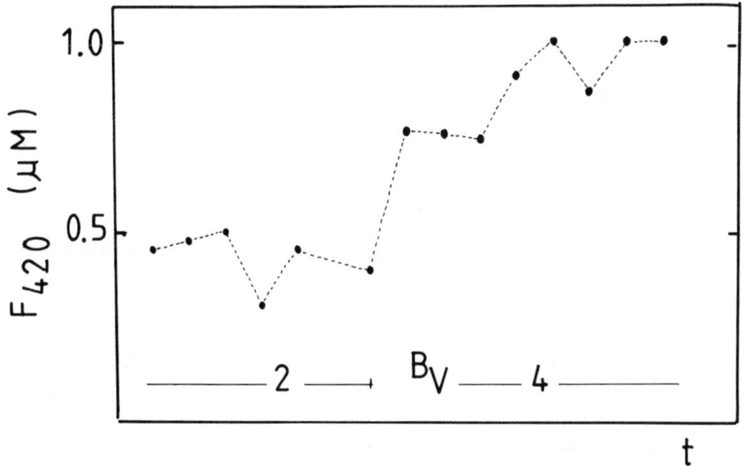

Figure 1. Evolution of the F_{420} content of a methanogenic digester of domestic refuse (35 °C)

3. THE UNRELIABILITY OF BIOMETHANATION OF DOMESTIC REFUSE

With the substrate described, the feasibility and the reliability of biomethanation were studied in CSTR maintained at 35 °C during 220 and 480 days. Figures 2 and 3 show the evolution of the methane production rate and of the total volatil acids concentration during the experiments; the running conditions are indicated under the figures. Tables 3 and 4 show the main results obtained.

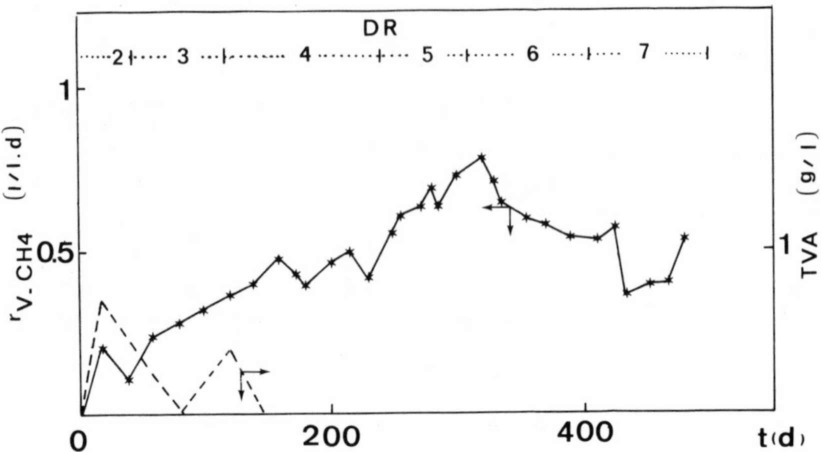

Figure 2. Evolutions of the methane production rate (✶——✶) and of the total volatil acids concentration (--------) during biomethanation of domestic refuse (DR) at 35 °C. B_V = 1 (d1 to d100) and = 1.5 (d140 to 482) g VS_o l^{-1} d^{-1}, θ = 20 d.

Table 3. Main results and yields obtained during biomethanation of different samples of domestic refuse (DR) at 35 °C in CSTR

	DR	B_V $(\frac{g\ VS_o}{l.d})$	θ (d)	TVA $(\frac{g}{l})$	$r_{V.CH4}$ $(\frac{lCH_4}{l.d})$	$Y_{VSr/VSo}$ $(\frac{g\ VS_r}{g\ VS_o})$	Y_{CH4/VS_o} $(\frac{lCH_4}{g\ VS_o})$
d160-d180	4	1.5	20	0	0.35	0.64	0.23
d300-d320	5	1.5	20	0.02	0.65	0.65	0.43
d340-d360	6	1.5	20	0.03	0.66	0.65	0.43
d430-d450	7	1.5	20	0	0.52	0.68	0.35

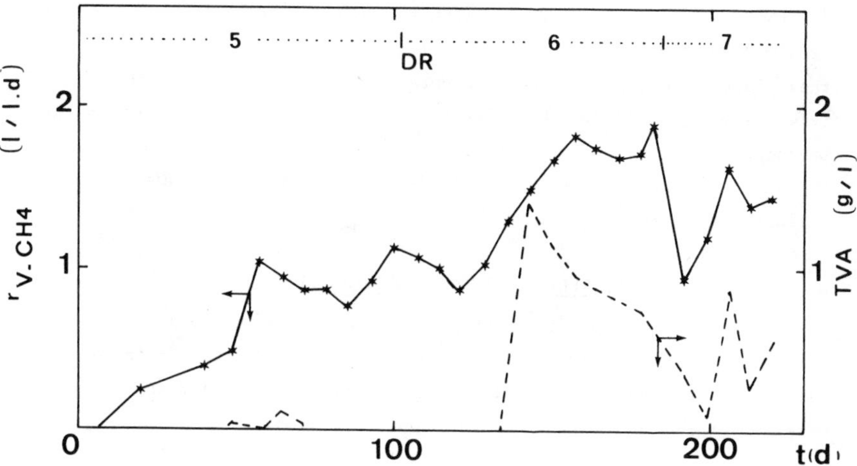

Figure 3. Evolutions of the methane production rate (✱——✱) and of the total volatil fatty acids concentration (--------) during biomethanation of different samples of domestic refuse (DR) at 35 °C. B_V = 1 (d1 to d40), = 2 (d50 to d122) and 4 (d137 to d220) g VS_o l^{-1} d^{-1}, θ = 20 (d1 to d58) and 14 (d65 to d220) d.

Table 4. Main results and yields obtained during biomethanation of different samples of domestic refuse (DR) at 35 °C.

	DR	B_V $(\frac{g\ VS_o}{l.d})$	θ (d)	TVA $(\frac{g}{l})$	$r_{V.CH4}$ $(\frac{lCH_4}{l.d})$	$Y_{VSr/VSo}$ $(\frac{g\ VS_r}{g\ VS_o})$	$Y_{CH4/VSo}$ $(\frac{l\ CH_4}{g\ VS_o})$
d20-d40	5	1	20	0	0.39	0.66	0.39
d66-d93	5	2	14	0.02	0.87	0.66	0.43
d165-d182	6	4	14	0.8	1.70	0.69	0.43
d255-d270	7	4	14	2.72	1.37	0.51	0.35

In conclusion one has confirmed in CSTR the feasibility and the yields as obtained during the biomethanation of non-defined refuse, that is a degradation of 65 % of the organic matter introduced, a methane production rate linearly correlated with the organic loading rate and a yield of about 90 Nm^3 of CH_4 per ton of refuse introduced. The comparison of these results with the analytical data on the various samples of domestic refuse used (see table 2), shows that the great variability of the gas production rate and of the yields is correlated with the composition of the refuse samples, especially the alkaline content (alkalinity) and the cellulose and hemicellulose content.

4. IMPROVEMENT OF THE RELIABILITY OF BIOMETHANATION

Since the reliability of the biomethanation seemed to be linked with its composition, two ways were used to study this hypothesis in more details : (1) influence of the alkalinity, eventually modified and (2) modification of the composition of the organic fraction.

4.1. Alkalinity content of the refuse

Figure 4 shows the methane volumetric production rate and the total volatile fatty acids concentration for a run with the samples DR 7 to DR 11 of different alkalinity. Table 5 shows the results obtained after 150 days of addition of Na_2CO_3 in comparison with the best results obtained with the sample DR 6.

Table 5. Results and yields obtained during biomethanation of domestic refuse with addition of Na_2CO_3 at 35 °C.

	B_V $(\frac{g\ VS_o}{l.d})$	θ (d)	TVA $(\frac{g}{l})$	$r_{V.CH4}$ $(\frac{l CH_4}{l.d})$	Y_{CH4/VS_o} $(\frac{l CH_4}{g\ VS_o})$
DR 6	4	14	0.82	1.70	0.43
DR 11	4	14	4.4	1.03	0.26

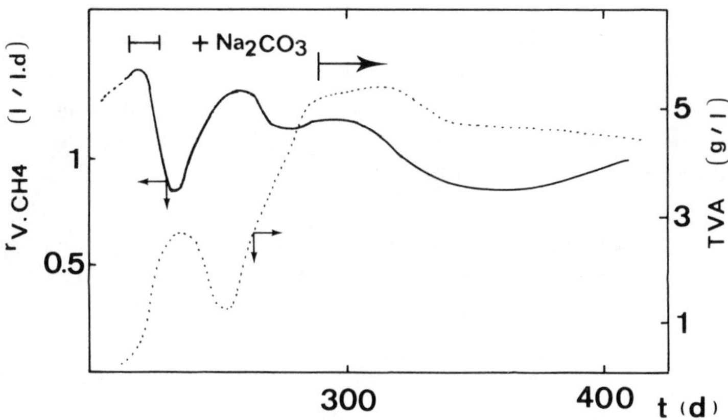

Figure 4. Evolutions of the methane production rate (———) and of the total fatty acids concentration (------) during biomethanation of different samples of domestic refuse with addition of Na_2CO_3 at 35 °C. θ = 14 d, B_V = 4 g VS_o l^{-1} d^{-1}.

With quite different alkaline content of the refuse and for the same running conditions, the best results are obtained with the higher alkalinity (16 g $CaCO_3$ per kg TS). Furthermore, feeding a digester with a sample of refuse with low alkalinity leads to the apparition of volatile fatty acids and to the fall of the methane yield and conversion yield. A punctual addition of Na_2CO_3 to this feed of refuse of low alkalinity leads to the stabilization of the pH and methane yield; however the situation then is not optimal and indeed, 4.4 g l^{-1} of volatile fatty acids (mainly propionic acid) still remain present. In conclusions, the alkalinity appears to be an important factor but not the only acting one.

4.2. Cellulose and hemicellulose content of the refuse

Influence of the cellulose and hemicellulose content of the refuse on biomethanation was studied in batch digesters during 100 days at 35 °C. Starting concentration in organic matter was 30 g COD_o l^{-1}, with various proportions of newspaper added to the refuse and hence various concentrations in cellulose and hemicellulose as described in Table 6. Ammonium carbonate was added to maintain a C/N ratio equal to 10 when the % newspapers was 50 % or more. All experiments were inoculated with a mixing of effluents of others digesters (3 per cent of COD_o).

Table 6. Experimental conditions of batch fermentors degrading domestic refuse combined with newspaper (35 °C, 30 g COD_o/l, eventually addition of $(NH_4)_2CO_3$).

Domestic Refuse (% COD_o)	Newspaper (% COD_o)	$(NH_4)_2CO_3$ (addition)	Cellulose + Hemicellulose (g/l)	(% COD_o)
100	0		3.9	15.5
80	20		6.1	23.4
50	50	+	9.5	37
20	80	+	13	50.5
0	100	+	15.2	60

Figure 5 shows the cumulative volumetric methane production after 10 days of experiment (in mmoles CH_4/l), as a function of the concentration in cellulose plus hemicellulose and figure 6 shows the volumetric methane production at the end of the experiments. A content of cellulose plus hemicellulose turning around 40 % of the COD introduced is shown to give the faster and the higher production of methane; furthermore a minimum cellulose plus hemicellulose content is necessary for a methane production.

Although some hypothesis may be made as the importance of this minimal content in cellulose plus hemicellulose in continuous processes, the respective and cumulative influences of this one and of the alkaline content of the refuse still need to be understood.

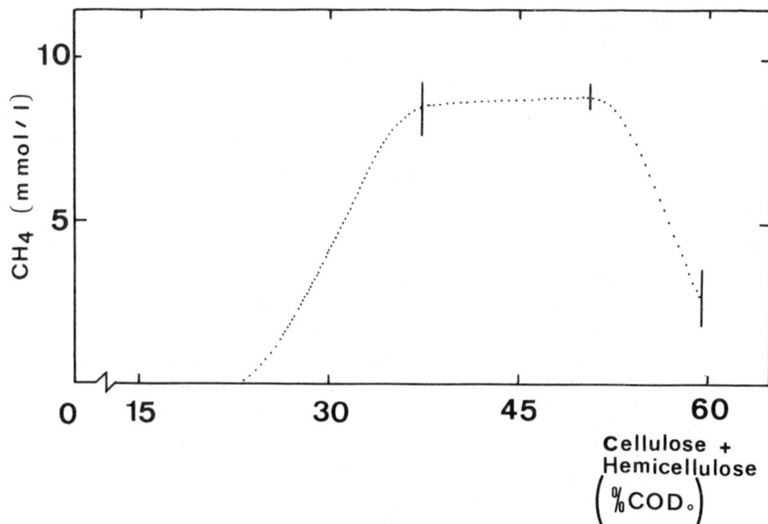

Figure 5. Cumulative volumetric methane production as function of the concentration of cellulose plus hemicellulose in batch digesters at 35 °C after 10 days of experiment.

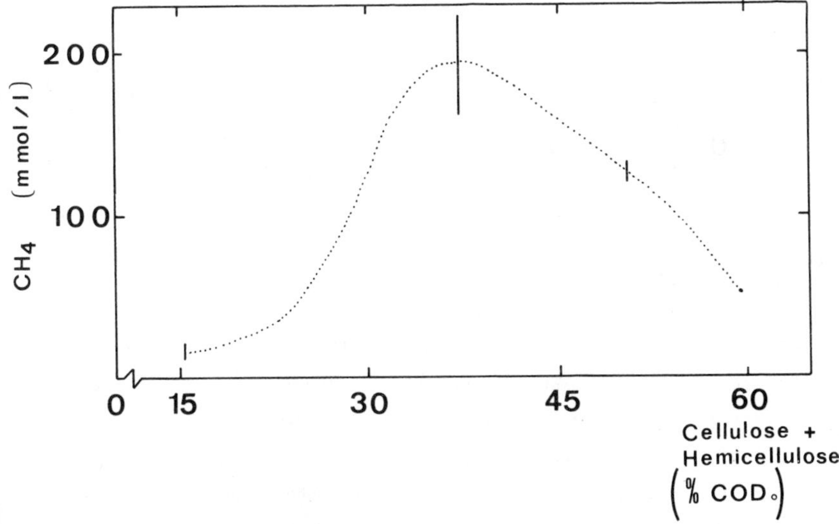

Figure 6. Cumulative volumetric methane production as function of the concentration of cellulose plus hemicellulose in batch digesters at 35 °C after 100 days of experiment.

5. DEVELOPMENT OF OTHER PROCESS

To biomethanize domestic refuse more efficiently, new processes are developed, based on the spontaneous fermentation pattern (b and c).

The fermentation pattern of the domestic refuse appears to be heavily acidic : 14 g l^{-1} of total volatil fatty acids (of which 50 % of butyric acid) are obtained in 20 days after introduction of 50 grams per liter of organic matter in batch fermentaters at 35 °C; all experiments transitorily present an important lactic acid phase (see Figure 7).

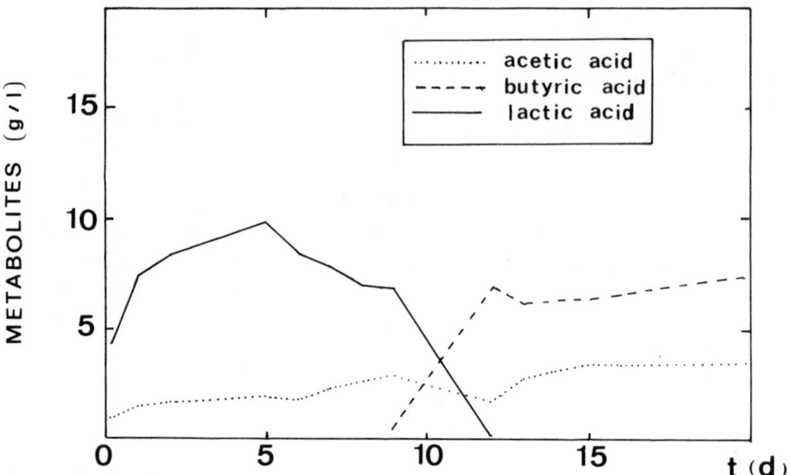

Figure 7. Evolution of metabolites concentration during fermentation of inoculated domestic refuse at 35 °C.

The presence of an inoculum does not influence the rate or metabolites production but well the final volatil fatty acids repartition.

Such an acid fermentation pattern has led to imagine two two-phase processes for the biomethanation of domestic refuse.

The first consists of a percolator reactor, in which the refuse is diluted and an up-flow digester. This process, developed in collaboration with an other CEC grant (n° ESE-R-025-B) has been optimized and patented. A methane production rate 5 times higher for a volumetric loading rate 7 times higher than in CSTR may be obtained as seen in Table 7 in comparison with results from a CSTR.

Table 7. Comparison between biomethanation of domestic refuse using two type of process (two-phases and CSTR)

	two phase	CSTR
B_V (g VS_o/l.d)	10.6	1.5
θ (d)	7	14
$r_{V.CH4}$ (lCH_4/l.d)	3.37	0.6
$Y_{CH4/VSo}$ (lCH_4/g VS_o)	0.32	0.24

The second process, not yet optimized, is composed of an other type of percolation reactor, in which the refuse is not much diluted, and a fluidized bed digester. Two type of percolator are tried; in the first, the refuse is sprayed with water and in the second, the water comes from the bottom of the reactor and completely submerges the refuse. The results for the first phase (percolation) are given in Figure 8 which represents the evolution of the cumulative COD extracted for the two reactors and in Table 8 which gives the results, in % of COD and of Kjeldahl nitrogen extracted of the refuse, after 12 and 26 days of experiment.

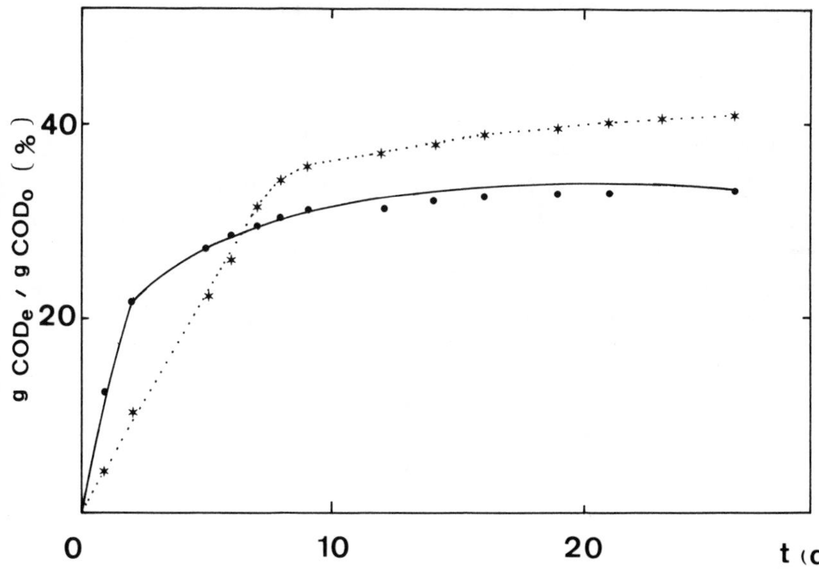

Figure 8. Evolution of the cumulative COD extracted (in per cent of COD introduced) during percolation processes of domestic refuse at 35 °C; aspersion process •——• and immersion process *·····*.

Table 8. COD and N_{TK} extraction yields in two types of percolator reactor of domestic refuse at 35 °C.

	COD_o (g O_2)	(g O_2)	COD_e (% COD_o)	N_{TKo} (g N)	(g N)	N_{TKe} (% N_{TKo})
12 days aspersion	394	123	31.3	-	-	-
immersion	393	146	37.2	-	-	-
26 days aspersion	394	130	33.1	8.9	6.7	76
immersion	393	162	41.5	8.9	7.0	79

Finally, this percolator effluent has been converted to methane in a fluidized bed digester with a COD conversion of 95 % using the following running conditions :

B_V (g COD/$l_{bed.d}$) : 7.89

θ (d) : 0.95

$r_{V.CH4}$ (l CH_4/$l_{bed.d}$) : 2.68

These results with two-phases systems need to be confirmed before developing a reliable industrial process able to degrade domestic refuse more efficiently.

6. BIOMETHANATION OF MIXTURES OF DOMESTIC REFUSE AND SEPTIC TANK RESIDUES

Residues of septic tanks, treated in a one-step completely mixed process, produce 0.4 liter of methane per gram of volatile solids introduced. Biomethanation of septic tank residues combined with domestic refuse is feasible and is more reliable and more productive than biomethanation of domestic refuse alone under similar conditions as shown in Table 9. The synergistic effect is about 19 %.

Table 9. Main results and yields obtained during biomethanation of domestic refuse (DR) and septic tank residues (S.T.R.) at 35 °C. The percentages are given in volatile solids introduced

	B_V ($\frac{g\ VS_o}{l.d}$)	θ (d)	TVA ($\frac{g}{l}$)	$r_{V.CH4}$ ($\frac{l CH_4}{l.d}$)	Y_{CH4/VS_o} ($\frac{l\ CH_4}{g\ VS_o}$)
S.T.R. 100 %	1	14	0.1	0.42	0.42
S.T.R. 35 % / DR 65 %	1	14	0.1	0.36	0.36
DR 100 %	1.5	14	0.2	0.36	0.24

7. CONCLUSIONS

The non-reliable character of the biomethanation of domestic refuse in completely mixed processes has been investigated.

(1) the alkaline content of the refuse is an important acting factor; a high alkaline refuse (16 g $CaCO_3$ per kg TS) yields the best results. Furthermore, the addition of Na_2CO_3 when feeding the digester with refuse of low alkalinity content leads to the stabilization of the process, but the situation is then not optimal, as shown by the presence of volatile fatty acids in the mixed liquor.

(2) experiments in batch fermenters have shown that a minimal content in cellulose plus hemicellulose of about 40 % of the COD of the refuse is necessary for a quick and a high production of methane.

(3) Answers to this unreliability have been tried in new digesters design : a two-step process has been developed and patented, which permits a methane production rate 5 times higher for a volumetric loading rate 7 times higher than in CSTR,

(4) an other two-step system, composed of a percolator and a fluidized bed digester is not yet optimized, but allows, at this time, an extraction of about 33 % of the COD and 80 % of the Kjeldahl nitrogen of the refuse and a conversion of the percolator effluent into methane with a yield of 95 %.

At least, the biomethanation of domestic refuse combined with septic tank residues has been shown feasible, more reliable and more productive than biomethanation of domestic refuse alone.

ACKNOWLEDGMENTS

This research has been supported by contracts RUW-057-B and RUW-099-B from the Commission of the European Communities (DG XII).

REFERENCES

a. FREROTTE, J., OMBREGT J.P. & PIPIJN P., 1982. La production de méthane à partir des ordures ménagères. La Technique de l'Eau et de l'Assainissement, 425, 13-28.

b. LEGROS, A., ASINARI DI SAN MARZANO, C.M., NAVEAU, H. & NYNS, E.-J., 1983. Fermentation profiles in bioconversion. Biotechnology Letters, 5 (1), 7-12.

c. COMPAGNION, D., NAVEAU, H. & NYNS, E.-J., 1982. Organigramme de décision en matière de digestion méthanique, plus particulièrement en milieu rural tropical. Revue de l'Agriculture, 6, 35, 3233-3241.

SCREENING OF CELLULOLYTIC ANAEROBIC BACTERIA, CELLULOLYTIC CO-FERMENTATIONS WITH METHANIC AND ACETOBUTYLIC FERMENTATION

PART ONE

FERMENTATION OF CELLULOSE BY A CO-CULTURE OF CLOSTRIDIUM CELLULOTYTICUM AND CLOSTRIDIUM ACETOBUTYLICUM

A. PETITDEMANGE, O. FOND, G. RAVEL, H. PETITDEMANGE AND R. GAY

Research carried out at the Laboratoire de Chimie Biologique 1,
Faculté des Sciences, Université de Nancy 1,
B.P. 239 / F - 54506 VANDOEUVRE-LES-NANCY CEDEX

and at the Laboratoire des Sciences du Génie Chimique - CNRS-ENSIC,
Institut National Polytechnique de Lorraine,
1, rue Grandville, F - 54042 NANCY CEDEX

SUMMARY

Two cellulolytic Clostridia, one thermophile and the other mesophile, were isolated. The first belongs to the species *Clostridium thermocellum*, while the second is a new species - *Clostridium cellulotyticum* (ATCC 35 319).

These two bacteria have a full range of cellulase activities (exocellulases and endocellulases) and a hemicellulase activity (xylanase). They break down more or less crystalline pure celluloses such as Solka Floc, MN 300 and Avicel and also various lignocellulosic compounds.

A co-culture (mixed culture of *Cl.cellulolyticum* and *Cl. acetobutylicum* was produced with a view to direct single-stage conversion of cellulose to the solvents acetone and butanol. It was hoped that the cellulolytic bacterium would provide the glucolytic bacterium with the cellobiose and glucose which its metabolism requires. This co-culture is effective but as yet only the first, acid phase of acetone-butanol fermentation has been observed. We have tried to determine the reasons.

INTRODUCTION

One of the possible ways of using biomass is to convert it into chemicals and/or energy products. Since our laboratory has some experience of acetone-butanol fermentation, we considered the possibility of converting the cellulose into solvents: acetone and butanol.

Acetone-butanol fermentaion is accomplished with *Cl.acetobutylicum*, an anaerobic bacterium which is capable of metabolising a large number of sugars including C_6 sugars (glucose and cellobiose) obtainable from cellulose and C_5 sugars (xylose and arabinose) obtainable from hemicelluloses. Such fermentation takes place in two phases (1) (2). The first or 'acid' phase consists largely in the production of acetic and butyrid acid while in the second 'solvent' phase the acids are partly remetabolsided and the solvents acetone and butanol are produced. Since *Cl.acetobutylicum* has no cellulase or hemicellulase ability, the sugars must be produced from the lignocellulosic compounds either by chemical acid hydrolysis or by enzymatic hydrolysis using e.g. the cellulase of 'Trichoderma reesei'. We considered an alternative route: a co-culture of a sugar-producing cellulolytic bacterium an *Cl.acetobutylicum*, which would ferment the sugars produced by the breakdown of the cellulose.

Since the number of cellulolytic Clostridia available is fairly low (when our work began there were only *Cl.thermocellum* and *Cl. cellobioparum*), we first tried to isolate new strains.

1. ISOLATION OF CELLULOLYTIC CLOSTRIDIA

A. Isolation of a thermophile Clostridium

A pure culture of a thermophile cellulolytic anaerobic bacterium was obtained from decomposing cow litter.

Isolation was effected by the Hungate method on a solid culture medium with cellulose MN 300 (Macherey and Nagel) as the sole carbon source.

Clearly distinct colonies surrounded by a zone of complete lysis of the cellulose were formed. We managed to make a culture from one well isolated colony.

Study of the thermophile cellulolytic anaerobic bacterium

. Morphological, cultural and physiological characteristics

The bacterium is in the form of fairly thin rods (0.5 to 0.6 μm thick) between 2.5 and 8 μm in length. In old cultures terminally located deforming spores appear, the cells which have sporulated measuring 10 to 15 μm. The bacterium is gram-negative. Cells cultivated in cellulose have a light yellow pigment.

The optimum gowth conditions for this thermophile bacterium are a temperature of 58° C and a pH of 7.8.

. Biochemical characteristics

In addtion to cellulose, the bacterium is capable of fermenting other sugars including cellobiose and glucose (though with a longer latent period), arabinose and xylose. The generation times vary from 7.5 h for growth on cellobiose to 14 h for growth on cellulose.

When the bacterium was cultivated for a period of 6 to 7 days in a fermenter on MN 300 cellulose (initial concentration 7.5 g/l), 50% of the substrate was degraded in a non-buffered culture medium and 78% in a medium buffered with MOPS (morpholino-propanesulfonic acid).

The cellulase activities - exoglucanase (Avicelase, filter paper activity and endoglucanase (carboxymethlycellulase) - and hemicellulase activity - xylanase - were detected in the culture medium, a β- glucosidase (arylglucosidase) activity being found in the cells.

The metabolic products are hydrogen, carbon dioxide, ethanol, acetic acid and small quantities of formic and lactic acid. Few reducing sugars accumulate as intermediates in the culture medium.

A detailed study of the cellulolytic action of this bacterium on various lignocellulosic substrates over a period of 20 days revealed a loss of 40 % by weight of untreated straw, 35 % of old newspaper and 35 % of delignified softwood sawdust.

This strain probably belongs to the species Clostridium thermocellum.

B. Isolation of a mesophile Clostridium H 10

This Clostridium was isolated from decomposing grass by the same techniques as the thermophile bacterium.

Study of the mesophile cellulolytic anaerobic bacterium

. Morphological, cultural and physiological characteristics

The cells are in the form of rods 3 to 6μ long and 0.6 to 1μ thick and are gram-positive (Fig. 1). The spores are terminally located and deforming (Fig. 1). This bacterium is motile and unpigmented. The optimum growth temperature is around 32-35°C.

. Biochemical characteristics

The bacterium grows at a moderate rate on arabinose, cellobiose, cellulose, fructose, glucose and xylose but never grows on lactose or saccharose.

The cellulolytic and hemicellulolytic activity of the strain was studied. The results are shown in Table I.

The main products of metabolism are hydrogen, carbon dioxide, ethanol and acetic acid with smaller quantities of formic and lactic acid.

The kinetics of a fermentation process (loss of the cellulosic substrate and appearance of the main products of metabolism) were observed as shown in Figure 2.

A final balance of this process was drawn up and is shown in Table II.

A study was made of the effect of strain H 10 on various lignocellulosic substrates which had undergone mild alkaline hydrolysis (at 80°C with NaOH 1%, for 3 hours). In 10-15 days, 45 % of the dry weight of maize stalks and leaves, 40 % of oat straw and 36 % of Jerusalem artichoke leaves were degraded.

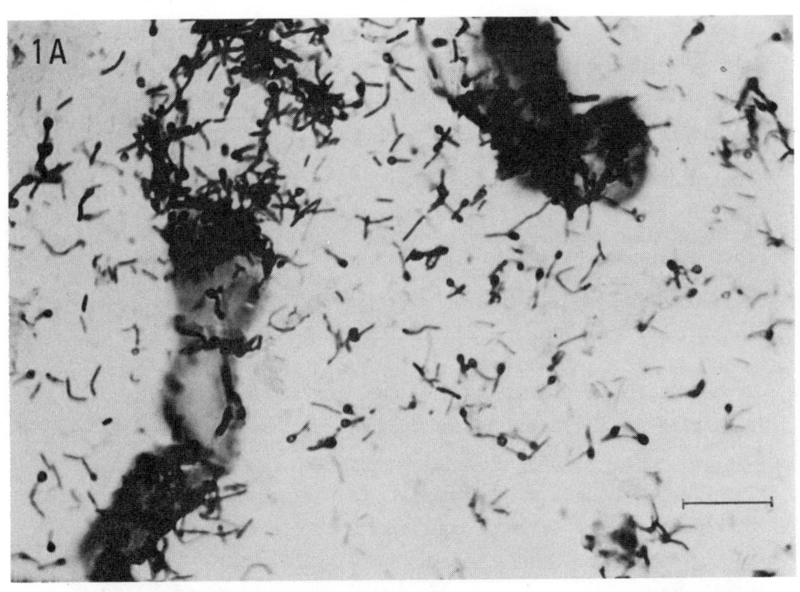

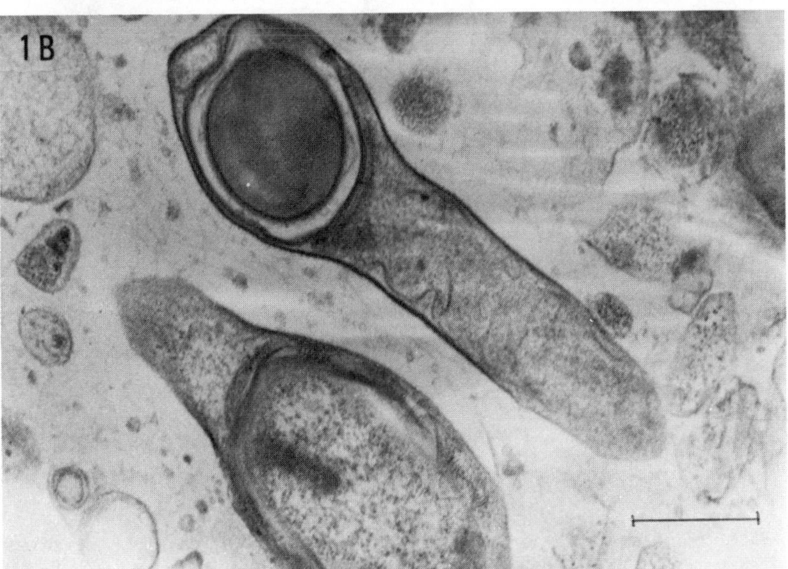

FIG. 1 : Vegetative and sporulating cells of _Clostridium cellulolyticum_

TABLE I - Cellulase and hemicellulase activity of Cl.cellulolyticum[a]

Cellulosic substrate	MN 300	Solka Floc	Avicel
% substrate loss	47	60	78
Cellulase activities			
- "cellulase" (filter paper)			
I.U. 10^{-2} ml [b]	0.93	1.3	1.5
- endoglucanase (carboxyme-			
thyl-cellulase) I.U. 10^{-2} ml	3	7	15
Hemicellulase activity			
- xylanase I.U. 10^{-2}/ml	12	15	17.5

a : These results were obtained with Hungate tube cultures, initial substrate concentration 7.5 g/l, duration 10 days at 35°C.

b : The I.U. is defined as the quantity of enzyme which released 1μmole of reducing sugar (glucose equivalent) in one minute.

TABLE II - Balance sheet for fermentation with Cl.cellulolyticum[a]

Products	g/l	Quantity μmol./ml	mol.of C/ml
Cellulose degraded[b]	4	24.69	148.14
Products formed:			
- CO_2 (estimate)		44.2	44.2
- acetate	2	33.33	66.66
- ethanol	0.5	10.87	21.74
- reducing sugars	0.45	2.5	15
C accounted for			99 %

a : This fermentation was carried out in a 2 l fermenter, the substrate being 6 % Solka Floc, with incubation at 35°C for 7 days.

b : For cellulose, the number of moles is expressed in terms of anhydrous glucose equivalent (MW162 g).

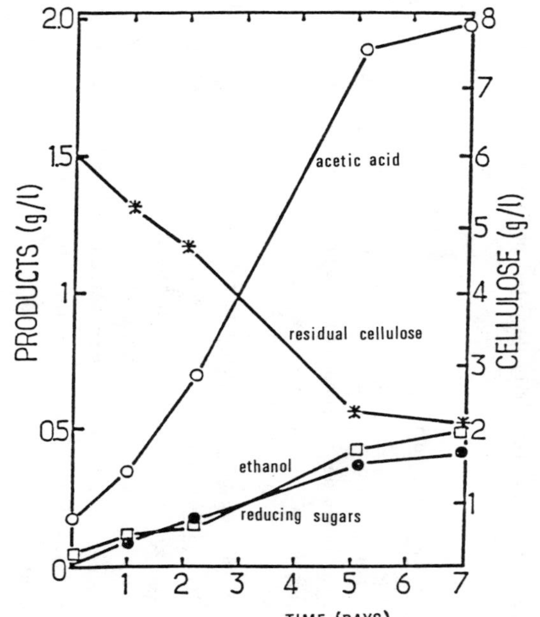

FIG. 2 : Kinetics of the utilization of Solka Floc cellulose and product formation by *Clostridium cellulolyticum*

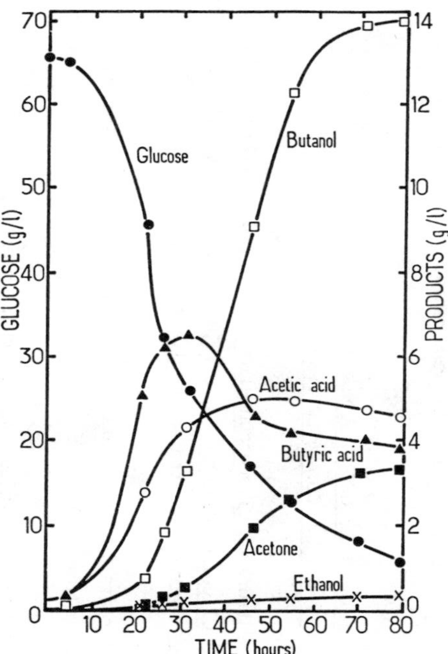

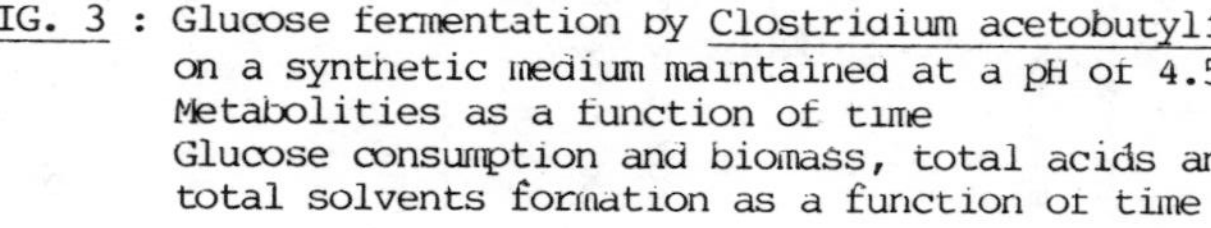
FIG. 3 : Glucose fermentation by *Clostridium acetobutylicum* on a synthetic medium maintained at a pH of 4.5
Metabolities as a function of time
Glucose consumption and biomass, total acids and total solvents formation as a function of time

. Identification of the strain

The bacterial strain H 10, which has a GC value of 41 %, represents a new species (3) Cl.cellulolyticum and was registered as ATCC No 35 319.

2. FERMENTATION OF CELLULOSE BY A MIXED CULTURE OF CL.CELLULOLYTICUM AND CL.ACETOBUTYLICUM

For a mixed culture of Cl.acetobutylicum and a cellulolytic Clostridium, we selected Cl.cellulolyticum, which is also a mesophile bacterium. (There are no thermophile strains of Cl.acetobutylicum).

The proposed route for utilization of cellulose is therefore mesophile, with a mixed culture at 35°C.

Acetone-butanol fermentation has a characteristic 2-phase breakdown profile. The phases are clearly apparent in Figure 3, which shows glucose consumption and metabolite production by Cl.acetobutylicum as a function of time. In the first phase, bacterial growth is accompanied by the production of acetic and butylic acid; in the second phase, the solvents are produced and the acids previously secreted are more or less entirely reconsumed.

A. Preparation of the mixed culture

The conditions permitting a mixed culture of Cl.cellulolyticum and Cl.acetobutylicum were determined (4). Figure 4 shows the product formation kinetics for a culture on Solka Floc.

As soon as the strain of Cl.acetobutylicum is added, the products of the metabolism of Cl.cellulolyticum cease to accumulate and the acetic acid is partially utilized. Cellulolysis is initiated since 27 g/l of Solka Floc cellulose are degraded. This primarily benefits the glucidolytic bacterium, which metabolizes the sugars formed, and only the metabolic products of this bacterium accumulate, mainly in the form of butyric acid (11 g/l) with traces of butanol.

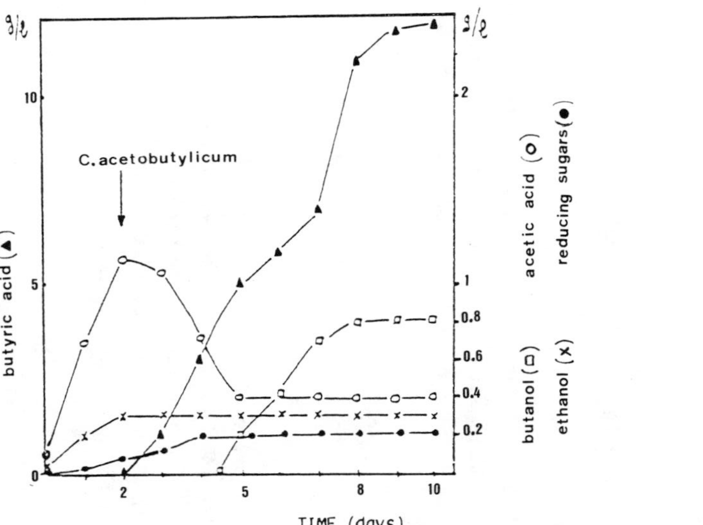

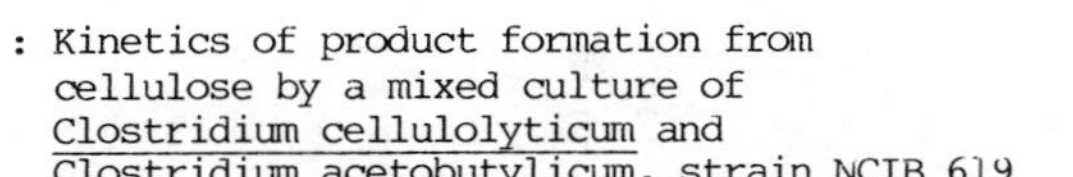

FIG. 4 : Kinetics of product formation from cellulose by a mixed culture of Clostridium cellulolyticum and Clostridium acetobutylicum, strain NCIB 619

As shown by the arrow, the NCIB strain is inoculated into the medium two days after Cl.cellulolyticum. The initial Solka Floc cellulose concentration is 18 g/l

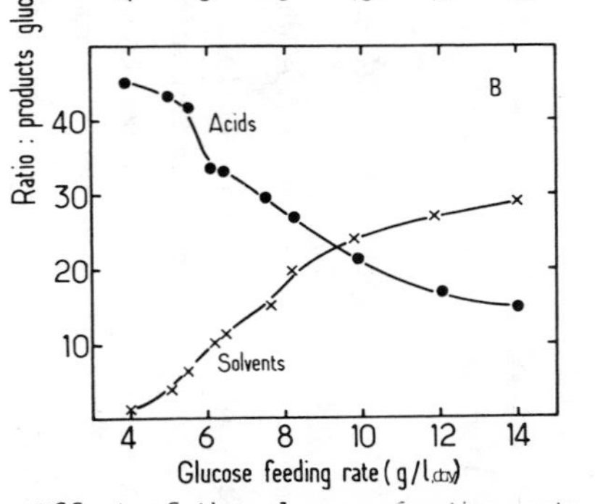

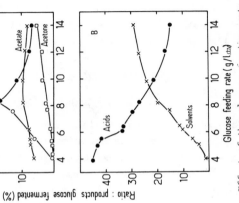

FIG. 5 : Effect of the glucose feeding rate on the nature of the products formed by Clostridium acetobutylicum (Ref. 5)

A. Percentages of glucose converted to butyrate, butanol, acetate, acetone.

B. Percentages of glucose converted to a total acids and total solvents.

This mesophile bacterial co-culture thus results in much greater cellulolysis (3 to 5 times as much cellulose being degraded). This is as efficient as hydrolysis with thermophile bacteria, but we were able to observe only the first phase of the acetone-butanol fermentation since the metabolic activity of the two bacteria ceases when the butyric acid concentration reaches a toxic level.

B. Explanation of the production of butyric acid by Cl. acetobutylicum in a bacterial co-culture (5)

The acetone-butanol fermentation profile in Figure 2 relates to a batch culture and an initial glucose concentration of 55 g/l. When bacterial co-cultures are used, the sugar concentration in the culture medium is always very low since Cl.acetobutylicum uses the sugars as soon as the Cl.cellulolyticum cellulases have produced them from the cellulose. These conditions may be compared with a continuous feed of sugars at a low rate.

We therefore studied acetone-butanol fermentation with a continuous feed of sugar solution.

Figure 5 shows the effect of various glucose feeding rates (expressed in grams of glucose per litre and per day introduced to the fermenter) on the nature and quantity of the products formed by Cl.-acetobutylicum.

At low glucose feeding rates the bacterium forms acids only and an increase in the feeding rate results in lower acid production and greater production of solvents.

The maximum rate of cellololysis by a co-culture occurs at a sugar feeding rate of 3-4 g/l/day. It is clear form Figure 5 that at this feeding rate only acids can be formed. The maximum conversion of glucose into solvents (30%), which is attained in a batch at 55 g/l (Figure 3), does not occur until the feed rate reaches 14 g/l/day.

CONCLUSION

In our attempts to obtain cellolytic anaerobic bacteria we succeeded in isolation two pure strains of the genus Clostridium; the first, which is thermophile and pigmented, belongs to the species Cl.thermocellum; the second, which is mesophile, is a new species Cl.cellulolyticum (registered as ATCC No 35 319).

Mixed cultures of Cl.cellulolyticum and Cl.acetobutylicum were produced; they make it possible to convert cellulose to C_4 compounds. Previous work has shown that C_2 compounds may be obtained by thermophile (6) or mesophile (7) (8) mixed cultures.

Our mixed culture degrades 25 g/l of Solka Floc in 8 days. NG et al. (6) report fermentation of 8 g/l of Solka Floc in 5 days by a co-culture of Cl.thermocellum and Cl.thermohydrosulfuricum, while KAHN and MURRAY (8) converted 17 g/l of cellulose in 10 days with a co-culture of 2 mesophile anaerobic bacteria. These findings demonstrate that mixed cultures of both mesophile and thermophile cellulolytic anaerobic bacteria are a promising approach to the production of chemicals from cellulosic materials. In our experimental acetone-butanol fermentation of cellulose by a mixed culture, however, only the first, acid, phase of the fermentation was observed. For the second phase, i.e. the production of solvent, to become feasible, the cellulose degradation rate would have to be increased by a factor of 2 to 3.

ACKNOWLEDGMENT

The authors thank Mr. G. KILBERTUS for preparing the electron micrographs.

REFERENCES

1. S.C. BEESCH (1952) Acetone Butanol Fermentation of Sugars. Ind. Eng. Chem. 44, 1677-1682.

2. F. MONOT, J.R. MARTIN, H. PETITDEMANGE and R. GAY - Acetone and butanol production by Clostridium acetobutylicum in a synthetic medium (1982). Appl. Environ. Microbiol. 44, 1318-1324.

3. E. PETITDEMANGE, F. CAILLET, J. GIALLO and C. GAUDIN - Clostridium cellulolyticum, sp. nov., a cellulolytic, mesophilic species from decayed grass. Int. J. Syst. Bacteriol. (Gone to press).

4. E. PETITDEMANGE, O. FOND, F. CAILLET, H. PETITDEMANGE and R. GAY - A novel one step process for cellulose fermentation using mesophilic cellulolytic and glycolytic Clostridia (1983) Biotechnol. Lett., 5, 119-124.

5. O. FOND, E. PETITDEMANGE, H. PETITDEMANGE and R. GAY - Effect of glucose flow on the acetone butanol fermentation in fed batch culture (1984). Biotechnol. Lett., 6, 13-18.

6. K.T. NG, A. BEN-BASSAT and J.G. ZEIKUS - Ethanol Production by Thermophilic Bacteria: Fermentation of Cellulosic Substrates by Co-cultures of Clostridium thermocellum and Clostridium thermohydrosulfuricum (1981) Appl. Environ. Microbiol. 41, 1337-1343.

7. J.N. SADDLER, K.H. CHAN and G. LOUIS SEIZE - A one step process for the conversion of cellulose to ethanol using anaerobic microorganisms in mono and co-culture (1981). Biotechnol. Lett. 3, 321-327.

8. A.W. KAHN and W.D. MURRAY - Single Step Conversion of cellulose to ethanol by a Mesophilic co-culture (1982) Biotechnol. Lett., 4, 177-180.

SCREENING OF CELLULOLYTIC ANAEROBIC BACTERIA, CELLULOLYTIC CO-FERMENTATIONS WITH METHANIC AND ACETOBUTYLIC FERMENTATION

SECOND PART

INFLUENCE OF AN ENRICHED INOCULUM WITH SELECTED ANAEROBIC BACTERIA (cellulolytic or methanogenic) ON THE METHANE FERMENTATION OF CELLULOSIC MATERIAL BY A WILD MICROBIAL POPULATION

J.S. BACHMAN*, E. PETITDEMANGE**, C. PROST*

* Laboratoire des Sciences du Génie Chimique, CNRS-ENSIC
1 Rue Grandville - 54042 NANCY Cédex - France

** Laboratoire de Chimie Biologique, Université de NANCY-1
B.P. 239 - 54506 VANDOEUVRE Cédex - France

Summary

Methane fermentation of two substrates i.e. pure cellulose and oat straw is carried out in batches in the presence of a wild microbial population and a medium more or less rich in nutrients. It is shown that adding a supplementary, pure, cellulolytic strain has no positive effect. On the contrary, adding a pure, methanogenic strain is beneficial whatever the substrate used. Finally a mixed inoculum with these two strains is favourable too.

1. INTRODUCTION

This study takes place within the general framework of research on how to increase methane production rate by fermentation of a solid ligno-cellulosic substrate.

In order to do this, three possible ways can be considered :

a - modifying the substrate by pretreating it,

b - modifying the process of fermentation (pH, temperature, stirring...)

c - modifying the bacteria population.

We chose experimentation (c) and showed the effect of a supplementary inoculum with selected bacteria acting simultaneously with a wild population coming from the "pied de cuve" of an anaerobic laboratory digester.

2. EQUIPMENT AND METHOD

Two substrates were tested

- solka-floc cellulose
- oat straw

The strain of the cellulolytic anaerobic bacteria used is mesophilic *Clostridium cellulolyticum* which has been isolated and described by a member of our research group (1).

The methanogenic bacteria strain is *Methanosarcina barkeri* DSM 800 coming from the Göttingen Collections (2-3)?

The wild microbial population, to which either or both above-mentioned bacteria are added, is taken from a digester which has been in operation in our laboratory for several months with an oat straw substrate.

The fermentations are to be carried out in batches for thirty days without stirring in two-litre reactors.

At the beginning of each experiment, the substrate is brought into contact with 1800 cm^3 wild population medium and 200 cm^3 culture medium of the bacteria to be tested.

Every day, during the fermentation process, we measure the pH, the flowrate and composition of bio-gas, the volatile fatty acids and reducing sugar contents in the digester.

3. RESULTS ACHIEVED WITH CELLULOSE

Having a *15 g/l cellulose concentration*, experiments are first carried out using a *wild microbial population alone*.

Figure 1 shows how the gas production varies with time : the flowrate is first low, then it passes through a maximum after the tenth day and finally decreases gradually.

Methane concentration settles at around 50 % from the very beginning and remains fairly constant in composition. As for the pH, it does not require any regulation and remains stable and almost neutral. Furthermore, no reducing sugar or building up of volatile fatty acids were noticed. Thus it seems that cellulolysis is the rate-limiting step of the process.

31. Influence of the "reference medium"

Our aim is to try speeding up the process of fermentation by modifying the microbial ecosystem of the wild population i.e. introducing cultures of selected bacteria either cellulolytic or methanogenic. Now, in this operation, not only a microbial population is added, but also the culture medium containing it. In order to show the exclusive influence of adding bacteria, it is necessary, first to evaluate the effect of the culture medium on the wild population.

The tested medium or "reference medium" is the mixture of the cellulolytic and methanogenic culture media.

At the beginning of the experiment, 200 ml of the reference medium are added to 1800 ml of the wild population medium in presence of 15 g/l cellulose.

Figure 1 shows that the reference culture medium has an important activating effect. The accumulated production of bio-gas increases from 4 to 7,2 l/l (table I).

32. Influence of an addition of cellulolytic bacteria

At the beginning of fermentation, 200 ml of the cellulolytic bacteria culture are added to 1800 ml of the wild population.

Figure 1 allows us to compare these results with those obtained with the reference culture medium. Contrary to our expectations, we see that adding cellulolytic bacteria does not bring about a rise in solka-floc conversion. On the contrary, it is less important and so is bio-gas production (Table I). There is still no reducing sugar but the concentration of acetic and propionic acids is no longer negligible (table II). However, this concentration is not high enough to have an inhibiting effect. It is therefore the process of cellulolysis itself which is less satisfactory than was expected.

Seeding	Bio-gas production in 30 days (cm3/1)	Methane production in 30 days (cm3/1)	Average % of methane	Cellulose Conversion in %
Wild population alone	4 060	2 390	58,9	40
Wild population + reference medium	7 225	3 830	53	63,3
Wild population + cellulolytic bacteria	3 330	1 725	51,8	58,7
Wild population + methanogenic bacteria	8 675	4 435	51,1	68,7

Table I : Results achieved with cellulose fermentation

33. Influence of the addition of methanogenic bacteria

The process is implemented in the same way as the one carried out with cellulolytic bacteria.

Figure 1 shows that adding methanogenic bacteria proves successful. Gas production goes from 7,2 to 8,7 1/1 while solka-floc conversion increases from 63,3 to 68,7 (table I). The curves for bio-gas flowrate are irregular and pass through several maxima. There is no latent period and bio-gas composition remains close to 50 %. There is still no reducing sugar and the amount of acid present in the medium is about the same as when using cellulolytic bacteria (table II).

Seeding / Concentration in g/l	Wild population alone	Wild population + reference medium	Wild population + cellulolytic bacteria	Wild population + methanogenic bacteria
Acetic acid	-	-	0,51	0,37
Propionic acid	-	-	0,61	0,53

Table II : Highest acid concentration with cellulose fermentation

4. RESULTS ACHIEVED WITH STRAW

In many agricultural wastes, cellulose is confined in a complex lignocellulosic structure. The straw studied here represents an example of these products. It is an oat straw with particle sizes ranging from 0.1 to 0.5 mm. Its composition determined by VAN SOEST'S method (5) is as follows :

cellulose 41 % ; hemicellulose 30,5 % ; lignin 7,5 % ; ashes 7 % ; micellaneous 14 %.

Straw concentration being 50 g/l, different processes of fermentation are carried out under the same conditions, as previously, i.e. first with the wild microbial population alone, then with the reference culture medium and finally with the addition of selected bacteria.

41. Fermentation carried out with the wild microbial population alone

Results are displayed in figure 2. They are quite different from those achieved with cellulose and show several phases :

- First phase : fermentation starts with an acidification phase of 5 days during which acetic acid builds up to over 3 g/l. During this phase, the bio-gas chiefly consists of carbon dioxide. A great deal of gas is produced at the beginning of this phase, then production decreases rapidly and the bio-gas flowrate falls down to a minimum. It is during this initial fermentation phase that the pH tends to drop rapidly and that the addition of ammonia is more necessary to maintain it near 7.

- Second phase : having reached a maximum value, the acetic acid concentration decreases rapidly while methane concentration and gas flowrate increase

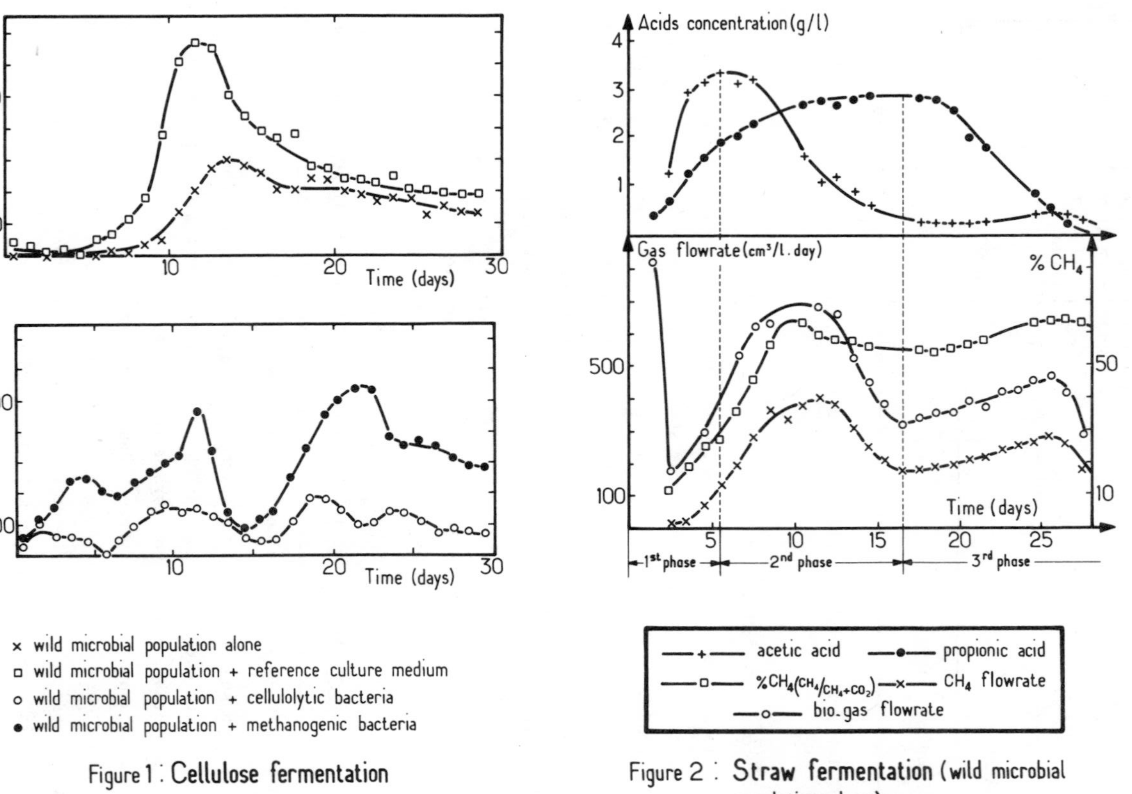

× wild microbial population alone
□ wild microbial population + reference culture medium
○ wild microbial population + cellulolytic bacteria
● wild microbial population + methanogenic bacteria

Figure 1 : **Cellulose fermentation**

Figure 2 : **Straw fermentation** (wild microbial population alone) rapidly. The methane flowrate passes through a first maximum value, then decreases and reaches a relative minimum at the end of this second phase that lasts about ten days and ends when acetic acid concentration no longer decreases and settles down to around 0,3 g/1. During this stage, the medium no longer tends to become acid and the pH remains spontaneously close to 7 without adding ammonia. From the moment fermentation started, the propionic acid has been building up slowly. At the end of the second phase, concentration reaches a value of more than 2,5 g/1

- Third phase : fermentation is characterised by the disappearance of the propionic acid together with a second increase in the production of methane slightly less important than the first.

In the course of these experiments, no reducing sugar could be detected.

All these results are in agreement with those, found in literature (6-7-8).

42. Influence of adding cellulolytic and methanogenic bacteria

In this last section, as we did with solka-floc cellulose, we shall study the influence of adding cellulolytic or methanogenic bacteria on oat straw fermentation using a wild population.

For this, a 50 g/1 straw sample was fermented using the four types of seeding mentioned in table III, with a regulated pH near 7. We noted the bio-gas production and carried out the analysis of the residues collected after thirty days of digestion. All the results achieved are put together in table III.

- It appears that adding the reference culture medium has a less activating effect on straw fermentation than on solka-floc cellulose fermentation. The reason of this can be found in the fact that straw already contains a great deal of complex substances useful for the growth of microorganisms.

- Adding *cellulolytic bacteria* does not bring about any increase in the gas production which even slightly decreases, while dry matter decay decreases by 23 % compared to the fermentation carried out with the reference medium. This confirms the negative effect of adding cellulolytic microorganisms which had already been noticed with cellulose.

Seeding	Bio-gas production in 30 days (cm3/1)	Methane production in 30 days (cm3/1)	average % of methane	Dry matter used in 30 days (g/1)	Dry matter conversion in %
Wild population alone	9 950	5 425	54,5	15,64	34
Wild population + reference medium	10 800	6 250	57,9	19,08	41,5
Wild population + cellulolytic bacteria	9 810	5 265	53,7	14,5	31
Wild population + methanogenic bacteria	14 325	7 700	53,8	16,4	35,6
Wild population + mixed seeding of cellulolytic and methanogenic bacteria	15 750	8 540	54,2	18,6	40,4

Table III : Straw fermentation : influence of adding reference culture medium and selected bacteria

- The effect of adding *methanogenic bacteria* is confirmed here : methane production is increased by 35 % and the decay of dry matter rises if compared to the previous fermentation with cellulolytic bacteria.

- The study of a *mixed culture of cellulolytic and methanogenic bacteria* has also been carried out. The results are collected in table III. It is clear that this mixed inoculum appreciably increases the production of bio-gas (+ 10 %) compared to that achieved with methanogenic bacteria. The conversion of dry matter is also higher.

5. CONCLUSION

The addition of a cellulolytic Clostridium culture to a methanisation wild population has no positive effect contrary to what we expected ; the cellulolytic bacteria do not seem to react when they are mixed with the complex wild population. Two hypotheses can be put forward :

- the cellulolytic bacteria that have been seeded are destroyed in this very heterogeneous population among which some microorganisms can produce substances such as antibiotics or bacteriocines reacting with strange cells.

- or the cellulolytic bacteria do survive but without producing active cellulases : either their production is hindered or their activity inhibited by some elements of the medium, or their destruction is caused by enzymes of the "protease" type.

Either of these hypotheses could explain thas C. cellulolyticum has no beneficial effect when added to a wild population producing methane although it is active in a pure or mixed culture with another Clostridium (9) and although cellulolysis is the rate-limiting step in the process of the methanisation of cellulolytic compounds.

Adding a Methanosarcina barkeri culture to a wild population has a positive effect which shows that the initial population of methanogenic bacteria of this wild population is not sufficient to exhaust the substrates of methanogenesis. More generally, a wild population which is not saturated with methanogenic bacteria could be profitably enriched with such bacteria, thus improving the performances of methane-production equipments.

REFERENCES

1 - PETITDEMANGE E., CAILLET F., GIALLO J., GAUDIN C., *"Clostridium cellulyticum sp. nov. a cellulolytic, mesophilic species from decayed grass"* Int. J. Syst. Bacterial, 1984 (to be published).

2 - BRYANT M.P. *"Methane producing bacteria, In Brergeg's Manual of derminative bacteriology, R.E. BUCHANAN and N.E. GIBBONS"*, 8th ed., William and William Co, 1974, 472-477

3 - BALCH W.E., FOX G.E., MAGRUM L.J., WOESE C.R., WOLFE R.S., *"Methanogens : reevaluation of a unique biological group"*. Microbiological Reviews, 1979, 43, n° 2, 260-296.

4 - BACHMAN J.S., Dr. Ing. Thesis, *"Contribution à l'étude de la fermentation méthanique de la cellulose et des ligno-celluloses"*, NANCY, nov. 1982.

5 - GOERING H.K., VAN SOEST P.J. *"Forage fiber analyses"*. A report of the Cornell University Agricultural Experiment Section. Agricultural Handbook, 1975, n° 379, Ed. Agricultural Research Service, U.S. Dep. of Agriculture.

6 - ANDREWS J.F., PEARSON E.A. *"Kinetics and characteristics of volatile acid. Production in anaerobic fermentation processes"*. Int. J. Air Wat. Poll., 1965, 9, 439-461.

7 - Mac CARTY P.L., JERIS J.S., MURDOCH W., *"The significance of individual volatile acids in anaerobic treatment"*. Proceedings of the 17th Annual Purdue Industrial Wastes Conférence, Purdue University, 1962, 421-439.

8 - KROEKER E.J., SCHULTE D.D., SPARLING A.B., LAPP H.M., *"Anaerobic treatment process stability"*. Journ. Water Poll. Control Fed. 1979, 51, n° 4, 718-727.

9 - PETITDEMANGE E., FOND O., CAILLET F., PETITDEMANGE H., GAY R., *"A novel one step process for cellulose fermentation using mesophilic cellulolytic and glycolytic Clostridia"*. Biotechn. Lett. (1983), 5, 119-124.

BIOGAS PRODUCTION FROM A DOMESTIC WASTE FRACTION

A.J. VAN DER VLUGT and W.H. RULKENS
Division of Technology for Society TNO
P.O.Box 342, NL-7300 AH APELDOORN

Summary

Research has been carried out into the possibilities of producing biogas from the fine organic waste fraction obtained from a mechanical separation system for municipal refuse. In particular the effect of an after-treatment of the digested organic fraction on the biogas production in a second digestion step has been investigated. These after-treatment methods are mechanical, thermal, chemical and combined methods.

The research is divided into four phases, phase 1 of which, the laboratory investigation, and phase 2, the preliminary technical-economic evaluation, have been completed.

The first digestion of the fine organic waste fraction gives a reduction in volatile solids content of about 50%. Biogas production at an organic load of 1.85 kg volatile solids/m³/day varied between 0.4 and 0.5 liter biogas/gram volatile solids added (55-60% CH_4).

The effect of the after-treatment methods used (mechanical, chemical, thermal and combined methods) on biogas production in the second digestion step is small. This is probably due to the substantial reduction in volatile solids content during the first digestion step. In terms of costs, mechanical treatment is the most economic method, compared with the other treatment methods.

According to the results and the preliminary technical-economic evaluation it can be concluded that the costs of after-treatment followed by a second digestion step are too high in relation to the profits of the extra biogas production. From the preliminary technical-economic evaluation it appears that one anaerobic digestion step of the fine organic waste fraction is attractive, provided the organic loading rate of the digestion can be increased.

According to this evaluation a modification of the research is proposed. A simple mechanical pre-treatment of the fine organic waste fraction and optimalization of the anaerobic digestion of the organic waste fraction are the main items. As a new substrate the organic fraction of source-selected collected municipal refuse is also going to be used.

This research project is carried out with the financial support of the E.C. and the Energy Research Projects Office of the Dutch Ministry of Economic Affairs.

1. INTRODUCTION

This research project is carried out with the financial support of the E.C. and the Energy Research Projects Office of the Dutch Ministry of Economic Affairs. It is an investigation into the effect of mechanical, thermal and chemical after-treatment methods on the production of biogas from the fine organic waste fraction which is obtained by the mechanical separation of municipal refuse. The project is divided into four phases, phase 1 of which, the laboratory research, and phase 2, the preliminary evaluation of the laboratory research, have been completed.

The fine organic fraction is obtained from a mechanical separation process consisting of the following successive steps:

. Bag shredding
. Electromagnetic separation
. Size reduction (shredding) with hammermill with 90 mm sieve
. Air classification
. Sieving of the light fraction with a 10 mm Spannwelle sieve.

The fall-through of the sieve is the fine organic fraction used. The mean particle diameter of the fine organic fraction varies between 0.6 mm and 1.4 mm. The total solids content (TS) is 65% to 55% and the volatile solids content (VS) is 45% to 60% (of TS). This means that 1 ton of fine organic fraction contains 300 to 330 kg of volatile solids (VS). A scheme of the process under investigation is given in the following figure.

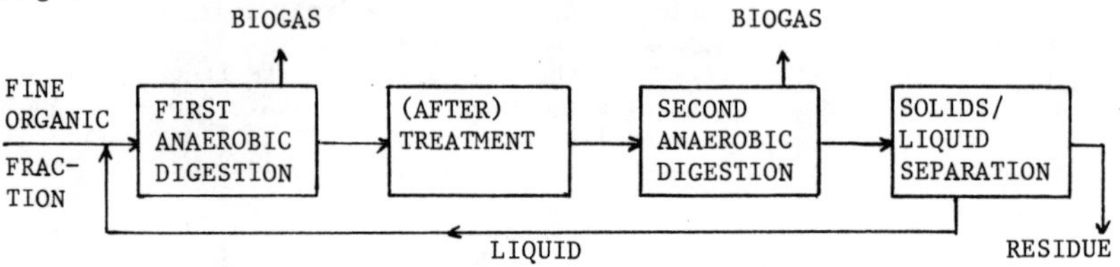

Figure 1. PROCESS SCHEME.

2. DIGESTION OF THE FINE ORGANIC FRACTION (FIRST DIGESTION)

The first digestion was carried out semi-continuously in 20-litre laboratory digesters (see figure 2). The fine organic fraction was added to the digester as a slurry.

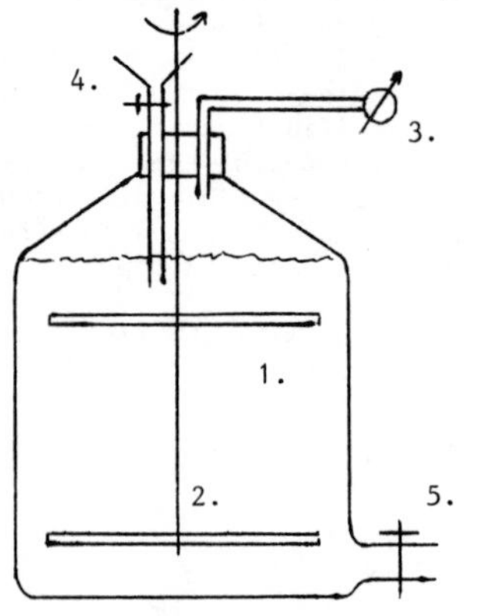

1. Digester
2. Stirrer
3. Wet precision gas meter
4. Feed inlet
5. Effluent outlet

Figure 2. 20-LITRE LABORATORY DIGESTER.

The process conditions of the digestion were:

. Temperature : 30°C ± 1°C
. Retention time : 40 days
. Total solids in feed : 6%
. Stirring at intervals : 30 sec/20 min.

The results are given in the following table.

Table I. SEMI-CONTINUOUS ANAEROBIC DIGESTION OF THE FINE ORGANIC FRACTION

	Mean	Range
Specific gas production (l/g VS added)	0,47	0,37 - 0,53
Biogas composition % CH_4	55	50 - 60
% CO_2	35	29 - 40
Volatile fatty acids (mg/l as acetic)	16,0	4,5 - 41,8
pH	6,8	6,7 - 7,0

The organic load of the digesters was 0,92 gram volatile solids per litre of digester volume per day (g VS/l_r day). Reduction of volatile solids was about 50%.

Also some experiments were carried out in shortening the retention time and in raising the total solids content of the feed. The aim of these experiments was to find out whether the specific gas production changes at higher organic loads. The higher the organic load, the smaller the digester volume can be chosen. In Table II the results of these experiments are given. The conclusion from these experiments is that doubling the organic load from 0.92 to 1.85 g VS/l_r day and shortening the retention time from 40 to 20 days has hardly any effect on specific gas production. From the figures in the table it appears that there is a tendency that the organic load can be increased still further.

Table II. SPECIFIC GAS PRODUCTION AT DIFFERENT ORGANIC LOAD AND RETENTION TIME

FEED TS %	FEED VS %	ORGANIC LOAD g VS/l_r day	RETENTION TIME days	SPECIFIC GAS PRODUCTION l/g VS added	NUMBER OF RETENTION TIMES
6.1	3.7	0.92	40	0.47	4.5
12.3	7.4	1.85	40	0.48	1.4
4.6	2.8	1.40	20	0.44	3.2
6.1	3.7	1.85	20	0.46	4.9

3. TREATMENT METHODS USED AND EFFECT ON BIOGAS PRODUCTION (SECOND DIGESTION)

Mechanical, thermal, chemical and combined treatment methods were applied to the digested fine organic fraction. The treatment methods selected on the basis of literature information and apparatus available used in the experiments were:

- . MECHANICAL - Homogeniser; 1 min, 5 min
  - Ball mill; ½ hr, 2 hrs
  - Colloid mill; 1x, 4x
- . THERMAL - 75°, 100°, 143°C; ½ hr, 6 hrs
- . CHEMICAL - Acid, alkali; ½ hr, 6 hrs
- . CHEMICAL-THERMAL - Acid, alkali; 75°, 100°, 143°C; ½ hr, 6 hrs
- . MECHANICAL-CHEMICAL - Ball mill 2 hrs, alkali ½ hr
  - Homogeniser 5 min, alkali ½ hr
- . MECHANICAL-THERMAL - Ball mill 2 hrs, 100°C, 143°C, ½ hr
  - Homogeniser 5 min, 100°C, 143°C, ½ hr.

The second digestion was carried out batch-wise in 1 litre digesters at a temperature of 30°C and with a retention time of 40 days.

Mechanical treatment with the homogeniser (5 min) and the ball mill (2 hrs), thermal treatment (143°C for ½ hr and 6 hrs) respectively chemical-thermal treatment (alkali, 143°C for ½ hr and 6 hrs) and mechanical-chemical treatment have a small positive effect on biogas production. The best treatment was mechanical-thermal, with an effect of 3% to 10% on biogas production. The negative effects of some treatment methods can be explained probably from the loss of material during the mechanical treatment (colloid mill) and from the fact that compounds were formed during the thermal and chemical treatment which were not or hardly digestible.

4. DEWATERABILITY AND COMPOSITION OF RESIDUES

By applying gravity sedimentation the digested fine organic fraction was thickened to about 5% total solids after two hours, which is about the same as obtained with digested activated sludge. Filtration with a filterpress after addition of polyelectrolyte gives a total solids content of the residue of about 30%. Centrifugation with a decanter resulted in a total solids content of the residue of 20% to 25%, also after addition of polyelectrolyte. The conclusion is that the residue of the digested fine organic fraction can well be dewatered by filtration and by centrifugation.

As regards the composition of the residue the contents of heavy metals of the residue were determined and compared with the contents in sewage sludge and compost from municipal waste (see Table III). In this table the Dutch guidelines from the "Unie van Waterschappen" are also given, regarding the heavy metals content in sludges for use on agricultural land.

Table III. HEAVY METAL CONTENT OF THE DIGESTED FINE ORGANIC FRACTION IN MG/KG TS

	DIGESTED FINE ORGANIC FRACTION	SEWAGE SLUDGE	COMPOST FROM MUNICIPAL WASTE	GUIDELINE
Cu	160	471	600	600
Pb	300	321	850	500
Cr	53	176	200	500
Ni	62	47	100	100
Cd	< 3	7	5	10
Zn	770	1359	1500	2000
Hg	< 3	5	-	10

From these figures it appears that the residue of the digested fine organic fraction has a heavy metals content which is lower than the heavy metals content of sewage sludge and compost obtained from municipal waste.

The C/N-ratio of the residue is about 2 which is too low for composting because nitrogen losses are to be expected through ammonia volatilization. The volatile solids content of the residue is above 50%.

5. FINANCIAL AND ECONOMIC ASPECTS

A rough calculation of the costs of the separate process steps has been made based on the following figures:

- 1 Ton of fine organic fraction contains 56% TS and 59% VS (of TS)
- Plant capacity 35,000 tons of fine organic fraction a year which means about 100 tons a day
- Specific gas production 0.47 m^3 biogas/kg VS added
- Gas production 5.5 million m^3 of biogas a year
- Selling price of 1 m^3 of biogas is 25 cents

In the following table the costs of digestion of the fine organic fraction (f.o.f.) are given as a function of the organic load applied.

Table IV. COSTS OF FIRST DIGESTION

ORGANIC LOAD kg VS/m_r^3.day	GROSS COSTS OF DIGESTION Dfl/ton f.o.f.	Dfl/m^3 biogas	NET COSTS Dfl/m^3 biogas
1.85	72.00	0.46	0.21
3.0	53.00	0.34	0.09
3.9	39.00	0.25	break-even

m_r^3 = m^3 of digester   Dfl = Dutch florin
f.o.f. = fine organic fraction

Assuming that the specific gas production is still 0.47 m^3 biogas per kg volatile solids added to the digester, the break-even point may be reached at an organic load of about 4 kg volatile solids per m^3 of digester a day.

The gross costs of the treatment methods used and the second digestion per m^3 of extra biogas are given in Table V.

Table V. GROSS COSTS OF TREATMENT AND SECOND DIGESTION PER M³ EXTRA BIOGAS (DFL)

TREATMENT METHOD	TOTAL COSTS OF TREATMENT AND SECOND DIGESTION* DFL/M³ EXTRA BIOGAS
No treatment	1.50 - 2.00
Mechanical	1.25 - 1.75
Chemical	5.00 - 7.50
Thermal	2.50 - 3.50
Chemical-thermal	2.75 - 4.00
Mechanical-chemical	2.50 - 3.50
Mechanical-thermal	1.50 - 2.00

* Costs of second digestion Dfl. 19.00 - 28.00/ton f.o.f.

From these costs it is evident that when the profits from the extra biogas are 25 cents, all the treatment methods used are too expensive, considering the small amount of extra biogas produced by these treatment methods.

The costs of dewatering the digested fine organic fraction to about 25% total solids with a filterpress or a decanter range from Dfl. 13.00 to 14.00 per ton of fine organic fraction. This means 8 to 9 cents per m^3 of biogas. About half these costs are attributed to the dosing of polyelectrolyte.

6. PRELIMINARY CONCLUSIONS

The following preliminary conclusions can be drawn from the research carried out so far:

. Good digestibility of the fine organic fraction with a biogas production of about 130 to 160 m^3 of biogas per ton of fine organic fraction.
. Gross costs of digestion (without treatment) are 46 cents/m^3 biogas at an organic load of 1.85 kg/m^3_r.day and can be reduced substantially at higher organic loads.
. Mechanical, chemical, thermal or combined after-treatment of digested organic fraction is of hardly any use, due to the small quantity of extra biogas produced.
. Costs of treatment are high, ranging from Dfl. 1.50 for mechanical treatment to Dfl. 5.00 to 7.50 for chemical treatment per m^3 of extra biogas.

Continuation of the research will be as follows:

. Optimalization of the anaerobic digestion of the fine organic fraction obtained from a mechanical separation system of municipal refuse.
. Investigation into the effect of mechanical pre-treatment of the fine organic fraction on biogas production.
. Use as a new substrate of the organic fraction from source-selected collected municipal refuse after mechanical pre-treatment.
. Phase 3 pilot-plant investigation.
. Phase 4 final technical-economic evaluation.

ANAEROBIC DIGESTION OF SLURRY FROM CATTLE FED HIGH ROUGHAGE DIETS

P.V. Kiely

An Foras Taluntais

Summary

Farm scale digestion of slurry from cattle fed a grass silage diet was studied in a 200m^3 digestor. Cattle slurries of different total solids content and fine size fractions of slurry were studied in laboratory scale batch digestors. The fertiliser value of digested slurry was compared with that of undigested slurry in a field experiment. Cattle slurry separates rapidly into a semi solid layer over a liquid layer during storage and regular or continuous mixing is necessary to provide a uniform feed to the digestor. The maximum gas production obtained was 0.8 vol./digestor vol./day and the average production was 0.5 vol./digestor vol./day or 240 to 300 l/kg volatile solids. Biogas production increased with total solids content of the slurry. Methane content of the biogas was 50 to 58 percent. Fine fractions of cattle slurry digest more rapidly. Anaerobic digestion reduced the volatile solids content by 20 percent, the BOD by 65 percent, the COD by 17 percent, the volatile fatty acids by 94 percent. Digestion causes a marked increase in fluidity of the slurry. Digested slurry is more easily pumped, is suitable for irrigation, does not require mixing prior to land spreading and does not coat pasture in the manner of undigested slurry.

INTRODUCTION

The substrate for this study was cattle slurry produced by animals fed almost completely on grass or grass silage. Approximately sixty six percent of the lowland area of Ireland is devoted to grassland most of which is utilised to feed six million livestock units for beef and milk production. The animals are grazed during summer and are housed for a four to six month winter period. As the animals are housed for nearly half the year almost half the faeces and urine produced is collected and forms a substantial source of biomass available for anaerobic digestion.

Disposal of the cattle slurry is by land spreading on the grassland area intended for silage cutting. As there is adequate area available for land spreading the object is to make optimum use of its fertiliser value. Cattle slurry contains an average 40 Kg N, 6 Kg P, 42 Kg K and 7 Kg S per 10 tonnes (Tunney, 1982). With the exception of nitrogen the availability of the nutrients in cattle slurry to grassland is generally high. Nitrogen availability from cattle slurry applied to grassland varies from zero to 45 percent depending on time of application and weather conditions following application (Kiely 1980). Also cattle slurry, because of its thick slimy consistency, has a tendency to coat grassland, occluding light and depressing grass growth. Any treatment process which would increase the availability of cattle slurry N and reduce the viscosity or slimy nature of the material must enhance the fertilizer value and handling properties of the slurry.

Most research on anaerobic digestion of animal waste has been concerned with pig slurry. Studies using cattle slurry generally report low levels of gas production relative to pig slurry, long retention times of 20 days or greater, and low levels of gas production per unit of digestor volume. Stewart (1980) found that cattle manure and sheep manure had the lowest biogas production per unit of dry matter of a wide range of agricultural crops and waste materials. This is not surprising since these materials have already been anaerobically digested in the rumen of these animals.

The effect of animal diet on biogas production from animal waste has not been widely studied. Hashimato et al (1981) found that manure from cattle fed a mainly maize grain diet produced 66 percent more methane per unit of volatile solids than manure from similar animals fed on a mainly maize silage diet.

The present study had three parts:

1. Evaluation of farm scale digestion of cattle slurry from animals fed a mainly grass silage diet.
2. A comparison of different cattle slurry fractions in laboratory scale digestors.
3. A study of the effects of digestion on the fertilizer properties of cattle slurry.

MATERIALS AND METHODS

Farm scale digestor

The farm scale digestor used was cylindrical with a diameter of 7.62m, a height of 4.72m and a slurry capacity of 202 m^3. The walls are constructed of steel plate vitreous enameled on both sides and the conical roof is built from epoxy coated aluminium segments. It is insulated with 6 cm polyurethane foam throughout. An internal mixer/heat exchanger unit consisting of a double layer tube of 32cm internal diameter and 7m long is located off centre in the digestor at an angle of 45° to horizontal.(Fig.1). The tube is heated by hot water at 50 to 65 °C circulated in the cavity between the layers of the tube. A propellor driven by a two speed reversible motor circulates the contents of the digestor through the tube thus heating and mixing the contents. Slurry was pumped from a level 25 cm above the floor of a subfloor tank by a 5.85 Kw Flygt submersible pump to a helical rotor and stator pump which was used to meter slurry into the digestor.(Fig.1). The slurry was pumped to a point at the base of the mixer/heat exchanger unit. Each time the digestor was being filled the mixer motor was activated and the cold influent slurry was passed through the heat exchanger before mixing with the contents of the digestor thus avoiding chilling the warm slurry and its bacterial population. The digested slurry overflowed over a weir at the top of the digestor. The biogas plant included a 140 m^3 diaphragm type gas holder, a heating boiler fueled by biogas, a gas meter and suitable control equipment.

Laboratory scale digestors consisted of 10 l glass bottles heated in a constant temperature waterbath with biogas measured by water displacement.

Analytical Techniques

Most of the analytical methods employed were standard chemical methods and are described by Byrne (1979). Volatile fatty acids were estimated by gas liquid chromatography (GLC) on Chromosorb 101. Hemicellulose was determined by hyrolysis to sugars in 1 $\underline{N}$ H_2SO_4 for 2 hours at 100 C and determination of the sugars by GLC following conversion to their alditol acetates. Cellulose was hydrolysed to glucose by treatment for 4 hours with $26\underline{N}H_2SO_4$. Neutral detergent fibre was determined by the method of Van Soest and Wine (1967). Methane content of the biogas was estimated by GLC.

RESULTS AND DISCUSSION

During storage cattle slurry separates into a semi solid layer over a liquid layer of approximately equal depth. This layer separation begins within 2 days after complete mixing and is complete after 3 weeks. It presents problems in feeding a digestor in that the liquid layer moves to the pump leaving behind the semi-solid layer which contains much of the organic material. This is reflected in the wide range in total solids content of the slurry fed to the farm scale digestor (table 1). The chemical composition of the cattle slurry used is presented in table 1. It had a high ash content of 25 percent resulting from the high ash content of the grass diet which was concentrated three-fold on passing through the animal. Also the organic components present indicate that the material has a low content of readily digestible organic matter and is unlikely to have a high biogas production capacity. Starch, sugars and other readily digestible material were not contained in the slurry.

Farm Scale Digestor

The operating conditions and output figures for the farm scale digestor are shown in table 2. Feeding the digestor was interrupted frequently by blockages and failures of the slurry pumping system. Gas production was generally in the region of 100 m^3/day. The highest gas production obtained was 161 m^3/day. This relatively low biogas production was attributed to the low total solids content of the slurry used and interruptions in feeding the digestor. The longer retention time of 25 days was found to give a higher nett biogas production during cold weather periods when much of the biogas was required to heat the incoming slurry feed to 35 °C. Highest biogas production was obtained when the slurry used had a high total solids content. Methane content of the biogas was low at 50 to 58 percent.

The effects of anaerobic digestion on the cattle slurry used are summarised in table 3. Smell and polluting potential of the slurry are significantly reduced as indicated by the reduction in volatile fatty acids and BOD. Most of the methane appears to be derived from breakdown of hemicellulose, cellulose and fatty acids. The principal benefit derived from anaerobic digestion is the marked increase in fluidity of the slurry. Digested slurry is:-

1. More easily pumped,

2. Is suitable for irrigation,

3. Does not coat pasture in the manner of undigested slurry and

4. Does not require mixing prior to land spreading.

These are significant benefits from a handling and fertiliser viewpoint.

After approximately 8 months operation the slurry outlet from the digestor became blocked occasionally by thick lumps of slurry forming within the digestor. The problem became progressively worse and biogas production tended to decrease during the following months. Efforts to stop the formation of solid material by operation of the mixing system for long periods failed.

Opening the digestor revealed a solid mass occupying approximately half the volume of the digestor over a liquid layer underneath. A high output air compressor was required to break the solid mass. Either the mixing system was unable to mix the contents of the digestor or the four 30-minute periods of operation used each day were not adequate.

The digestor has a high capital cost and a high maintenance requirement. Biogas production per se from cattle slurry is not economically viable; other benefits may accrue in the form of increased fertiliser value of the slurry, easier handling of the slurry and a reduction in polluting power and disease content of slurry.

Laboratory scale digestors.

Biogas production from four different cattle slurry types was studied in batch digestors operated at 35 °C for a 66 day period. Each of the digestors was seeded with 10 percent by volume of slurry from a digestor operating on cattle slurry at 35 °C. The slurries studied were as follows:

Constituent	In slurry	In total solids
Total solids %	3.0 - 11.5	
Volatile solids %		72 - 80
Total organic carbon %		22 - 29
Cellulose %		8 - 16
Hemicellulose %		14 - 15
Lignin %		2 - 4
Starch %		0
Soluble sugars %		0
Lipids %		2
Organic N %		2.2 - 2.6
BOD	13,000 - 17,000	
COD	60,000 - 90,000	
Acid soluble P mg/l	380 - 580	
Ammonia N mg/l	1,700 - 1,900	
Volatile fatty acids mg/l		
C_2	1,300 - 2,100	
C_5	180 - 400	
C_4	0 - 40	
pH	7.6 - 8.2	

Table 1. Chemical compostion of the cattle slurry used

Temperature	35°C
Retention time	20 and 25 days
Slurry feed	3.0 - 11.5% D.M.
Maximum gas produced	161 m³/day
Gas required for heating	45 to 95 m³/day
Gas production/Kg volatile solids	240 - 300 l
CH_4 content of gas%	50 - 58

Table 2. Operating conditions and gas output figures of farmscale digestor

Volatile solids reduction %	20
BOD reduction %	65
COD reduction %	17
VFA reduction %	94
Hemicellulose reduction %	35-40
Cellulose reduction %	35-40
Thick slurry → free flowing liquid	

Table 3. Effects of digestion on the cattle slurry

	A	B	C	D
Slurry fraction	Total	Total	1 mm	0.3 mm
Total Solids (TS) %	9.2	7.6	5.7	5.7
Volatile solids in TS %	75	76	66	68
Cellulose in TS %	16.0	8.3	3.3	1.6
Hemicellulose in TS %	15.0	14.6	7.1	4.8
Lignin in TS %	3.7	2.8	3.4	2.5
Gas production 1/1 digestor volume in 66 days	20.9	19.6	12.3	12.4
Percent of gas produced in 21 days	62	59	74	77

Table 4. Digestion of cattle slurry fractions

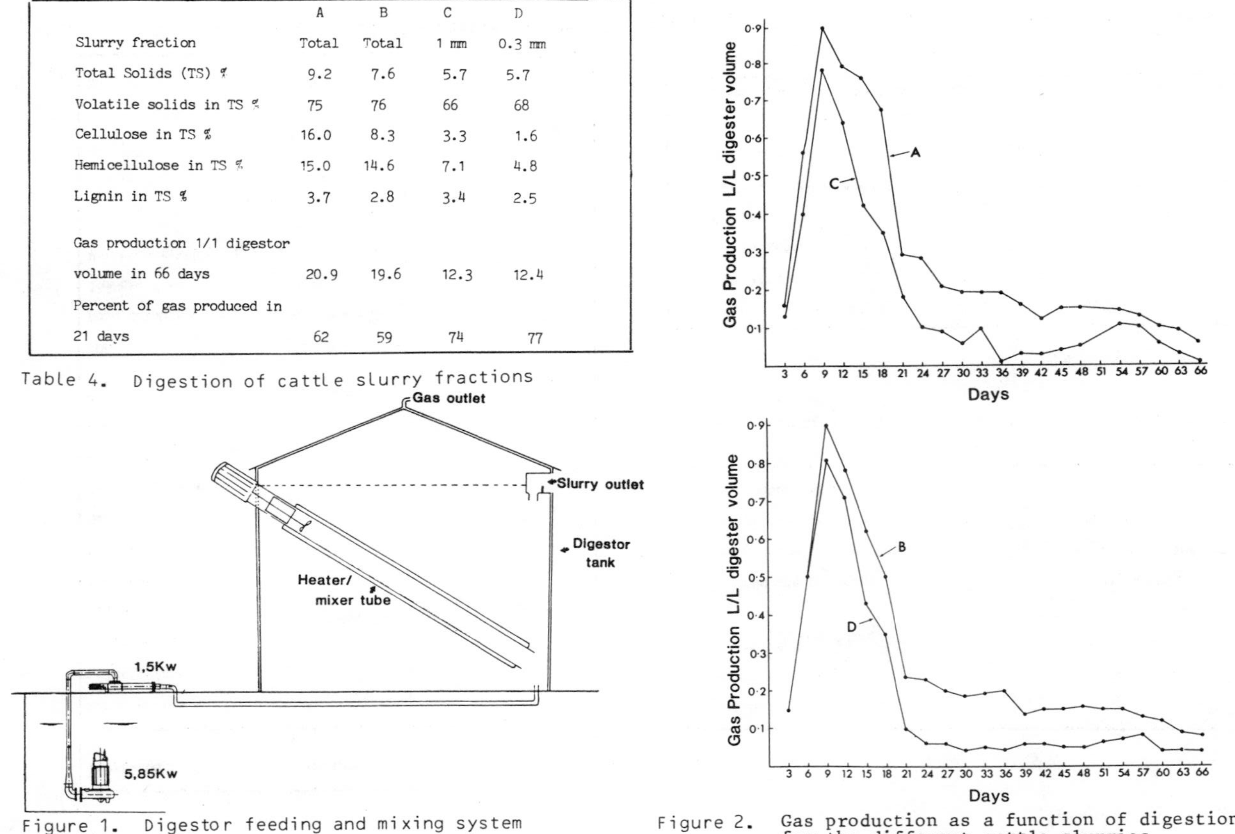

Figure 1. Digestor feeding and mixing system

Figure 2. Gas production as a function of digestion time for the different cattle slurries

A Cattle slurry containing 9.2 percent dry matter obtained from cattle fed grass silage and 2 kg concentrates per day.

B Cattle slurry from a similar source containing 7.6 percent dry matter.

C Cattle slurry as in B which had passed through a 1 mm seive and contained 5.7 percent dry matter.

D Cattle slurry as in B which had passed through a 0.3 mm seive and contained 5.7 percent dry matter.

The rate of biogas production as a function of time for the diffferent slurries is presented in figure 2 and the results of digestion are summarized in table 4. Gas production increased with increasing dry matter content of the slurry. The finer fractions of cattle slurry produced a higher proportion of gas in the first 21 days of digestion and require a shorter retention time. The results suggest that the minimum retention time for cattle slurry is in the 18 to 21 day range and that a longer retention time is required for cattle slurry of high dry matter content containing more cellulosic material. Only 20 to 30 percent of volatile solids in cattle slurry were destroyed during a 66 day digestion. Aproximately 35 percent of hemicellulose and cellulose and all the fatty acids of the slurry were destroyed during digestion.

Field Experiment

The availability of N in pig slurry is approximately twice that of cattle slurry. Anaerobically digested cattle slurry has a physical consistency similar to that of pig slurry in that the slime and much of the fibrous material has been destroyed resulting in a low viscosity slurry which is unlikely to coat grassland in the manner of raw cattle slurry.

In order to determine any beneficial effect on the N availability of cattle slurry which may accrue from anaerobic digestion, a field experiment was established to compare the N availability to grassland of anaerobically digested cattle slurry, similar undigested cattle slurry and fertiliser nitrogen. Three rates of each slurry type and three rates of fertiliser nitrogen were compared in a randomised block experimental design in which there were five replications of each treatment. Slurry and fertiliser treatments were applied on April 5. The rates of application were as follows:

	Slurry volume m^3/ha	Cattle slurry N kg/ha	Digested slurry N kg/ha	Fertiliser N kg/ha
Control	0	0	0	0
Rate 1	27	111	95	50
Rate 2	45	188	160	100
Rate 3	54	228	200	150

Mean yield of dry matter in tonnes/ha obtained in the first silage cut taken on May 31 were as follows:

Rate	Undigested cattle slurry	Digested cattle slurry	Fertiliser
0	4.169	4.169	4.169
1	4.777	4.666	5.358
2	3.986	4.352	5.350
3	3.701	4.675	5.235

The response to increasing amounts of fertiliser nitrogen was low presumably because of high soil nitrogen levels. A marked depression in yield was obtained from applications of undigested cattle slurry in excess of 27m^3/ha applied under wet weather conditions. No depression in yield was obtained from high rates of application of digested slurry. Grass samples taken from undigested slurry treated plots contained on average more than 5 times as much slurry contamination as samples taken from the digested slurry treated plots. Digested slurry did not require mixing, was more easily pumped and did not coat grassland in the manner of undigested slurry.

ACKNOWLEDGEMENTS

I wish to acknowledge the following:

Dr. E. O'Riordan for carrying out the field experiment.
Mr. T. Roche for expert technical assitance throughout the project.
Dr. D. McGrath for analysis of cellulose, hemicellulose, lignin, starch soluble sugars and fatty acids.
Mr. E. Byrne and his staff for analysis of P, N, BOD and COD.

REFERENCES

1. BYRNE, E. (1979). Chemical analyses of agricultural materials. Published by An Foras Taluntais, Dublin.
2. HASHIMATO, A. G., VAREL, V. H., and CHEN, Y. R. (1981). Ultimate methane yield from beef cattle manure. Effect of temperature, ration constituents, antibiotics and manure age. Agricultural Wastes, 3: 241-256.
3. KIELY, P. V. (1980). Time and rate of application of animal manures. In Effluents from Livestock Published by Applied Science Publishers Ltd., Barking, Essex, England.
4. STEWART, D. J., (1980). Energy crops to methane. In Anaerobic Digestion Published by Applied Science Publishers Ltd., London.
5. TUNNEY, H. (1982). Animal waste utilisation in energy management and agriculture. Proc. 1st. Int. Summer School in Agriculture. RDS/Kellogg Fund. Dublin. pp 351-66.
6. VAN SOEST, P.J., and WINE, R. H. (1967). Use of detergent in the analysis of fibrous feed. IV Determination of plant cell wall constituents. J. Assoc. Off. Anal. Chem. 50(1): 50-55.

Summary of the discussion of Session II :

PART 1 - ANAEROBIC DIGESTION

Chairman : G. LETTINGA

Rapporteur : E. COLLERAN

The discussion following the first seven papers of Session 11: Anaerobic Digestion was wide-ranging and comprehensive, encompassing both research and full-scale aspects of the various processes described by the authors. For convenience, the report of the various questions and comments covered by the discussion follows the order of formal presentation of the papers.

The advisability of phase separation for digestion of solid wastes was raised by Rijkens of the IBVL at Wageningen. Verstraete pointed out that a considerable body of literature supports the view that hydrolysis/ liquefaction is the rate-limiting step in the digestion of particulate organic matter. From a straight kinetic point of view, therefore, phase separation is not desirable and would be difficult to achieve for cellulosic substrate digestion, although it may be beneficial in the case of readily digestible particulate wastes such as sugar beet residues or cannery wastes. In response to questions from Colleran (Ireland) on the operation of the dry solids digester under study at Gent University, Verstraete stated that the development of an adequate innoculum was essential, though not problematical, and that nutrient addition (N or P) may be necessary, in certain cases, in order to maximise biogas production. Dewatering of the solid residue after digestion was accepted by Verstraete as being a major problem area in solid substrate digestion. In his opinion, the utilization of a dry solids fermentation system was preferable since the digested residue was produced at 30% TS, thereby reducing the final dewatering volume with attendant reduction in costs. A combination of various mechanical and drying procedures, using the waste heat output of an electricity generator envisaged for biogas utilization, is under investigation in the Gent laboratory. Rijkens commented that, in his experience, the solid residues obtained from his two-phase system are of a coarse nature and dewater or drip-dry readily, residual water being subsequently removed during an aerobic after-composting step. Addition of water, as practiced in two-phase systems, has certain advantages in that the likelihood of souring is minimized and the concentration of potentially inhibitory materials, such as NH_4+, can be reduced below toxic levels. In addition, heavy metal contaminants may transfer into the recirculation liquid phase from which they can be removed, thereby reducing the degree of contamination of the final solid residue. Verstraete cautioned against the addition of excessive volumes of water and suggested that dilution of particulate substrates to a 1/1 or 1/2 basis should be regarded as the maximum acceptable.

The rate-limiting nature of the cellulose liquefaction step in the CSTR digestion studies of domestic refuse by Nyns and Pauss at Louvain-la-Neuve was also commented on by Verrier of INRA who queried whether thermophilic operation increased the rate of liquefaction. Although not so far investigated at Louvain-la-Neuve, Pauss agreed that thermophilic operation could increase both the rate and the extent of MSW liquefaction but cautioned that the improvement gained would have to be sufficient to justify the increased heating costs associated with thermophilic systems. In addition, Pauss commented that it was essential to investigate whether the degree of cellulose and hemicellulose hydrolysis obtained in the batch studies was reproducible under continuous operation.

In reply to questions posed by Verrier of INRA and Breure of the University of Amsterdam, the basis for the changeover from acids to solvent production by Clostridium acetobutylicum was discussed. The pH of the culture medium is of critical importance since the inhibitory effect of the acids, notably butyric, depends on the concentration of the unionised form. The pH chosen for the co-culture experiments involving growth of Cl. cellulolyticum and Cl. acetobutylicum on cellulose was pH 5.8. A lower pH would be preferable in order to promote solventogenesis but the limiting factor was the pH tolerance of the cellulolytic species. Ongoing experiments at the University of Nancy are utilising other strains at pH values between 4 and 5. The switch from acid production to solventogenesis was stated by Gay to be a highly complex process which has been under study in his laboratory for the past 16 years. The carbon flux is undoubtedly an important factor since it was clearly demonstrated that a high flux of carbonaceous material promotes butanol formation whereas a low flux results mainly in acids production. From an energetic point of view (i.e., ATP production) acid formation is the preferred route for Cl. acetobutylicum. The concentration of unionised acids, in turn, may then trigger the changeover to solventogenesis. Gay suggested that considerable research is still required in order to understand the factors involved in the pattern of enzyme induction and repression associated with the changeover.

With respect to the co-culture system, the extent of sugar utilisation by Cl. cellulolyticum was not determined although the data obtained indicated that the metabolic end-products of Cl. cellulolyticum did not accumulate further on addition of Cl. acetobutylicum and that the acetic acid level actually decreased. The maintenance of Cl. cellulolyticum after introduction to a mixed culture such as is present in operational digesters was not investigated at the University of Nancy. In response to a query from Breure, Gay suggested that it was unlikely that the organism could compete and survive in such a mixed culture system. Arising from the paper presented by Proust on the survival of introduced methanogenic strains in batch digestion systems, Campagna of Italy queried the inorganic nutrient concentration and methanogenic population levels in the test system. The presence of straw particles in the batch reactor prevented accurate estimation of the methanogen population by epifluorescence techniques. The nutrient concentrations used were cited by Proust as KH_2PO_4 1.5 g.l^{-1}; K_2HPO_4 2 g.l^{-1}; NH_4Cl 0.5 g.l^{-1}; $(NH_4)_2SO_4$ 1.3 g.l^{-1}; $MgCl_2.6H_2O$ 1 g.l^{-1}; NaCL 2.25 g.l^{-1}; $NaHCO_3$ 0.25 g.l^{-1}; $CaCl_2$ 0.15 g.l^{-1} and $FeSO_4$ 1.25 g.l^{-1}. The time period of 30 days for the experiments were, in his opinion, sufficient to conclude that the introduced methanogenic strain disappeared quickly from the fermentation system and did not exert any appreciable effect on the biomethanation process. Stirring of the fermentation vessel did not appear to significantly affect the results obtained. Hobson described the disappearance, within a few hours, of a highly active cellulolytic rumen species after introduction into a digester treating pig slurry. It appears from a variety of studies in the Rowett Institute that it is difficult, if not impossible, to add extra bacteria to mixed, stable cultures such as exist within digesters or in the rumen.

The energetic balance of the process described by van der Vlugt of the Netherlands for the digestion of the fine organic fraction of MSW; the economic value of the digested residue and the feasibility of co-digestion of solid wastes, such as municipal wastewater sludges, was queried by Gerletti of Italy. Van der Vlugt replied that the scope of the project did not allow, as yet, investigation of mixed substrate digestion although the possibility of such mixtures had already been commented on in the papers presented by Verstraete and Pauss. It was proposed to include the energetic balance in the final technical/economic evaluation at pilot scale. Zero economic value was attributed to the digested residue because of its inherently low nutritional value. Van Landuyt of Belgium raised the problem of subsequent treatment of the liquor eliminated from the digested solids during dewatering and queried the COD concentration of this liquor and the anticipated cost of treatment. In reply, van der Vlugt stated that the liquid removed during dewatering had a COD concentration of 1,600 $mg.l^{-1}$ and a Kjeldahl nitrogen of c. 130 $mg.l^{-1}$. No treatment costs were included since the process involves complete recycling of dewatering liquor for slurrying the incoming solids. In practice, the volume of water removed during solids drying provides 90% of the volume required for dilution of influent solids. This could, in theory, give rise to problems during long-term operation, due to salts or toxicant accumulation and this will be investigated during pilot-scale trials.

The difficulties associated with solids settlement, both in under-floor slurry holding tanks and in the digester itself, were raised by Ferrera of Portugal in relation to the paper presented by Kiely of Ireland on the operation of a mixed digester treating slurry from cattle fed high roughage diets. Although solid settlement in the digester was noted, the problem was overcome by the removal of 10 m^3 of bottom solids every two months via a specially-fitted removal outlet located at the base of the reactor. Solids settlement in holding tanks was more problematical and the heavy settled fraction was eventually removed by the addition, in season, of silage effluent to the holding tanks, followed by vigorous mixing by tractor-operated pumps and eventual disposal by land-spreading.

THE PROSPECTS OF ANAEROBIC WASTE WATER TREATMENT

Gatze Lettinga
Agricultural University
Department of Water Pollution Control
De Dreijen 12
Wageningen

INTRODUCTION

In view of its significant advantages over conventional aerobic methods anaerobic waste water treatment is in many ways an ideal alternative for waste water treatment. Limitations and drawbacks of anaerobic treatment are presented in Table 1.

Table 1. Benefits and drawbacks of anaerobic waste water treatment over conventional aerobic methods

Benefits

1. low production of stabilized excess sludge
2. low nutrient requirements
3. no energy requirements for aeration
4. production of methane
5. frequently high organic and hydraulic space loads can be applied
6. adapted anaerobic sludge can be preserved unfed for long periods of time without dramatic detoriation.
7. valuable compounds like ammonia are conserved, which in specific cases might represent an important benefit (e.g. in the case the effluent can be applied for irrigation and fertilization)

Drawbacks

1. anaerobic bacteria (particularly methanogens) are susceptible for a large number of compounds
2. the first start-up of the process is slow
3. anaerobic treatment requires generally aerobic polishing
4. there still exists little practical experience with most of the high rate treatment systems

Very significant progress has been made in the last two decades both in understanding the fundamentals of the process as well as in the field of engineering. Despite that only approximately 40 high rate anaerobic treatment processes (viz. mainly UASB-plants) have been installed in recent years. It still remains rather difficult to convince potential users and established consultancies of the great economic value of the anaerobic treatment alternative. Nevertheless the definite breakthrough of anaerobic waste water is at hand, because the feasibility of the process for various types of liquid wastes has been clearly demonstrated.

BASIC PRINCIPLES OF THE MODERN HIGH RATE PROCESSES

Both the conventional -medium rate- anaerobic treatment systems and the modern -high rate- anaerobic systems are based on the principle of a high viable biomass retention.

Figure 1 illustrates a number of available anaerobic treatment systems.

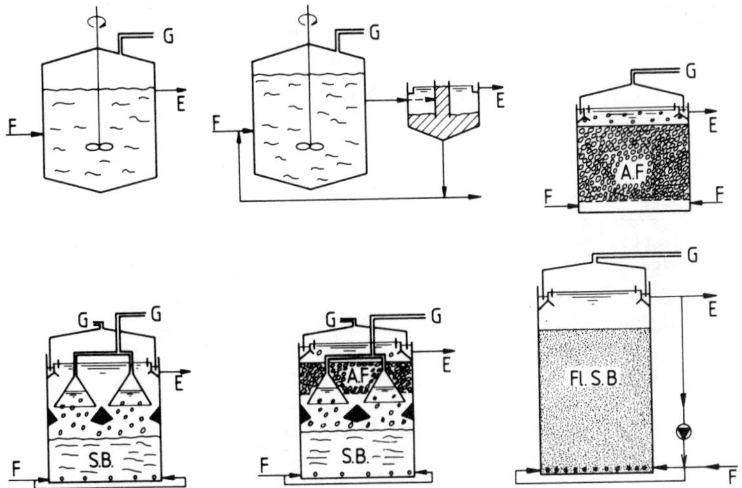

Figure 1.: Various anaerobic treatment systems.
Conventional sludge digester, the Contact process, the Anaerobic Filter (A.F.), the UASB-process, a combined UASB-A.F.-process and a Fluidized bed system.

In the conventional processes the required biomass retention is achieved by combining the -continuously or intermittently mixed reactor- with an internal or an external sludge separation and return sytem (e.g. the Anaerobic Contact process). Despite the large difficulties frequently experienced in achieving the required separation and thickening of the sludge, various modes of the Contact concept are beneficially applied in practice, although this generally merely is the case at temperatures exceeding 25°C.

All the modern 'high rate' processes are based on some mode of bacterial sludge immobilization (1), viz.

1. the formation of highly settleable sludge aggregates, combined with early gas separation and sludge settling
2. bacterial attachment to high density particulate carrier materials or to immobile support structures supplied to the reactor
3. entrapment of sludge aggregates between packing material supplied to the reactor.

MERITS AND LIMITATIONS OF THE MODERN HIGH RATE PROCESSES

The loading potentials

Each of the high rate processes has its own specific merits and limitations, depending on the extent and the ease in which the primary conditions underlying the process can be met and on the operational conditions that should be fulfilled for their proper performance.

As a matter of fact considerable differences exist between the various 'high rate' processes with respect to their maximum hydraulic and organic loading potentials. In order to make this clear the main factors governing the loading potentials of a system should be reviewed.

1. The total amount of active biomass which can be retained under high loading conditions.

Positive and negative factors determining the retention of active biomass are presented in Table 2.

Table 2. Factors determining the retention of active biomass

Factor	effect
redispersion of sludge	diminishes the sludge retention (at high loads in UASB, perhaps in Upflow A.F.systems)
desintegration of sludge aggregates	diminishes the sludge retention (perhaps in granular UASB-systems)
de-attachment of biofilms	diminishes the sludge retention (at incomplete acidification and at HRT > sludge generation rate in Fluid bed systems)
space occupied by support/carrier materials	diminishes the overall sludge retention (particularly important in AFFEB, AF, AFF)
surface area of carrier/packing material	larger surface areas are beneficial (particularly important for Fluid Bed, AFFEB and AFF-processes)
film thickness	thicker films are beneficial with respect to the sludge retention
bed expansion pursued	more bed expansion results in lower sludge hold-up (particularly important for Fluid bed and AFFEB-systems)

In granular sludge UASB-reactors (2, 3, 4, 5) the sludge retention at very high organic loading rates is mainly limited by the redispersion of the sludge granules due to the high turbulence prevailing in the digester compartment of the reactor and the increasing tendency of the granules for flotation. Desintegration of the granules at high loading rates due to shear presumably is insignificant.

In Upflow Anaerobic Filters (6, 7) the main limiting factors with respect to the overall sludge retention frequently is the void space left by the packing material, although redispersion of the sludge due to the high turbulence and flotation of the sludge at high loading rates may also represent an important limiting factor. In downflow Fixed Film reactors (8, 9) the surface area available for bacterial attachment is the limiting factor, although also here the void space left by the packing material is important.

Apart from these factors, the film thickness is of eminent importance with respect to the overall sludge retention in expanded/fluidized bed reactors (10, 11), together with the sludge bed expansion pursued. As the surface area available for attachment is much larger in the modern fluidized bed systems as compared to the AFFEB-process (12, 13) the fluidized bed systems compete favourably with AFFEB-systems with respect to the amount of sludge which can be retained.

However, in the latter system only 10-20% sludge bed expansion is pursued, whereas this is up to 50% in the Fluidized bed system.

2. The amount of contact achieved between retained sludge and the incoming waste water.

In fluidized bed and downflow AFF reactors a sufficient contact is guaranteed. In fact this is also the case for UASB- and AAFEB-systems, although the incoming waste should be distributed as evenly as possible over the bottom of the reactor, particularly when lower organic loads are applied, viz. in the case of low strength cold waste waters. Upflow anaerobic filter systems suffer from the problem of clogging, particularly in treating wastes with a higher fraction of suspended solids.

3. Kinetic factors, such as film and/or particle diffusion limitation.

So far little relevant information has become available about the importance of film diffusion limitation, although in experiments with granular sludge evidence has been obtained that the particle size has little if any effect at higher substrate concentrations (4).

However, a significant drop in the substrate utilization rate may occur when the substrate level in the reactor is maintained at a relatively low value, as frequently will be the case for systems requiring a high effluent recycle factor, e.g. fluidized bed systems, where supercial velocities of 7-12 m/hr are maintained.

In view of the above considerations, the maximum achievable organic loading rate will probably diminish in the sequence:

granular sludge UASB >
fluidized bed >>
fixed film expanded bed (AAFEB) >
flocculant sludge UASB, upflow AF, downflow AFF

With respect to the maximum achievable hydraulic loading rate under conditions that the organic loading rate is not limitative, the attached film systems presumably are in favour of the 'non-attached' systems, although it is likely that the organic loading rate for most of these processes becomes much sooner restrictive than the hydraulic loading rate. Nevertheless, for conditions in which the organic loading rate is not restrictive, consequently generally with very low strength (cold) wastes, the maximum achievable hydraulic loading presumably will diminish in the sequence:

downflow AFF >
fluidized bed (depending on the specific density of the inert carrier material) >
AAFEB >
granular sludge UASB >>
upflow AF >
flocculant sludge UASB

So far comparable experiments between the various high rate processes on a relevant scale have not been performed. It should be recognized that in practice the primary conditions underlying the various processes can not always sufficiently be met. Moreover, apart from the maximum achievable loading rates, there are a number of other selection criteria, such as

1. the period of time required for starting up the process
2. stability of the process with respect to fluctuations in loading rate, in the composition and the strength of the waste, in the temperature etc.
3. capability of the process for treating partially soluble wastes.
4. capital and running costs, e.g. energy and chemical requirements, labour need for phase separation, control equipment required
5. area of land required
6. requirement of skilled man power
7. the need of phase separation

The start-up of high rate reactors

As far as the first start-up of the process concerned for most of the processes a period of time exceeding 3-4 weeks will be required to achieve the design loading capacity. It will be evident that, when moderate goals are pursued with respect to the ultimate loads of the process, the period of time required for the first start-up will be significantly shorter than in the case very high organic space loads ultimately should be accommodated. In comparing the various processes presumably the rate of start-up will be slowest for UASB-systems, unless a sufficient amount of viable granular seed sludge from an existing full scale plant is available. In view of the rapidly increasing number of full scale plants there is no doubt that indeed sufficient amounts of high quality seed sludge will become available.

However, if a granular seed sludge is not available, the first start-up of a UASB-plant may be fairly time consuming, mainly because of the slow rate at which sludge granulation proceeds with various types of liquid wastes. The mechanism of sludge granulation is still not fully understood, although considerable progress already has been made on this topic (14, 15, 16). In summarizing the results obtained so far the following statements can be made:

1. In using digested sewage sludge as seed and VFA-solutions as substrate, the best results are obtained with sludges characterized by a relatively low residual methanogenic activity, viz. approximately less than 0.6 kg CH_4-COD.m^{-3}.day^{-1}, as well as a good settleability (SVI sludge < appr. 50 ml.g DSS^{-1}) of the residual sludge remaining after the wash-out of the finely dispersed fraction of the sludge (17). The presence of inert carrier materials in the sludge has a distinct beneficial effect. The start-up of the process is also considerably enhanced by supplying a small amount of well adapted granular seed sludge to the digested sewage sludge (18).
2. Using VFA-substrates a satisfactory granulation of the sludge can be achieved, although the process of granulation proceeds rather slowly and a filamentous type of granule developes (unless a small amount (1-2%) of rod-shaped granules is supplied to the seed sludge!).
   The granulation process proceeds easier (faster) on slightly acidified carbohydrate substrates, such as sucrose solutions, potato processing wastes and liquid sugar wastes.
   In all cases the waste water should contain all essential growth ingredients in sufficient amounts, whereas inhibitory concentrations of toxic compounds should be absent.
3. With respect to the mode of start-up the following factors are important:
   a. The initial sludge load should be in the range 0.05-0.10 kg COD.kg VSS^{-1}.day^{-1} and the space load of the reactor should not be increased unless approximately 80% of the biodegradable COD is removed.
   b. Voluminous ingredients present in the seed sludge and/or developing during the first phases of start-up (particularly dispersed filamentous bacterial matter) should be allowed to rinse out from the reactor.

c. Temperatures near 40 °C are beneficial (18).
d. During the initial phases of start-up the influent COD should be maintained below approximately 3000 mg/l. In the case effluent recycle has to be applied for this purpose, the recycled effluent should be stripped of dispersed matter, because finely dispersed matter -particularly filamentous bacterial matter- is detrimental with respect to sludge granulation.

Results of a comparative start-up experiment between a 10.7 liter UASB-reactor (h = 1.1 m) and a fibrous plastic medium packed 10.7 liter A.F. (porosity of the medium > 0.90; the plastic medium cut in pieces of approximately 10 x 20 x 20 mm, which occupied approximately half of the reactor volume in two layers of 30 cm, leaving an unpacked volume of approximately 3.5 liter), show a slightly faster start-up of the A.F. during the very initial stages of the process. Ultimately considerably higher space loads could be accommodated in the UASB-system as shown in the Figures 2 en 3. The experiment was conducted at 30 °C with a VFA-mixture consisting of 1250 mg C_2/l and 1000 mg C_4/l. A distinct granulation was observed in the UASB-reactor after 5 weeks. To our experience starting from that instant the space load in fact can be increased considerably faster than was practiced in the concerning experiment.

It also should be mentioned that only approximately 40 % of the UASB-reactor volume was occupied with sludge at day 114, indicating that ultimately much higher loading rates can be accommodated in the UASB-system. This is confirmed by other (30 liter) UASB-experiments with VFA-substrates and VFA-cultivated granular sludge, in which space loads were applied of 60 $kg.m^{-3}.day^{-1}$ at 30 °C, despite of the fact that the sludge bed also here only occupied approximately 30 % of the reactor volume (14).

Results of other start-up experiments with A.F.-systems are summarized in Table 3.

Table 3.: Period of time required for the first start-up of anaerobic filter processes.
(Data collected from the literature).

A.F. Process investigated	Wastewater used in the investigation	Initial space load ($kg\ COD.m^{-3}.day^{-1}$)	Inoculum used in the start-up	Results obtained (days for 90 % red.)	Reference
28.5 L quarz-stone 28.5 L quarz-stone	3.0 g/l VFA-mixture 3.0 g/l VFA-mixture	0.42 (30 °C) 0.84 (30 °C)	2 x 30 g VSS Dig.Sew.Sl. light seed Dig.Sew.Sl.	40 days 180 days	Young & McCarty
14.2 L quarz gravel	(glucose) pharmaceutical, 1-16 g COD/l	0.37 (30 °C) 1.76 (30 °C)	30 g VSS Dig.Sew.Sl. 30 g VSS Dig.Sew.Sl.	38 - 40 days 100 days	Jennett & Dennis
33.4 limestone	Brewery press liquor waste (6 g/l)	0.8	supernatant An.Digester	25 days	Lovan & Foree
24 L coke packed	Potato starch waste 4-6 g/l	1.0 (30 °C) 0.5 (30 °C)	14 g VSS slightly adapted Dig.Sew.Sl.	36 days 24 days	Lettinga & Janssen
6.7 L plastic porous media packed	Soluble starch COD: 8.7 g/l	0.7 first week 1.4 second week 2.5 week 3-5 5 week 5-8	63 g VSS Dig.Sew.Sl.	first week second week week 4-5 week 6-8	Frostell

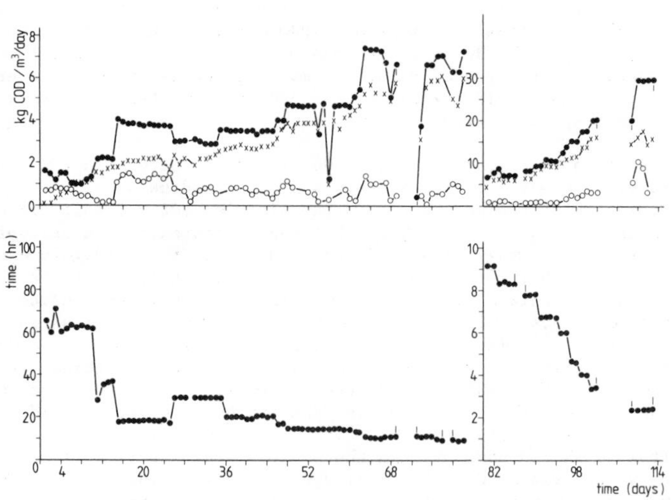

Figure 2.: Results obtained in a 10.7 liter UASB-experiment (height 110 cm, diamter 12 cm) using a VFA-solution (C_2: 1250 mg/l; C4: 1000 mg/l) as substrate and 4.2 liter digested sewage sludge (TSS: 4.02 %; VSS: 26.0 g/l) as seed. (Temperature 30 °C). (Upper figure: ... COD-load; xxx CH_4-COD production in kg.m^{-3} . day^{-1}; ooo VFA-COD load in the effluent expressed as kg VFA. m^{-3} .day ; upper figure: hydraulic detention time.

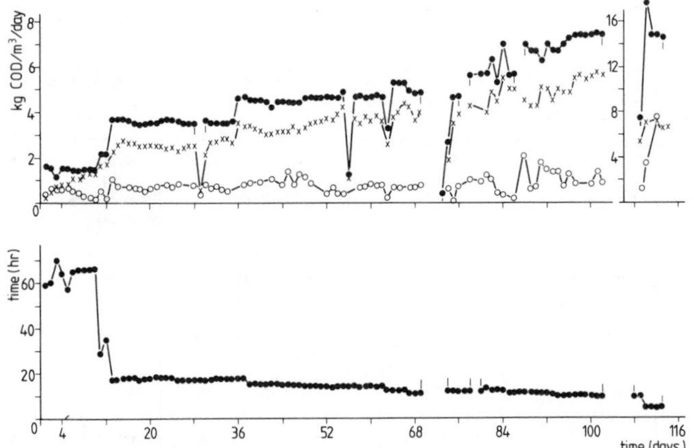

Figure 3.: Results obtained in a 10.7 liter A.F.-experiment (height 110 cm; diameter 12 cm; reactor equipped with GSS-device) using the same feed-solution (symbols, see Figure 2.).

As far as the <u>fluid bed</u> process concerned the period of time required for the first start-up is approximately 7 weeks to achieve a space loading rate of 30 kg COD.m^{-3}.day^{-1} in using an acidified yeast-waste water and at a temperature of 37°C. This certainly is significantly faster than the first start-up of a UASB-reactor at 30°C. However as mentioned before the rate of start-up, including the granulation process, is significantly faster at 38°C, viz. at least a factor twice as high.

Unlike most of the other high rate systems the UASB- and the upflow AF-system frequently produce a relatively large quantity of highly active sludge, which really represents a very important benefit of these systems. Moreover a secondary start-up of these two reactor-concepts after a prolonged period of standstill (e.g. 9-12 months) generally will take little if any time, whereas this is still uncertain for the other high rate systems.

<u>The stability of the process</u>

The stability of the process with respect to suboptimal conditions, fluctuations in the composition and the strength of the waste water, the presence of inhibitory compounds depends strongly on the degree of underloading at which the process will be applied under practical conditions. Processes operated near their maximal loading potentials obviously will be fairly susceptible. So far most of the UASB-plant installed are designed far below their ultimate maximal loading capacity, and consequently are capable to accomodate satisfactorily stress situations. This presumably holds less for the fluidized bed system recently installed at a yeast-factory in Delft, the Netherlands, because the performance of this system depends strongly on the extent of acidification of the organic pollutants achieved in the fluid bed acidification step, whereas it presumably has been designed close to its maximum loading potentials. In order to prevent any severe upset, this system requires careful monitoring and control, and for that reason it presumably is relatively expensive, as it requires skilled escort and a relatively high investment for monitoring and control equipment.

<u>The feasibility of high rate processes for removing suspended solids</u>

As far as the feasibility of the various systems concerned in removing and stabilising dispersed organic matter from the waste water, considerable differences exist between the various systems. So the efficiency of fluidized bed, downflow AFF, AAFEB- and high rate granular sludge UASB-reactors will be low to insignificant in this respect, whereas an upflow AF-system under certain conditions is efficient in this respect, viz. in the absence of clogging, which means that it should not be applied to wastes containing a high concentration of dispersed matter.

Although the efficiency of fluidized bed and the high rate granular sludge UASB is low in removing SS, the performance of these processes is not seriously affected by the presence of SS, although there certainly is some competition for bacterial attachment between the supplied carrier material and the dispersed matter. Medium loaded granular sludge UASB- and flocculant sludge UASB-reactors -which principally is a medium rate system- are fairly efficient in removing and stabilizing dispersed organic matter, as will be the case for medium rate AAFEB-reactors. The entrapment of dispersed inert matter in the sludge bed may result in a certain drop of the specific activity of the sludge. Such a drop in specific activity can easily be prevented for granular sludge UASB- and AAFEB-reactors, because the entrapped -generally flocculant inert sludge- can be selectively rinsed out from the reactor.

The feature of fluidized bed -and to some extent of granular sludge UASB and AAFEB-reactors- that it can handle high superficial velocities represents an important benefit for those specific cases, where little land is available for installing the plant.

The need of phase separation

In treating complex soluble wastes, under specific conditions phase separation might be an attractive option, e.g.

1. for the elimination of toxic/inhibitory compounds,
2. for the removal of sulphate (sulphite) or nitrate, if present in the waste,
3. for preventing a detoriation of granular seed sludge.

Whether or not phase separation is attractive, it should be taken into account that the acidification step generally proceeds very fast, e.g. in some cases it may be already completed in feed supply lines or in equalization tanks.

As mentioned before a sufficient pre-acidification is required in applying fluid bed anaerobic reactors, unless high loading rates are not pursued.

In treating largely insoluble wastes phase separation may be sometimes also an attractive solution, although it is still uncertain whether or not it is more attractive than a simple one step process. Latter certainly applies for those cases where the separation of dispersed solids after the liquefaction step -which generally is the rate limiting step (particular at lower temperature)- cannot be accomplished in a simple and inexpensive manner.

However, in specific cases, such as in the digestion of rigid agricultural residues like stalks of tomatoes, hay etc., phase separation combined with effluent recycle from the methanogenic step certainly may represent an attractive solution.

THE PRACTICAL IMPORTANCE OF HIGH RATE PROCESSES

Although the loading potentials of the modern high rate processes exceed considerably those of conventional systems, certainly a place will be left for the latter processes, particularly in the temperature range 30-40°C. In this connection it should be understood that the applicability of high loading rates is only one of the principal benefits of anaerobic treatment over aerobic activated sludge processes.

Nevertheless the high loading potentials of the modern high rate processes remain extremely interesting for a number of reasons:

1. the anaerobic treatment process becomes available for treatment of low strength waste waters,
2. unless not operated at the limits of its loading potentials the anaerobic treatment process can accommodate satisfactorily stress situations,
3. the anaerobic treatment process will be applicable at low temperatures.

Consequently, at the present state of process technology it would be a serious underestimation to restrict the use of anaerobic treatment to the optimal mesophilic temperature range or to medium and high strength wastes. With the modern high rate systems, considering also possible improvements in these systems, a tremendous field of application lies open, both for technologically developed countries as well as for developing countries.

Experiments performed with raw domestic sewage in a granular sludge UASB-reactor reveal the feasibility of these systems for moderate climatic areas. Under dry weather conditions 65-85% COD-reduction can be achieved at temperatures in the range 8-20°C at 12 hrs liquid detention time (19).

Recent experiments performed with settled domestic sewage even indicate the feasibility of granular sludge bed systems at liquid detention times of 30-60 min at temperatures in the range 15-22°C. The COD-removal-efficiency achieved was in the range 50-70%.

Flocculant sludge UASB-reactors look particularly attractive under tropical conditions. Preliminary experiments obtained in a 50 m^3 demonstration plant in Cali, Colombia, indicate 75-85% COD-reduction at detention times of approximately 6 hrs (waste temperature: 26-28°C) (20). Flocculant sludge bed systems perhaps even might be applicable at temperatures as low as 10-12°C, although the liquid detention time presumably should exceed 12 hrs in that case.

In various cases, also thermophilic high rate systems might become an extremely attractive alternative. Recent experimental results with thermophilic UASB (21, 22) and AAFEB-experiments (10) clearly indicate the enormous potentials of thermophilic systems. Results with UASB-reactors in using an unacidified glucose solution (10-25 kg COD/m^3) reveal (21) that in using fresh cow manure as seed material a space load of 33 kg COD.m^{-3}.day^{-1} can be well accommodated within a period of 7 weeks. The first granules appeared approximately four weeks after the start of the experiment. Ultimately space loads of 60 kg COD.m^{-3}.day^{-1} were well accommodated (day 87). On the other hand it was also observed that granulation didn't proceed easily on a pure VFA-substrate. However, in using a granular thermophilic seed sludge, the sludge doesn't detoriate on a VFA-substrate. The same applies for treating vinasse solutions.

Thermophilic AAFEB-systems using soluble (sucrose) substrates (1.5 - 16 kg COD.m^{-3}) require a period of start-up of 2-3 months, and then a 70% COD-removal efficiency can be achieved at 30 kg.$m^{-3}_{expanded\ bed}\cdot day^{-1}$.

CONCLUSIONS

In conclusion it may be stated that anaerobic treatment systems increasingly will replace conventional aerobic systems as the secondary biological treatment step, leaving the aerobic process more and more as a polishing method for the secondary effluent. Ultimately this will lead to a considerable reduction in the costs of waste water treatment, whereas it makes waste water treatment economically and practically feasible for third world countries.

As explained in the paper, each of the high rate anaerobic treatment systems has its own typical merits, limitations and potentials and it will therefore greatly depend on the local situation, the type of waste water to be treated, the specific experience available on the various systems, which of the systems will be selected. So far, most experience at full scale size has been obtained for the UASB-plant and the upflow A.F. process, and in most cases the results obtained were very satisfactory, although the UASB-process still suffers on incomplete understanding of the granulation process whereas the A.F.-system suffers on problems with clogging. As far as the granulation process concerned, there is good hope that the mechanism can be cleared up adequately in the near future. The same will be true with respect to the mechanism underlying the attachment of bacterial mass to mobile and stationary carrier materials. Once this situation has been attained, each of the processes can be properly designed and operated. Then it presumably will also turn out that the differences between systems such as the UASB, AAFEB and the fluid bed process are considerably smaller than they look at first sight at this stage of development.

LITERATURE

1. Lettinga, G., S.W. Hobma, L.W. Hulshoff Pol, W. de Zeeuw, P. de Jong, P. Grin, R. Roersma.
Design, operation and economy of anaerobic treatment.
Wat. Sci. Techn. 15, 177-195, 1983.

2. Lettinga, G., A.F.M. van Velsen, S.W. Hobma, W. de Zeeuw, A. Klapwijk.
Use of the Upflow Sludge Blanket (USB) reactor concept for biological waste water treatment.
Biotechn. and Bioeng. 22, 699-734, 1980.

3. Lettinga, G.
Feasibility of anaerobic digestion for the purification of industrial waste waters.
4th European Sewage and Refuge Symposium, EAS, Munich 1978.

4. Lettinga, G., L.W. Hulshoff Pol, S.W. Hobma, P. Grin, P. de Jong, R. Roersma, P. IJspeert.
The use of a floating settling granular sludge bed reactor in anaerobic treatment.
Preprints of the 'Anaerobic Waste Water Treatment' Symposium, Noordwijkerhout, Nov. 1983.

5. Lettinga, G., L.W. Hulshoff Pol, W. Wiegant, W. de Zeeuw, S.W. Hobma, P. Grin, R. Roersma, S. Sayed, A.F.M. van Velsen.
Upflow sludge blanket process.
Proceedings 3rd Int. Symp. on Anaerobic Digestion.
Boston, August 1983, 139-158.

6. Young, J.C.
The anaerobic Filter, past, present and future.
Proc. 3rd Int. Symposium Anaerobic Digestion, Boston, August 1983.

7. Young, J.C., P.L. McCarty.
The anaerobic filter for waste water treatment.
Jour. Water Poll. Control Fed. 41, R160-171, 1969.

8. Berg, L. van der, K.J. Kennedy.
Comparison of advanced anaerobic reactors.
Proc. 3rd Int. Symposium Anaerobic Digestion, Boston, August 1983.

9. Berg, L. van der, K.J. Kennedy.
Support Materials for stationary fixed film reactors for high rate methanogenic fermentations.
Biotechn. Letters, 3 (4), 165-170, 1980.

10. Schraa, G., W.J. Jewell.
Conversion of soluble organics with the thermophilic anaerobic attached film expanded bed process.
Preprints AWWT-symposium, Noordwijkerhout, Nov. 1983.

11. Heijnen, J.J.
Anaerobic Waste Water treatment.
Preprints AWWT-symposium, Noordwijkerhout, Nov. 1983.

12. Switzenbaum, M.S., W.J. Jewell.
Anaerobic attached film expanded bed reactor for the treatment of dilute organics.
51th Manual Water Poll. Control Fed. Conf., Anaheim, California (1978).

13. Jewell, J.W., M.S. Switzenbaum, J.W. Morris.
Sewage treatment with the anaerobic attached microbial film expanded bed process.
52nd Annual Water Poll. Control Fed. Conf., October 1979.

14. Hulshoff Pol, L.W., W.J. de Zeeuw, C.T.M. Velzeboer, G. Lettinga.
Granulation in UASB-reactors.
Water Sci. Techn. 15, 291-304, 1983.

15. Hulshoff Pol, L.W., H.A.A.M. Webers.
The effect of the addition of small amounts of granular sludge to the seed sludge on the start-up of UASB-reactors.
Preprints AWWT-symposium, Noordwijkerhout, Nov. 1983.

16. Hulshoff Pol, L.W., J. Dolfing, K. v. Straaten, W.J. de Zeeuw, G. Lettinga.
Pelletization of anaerobic sludge in UASB-reactors on sucrose containing waste water.
3rd Int. Symposium on Microbial Ecology, August 1983, East Lansing.

17. Zeeuw, W. de, G. Lettinga.
Start-up of UASB-reactors.
Preprints AWWT-symposium, Noordwijkerhout, Nov. 1983.

18. Hulshoff Pol, L.W., E. ten Brummeler, G. Lettinga.
Unpublished results.

19. Lettinga, G., P. Grin, R. Roersma.
Anaerobic treatment of raw domestic sewage at ambient temperatures using a granular bed UASB-reactor.
Biotechn. Bioeng. 24, 1701-1724, 1983.

20. Schellinkhout, A., G. Lettinga, A.F.M. van Velsen, G. Rodriguez.
Unpublished results.

21. Wiegant, W.M., J.A. Claassen, A.J.M.L. Borghans, G. Lettinga.
High rate thermophilic anaerobic digestion for the generation of methane from organic wastes.
Preprints AWWT-symposium, Noordwijkerhout, Nov. 1983.

22. Wiegant, W.M., G. Lettinga.
Solar Energy in the European Countries, Series E., vol. 5, 323-330, June 1983.

23. Jennett, J.C. N.D. Dennis.
Anaerobic Filter Treatment of pharmaceutical waste.
J. Water Poll. Control Fed, 47 (1), 104-121, 1975.

24. Lovan, L.R., E.G. Force.
The anaerobic Filter for the treatment of brewery press liquor waste.
Proc. 26th Int. Waste Conf., Purdue Univ., 1974.

25. Lettinga, G., P.G. Fohr, G.G.W. Janssen.
Toepassing van methaangisting voor de behandeling van geconcentreerd afvalwater.
H_2O 5, (22), 510-517, 1972.

26. Frostell, B.
Anaerobic treatment in a sludge bed system compared with a filter system.
J. Water Poll. Control Fed., 53, 217-222, 1981.

ENERGY RECOVERY FROM AGRO-INDUSTRIAL WASTES USING FIXED BED ANAEROBIC REACTORS

P. Sanna, M. Camilli and L. Degen

ASSORENI, ENI Group Research Association, 00015 Monterotondo (Roma), Italy.

Summary

Agroindustrial wastes, generally marked by high flow rates and soluble pollutants, call for the use of low-HRT and highly efficient reactors to ensure economy in their treatment.
Fixed-bed anaerobic reactors have been selected.
20-and 300-liters laboratory scale fixed bed anaerobic reactors have been utilized for several years to optimize the process and the start-up and re-start procedures, and to examinate both various filling materials and agroindustrial wastes.

Among sugar refinery wastes, the eluates coming from regenerating ion-exchange columns, utilised for purification of the sugar juices, immediately appeared especially interesting.

Based upon preliminary laboratory results, a pilot plant was built, early in 1981, at a sugar refinery at Argelato, near Bologna (Italy).
The main features were two reactors, in parallel, respectively of 10 and 50 m^3 operating volume.
With an hydraulic retention time of less than one day, a maximum gas production of 11.5 volumes / reactor operating volume / day was obtained together with a reduction of BOD_5 and volatile solids up to 80% and of COD up to 70%.
The gas contained from 70 to 75% methane.
On the basis of these results a full-scale industrial plant was built at the same sugar refinery in 1982.

1. Introduction

Effluents coming from industries which process farm products often contain rather large amounts of pollutants generally of the soluble type (1), (2), (3).

This brings up more than a few treatment problems, in that the elimination of such pollutants using conventional aerobic systems calls for construction of large, costly plants whose operation weighs heavily upon the company's economics.

Using anaerobic digestion to treat these kinds of wastes, it is possible to recover large part of the energy potentially remaining in them.

Effluents coming from the digester can be treated, finally, in an aerobic section, which has only the function of effluent finishing for obtaining clarified water at a lesser cost and with the least possible consumption of energy.

2. Fixed bed anaerobic reactors

Agro-industrial wastes are generally marked by high flow rates and soluble pollutants.

Therefore they call for the use of low-HRT and higly efficient reactors to ensure economy in their treatment.

For this purpose, fixed bed anaerobic reactors have been investigated.

These are reactor vessels filled with inert material which has the purpose of entrapping the bacteria inside the system, both through adhesion of such microorganisms to the surfaces of the filler and through simple mechanical effects, allowing the wastes under treatment to flow freely trough the reactor (4).

In this way any wash-out of the microbial flora is obstructed and it becomes possible to feed the reactor at flow rates corresponding to HRT values which are lower than duplication times of anaerobic microorganisms.

3. Laboratory research: materials and methods

20-and 300 liters fixed bed anaerobic reactors have been utilized for several years to optimize the process and to examinate both various filling materials (stones, plastics, ligno-cellulosic residues) and agro-industrial wastes (whey and effluents from sugar, alcohol and cannery industries).

Start-up procedures have been optimized.

One very important characteristic of agroindustries is that most of them are seasonally operated.

Therefore, in our laboratories we repeatedly verified that, if suitably re-started, fixed bed anaerobic reactors can return to full operation condition within a few days after they have been idle for many months.

Waste analyses were made in accordance to Standard Methods (5).

For measuring H_2S in the gas Draeger ampules were used.

The other components of the gas were determined using a gas chromatography method.

Gas flow was measured with a wet gas meter specially suitable for corrosive gases.

4. Selection of the waste

Among sugar refinery wastes, the eluates coming from regenerating ion-exchange colums, utilized for purfication of the sugar juices, immediately appeared specially interesting. In fact they are difficult to clarify as well as they constitute about half of the effluent pollutant load from a sugar refinery which uses this processing technique.

The experiment was carried out on waste from a factory processing sugar beets at Argelato, near Bologna (Italy).

The main characteristics of these wastes and their variables are shown in Table I.

5. The pilot plant

Pilot plant was built early in 1981, within the CEC contract n°

RUW-045-I, in accordance with the flow-sheet design shown in Figure 1.(6), (7).

The effluent coming from anionic ion exchange resins was collected in two equalization tanks, and then was preheated to 35°C by means of tube-sheet heat exchangers and feedto two separate fixed-bed anaerobic reactors installed on process lines which were parallel and indipendent.

The first reactor, indicated as R-1, is cylindrical with a conic roof, was made of carbon steel and had an operating volume of 10 m^3.

The second, indicated as R-2, was rectangular with a flat roof. The sides of reactors were made of prefabricated self-supporting concrete and the roof was of carbon steel. The operating volume was 50 m^3.

The term "operating volume" is calculated by subtracting from the total volume the space at the top occupied by the gas as well as the effective space taken up by the filler material.

The exterior surfaces of both reactors were insulated with sprayed-on polyuretane.

Because of its small size, reactor R-1 was also provided with an external thermal guard consisting of a serpentine coil of tubing firmly attached to the wall connected which hot wastewater from the factory.

The biogas produced by the anaerobic digestion process was measured by two separated gas meters, dry tipe, and its CO_2 content was determined in the field using a Gas Fyrite instrument.

6. Results

The pilot plant was operated from September to December 1981, during the period when sugar manufacturing was in progress.

Table II shows the main results obtained in the plant after start-up period, during continuous operation, at steady-state conditions and HRT less than one day.

Figures 2 to 5 show COD values at the feed input and waste - discharge output, as well as daily production of biogas and its

Table I - Sugar beet refinery effluents composition (ion exchange resin eluates)

pH	6	to	12
COD (g/l)	9	to	40
BOD_5 "	5	to	25
N tot "	0.4	to	1.8
NH_4^+ "	0.01	to	0.04
TS "	12	to	60
VS "	6.5	to	32

where:

COD = chemical oxygen demand

BOD_5 = biological oxygen demand, five days

N tot = total nitrogen

NH_4^+ = ammonia nitrogen

TS = total solids

VS = volatile solids

Table II - Main results

COD reduction : up to 70%

BOD_5 reduction : up to 80%

Volatile solids reduction : 70 - 80%

NH_4^+ in % of total N (average) : 90%

Biogas production : up to 11,5 volumes/reactor operating volume/day

Biogas composition : CH_4 70 - 75%

CO_2 25 - 30%

H_2S 0,2 - 0,3%

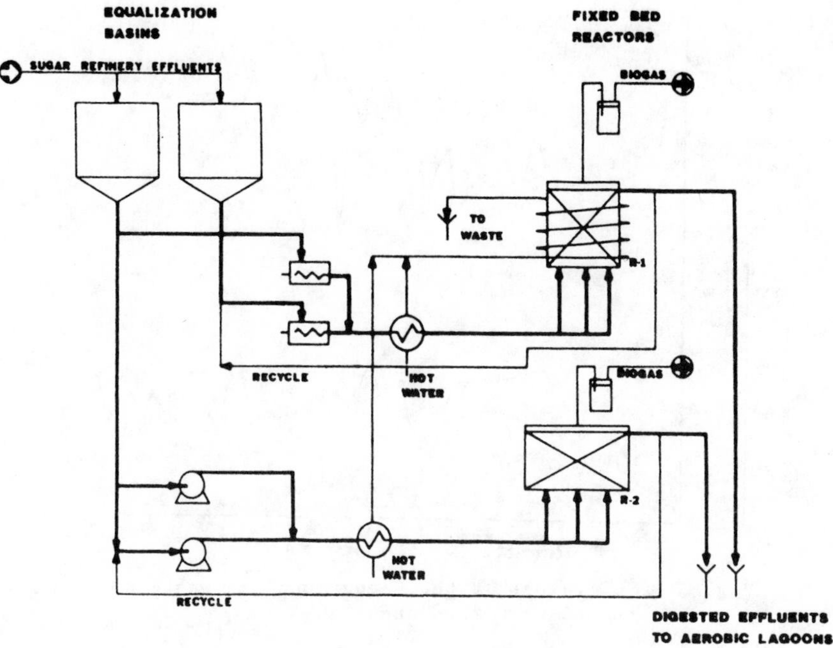

Fig. 1 - Flow - sheet of the pilot plant

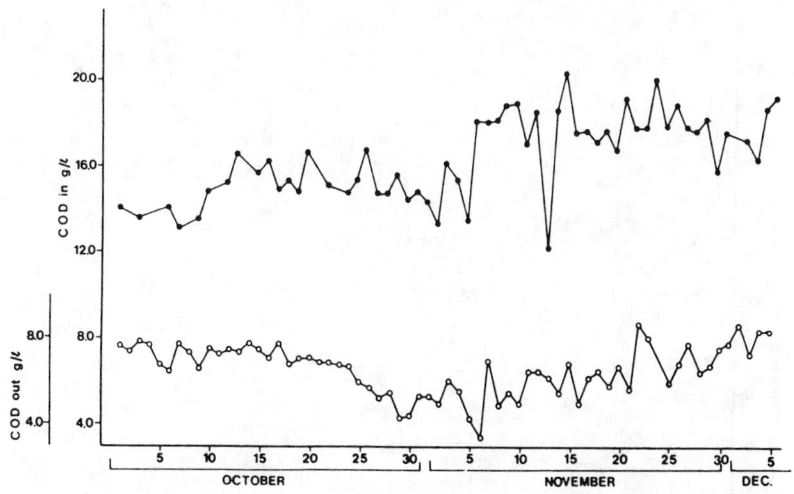

Fig. 2 - COD values at the feed input and waste - discharge output of the R-1 reactor

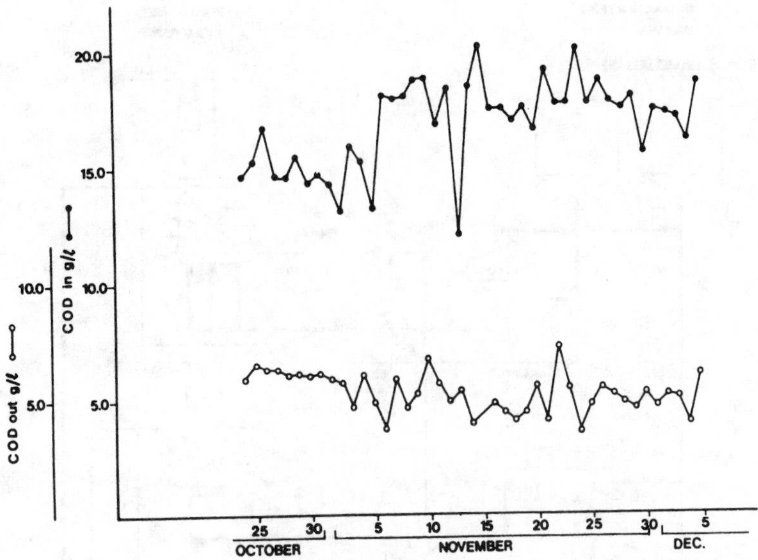

Fig. 3 - COD values at the feed input and waste-discharge output of R-2 reactor

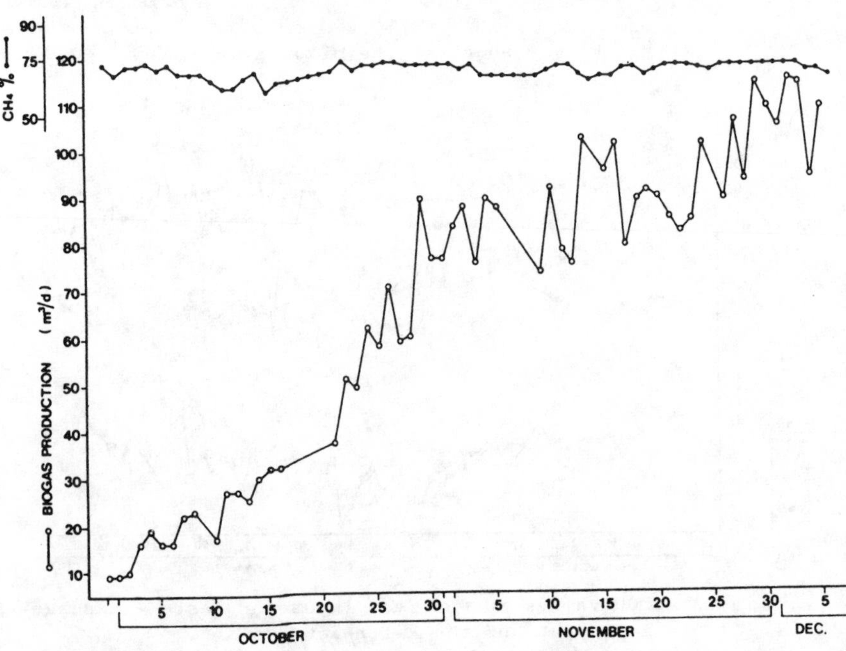

Fig. 4 - Production of biogas and its methane content for R-1 reactor

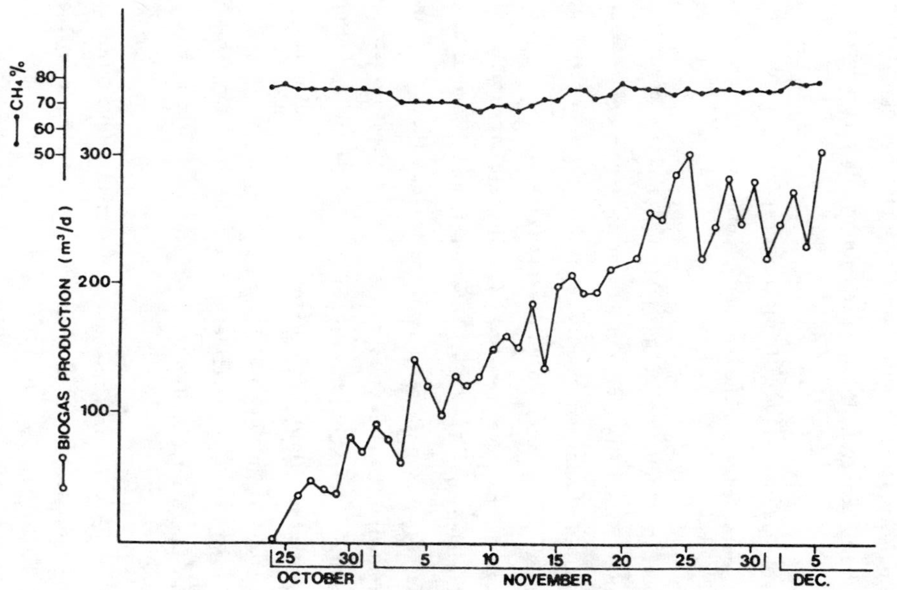

Fig. 5 - Production of biogas and its methane content for R-2 reactor

methane content, respectively for reactors R-1 and R-2.
Generally the maximum efficiency values where shown by reactor R-1, but probably this was because reactor R-2 was operated for a shorter period of time because of some delays in its construction.

7. Discussion

Despite its brevity, the results obtained during operation of the pilot plant fully confirmed the preliminary data produced in the laboratory.
For this waste, particularly difficult to treat using aerobic methods, the results obtained in the pilot plant are to be considered satisfactory.
The high specific production of gas obtained and its optimum quality confirmed the possibility of an important energy recovery. An experiment originally aimed chiefly at solving an ambient pollution problem, turned out to be a valuable energy source.
Because of these positive results, a full-scale anaerobic digestion plant, engineeried by AgipGiza, was built and started up during 1982 at the same sugar refinery, confirming that this treatment system for industrial wastewaters can also be a good investment.
The successful re-start of this plant in 1983 confirmed that it is possible to interrupt flows of feed to reactors for quite long periods without compromising process efficiency.
In the case of the sugar refinery cited herein, for example, the factory is in operation for five months and is idle for seven months each year.
A condition of this nature can quite decisively favor the choice of an anaerobic system for those industries where certain effluents are characteristically seasonal.

8. Conclusion

The behavior of both pilot and full scale plant had shown that

suitably designed fixed-bed anaerobic reactors may be used advantageously in the treatment of wastes containing soluble pollutants.
At the present state of the art, anaerobic digestion is clearly a valid alternative to aerobic treatment of industrial wastes, particularly when the availability of such wastes is seasonal or when they are difficult to treat with an aerobic system.

REFERENCES

(1) INRA, Pollution by the Food Processing Industries in the EEC, Graham & Trotman Ltd., London, 1977

(2) Litchfield, J.H. in "Food and bioengineering: fundamental and industrial aspects", AICHE Symposium Series, Vol. 67, 1971, pp. 164-172.

(3) Sanna, P., Camilli, M., Anaerobic treatment of agroindustrial wastes by means of fixed bed reactors, 1[a] Conferenza Internazionale Energia e Agricoltura, Milano, 27-29 aprile 1983, Vol. 2, pp. 92/1 - 14

(4) Young, I.C. and Mc Carty, P.L., The anaerobic filter for waste treatment, Technical Report N. 87, Stanford University, Stanford, California, 1968.

(5) APHA, AWWA, WPCF, Standard Methods for the Examination of Water and Waste Water, 15th Edition, Washington, 1980.

(6) Sanna, P., Camilli M., Energy recovery from agroindustrial wastes with prevalently solute pollutants using fixed-bed anaerobic reactors, ASTM 3rd International Symposium on Industrial and Hazardous Solid Wastes, Philadelphia, Pa., 7-10 March 1983 (In course of publication)

(7) Sanna, P., Camilli, M., Agroindustrial wastes treatment using fixed bed anaerobic reactors, Proceedings of the AWWT European Symposium, Noordwijkerhout, Netherlands, 23-25 November 1983.

ENERGY RECOVERY AND EFFLUENT TREATMENT OF STRONG INDUSTRIAL WASTES BY ANAEROBIC BIOFILTRATION

Wheatley, A.D., Ph.D., Cassell, L., M.Sc. and Winstanley, C.I., LRSC

Environmental Biotechnology Group, Department of Chemical Engineering,
University of Manchester Institute of Science & Technology,
P.O. Box 88, Manchester, M60 1QD.

Abstract

A process for utilising waste organic materials in industrial wash and waste waters is described. The process relies on the continuous anaerobic fermentation of the waste by mixed culture. The bacteria are immobilised in the reactor on a high surface area plastic support media.

Laboratory research, on a range of effluents, has established the feasibility of using anaerobic filtration for effluent treatment and gas recovery from warm, strong, food, drinks and fermentation industry wastes. This type of effluent is not easily treated by conventional techniques because supplying the necessary air for bio-oxidation (compressors, pumps, aerators) and sludge disposal is now very expensive.

A neutral pH was found to be necessary for successful anaerobic treatment but difficult to achieve with some types of waste low in alkalinity. It was possible to control pH by recycle, chemical dosing and addition of other wastes.

Experiments on upflow anaerobic biofiltration were extended to a $10m^3$ test plant, to treat the waste from a local sweet factory. The plant ran for 8 months and demonstrated the potential to save 70% of the effluent treatment costs and make a 5% contribution to the fuel costs. COD removals of up to 80% can be achieved at economic residence times (24-30 hours) at loads of 5-15 kg $CODm^{-3}$ day^{-1}. Gas yields are between $0.5\text{-}0.7m^3$/kg COD removed and can contain between 50-70% methane. On this basis a full scale plant would pay for itself in 2½ years but an economic constraint on the system might be the chemical conditioning costs necessary to control fermentation. The best gas production is at neutral pH, but the sweet waste in this case is very acidic with no buffering capacity, and experiments are continuing to reduce this cost on this system. A $50m^3$ plant has now been built.

1. INTRODUCTION

The most common type of industrial effluent treatment is biological oxidation (biofilters/activated sludge), but supplying the necessary air and equipment for this aerobic process is so expensive that discharge to sewer is normally more attractive. Table i shows the Mogden formula, the basis of effluent treatment charges in the UK.

TABLE i - Trade Effluent Charge Formula Costs North West Area 1982

$$C = R + V + \frac{Ot}{Os} B + \frac{St}{Ss} S$$

where C = Cost in p/m^3 trade effluent/day

R is the reception and conveyance costs of sewage (2.45p/m^3)

V is the volumetric and primary treatment cost of sewage (1.95)

Ot is the COD of the trade effluent after 1 hour settlement

Os is the average COD of settled sewage (385 mg/L)

B is the unit biological oxidation cost for settled sewage (3.01p/m^3)

St is the total suspended solids mg/l of trade effluent

Ss is the total suspended solids mg/L of crude sewage (259 mg/l)

S is the treatment and disposal cost of primary sludges (1.932p/m^3)

Trade effluent discharge costs vary from area to area, but typically they are between 100 to 300p a m^3 for food, drinks and pharmaceutical industry wastes with CODs of between 10,000 and 20,000 (1). Table ii shows some typical strengths of these types of wastes.

Table ii - Some strengths of common food and drinks wastes, they are approximate figures and actual values vary from site to site, depending on the type process.

Industry	Strength of Waste COD mg/L
Pharmaceutical Citric acid	20,000
Pharmaceutical Antibiotics	10,000
Distillery waste	25,000
Confectionery	15,000
Cheese, butter, cream	5,000
Brewery	5,000

If the formula is used to calculate the charges for a small sweet factory employing 600 people producing 12,250m^3 of effluent per annum at 10,000 mg/l COD then the charge is £11,000 per annum. This is insufficient to justify capital expenditure on an aerobic effluent plant to remove 70-80% of the organic load and costing £100,000.

A possible alternative treatment for these strong wastes is anaerobic digestion. This process has been successfully applied since 1900 to:- reduce sludge volume, eliminate pathogens, prevent smell nuisance and generate methane as a by-product. There are however difficulties with this process associated with the slow growth rate of the methanogenic bacteria and the solid, recalcitrant nature of the substrate sludges. Until recently 30 days was a typical reactor residence time for the conversion of 50% of the organic material, into inorganic salts. The organic waste in strong industrial effluents however is mostly in solution and is thus more amenable to treatment.

In a completely mixed system such as the traditional sludge digester Hydraulic Retention Time (HRT) and Solids Retention Time (SRT) are the same and the minimum HRT is defined by microbial growth rate. There have recently been some major developments in reactor design to retain the active biomass and to uncouple SRT from HRT. The biomass can be held in the reactor; by flocculation (Upflow Anaerobic Sludge Blanket UASB), or by immobilisation to an internal support media (Fluidised beds, Expanded beds or Anaerobic Filtration), or by downstream separation of the biomass for recycling back to the digester (Anaerobic Contact System). Both filtration and assisted sedimentation have been used for biomass separation (Fig.1). An additional economic motivation for the utilisation of any anaerobic treatment system is the generation of CH_4 gas as a by-product. The rising costs of aerobic treatment and energy has revived interest in anaerobic treatment as an alternative and this has led to important developments in the understanding and control of the anaerobic process. A recently completed survey (2) on anaerobic waste treatment in Europe includes data on 400 operating plants.

Biological filters have been shown to be more reliable, easier to operate and more resilient than other fermenter systems (3). They are therefore ideally suited to treat variable industrial effluents.

At the time this research was started in 1975 there was little published information on anaerobic industrial effluent treatment (4) in the UK, and an experimental programme was instigated to determine the feasibility of Upward Flow Anaerobic Biological Filtration on various types of effluent. The apparatus was based on that reported by Young and McCarty (1967)(5). The next phase of the work, started in 1981, utilised a $10m^3$ pilot plant operating in an industrial environment to treat confectionery waste. A small full scale plant was subsequently designed and built in 1983.

2.1 Laboratory Apparatus and Methods

The laboratory fermenters were based on standard 10 litre glass aspirator bottles (Fisons Scientific Apparatus) packed with 25mm Flocor R (ICI Pollution Control Systems) Figure 2. Up to four filters were operated simultaneously. Early in the experiments it was noted that a large number of photo-synthetic bacteria were growing the in anaerobic filters, these may have been affecting performance by occupying space and regenerating organic matter within the filter. One of the filters was painted black to measure any differences attributable to photo-synthesis. A fermenter temperature of 35-37°C was maintained by a water bath (Grant Instruments).

Effluent was circulated by a variable speed peristaltic pump (Watson

Marlow) using 5mm (i.d.) neoprene tubing. Flow rates of between 4-15 mls a minute were employed which equated to hydraulic retention times of between 10-40 hours.

Some difficulties with pipe wear and flow control were experienced and the peristaltic pumps were replaced with positive displacement pumps (Metering Pumps). Flow control and pipe wear were very much improved.

The feed pipe discharged into the centre of the reactor and was turned down to dissipate the flow at the base of the packing. Gas was disengaged from the effluent at the top of the reactor using a 'T' piece syphon arrangement as shown in Figure 2. The gas was collected over water in a 10 litre glass tube (QVF id. 40mm x 2.0m) which was normally sufficient for half a days storage. The gas was collected over water rather than a salt or acid, since this was likely to be the cheapest and simplest method for large scale use. If the bio gas is to be used as boiler fuel, no scrubbing should be necessary. The volumes of gas produced were not corrected to STP as a matter of routine. A tap was fitted to the top of the tube, via a bung, to allow evacuation and sampling of the gas. A Bird and Tole tipping trough gas flow meter was also used but fouling problems were encountered which affected its reliability.

The methane content of the gas was measured by infra-red specroscopy (Miran 104 Foxboro Insturments Ltd.) at a wavelength of 7.7 μm and a pathlength of 10.5m. Normally adjustments to wavelength and pathlength were made with every determination to ensure maximum response from the meter. Mains natural gas (95% methane) was used for calibration. Calibration and anlaysis was carried out twice daily on gas samples obtained from the storage tubes using a 10ml syringe injected into a clamped rubber connecting tube. The samples were transferred to the analyser and the results of 3/4 determinations averaged. Influent and effluent pH were measured periodically throughout the day.

Samples of the feed and effluent were taken on most working days and the COD and suspended solids determined by the standard analytical methods (6). Less frequent determination of NH_3N, BOD, PO_4P, total solids and volatile acids were made to check treatability. Occasional sludge samples were also taken from the reactor and their total and volatile solids content determined.

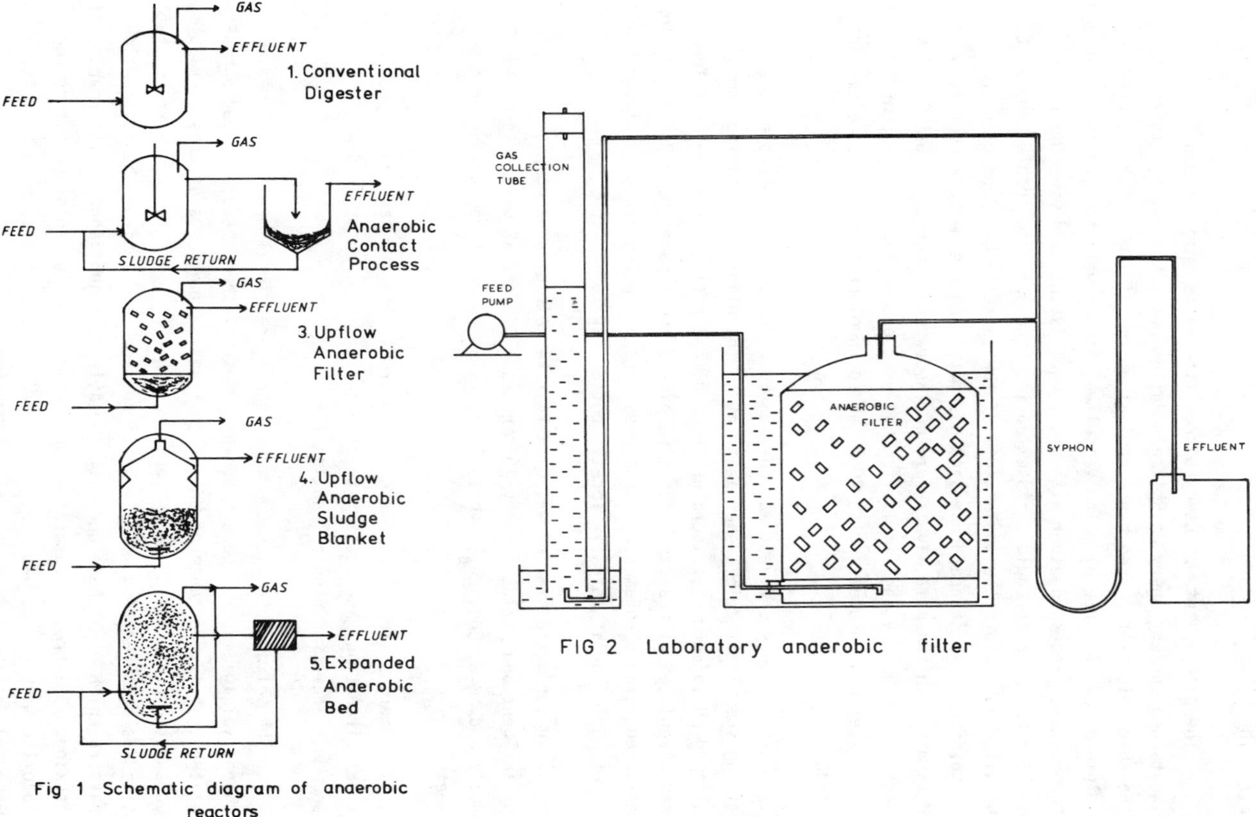

Fig 1 Schematic diagram of anaerobic reactors

FIG 2 Laboratory anaerobic filter

2.2 The Pilot Plant

The pilot plant was used to treat the waste from the manufacture of sweets and confectionery at Swizzels Matlow Ltd., New Mills, Stockport. The anaerobic filter used had a nominal volume of $10m^3$, with a packed bed volume of $6m^3$ (Figure 3). The vessel was fitted with top and bottom pH and temperature probes together with eight equidistant side sampling ports. Smaller bore tubing could be inserted into the ports to enable samples to be withdrawn across the whole width of the vessel. Gas production was measured using a standard domestic gas meter and its methane content determined in the same way as for the laboratory apparatus. Gas was disengaged using a syphon similar to that used on the laboratory apparatus. The vessel was insulated with 200mm of rockwool protected by an aluminium cladding.

The factory effluent was pumped (Mono Pump) from a manhole to a mixing sump located in a shed adjacent to the plant. The mixing sump allowed pH adjustment temperature control and effluent recycle. Effluent was circulated with another positive displacement pump (Mono Pump) and flow measured using a rotameter with bypass, and a tipping trough. Sludge could be withdrawn by a centrally located bottom take off valve.

An automatic sampler was used to take refrigerated composite samples of influent and effluent every 40 minutes. COD and SS were determined daily and NH_3H, BOD, PO_4P, volatile fatty acids and total solids once a week.

Numerous studies on fixed bed anaerobic filters have noted possible short-circuiting, and loss of treatment efficiency associated with accumulations of biological solids (7).

The solids retention characteristics are closely related to media shape, voidage and to reactor design. Most of these studies have concentrated on the importance of HRT which ignores the effects of these variables on substrate contact or detention times (8). In this study tracer contact time was measured on a monthly basis to check and monitor changes in flow pattern which might have been attributable to channeling or disturbances of packing. 3 litres of saturated (357 g/litre) sodium chloride introduced into the mixing sump was used as a tracer. Concentrations in the effluent were (6) monitored using the Mohr titration.

2.3 The Full Scale Plant

The full scale plant has been built to treat the same confectionery waste tested at pilot scale. In this case a balance/1st stage tank was combined with the methanogenic biofilter in an integral structure (Figure 4). The tanks were fabricated in glass fibre, reinforced plastic. The biofilter is 21m^3 in volume and contains 15m^3 of plastic filter media (average 50 x 50mm. ETA Ltd., Rugeley). The balance tank is a similar volume to the reactor (23m^3) built as an annulus to the biofilter. The vessel was insulated with 200mm of rockwool protected by an aluminium cladding. The pilot study showed that heating of the vessel was unnecessary (effluent temperature 35-50°C) and no heat exchanger is fitted to the plant. If supplementary heating is required, direct steam injection into the balance tank is used. Gas is collected in a 5m^3 floating gasometer also made from glass fibre reinforced plastic (Farm Gas Ltd.), this is to be connected to a boiler. The capital costs of the plant are shown in Table 3.

The earlier work on the effluent has shown that 80% of the COD load occurs in 3 flow peaks lasting no more than an hour each, these are pumped into the balance tank by a positive displacement pump (Mono Pumps Ltd.). The remaining effluent, which has a COD of <2000 mg/l is discharged to sewer in the normal way. The forward feed to the reactor is by a similar, but smaller, positive displacement pump. The domestic sewage from the factory is pumped intermittently into the balance tank as required using a submersible pump in the sewer (Flygt pump).

Monitoring of the plant is similar to that used on the pilot plant, but the pH and temperature probes inside the reactor have been omitted. Experience showed that these probes were rapidly contaminated and could not easily be restored to give meaningful results. The biofilter effluent pH and temperature are continuously monitored by probes in the effluent line immediately prior to discharge to the sewer. The pH probe can automatically control pH by a variable speed chemical addition pump discharging into the feed line prior to the pump.

Filter Start Up

Less energy is derived from anaerobic than aerobic metabolism and, in general, microbial growth rate is slower. Start up of an anaerobic system,

FIG 3 Schematic diagram of the pilot plant

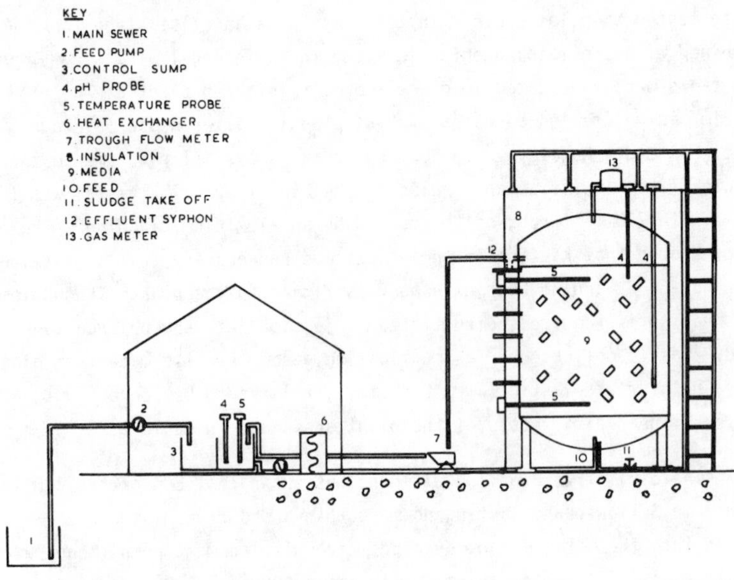

Fig 4 Schematic diagram of anaerobic filter plant treating sweet waste

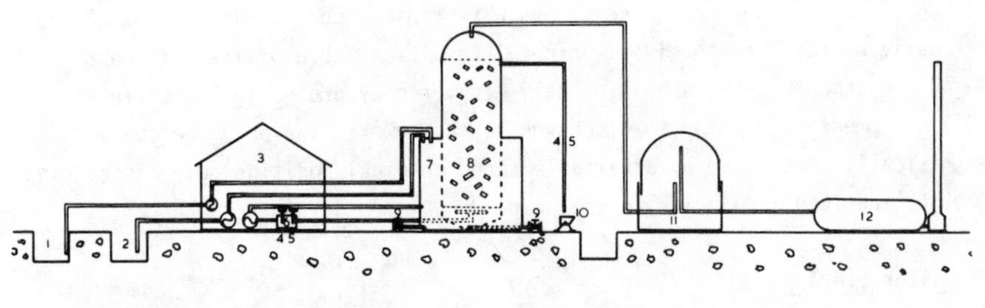

KEY

1. DOMESTIC SEWAGE
2. FACTORY WASTE
3. EQUIPMENT HUT
4. TEMPERATURE PROBE
5. pH PROBE
6. MIXING CHAMBER
7. BALANCE TANK
8. REACTOR
9. SLUDGE VALVES
10. TROUGH FLOW METER
11. GASOMETER
12. BOILER

is as a consequence more difficult than an aerobic system and seeding is a necessity. Similarly although anaerobic systems can operate steadily at lower temperatures by a slow acclimatisation of the culture, growth rate is too slow to allow adequate start up. Normally, therefore, anaerobic systems are initiated with an inoculum of actively digesting sludge.

Table iii - Capital Costs Fill scale anaerobic filter treating sweet wastes

Reactor and	
Balance tank combined structure in GRP	10,000
Insulation	2,000
Base slab	2,000
Pumps/pipes	1,000
Gasometer $5m^3$ GRP	2,500
Instrumentation	2,000
Boiler modifications	1,000
Extras at 10%	2,500
	£23,000

The pilot and full scale plants were started with a 30% by volume inoculum of actively digesting sludge. An initial load of 1.0 kg COD/m^3 was applied by dilution of the strong confectionery waste with water. Chemical nutrients and pH control was used to ensure effluent pH was about neutral. The load was then increased by 0.5 kg COD/m^3/week during an acclimiatisation period lasting 4 months. Continuous measurements of effluent pH and the methane content of the biogas were made and the increases in load abandoned if pH deviated from neutral or the methane content fell below 60%.

RESULTS OF LABORATORY TREATABILITY TRIALS

3.1 Treatment of sewage

Sewage is one of the most common liquid wastes, but in many respects represents the most difficult challenge to anaerobic treatment, since it is cold and dilute (500 mg/l COD). Thus, the first treatability trial was conducted on sewage at room temperature. Table iv shows the average results (COD, SS) of filters operated at different hydraulic loads. Gas production during these first experiments was poor and variable averaging 100-200 mls/ day and 10-20% CH_4. This was attributed to the low temperature and the low concentration of the substrate in the waste.

In further experiments the concentration of the feed was increased by the gradual addition of abattoir waste to the sewage. The HRT was kept constant while the load was increased to twice that of the first experiments (Table iv). The performance of the filters improved in terms of COD removal, but gas production was variable and the gas produced contained only small amounts of methane (10-30%). These results are similar to other reports in the literature (8) but the major problem appears to be the loss of activity of methanogenic bacteria at lower temperatures. Removal of organic materials is substantially more efficient by aerobic filtration at these loadings (Figure 5).

3.2 Removal of Nutrients

Another possible application of anaerobic techniques is the removal of inorganic nutrients after conventional sewage treatment. Final effluents from 3 sewage works were passed at ambient temperature through one of the anaerobic filters used previously to treat sewage. The average results of a 3 month trial are shown in Table 5. Suspended solids and BOD are removed, but the COD increased. This was attributed to some digestion of solids material. The reduction in nitrate through the filter was directly equivalent tc an increase in ammonia. The overall change in inorganic nitrogen and phosphorous was negligible and it was assumed that little bacterial activity occurred in the filter. Addition of a complex substrate such as sewage may have stimulated bacterial growth; methanol, glycol and glucose have been used in other trials with facultative anaerobes (9). The nitrate however was not reduced to nitrogen gas and the effluent produced unsuitable for discharge to a water course since it contained both ammonia and sulphides. No further tests were carried out.

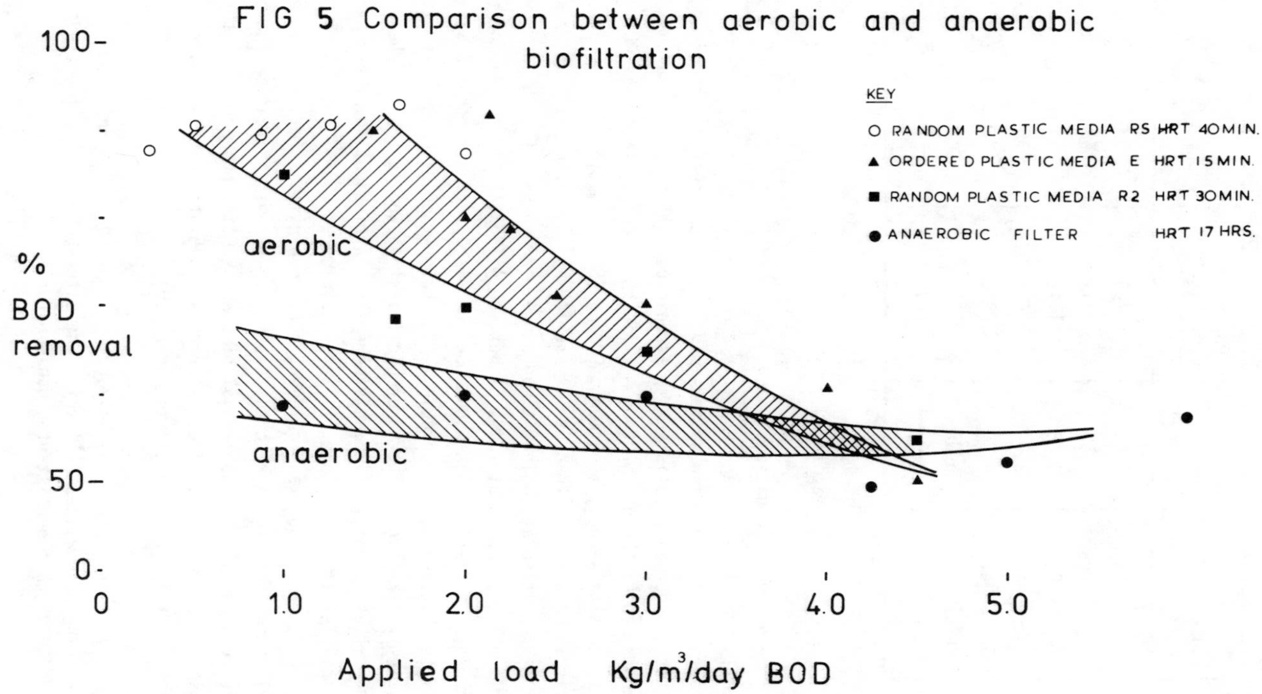

FIG 5 Comparison between aerobic and anaerobic biofiltration

Table iv - The anaerobic treatment of domestic sewage at ambient temperature

Date	COD load kg/m^3/day	COD concentration mgl^{-1}	COD % removed	BOD % removed
Jan - May	1.05	704	88	45
May - Aug	1.55	1111	40	55
Sept - Dec	3.15	1142	46	62
Dec - Jan	2.9	656	35	-
Jan - March	4.3	2200	66	74
April - June	5.0	2210	41	-
June - July	6.0	3878	74	-

3.3 Recalcitrant Materials

The redox conditions in anaerobic treatment systems differ substantially from those in an aerobic system. It is possible therefore that these anaerobic reducing systems may promote the degradation of organic materials normally resistant to aerobic oxidation. Data was available on such a waste: the effluent from a finishing and dyeing works which was resistant to aerobic treatment, COD and BOD reduction by low rate biofiltration was less than 5%. A previously inoculated laboratory anaerobic filter operated at ambient temperature, and an HRT of 18.5 hours was used to treat this waste at loads of between 4-13 kg COD/m^3/day. The results showed an average of 31% COD reduction, Table vi. If recycle was used to dilute the feed to the aerobic filter however then the toxicity to the system was removed. The anaerobic filter showed no benefit when operated under these conditions. Tests were also carried out to determine colour removal efficiencies using filtered samples at 385-470 nm (6). Some visible reduction in colour was achieved, but the persistant cloudiness, even after filtration interfered with the tests. In aerobic filtration the protozoans are responsible for the removal of a substantial proportion of the fine solids, no protozoans were observed in the biofilm of the anaerobic filters, although they are reported to occur (10).

Similar experiments have been conducted in Sweden and Finland on bleaching pulp wastes (11) which indicate anaerobic treatment system can be used to degrade persistant organic materials.

3.4 Strong Food and Drinks Wastes

Seven different food processing wastes with COD concentrations above 2000 mgl have now been tested; a maltings, a distillery, a brewery, potato wastes, the effluent from the production of pharmaceuticals by fermentation, ice cream waste, and a sweet factory waste. All the effluents are warm (30-40°C) and the trials carried out using inoculated filters and a 24 hour or below HRT. The results are summarised in terms of COD removals and gas production in Table vii, other results have been omitted for brevity. There is as yet insufficient data from the brewery or pharmaceutical wastes, but indications are that they are easily treated. Problems were encountered with the maltings and confectionery wastes, filter effluent pH was acid and COD removal was often poor. The largest amounts of methane from these two wastes, were produced early on in the tests, and this was thought to be carried over from the original inoculum. There was an eventual substantial improvement in COD and BOD removal from the maltings waste, but this was not associated with any biogas production and was attributed to the precipitation and accumulation of solids by filter. A combination of acid pH and fluctuating loads were assumed to be responsible for the failures. The filter treating the distillery waste, although at a similar initial pH and conditioned in the same way, encountered no similar problems and loads in excess of 20 kg COD/m^{3}/day have now been attained. This filter was operated continuously for 24 months with a neutral effluent pH.

Most investigations into anaerobic treatment (12) have noted problems associated with pH control. The difficulty arises from differences in the optimal pH of the two synergistic bacterial populations in anaerobic fermentation. The acid forming group of bacteria grow relatively quickly (doubling time 2-3 hours) and have a pH optimum between 6 and 8. The methane formation is the rate limiting step. Under stable operating conditions a balance is established between the methanogenic and acidogenic bacteria and the combined population are able to operate without external pH control. The differential rates of growth of the bacterial populations however make the system susceptible to overload, especially when treating industrial wastes with insufficient alkalinity or additional nutrients. A potential method of controlling pH is to recycle effluent. Thus the filter treating the maltings waste was reseeded and the hydraulic load increased to give an HRT of 17 hours. The recycle ratio used was 20:1. pH control was better although not sufficient to encourage gas production.

Table v - Nutrient Removals one months results mg/l. (Hydraulic Retention Time 7 hours)

	Inf.	Eff.	Changes
COD	97	111	+ 14
BOD	10	8	- 2
SS	37	27	- 10
NH_3	6	12	+ 6
NO_3	15	7	- 8
PO_4	25	24	
pH	7.1	7.2	

Table vi - A comparison between the aerobic and anaerobic treatment of textile wastes

Type of Treatment	COD			BOD		
	inf.	eff.	%R	inf.	eff.	%R
Aerobic	2035	1990	3	398	374	6
Anaerobic	1990	1353	32	396	242	39
Aerobic (with recycle)	466	248	39	225	105	54
Anaerobic (single pass)	6.6	431	30	296	180	39

High rates of recycle are an integral part of more recent anaerobic reactor designs, such as the expanded and fluidised beds and is reported to improve contact between substrate and organisms, reduce toxicity and improve pH stability (13). It is inexpensive compared to chemicals but in these experiments was insufficient to control pH and counteract nutrient deficiency. pH can also be controlled by adding alkalinity and buffer.

The anaerobic filter treating the confectionery waste was re-seeded and pH controlled by chemical addition. Laboratory titrations were used to establish a buffering mixture. The effluent required 1 grm diammonium phosphate, 3 grms of sodium bicarbonate and 0.5 grms of sodium hydroxide to stabilise the pH to 7.7. It was possible to control pH using this method but at substantial cost estimated as £1.45/m^3. 80% of this cost was for the phosphate and further laboratory work may establish suitable substitutes.

COD removal, gas production and methane content of the biogas deteriorated even though it was possible to control pH using the buffer. This was attributed to the fluctuating loads, and lack of additional nutrients.

Table vii - Summary of Laboratory Results from various types of industrial wastes. Reactors operated at mesophilic temperature 35-37°C

Period	Load Applied kg COD/m^3/day	COD removal %	Gas Ratio m^3/kg COD removed	Methane content % not corrected
Maltings Wastes 24h HRT Tests carried out 1976/77				
May - July	4	65	0.23	68
Aug - Oct	6	72	0.16	70
Nov - Jan	8	70	0.10	65
Confectionery Wastes 24h HRT Tests carried out 1982				
May	5.0	40	0.23	74
June	7.5	72	0.13	70
July	5.4	78	0.10	59
August	4.6	78	0.15	61
September	4.2	75	0.18	58
October	12.5	33	0.13	66
November	10.0	58	0.19	62
December	12.0	52	0.04	43
Distillery Wastes 24h HRT Tests carried out 1982				
Jan - March	13.5	45	0.3	57
April - June	22.1	69	0.6	57
July - Sept	23.1	63	0.5	59
Oct - Dec	25.6	56	0.5	59
Potato Wastes 7h HRT Tests carried out 1983				
September	5.0	74	0.31	80
October	7.0	72	0.28	79
November	7.0	70	0.30	77
Ice Cream Wastes 24h HRT Carried out 1983				
May	4.0	78	0.18	56
June	4.4	84	0.32	66
July	5.7	82	0.32	70

The anaerobic filter treating the maltings waste was also reseeded and the strength of the waste deliberately reduced to 2000 mg/l COD by dilution. Domestic sewage (25% by volume) was also introduced to provide supplementary nutrients. No pH control problems were encountered and the load was increased to 6000 mg/l COD. Gas production was proportional to the COD load applied and the gas was normally about 70% methane. Some inhibition of gas production and COD removal occured at loads above 5000 mg/l COD, and this was attributed to a fall in the reactor pH. Methanogenic activity was found to be dependant on effluent pH.

A similar trial was carried out with the confectionery waste, the strength and pH were controlled by adding domestic sewage. For the first two weeks of the trial the COD strength was kept at 5000 mg/l by a 50% dilution with sewage and no pH adjustment was required. Gas production and methane content were stable and the strength was increased to 7000 mg/l COD by reducing the sewage content to 30%. Stable operating conditions continued and the gas produced was close to the theoretical value. The load was further increased to 10,000 mg/l by a reduction of the sewage content to 10%. The early results were good but a pump malfunction reduced the HRT to 8 hours. This overloaded the filter and the pH dropped to 4.9 when the filter failed.

The distillery, ice cream and potato wastes could be treated without these attendant difficulties and this was attributed to a better balance of nutrients and a more even strength of substrate (table vii). Gas production from the distillery waste did however, vary and this was linked to changes in character and type of suspended solids in the waste. Yeast and other larger debris from the malting and fermentation process are physically separated from the waste.

The efficiency of this separation process effects the amount of suspended solid in the effluent. There is also carry over of a flocculant which increases the viscosity of the waste. Increases in suspended matter from either of these conditions was found to reduce gas production and quality, although COD removal was not as seriously affected.

Gas chromatography was used to try and establish what metabolic intermediates were formed. Recently published work on the microbiology of the anaerobic process has suggested that the quantity and type of volatile acid can control methanogenesis (13). Acids other than acetic and formic may not be utilized by the methanogens and can cause inhibition of process by a reduction in the pH. The results indicated important differences between the wastes. The confectionery and distillery wastes contained a very high concentration of of acids. In the distillery waste the concentration of acids were relatively constant, but in the confectionery waste fluctuated widely. In contrast the potato effluent had a very low and constant concentration of volatile acids. Thus the very good performance from the potato waste may result from a different series of metabolites which do not affect pH. Further tests on the potato wastes showed that the sugars and starches did not play an important intermediate role in the fermentation either and further work is required on other possible metabolites.

4. OPERATION OF THE PILOT PLANT

The basic HRT of the plant was tested using mains water and NaCl as a tracer. Similar tests were conducted at two different flow rates 4 litres/minute and 13 litres/minute. These flows represented the maximum and minimum rates of the pump. They equated theoretical HRTs of 461 minutes and 1500 minutes. The measured median equivalent values were 550 minutes and 1800 minutes respectively.

The average COD results are shown in Table viii. There were major difficulties in controlling sudden increases in flow and strength. These could cause a significant fall in pH within the reactor (- 1 to 2 pH units) which is subsequently difficult to recover. A study of the factory effluent showed that there were 3 major coincident peaks of flow and strength. These peaks correspond to production wash downs and in combination they represent 80% of the total daily COD discharge. The sugary warm waste is an ideal fermentation medium and although it has a nearly neutral pH at source it rapidly becomes acidic and septic even during the short retention period in the drainage system. Trials were carried out using chemical addition and high recycle rates to try and buffer the pH of the feed to the reactor. A recycle ratio 5:1 was insufficient to control pH and the chemicals too expensive (£1m^{-3}). Flow balancing and treatment of these peaks alone could lead to substantial savings on the size of plant installed and it was decided to build a larger plant with better balancing capacity.

Table viii - Pilot plant results March - June 1982

Month	Load Applied kg CODm3 day^{-1}	Removed % COD	Gas Ratio m^3/kg removed	%CH_4
1982				
March	3.8	65	0.19	41
April	6.5	70	0.14	63
May	10.7	67	0.16	75
June	12.5	65	0.13	53

5. FULL SCALE PLANT

This was built with balancing capacity of about one days flow ($25m^3$). It was hoped this might even out the concentration of the waste to between 5-10000mgl^{-1} COD. The nature of the waste however means that it deteriorates with the formation of acid very quickly. Crude pH control was fitted to the balance tank to maintain the pH at about 6.5 with lime, without this control the pH falls rapidly to 3.5. The cost of this lime is 20pm^{-3}. Fine pH control is then carried out with the pH control equipment using a 10% caustic solution. The results so far are shown in table ix. The capital cost of the plant is shown against the expected savings, table x. These figures do not include running costs and further work will be required to establish the most cost effective solution to the pH problem.

Table ix - Results from the full scale plant

Date	Load kg CODm^{3}day^{-1}	% COD removal	HRT hrs	pH	Temp ^oC	Gas m^3day	% CH_4
May	0.5	70	500	7.1	35	0.5	54
June	1.0	68	500	7.2	34	0.5	62
July	1.6	88	200	7.2	34	1.0	70
August	3.0	86	100	7.0	33	1.0	70
September	6.1	84	60	7.2	32	5.7	55
October	4.2	36	60	6.3	34	4.2	51
October	0.5	30	250	6.0	32	0.3	48
November	0.6	65	250	6.8	30	1.9	58
December	3.0	70	250	6.0	32	4.2	66

Table x - Summary of Pilot Plant results and possible savings at full-scale

Effluent characteristics	strength	10-15,000 mg/l
	flow	12,250 m^3 annum
	load	122.5 tonnes COD annum
Present Charge at 88p/m^3		£11000
New charge assuming 71% COD removal 73% SS removal		
	28p/m^3	£ 3430
	Saving	£ 7600
Biogas		
Gas production 0.35m^3/kg COD removed at 70% methane from applied load 122.5 tonnes is 42,875 m^3		
Energy equivalent biogas at 25Mj/m^3 fuel oil 43 Mj/l		
25,000 litres at 11.12p litre		
	Saving	£ 2771

6. DISCUSSION

From the treatability studies conducted, anaerobic filtration is most useful at high organic loads compared to aerobic systems, i.e. 5-25kg/COD/m^3/day. Under ideal conditions 0.3-0.35m^3 methane per kilogram of COD removed can be produced. There are, however, several difficulties which still have to be resolved if the process is to be adopted for an extensive range of wastes. The reliability and stability of gas production is likely to be a problem until the exact nutrient requirements of the methanogenic bacteria are identified. There were major differences between industrial wastes containing different amounts of additional nutrient.

More work is required on the metabolic intermediates in anaerobic treatment some wastes produce large quantities of acids others only small quantities, and this exerts an important influence over the control of the process.

The character of the waste organic matter also has an important effect on performance, the results show that suspended solids are only slowly degraded within the primary balancing stage and the anaerobic filter. Further work is required on the design and performance of a primary non methanogenic stage to release simple materials from complex substrates. Acid formation from the relatively soluble substrates is very rapid.

Little information has been derived from this study on the nature and quantity of the sludge produced. Sludge deliberately withdrawn from the bottom of the reactors has been highly mineralised between 50-70% depending on the inorganic content of the feed; this is likely to be an asset when disposing of this sludge. Blockage of the laboratory filters has occurred occasionally but this has been overcome by temporary increases in flow, rather than backwashing. Settlement of suspended solids from the effluent of the anaerobic filters does occur, but no attempt has been made at quantifying this material. Observations on these settleable solids indicates that they are different from the original waste suspended matter.

Both the biogas and the effluents from high rate anaerobic systems are malodorous and special precautions were necessary to avoid agitation and release of these vapours. No problems were encountered with smell when the effluent from the pilot and large scale plants were discharged directly to sewer. The anaerobic filter effluent was however also found to be easily treated aerobically with in some cases better than 95% COD removal overall. If the biogas is used as a boiler fuel then the malodorous sulphide compounds should be oxidised in combustion.

7. CONCLUSIONS

1. The most successful application of anaerobic filtration to date is the treatment of soluble, warm, strong, food and drink industry wastes. Loads up to 25 kg COD/m^3/day can be continuously applied with up to 70% COD removal at 24 hours HRT. The inherent buffering capacity of the waste, i.e. alkalinity, dissolved solids and nutrients affected waste treatability.

2. The optimal operational conditions for gas production were a pH between 6.8 and 7.8 and a temperature of 37°C. No gas production

occurred below 27°C. Gas production was directly proportional to the COD concentration of the feed.

3. The production of volatile acids from simple substrates proceeds very quickly, but the breakdown of complex polymeric substrate is slow under anaerobic and microaerobic conditions.

4. The chemical composition of the waste exerts a crucial influence on treatability and full scale plant design is not recommended without laboratory or pilot scale trials.

5. Some wastes will require high recycle ratios, nutrient addition and pH control to ensure good removals of COD.

6. Slow acclimatisation with small increases in load is essential for successful filter operation. Start up is even more important if the waste is nutrient deficient. Constant monitoring of effluent pH gas production rates and gas quality are critical.

7. High rate anaerobic treatment can be malodorous.

ACKNOWLEDGEMENTS

The authors gratefully acknowledge the help, interest and support given to the project by Swizzels Matlow, New Mills Ltd. Thanks are also due to ETA Rugeley Ltd. for providing the reactor and biosupport media, and to the Commission of the European Community Recycling of Urban and Industrial Waste Programme DG12 for financial support.

REFERENCES

1. Treagus, M.J. (1982), "A survey of Anaerobic Processes for the Treatment of Food Factory Effluents", Special Report issued by the Food Research Association, Leatherhead.

2. Demuynk, M.G. , Naveau, H.P. and Nyns, E.F., Anaerobic Fermentation Technology in Europe, Environmental Biotechnology: Future Prospects pp64-70, Edited Holdom, R.S., Sidwick, J.M. and Wheatley, A.D., FAST project CEC Brussels, 1983

3. Hawkes, H.A. (1963), "The ecology of waste water treatment" 1st Edition - Pergamon, Oxford.

4. Department of Scientfiic and Industrial Research, "Water Pollution Research", The report of the Water Pollution Research Board with the

Report of the Director of the Water Pollution Research Laboratory. 1955-1959, Annual Reports, HMSO London, 1956-1960.

5. Young, J.C., and McCarty, P.L., "The Anaerobic filter for waste treatment", JWPCF (1969), 41, 5, R160-R173.

6. Department of the Environment, "Analysis of raw potable and waste waters", 1st Edition, HMSO, London, 1972.

7. Lettinga, G., Fohr, P.G., and Janssen, G.G.W., "Anaerobic digestion of concentrated waste H_2O", (1972), 5, 22, 510-517.

8. Henze, M., and Harremoes, P., "Anaerobic treatment of waste water in fixed film reactors: a review", IAWPR Specialised Seminar: Anaerobic treatment of waste water in fixed film reactors, Copenhagen, 1981.

9. Gauntlett, R.B., "Removal of ammonia and nitrate in the treatment of potable water", "Biological fluidised bed treatment of water and waste water", p48-60. Ed., P.F. Cooper and B. Atkinson, Ellis and Horwood, Chichester, 1981.

10. Hobson, P.N., "The bacteriology of anaerobic sewage digestion", Process Biochemistry (1973), p19-25.

11. Salkinoja-Salonen, M., Hakulinen, R., and Apajalhti, J., "Biodegradation of recalcitrant organochlorine compounds in fixed film reactors", IAWPR Specialised Seminar, Cogenhagen, 1981, p145-159.

12. Mosey, F.E., (1981), "Anaerobic Biological Treatment of Food Industry waste waters", J. Inst. Wat. Pollut. Cont., 80 (2), 273-291.

13. Mosey, F.E., (1982), "New Developments in Anaerobic Treatment of Industrial Wastes", J. Inst. Wat. Pollut. Contr., 81 (4), 540-552.

ANAEROBIC TREATMENT OF OLIVE PROCESSING WASTES

Prof. Dr. rer.nat. L. Hartmann
Institut für Ingenieurbiologie und Biotechnologie des Abwassers, Universität Karlsruhe

Dipl.-Biol. D. Ntalis
Institut für Ökologie und Taxonomy, Universität Athen

Prof. Dr. rer. nat. K. Anagnostidis
Institut für Ökologie und Taxonomy, Universität Athen

1. INTRODUCTION

The European Community has sponsored a joint research project, Contract No. RUW-056-D, performed by the Institute of Bioengineering and Biotechnology of Waste Water of the University of Karlsruhe and the Institute of Systematic Botany of the University of Athens. The purpose of this project was to study methane production by anaerobic treatment of liquid wastes stemming from olive oil mills. The goal of this project was to develop a stable process which will not only reduce the organic content of the olive processing but will also serve for energy creation.

Greece, with 15% of the total production ranking third of the olive oil producers after Spain and Italy, has about 350,000 farmers occupied in this enterprise, which corresponds with one third of the total farmer population. The olive oil production has increased from 130,000 tn in the mid-sixties to 236,000 tn in the season of 1981/1982. On the other hand, at the same time, the number of olive mills has decreased from 5,300 to 3,500. The average mill processes 8 to 10 tons of olives in eight hours with an output of 300 to 500 cubic meters of waste water (the total production per season is 1,000,000 m^3). This waste water has an average DOC concentration of 42,000 mgC/l; COD and BOD_5 values reach 156,800 mgO_2/l resp. 50,500 mgO_2/l. The total pollution produced by this industry is equal to the amount of pollution produced by one half of the Greek population per year, if the output per person and day is assumed to be 60 g BOD_5.

Work on this project was started in January 1981. The lab experiments took place in the Institute of Bioengineering and Biotechnology of Waste Waters of the University of Karlsruhe and the pilot plant experiments were performed in the Institute for Systematic Botany of the University of Athens. These experiments are still going on.

2. MATERIALS AND METHODS

2.1 Lab bioreactors

The laboratory system consisted of five reactors (Fig. 1) with each of them having a volume of 4.3 l totalling up 21.5 l. These five reactors were operated in sequence (Fig. 2) to permit separate evaluation of the biochemical steps in the process.

2.2 Aerobic fixed bed reactor

The aerobic fixed bed lab reactor (trickling filter) had the following dimensions: h=184 cm, d=22cm; the fixed bed was "Blähton" (brick pebbles with a high inner surface).

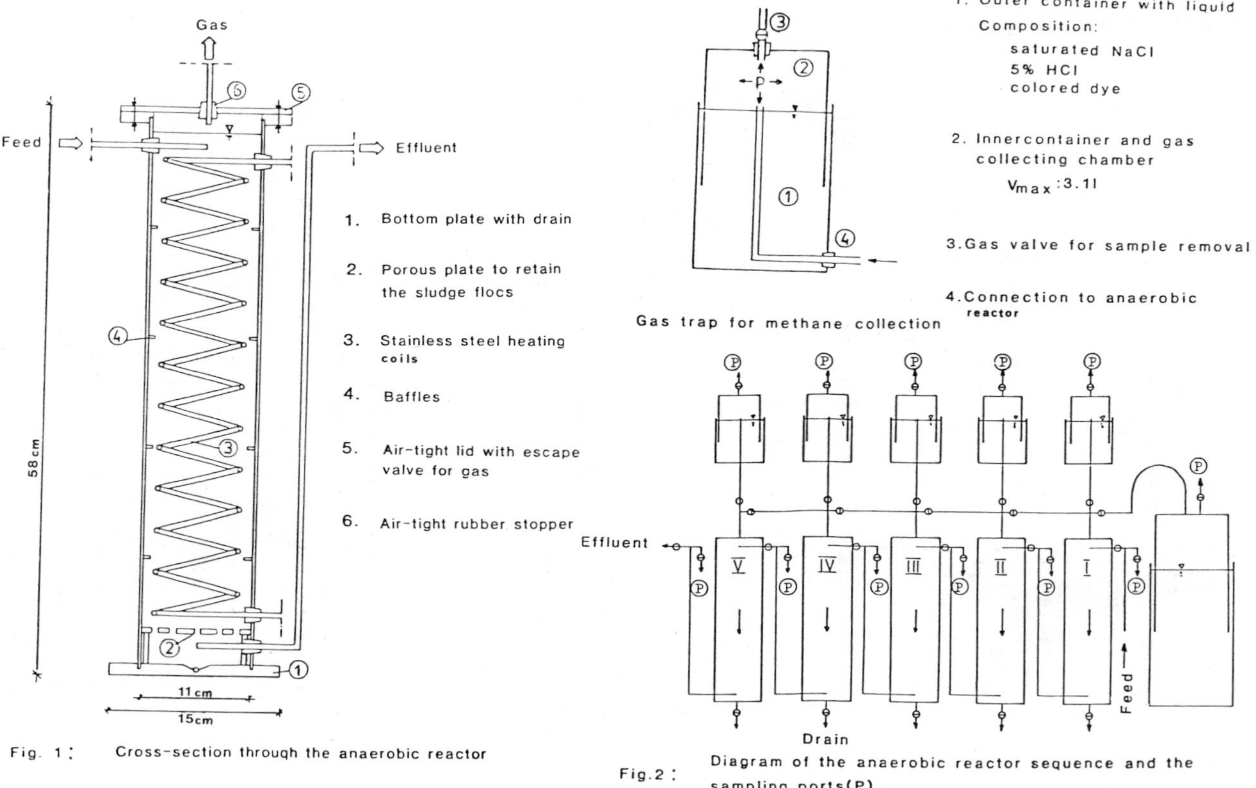

Fig. 1: Cross-section through the anaerobic reactor

Fig.2: Diagram of the anaerobic reactor sequence and the sampling ports(P)

2.3 Pilot plant bioreactor

The pilot plant is a continuously stirred tank reactor with a total volume of 520 l with an iron sheet mantle and a central axis for stirring.

2.4 Liquid waste composition

Six hundred liters of a concentrated fraction of liquid waste from one olive mill in Greece (Korinthos) was transported to Karlsruhe, where it was stored at -27°C. Details of the olive waste production process are shown in Fig.3.

Only fraction 2 could be used for the experiment; this waste water fraction was further separated into three layers by centrifugation in the laboratory. The top layer (oil phase) contained primarily the small amount of olive oil remaining in the waste water. The middle layer (aqueous phase) contained most of the dissolved organic matter in this fraction, and had a high Chemical Oxygen Demand of around 56,000 mgO_2/l as well as high BOD_5 values and a C:N ratio of 51:1. The bottom layer consisted of particulate matter. The chemical composition of all these materials is described in Table 1.

3. RESULTS

3.1 Anaerobic treatment in the lab experiments

Experiences from previous experiments indicated a relatively slow establishment of the biocommunity. To get better starting conditions for the reactor, the liquid wastes were initially enriched with molasses. The experiments were continued with pure olive oil waste without any addition of molasses. The system was buffered to hold a pH of 7.5 by using $NaHCO_3$ solution as a buffer. The waste composition was:

BOD_5 (filtered)	:	5,963 mg/l
BOD_5 (unfiltered)	:	5,275 mg/l
COD (filtered)	:	21,248 mg/l
COD (unfiltered)	:	33,558 mg/l
DOC	:	7,380 mg/l

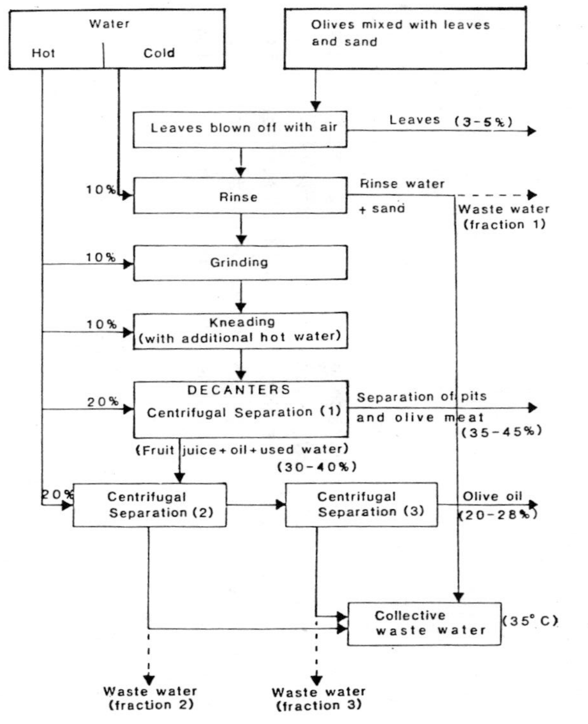

Fig.3: Flow diagram for water use and production of waste water fractions

	Oil phase (4-8% v/v)	Aqueous ph. (75-85%v/v)	Residue (11-17%v/v)	Collective Waste water
BOD_5 (Winkler) mgO_2/l	9500	18500	15600	50500
BOD_5 (with Sapr.) mgO_2/l	900	7900	3300	12800
$BOD_{pl.}$ mgO_2/l				6360
COD mgO_2/l	35650	55840	58460	156830
DOC mgC/l	6200	23200	12400	42400
TKN mgN/l	94	453	623	1170
C:N	66:1	51:1	20:1	36:1
NH_4^+ $mgNH_4^+/l$	1.4	49	73.5	132
Total P mgP/l				225
C:N:P				36:1:0.2
pH				5.5
Dry weight by 105°C g/l				89.8
Fat g/l				23.4
Fat %				2.5
Sucrose (Fehlings) g/l				20
Sucrose (Fehlings) %				2.1

Table 1: Chemical Analyses of the Centrifuged Waste water Fraction 2 Phases

Kj-N (filtered) : 165 mg/l
Kj-N (unfiltered) : 293 mg/l
NH_4^+ : 25 mg/l
dry solids (centr.) : 8,600 mg/l
dry solids (filter) : 10,100 mg/l

Observations and analysis were performed for a total period of 42 days to study:

- concentrations of pollutant of the original waste
- stability (or changes) in the reactor systems with respect to:
  - gas production
  - gas composition
  - pH value and
  - purification efficiency.

The system could be optimized at a total detention time of 12.5 days with a volumetric load of 0.59 g DOC/l/d, resp. 0.48 g BOD_5/l/d, as listed in Tab.2. Under these conditions the DOC concentration could be reduced by 65-70%. Biogas yield averaged 1.6-1.7 l/g $DOC_{elim.}$; methane concentration in the biogas was 70-75%.
The results after 42 days of operation are summarized in Fig.4; they indicate that the most significant changes with respect to gas production and DOC elimination occur in reactor III (resp. III and IV). It is also interesting to note (Fig.5) that the gas quality (percentage of CH_4 of total gas) is above 80% in the reactors III to V, where rest pollution in acids (expressed as acetic acid) is diminishing. Of importance is also the relatively high gas yield which amounts to roughly 1.6 to 1.7 liters of gas per g DOC eliminated; theoretically a value of 1.8 to 2.0 can be reached.

3.2 Aerobic post treatment of the anaerobic pretreated liquid waste

The anaerobic treatment does, however, not as could be shown above result in the satisfactory reduction of the organic load

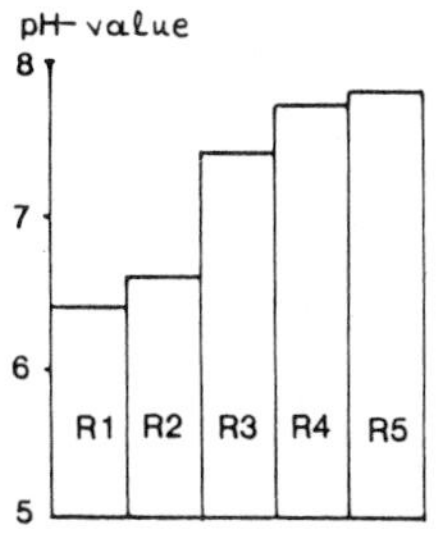

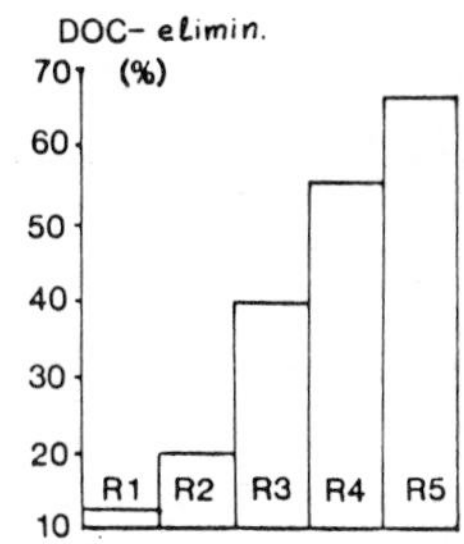

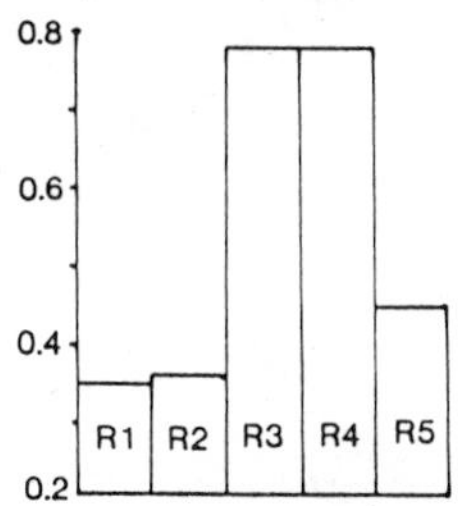

Fig. 4: pH - value, DOC - elimination and specific Gasproduction in each Reactor at the end of the experiment.

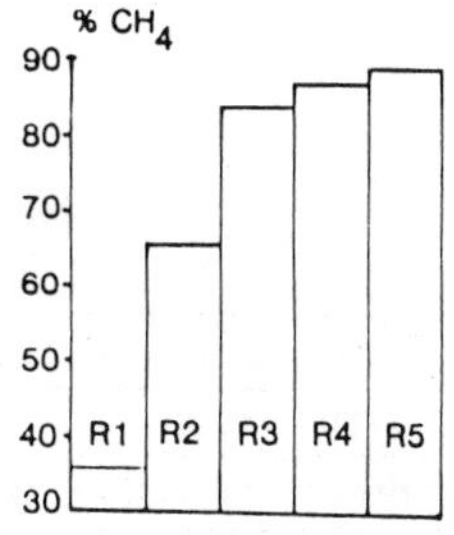

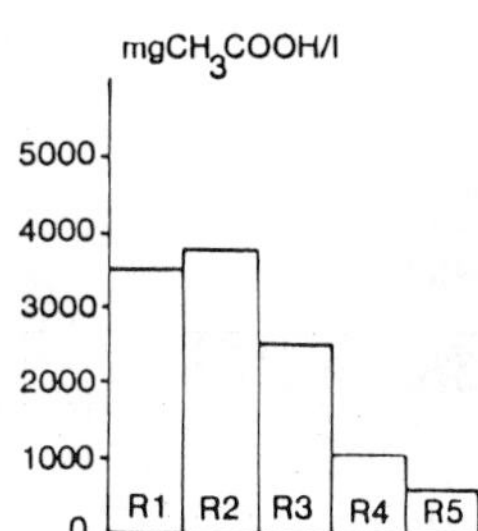

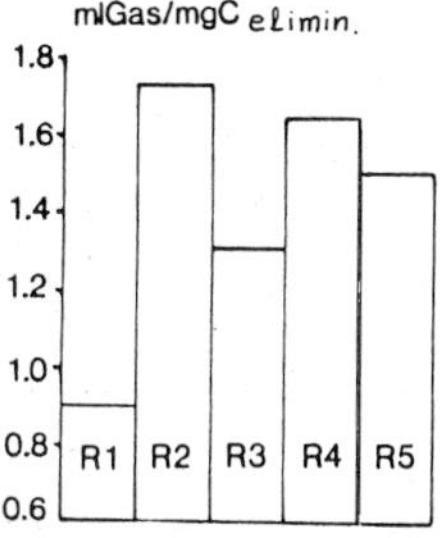

Fig. 5: Methane, acetic acid production and Biogas yield in each Reactor at the end of the experiment

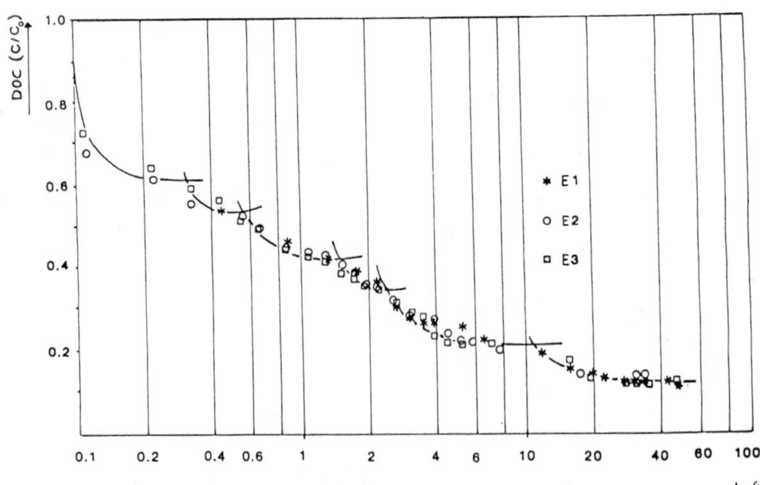

Fig. 6: Aerobic post treatment of anaerobically pretreated liquid wastes. (Results of three experiments)

1 : mg O_2/l
2 : mg C/l
3 : mg N/l
4 : mg NH_4^+/l
5 : g/l

	BOD_5 [1]		η	COD [1]		η	DOC [2]	η	BOD_5:COD:DOC	Kj-N [3]		NH_4^+ [4]		org. N [3]		C:N	dry solids [5]	
	unf.	fil.	unf./fil.	unf.	fil.	unf./fil.				unf.	fil.		fil.	unf.	fil.		centr	filter
Feed	5,963	5,275		33,558	21,248		7,380		1:4:1.4	293	165		25	253	140	53:1	8.6	10.1
Effluent	3,187	2,575	47%/51%	11,041	7,310	67%/66%	2,512	66%	1:2.8:1	193	95		15	167	80	31:1	3.3	3.4

Table 2: Feed-and Effluent values of filtered and unfiltered sambles at the end of the experiment

in the wastes. The effluent rest pollution in organics has a COD of 7,000 mg/l and a BOD_5 of 2,500 mg/l, and is therefore under environmental considerations not tolerable. Therefore aerobic post treatment had to be considered. For this experiments were performed using an aerobic fixed bed lab reactor (trickling filter). The experiments (batch experiments) were carried out under the following conditions:

- waste quantity : 5 l
- period of experiment : 50 h (each experiment)
- B_v (volumetric load) : 0.22 g BOD_5/l/d
- Q (influent) : 0.5 l/min
- q_F (surface load) : 0.8 $m^3/m^2/h$
- t_m (mean detention time) : 6.5 min
- t_{min} : 3.5 min
- t_{cycle} : 10.0 min

and yielded the following results:

- DOC elimination reached 85-89%
- BOD elimination reached 96-98%

The results are summarized in Fig.6; they show that there is a sequential reduction of the organic load as a result of fractional degradation of the different pollutants, and the time lag from one step to the next, is increasing from step to step. It can, therefore, be concluded that each stage of biodegradation is a result of specific adoptation of the biocommunity involved. That means that for each group of organic substances, there must be a specific biocommunity.

3.3 Pilot plant studies

Due to difficulties with the lab equipment after continuing the research in Athens, the pilot plant studies suffered a time lag. The pilot plant, though installed in 1983, is only now starting to yield results which can worth to be discussed. Therefore we can give only a few informations.

The first problem was the low amount of seed available. The plant fell out of equilibrium with high acidification (ph=5.7) and almost no gas production. By putting in lime, the ph was increased to ph=7.6. As Fig.7 shows, this addition of lime stimulated acid producing bacteria and had no effect on methane production. It needed a second addition of lime and an additional time span of roughly two weeks to also stimulate the methane production. The ph increase after 30 days, as the result of increasing equilibrium conditions between acidification and methanisation. Reduction of DOC concentration is a direct result of this process.

The next step is to transform the batch operated plant in a continuously fed plant, to reach steady state conditions. This transformation is expected to take roughly two to three months, and then we get data, which can be used for practical purposes.

With the results already gained, some conclusions can be made with respect to practical application: that for the real installment of technical plants it will be necessary to build up some advisory group to help the farmers in the initial phase as installment of steady state conditions in all case will need several months if large amounts of seed are not available. On the other hand, experiences with other types of wastes show, that the start up in the seasons after a long time of rest will not be accompanied by such difficulties.

(Part of this research is performed within the Ph.D. thesis of Mr. D. Ntalis)

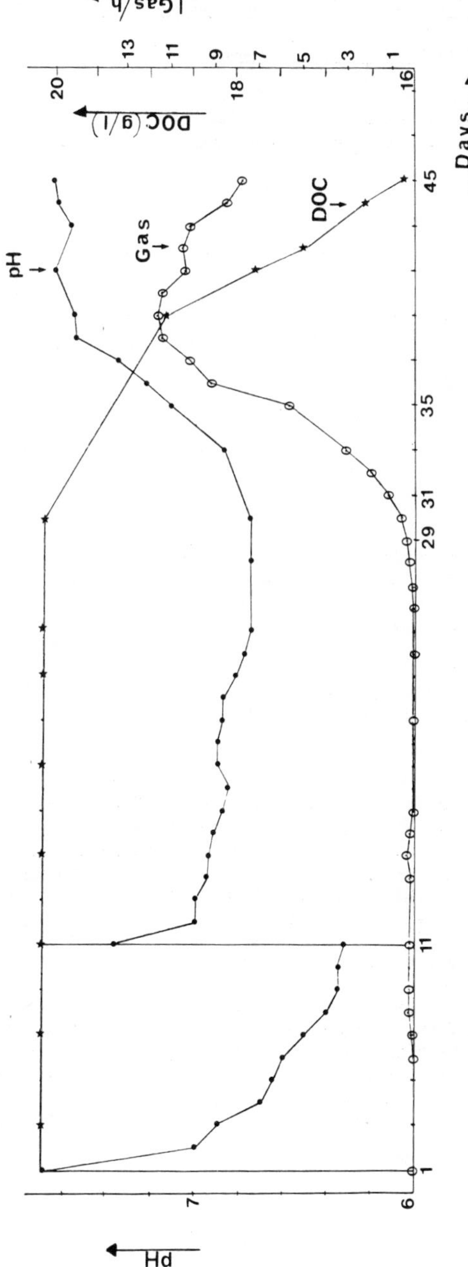

Fig: 7: Results of first pilot plant batch experiment.

DEVELOPMENT OF INSTALLATIONS FOR THE PRODUCTION OF BIOGAS FROM STOCK-FARMING WASTE

C. Aubart and F. Bully
Laboratoire de Recherches sur les Fermentations (EMC Group)

Summary

The yearly results of an anaerobic digestion unit for pig slurry at a stock-breeding farm with 3 000 pigs are presented as follows:

daily production - 208 m^3/day;

methane content - 69.5%;

gas yield - 320 m^3 CH_4/t V.S. introduced;

COD reduction - 43%.

The climate, which is particularly harsh in winter, is a major factor since 94% of daily biogas production is utilized for five months of the year (May to September), while only 70% is utilized during the winter months. Yearly electricity production is approximately 72 000 KWh and since the slurry is deodorized during the digestion process it is possible to save the 94 900 KWh consumed by the previous aerobic treatment process.

The following are the main technical observations:

- mechanical agitation is needed in the buffer tank of the digester;
- the buffer tank should be fitted with a cover to reduce heat loss during winter and to avoid the intrusion of rainwater;
- heat recovery is necessary from the exchanger of the generator set and from the waste gases;
- the highly corrosive condensates in the fume ducts should be eliminated.

1. INTRODUCTION

The EMC Group (and its subsidiaries PEC Engineering, the SCPA and Sanders) has since 1979 been working on a research programme on the utilization for energy purposes of stock-farming waste and waste from the foodstuffs industries by means of anaerobic digestion.

The development of industrial-scale installations is discussed in the paper, which follows on from those read at Wageningen in 1982 and at Pavia in 1983.

A general description of the process and the main features of the installation in question have already been given, as have the energy results for the first nine months of 1982.

The farm in question is a pig-breeding farm with an average of 3 000 pigs.

Details of a technical nature or energy data derived from three years of study are provided for certain features of the installation.

2. BUFFER TANK

The cylindrical shape of the structure is perfectly suitable. The system for homogenizing the pig slurry, which involves recirculation, has proved totally inefficient because of the considerable sedimentation of suspended matter which occurs. This problem was overcome by installing a slow-operating paddle homogenizer. The electricity consumption of this device, which operates 24 hours a day, is low (less than 10 KWh/day). However, the fact that the tank has no cover results in two major drawbacks:

- substantial heat loss in winter;
- intrusion of rainwater.

The buffer tank, which has a useful capacity of 70 m^3, is filled every three to four days, depending on the needs of the farm. The characteristics of the slurry, samples of which were taken from a tap in the digester supply circuit, are as follows:

Table I : Slurry composition

T.S.	35 g/l
% V.S. in the T.S.	69.7 %
COD	36.5 gO_2/l
pH	7.2
supernatant solids	10.8 g/l
% V.S. in supernatant solids	30.8 %
N_T	3.1 g/l
NH_4	2.8 g/l

K	1.2	g/l
P	878	mg/l
Ca	1.18	g/l
Mg	328	mg/l
Na	302	mg/l
Cu	22	mg/l
Fe	72	mg/l
Zn	43	mg/l
Mn	19	mg/l

3. DIGESTER

The polyester structure is completely satisfactory. Since its installation (in July 1981), it has operated without interruption. The sludge which accumulates at the base of the digester has been drained off periodically since July 1983. The problem is caused mainly by the sawdust which is used in stock-farming and which causes the pig slurry to clog and does not decompose (owing to the presence of lignin). More effective agitation would obviate the need for periodic draining but would substantially increase power consumption (this is now around 30 KWh/day).

Average daily biogas production is 208 m^3, with a methane content of 69.5%, and an H_2S content of 0.3%.

The number and age of the animals vary over the year, and there is therefore a slight variation in the amount and quality of slurry produced. Consequently, average biogas production for May, June, July, August and September is 223 m^3 per day, while for the rest of the year it is 200 m^3 (Fig. 1). The gas yield, however, remains constant at 320 l CH_4/kg V.S. introduced.

4. HEATING OF THE DIGESTER

A number of problems arose after the installation of the boiler:

- it was necessary to fit a pressure regulator to the boiler's biogas supply circuit;
- the mild steel fittings (bolts, casings and screws) are subject to corrosion, but so far no corrosion has occurred in the cast iron heating chamber of the boiler;
- the combustion of gases gives rise to substantial and highly corrosive condensation in the fume ducts (mild steel sheet and asbestos cement). These were replaced by a stainless steel vent.

Heat recovery from the heat exchanger of the generator set and from the waste gases, which proved reliable after some adjustments, makes it possible to reduce consumption of the boiler to 30% of the daily production of biogas during the seven "winter" months (average temperature 7.4°C), or 306 304 Kcal/day. The daily heat saving from the heat exchangers is 217 236 Kcal. During the "summer" months (average temperature 22.8°C) only 6% (70 805 Kcal) of daily biogas production is used by the boiler. A daily average of 325 768 Kcal is recovered from the heat exchangers of the generator set. The average heat requirement for the year is 470 637 Kcal/day.

In theory, about 440 000 Kcal is required daily to heat 18.5 m^3 of pig slurry from 13°C to 37°C. Heat losses appear relatively slight at about 30 000 Kcal/day. This shows the effectiveness of the digester's insulation (Table II)

SEASON	"WINTER" MONTHS	"SUMMER" MONTHS
Mean temperature °C	7,4	22,8
Daily biogas production m^3	200	223
Heating of digester - boiler - heat exchangers (Kcal/day)	 306.304 217.236	 70.805 325.768
Daily electricity production Kwh	159	250

Table II : Yearly results

5. ELECTRICITY PRODUCTION

Various components and accessories of the electronic part of the alternator's excitation unit were replaced after two years of operation.

Total yearly electricity production is 71 958 KWh. In summer, when 94% of biogas is utilized by the generator set, daily electricity production is 250 KWh, which supplies the ventilation system of the farm buildings. In winter, when 70% of biogas is utilized by the generator set, daily power production is 159 KWh, which is used for the farm's power requirements.

6. PURIFICATION

The daily COD load to be treated is 680 kg. The numerous tests which have been carried out indicate 43% COD removal.

The pig slurry is completely deodorized and can be used on cultivated land with no unpleasant effects for the immediate surroundings.

The fermentation unit has made it possible to abandon the previous aerobic process, the yearly power consumption of which was 94 900 KWh.

7. CONCLUSION

The biogas production unit in question, which is located in a part of France where the winter is particularly harsh, is completely satisfactory for the following reasons:

- COD is reduced by about 45%;
- the pig slurry is completely deodorized;
- more than 200 m^3 of biogas is produced per day;
- yearly power production is approximately 72 000 KWh.

Power production matches consumption during the summer because of the power needed to drive the ventilators in the farm buildings. However, a study is now being carried out with a view to the direct harnessing of biogas for winter heating.

All the equipment in the slurry circuit and in the gas circuit has so far proved entirely satisfactory.

Figure 1

GROSS BIOGAS PRODUCTION

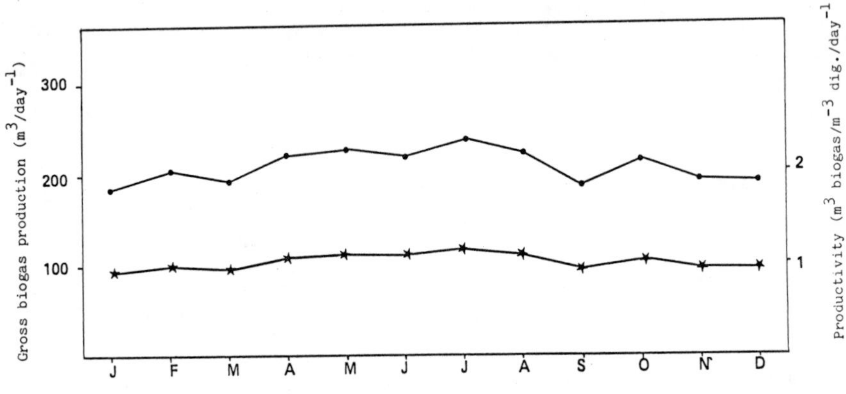

BIOLOGICAL WASTE DISPOSAL FROM SLAUGHTERHOUSES

H. POLLACK

Schmidt Reuter Ingenieurgesellschaft

Summary

During the research project RUW-041 several lab tests for anaerobic fermentation have been made. 16 liter fermenters of the horizontal type with mechanical stirrers were used. The resulting gas yields were slightly lower than tests with completely mixed fermenters.

In the second stage, energy consumption and waste disposal conditions for slaughterhouses were analyzed. The model of a modern slaughterhouse is described along the given data, and the economics of the use of methane digesters are examined.

With the presently used single-stage digesters the anaerobic fermentation seems to be no economic option for abbatoir wastes, if it is only used for energy production. Anaerobic fermentation can be economical, if additional benefits regarding the necessary waste disposal can be used.

1. SCOPE OF THE PROJECT

The research project RUW-041 was to reach two goals. One was to confirm data known from literature about methane production from slaughterhouse wastes by own lab tests. The other was to analyze energy and waste disposal conditions in order to get first hints for technical and economic feasibility of methane digesters.

2. EXPERIMENTAL RESULTS

The experimental results have been reported earlier. Thus they only will be concluded in the following.

In small fermenters with 16 liters volume, four batches in the continous flow system have been tested (Fig. 1). The substrate is agitated by a mechanical stirrer. This type of reactor has been frequently used for applications in agriculture. For these substrate it has been demonstrated, that the specific gas production rates of lab size and of large fermenters fit very well.

The batches have been fermented mesophilic and thermophilic during 10 and 20 days residence time. The goal was to operate without dilution of the substrate. Therefore the volumetric loads with respect to Volatile Solids were in the range between 4 and 9 kg VS/m^3 d (table 1). During the tests F 2 and F 4 no stable gas production could be reached. Despite special precautions during start-up the transition to high acid contents in the reactor and high CO_2 contents in the gas could not be avoided.

Stable gas production was reached for the mesophilic tests at 4.48 kg VS/m^3 d with a gas production rate of 0.263 m^3/kg VS and for the thermophilic tests at 8.96 kg VS/m^3 d with a gas production rate of 0.275 m^3/kg VS. With CO_2 contents of about 40 % this corresponds to methane yields of about 0.16 m^3 CH_4/kg VS or 1.6 KWh/kg VS as an energy equivalent. According to experiments of K. Dahl the contents of paunch manure yield 0.2 m^3 CH_4/kg VS at 10 days residence time and 8 % VS in the mesophilic as well as in the thermophilic range.

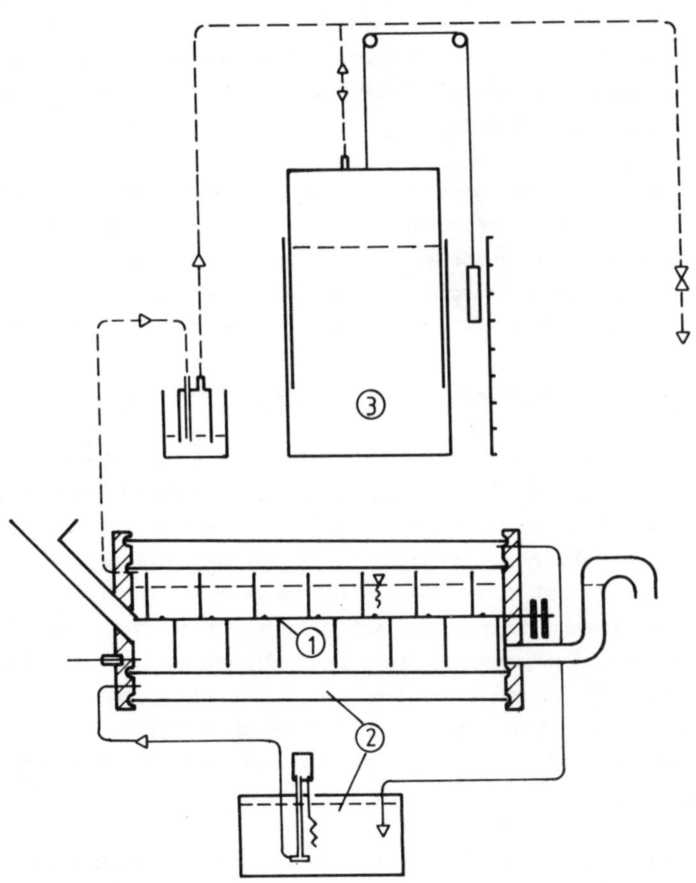

<u>Fermenter</u> :

Length	:	690 mm	①	mechanical stirrer
Diameter	:	200 mm	②	heating device
Volumen	:	16 l	③	gasometer

Figure 1: reactor - type (n Dürr/Rüprich)

The test result show, that similar gas production rates can be reached both for horizontal cylinders with mechanical stirrers and for small fermenters of the completely mixed type. A greater number of tests was planned based on the experiences with agricultural substrates. However, the paunch contents turned out to be substrates which were difficult to handel mechnaically. Start-up and changing of the parameters took longer than expected.

Based on the experience of the EC project, we are presently running a 2 m^3 reactor with the content of the intestins of cattle. Also in this case it has to be stated, that the mesophilic process, which is commonly regarded as uncritical, suffers from long start-up periods and permits only small velocities in changing the volumetric load.

3. ANALYSIS OF THE ENERGY AND WASTE DISPOSAL CONDITIONS

Another focus was the description of the technical and economic feasibility of the anaerobic fermentation in slaughterhouses. Energy consumption, the situation of waste disposal and costs were examined in detail. Even though the questions are simple, there are no answers which can be generalized. Operational conditions vary too much between different slaughterhouses. They are influenced by the number of killings, internal organisation and local conditions for waste disposal, the form of discharge and the final use of the organic residues. Furthermore a great number of legislative conditions have to be observed for waste disposal. They greatly restrict the choice of substrates which can be used.

Additional problems during the evaluation of the actual situation also have to be considered. The companies are reluctant when they are asked for quantities and costs. From the data obtained we developped a model of a large modern slaughterhouse in the F. R. of Germany, at which we will demonstrate some interesting aspects.

In the large slaughterhouse around 600.000 pigs and 30.000 cattle are killed annually. The slaughterhouse is outside the densely populated areas and disposes of its own waste water treatment plant or reduces the water pollution load as far as possible. In the F.R.G., bodies or parts of the bodies of the animals have to disposed of after legislative regulations, which can not be met by anaerobic fermentation. The contents of the intestines and sludge from waste water treatment are not counted as parts of the animals bodies (see table 2).

Paunch manure occurs only during two slaughter days per week. They are collected separately. The contents of pigs stomachs are also collected separately. Only a small part of the intestines is pressed. The contents flow with the waste water and are sieved later. The rest of the intestines is regarded like animal bodies which cannot be used further for anaerobic fermentation. All sieved residues are taken out of the slaughterhouse waste water. A part of it occurs at the truck washing station. The straw or saw dust regarded as other wastes may be used in methane digesters only to a certain degree.

The waste water is treated aerobically and thickened in gravity separators up to a solids content of 1.5 % dry substance. From this substrate after stabilization only low gas yields can be expected. Therefore this substrate may be only important for the necessary dilution.

Gas energy is used to burn the bristles of the pigs and to heat up water, electric energy is mainly used to run the cooling stations. The plant has a warm water storage tank to reduce the maximum consumption load for natural gas and heat recovery systems. The warm water demand is much higher than the possible supply from biogas. So the total consumption is ensured.

4. BASIC DESIGN OF A METHANE DIGESTER

For these conditions the economic feasibility has been studied (table 3). The substrate supply of about 15 t/d is composed of paunch manure and pig stomachs content (8.3 t/d) and 7.5 t/d waste water, which is blended to dilute down to a total solid of 8 %. A mesophilic process is assumed, where about 20 % of the methane production have to be used to heat up the substrate.

For fermenter sizes of about 300 m^3 a double wall concrete container seems to be the most favorable system. It is planned to burn the methane in a separate boiler and to feed the warm water into the reflux of the warm water supply system. This is more economical than to feed the biogas into the conventional gas supply system, which would comprise higher piping costs and complex control systems. Capital costs are calculated according to common methods. With an assumed period of 15 years technical use and 8 % interest, annual capital costs are about 11.7 %. This assumption seems to be a favourable one, actually annual capital costs may even be higher.

These costs have been confirmed by offers. The resulting price of the methane is around 92 DM/MWh or 76 DM/MWh, if the investment for the final storage is not taken into account. The comparable price for natural gas is between 55 and 70 DM/MWh.

The energy yield alone can not ensure the economic feasibility. Economic operation can be attained, if

a) Investment costs are reduced,
b) methane production is increased by about 40 % or
c) additional benefits occur by anaerobic fermentation.

There seems to be only a restricted potential for improvements regarding the first two possibilities. Additional benefits by anaerobic fermentation seem to be feasible, if waste disposal costs can be reduced thereby. According to the present legislative regulations direct disposal on agricultural areas is restricted. If the hygienic conditions of the substrate are improved by anaerobic fermentation, direct disposal might be possible. This could result in economic advantages over the disposal methods currently applied, because the fermented substrate also has a certain value as a fertilizer. The necessary dilution is a disadvantage in this respect, because of the volume and therefore transport costs are increased.

Further experiments to determine hygienic conditions are needed to confirm this potential. It can be expected that there will be even stricter regulations for waste disposal in the future and that energy costs will rise generally. Both tendencies will improve the economic conditions for anaerobic fermentation and justify further efforts.

TABLE 1: SUMMARY OF TESTS RESULTS

No.	F1	F2	F3	F4
temperature (^{o}C)	55°C	35°C	35°C	35°C
residence time (d)	10	10	20	20
substrat paunch contents	diluted	diluted	diluted	not diluted
MEASURED VALUES:				
daily introduced feed(e)	1,6	1,6	0,8	0,8
introduced: TS_e (%)	9,63	9,63	9,63	13,89
VS_e (%)	8,96	8,96	8,96	12,83
$pH\text{-}_e$	6,70	7,04	6,76	6,99
effluent: TS_a (%)	6,90	7,51	7,48	8,55
VS_a (%)	5,88	6,05	6,40	7,32
$pH\text{-}_a$	7,98	7,65	7,93	8,01
gas production:				
quantity (l/d)	39,41	-	18,87	-
CO_2-portion (Vol.%)	38,93	-	40,56	-
CALCULATED VALUES:				
introduced volatile solids (g)	143,36	143,36	71,68	102,64
volumetric load (g VS/l d)	8,96	8,96	4,48	6,42
volumetric gas production (l/l)	2,46	-	1,18	-
gas production related to introduced total solids (l/g TS d)	0,256	-	0,245	-
gas production related to introduced volatile solids (l/g VS d)	0,275	-	0,263	-
reduction of volatile solids (%)	34,38	-	28,57	-

TABLE 2: DATAS OF THE MODEL SLAUGHTERHOUSE

number of killings:	pigs 600.000 per year		
	cattle 30.000 per year		
RESIDUES:			
paunch manure			
70 Kg/animal, 14 % TS, 12 % VS		5.800 kg/d	700 kg VS/d
contents of pigs stomachs			
1 kg/animal, 26 % TS, 24 % VS		1.670 kg/d	400 kg VS/d
sieve residues and others			
8 % TS, 7 % VS		2.000 kg/d	140 kg VS/d
waste Water			
200 l/animal, 0,5 % TS, 0,4 % VS		350 m^3/d	1.750 kg/d
ENERGY CONSUMPTION:			
gas consumption	26,25 MWh/d	40 % for flaming of pigs	
		60 % for warm water	
electric consumption	14,00 MWh/d		

TABLE 3: BASIC DESIGN OF A METHANE DIGESTER FOR SLAUGHTERHOUSE SUBSTRATES

DESIGN DATA:

substrate supply: (8 % TS, 7,2 % VS)	15,775 kg/d	1,136 kg VS/d
residence time:	20 d	
volumetric load:	3,8 kg VS/m^3 d	
methane yield: (treating value 10 KWh/m^3)	0,2 m^3 CH_4/kg VS	
internal consumption: (heating from 10 to 35°C)	20 %	

INVESTMENT:

a)	methane digester, complete with double wall concrete container of 300 m^3, gas storage 50 m^3, gas boiler 140 kW, controls	240.000,— DM
b)	first dump pit 50 m^3 with mixing and cutting device	30.000,— DM
c)	final storage	(90.000,— DM)
d)	integration into the plant	40.000,— DM
e)	engineering and licensing	30.000,— DM
	total	430.000,— DM (340.000,— DM)

COSTS:

annual capital costs 11,7 %/y (15 years technical use, 8 % interest)	50.310,— DM (39.780,— DM)
annual operating costs	10.000,— DM
total costs	60.310,— DM (49.780,— DM)

energy yield	654 MWh/a	that means	92 DM/MWh (76 DM/MWh)
or heat supply	589 MWh/a	that means	102 DM/MWh (84 DM/MWh)
cost comparison:	natural gas	50 - 70 DM/MWh	
	heat	55 - 80 DM/MWh	

MESOPHILIC AND THERMOPHILIC ANAEROBIC DIGESTION OF ABATTOIR WASTE AND SEWAGE SLUDGE

K. DAHL and P. NØRGAARD
Environmental Engineering Laboratory
Department of Civil Engineering, University of Aalborg
DK-9000 Aalborg, Denmark

Summary

Digestions of sewage sludge of 3% volatile solids (VS) content and sewage sludge supplemented with either 1.5 or 3% VS of abattoir wastes were performed at 24, 16, 10.7 and 7.1 days retention times (RTs) at 37° C. Bovine waste (bw) was digested alone at 24, 16, 10.7, 7.1 and 4.7 days RT at 37° C and at 12, 8, 5.3 and 3.6 days RT at 55° C. Pig waste (pw) was digested alone at 24, 16 and 10.7 days RT at 37° C and at 12 days RT at 55° C.

Addition of abattoir wastes increased the methane production with 84-144% (pw, 1.5% VS added), 154-225% (pw, 3% VS added), 48-76% (bw, 1.5% VS added) and 92-148% (bw, 3% VS added), depending on the RT and the sewage sludge character.

Digestion of abattoir wastes appeared to be better in mixed digestion. The methane yield increased with about 21% (1.5% VS added) and 13% (3% VS added) when compared to abattoir wastes digested alone. The highest specific mesophilic methane production rate ($r_{v.CH4}$ = 2.3 l l^{-1} day^{-1}) was obtained by the digestion of a mixture of 3% VS of and 3% VS of sewage sludge at 7.1 days RT.

The highest possible loading rate of mesophilic pw digestion (8% VS) was 7.5 g VS l^{-1} day^{-1} with a $r_{v.CH4}$ of 2.1 l l^{-1} $days^{-1}$ (10.7 days RT). A $r_{v.CH4}$ of 3.7 l l^{-1} was obtained at 12 days RT, 55° C. The ultimate methane yield of pw was estimated to 0.44 l g VS^{-1}.

Mesophilic digestion of bw was possible at a RT down to 4.7 days (loading rate = 17 g VS l^{-1} day^{-1}) with 21% digestion of VS. VS digestion increased to 37% at 16 and 24 days RT. The maximum $r_{v.CH4}$ of RTs shorter than 8 days was greatly enhanced at thermophilic conditions. The $r_{v.CH4}$ was 3.3 l l^{-1} day^{-1} at 3.6 days RT, 55° C. The ultimate methane yield was estimated to 0.27 ± 0.02 l g VS^{-1}.

1. INTRODUCTION

Slaughterhouses produce abattoir wastes of strong polluting power. Slaughterhouses are often placed in urban areas nearby municipal sewage works. To decide whether the abattoir wastes and the sewage sludge should be treated separately or in mixture in existing digesters, the process efficiency was examined in both cases.

2. MATERIALS AND METHODS

The digesters had a working volume of 4 liters and a head space of 1.6 liters. The content was stirred by an assymetric two-bladed stirrer driven at a speed of 11 rpm. Gas was collected in calibrated 25 liter bottles placed in inverted position over a brine containing 20% sodium chloride and 0.5% citric acid.

The total solids were determined by drying overnight at 103-105° C. Volatile solids (VS) were determined by ignition at 550° for 2 hours. Chemical oxygen demand (COD) was determined by a modification of the method given by the American Public Health Association (1971). Ammonia was determined by steam destillation and titration with mineral acid. pH was determined by inserting a pH-probe directly into the digesting mass via the feed line. Methane was determined by a gas chromatograph (Carlo Erba 2350) equipped with flame ionization detectors. A column of glass 2 m in length and 4 mm in inner diameter packed with chromosorb G AW-DMCS 80-100 mesh coated with 2.5% silicone gum rubber E301 was used. Volatile fatty acids were determined according to Jensen (1973). Total concentration of volatile fatty acids was also determined by direct titration after Nørgaard (1983). A minimum of 3 RTs were passed before measurements were made.

3. WASTE INPUT

The composition of the feeds employed in the study is shown in table I.

Table I

Feed		A	B	C	D	E	F	G
sewage sludge	% VS (w/w)	3.0	3.0	3.0	3.0	3.0	0.0	0.0
pig wastes	% VS (w/w)	0.0	1.5	3.0	0.0	0.0	8.0	0.0
bovine wastes	% VS (w/w)	0.0	0.0	0.0	1.5	3.0	0.0	8.0
Total organic matter	% VS (w/w)	3.0	4.5	6.0	4.5	6.0	8.0	8.0
Total dry matter	% TS (w/w)	4.2	5.9	7.5	6.0	7.7	8.8	9.4
COD* (g O_2/kg slurry)		48.8	77.1	105.5	72.6	96.4	151.4	127.0
Density (25° C)		1.00	1.01	1.01	1.01	1.02	1.03	1.03

*calculated on the basis of the relationship between COD and VS contents of the wastes. The value is based on 8 samples from each of the slaughterhouse wastes and 4 samples of sewage sludge.

Due to the crude character of the abattoir wastes they were passed three times through a mincing machine (5 mm bores). Batches of approx. 1 kg were stored in polythene bags at -20° C.

A centrifuged mixture of primary and secondary activated sludge was obtained fresh from a sewage work every two weeks.

Batches of the concentrated wastes were analysed for VS and TS prior to mixing and diluted with tap water in order to give the appropriate final content of VS. Feed was made for a two week demand and kept in closed 1 liter bottles at +5° C. Due to an initial pH of approx. 4.5, bottles containing 8% VS of pig waste were adjusted to pH 7-7.5 with approximately 8 mℓ 40% Sodium Hydroxide.

4. EXPERIMENTS

Digestions of sewage sludge of 3% volatile solids (VS) content and sewage sludge supplemented with either 1.5 or 3% VS of abattoir wastes were performed at 24, 16, 10.7 and 7.1 days retention times (RTs) at 37° C. Bovine waste (bw) was digested alone at 24, 16, 10.7, 7.1 and 4.7 days RT at 37° C and at 12, 8, 5.3 and 3.6 days RT at 55° C. Pig waste (pw) was digested alone at 24, 16 and 10.7 days RT at 37° C and at 12 days RT at 55° C.

5. RESULTS AND DISCUSSION

The methane yield and VS destruction of sewage sludge digestion were expected to decrease by decreased RT.

This has not been an unequivocal finding (table III). This could be caused by variation of composition and degradeability of the sewage sludge

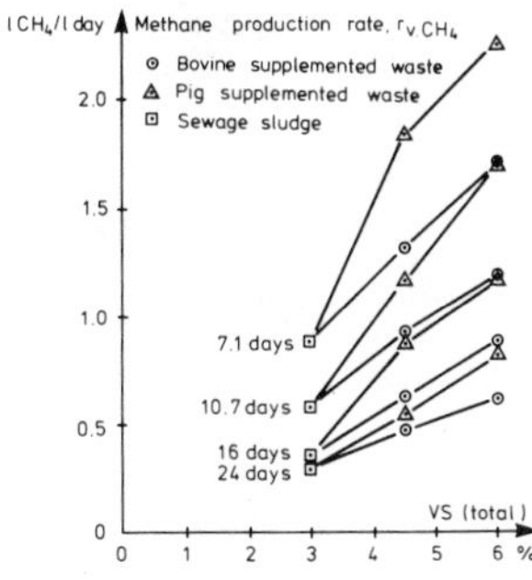

Fig. 1 Influence of waste supplementation on methane production rate. 37° C.

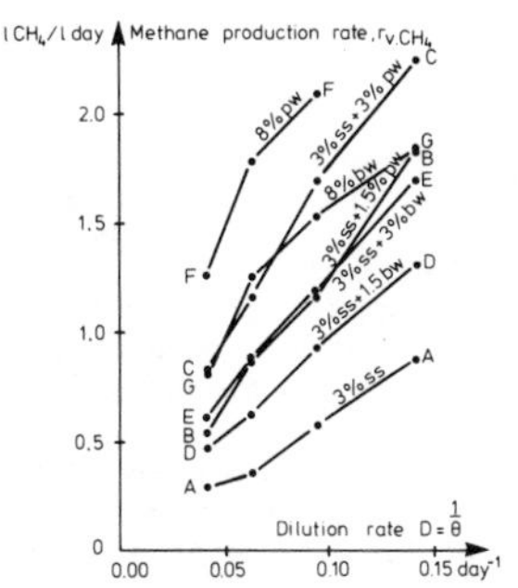

Fig. 2 Relation between dilution rate and methane production rate. pw = pig waste, bw = bovine waste, ss = sewage sludge. 37° C.

VS fed during the experimental period. Variation in sludge character due to varying performance of the activated sludge plant during the year is also likely to occur. Consequently, the finding that methane production from sewage sludge is related to the dilution rate in a linear way (fig. 2) in the range of RTS times examined, may be treated with some reservation. When the loading rate was increased by waste addition the methane production rate also increased in a rather linear way (fig. 1).

The pH of mesophilic sewage sludge digestions was 7.2-7.3. Abattoir waste supplementation caused a slightly higher pH of 7.3-7.5 due to the higher cation concentration as exemplified by ammonia ions (table III). The ammonia concentration of the sewage sludge digestion did almost correspond to the alkalinity measured. From the principle of electric neutrality, it can be calculated, that ammonia ions are the dominant cation species present, if bicarbinate is assumed to be the dominant anion species. The same conclusion could have been drawn with respect to the pig waste digestion, if sodium ions had not been artifically added to the feed. In contrast ammonia ions can only account for half of the negative charges present in bovine waste digestion. The rest of the cations are presumeably mainly sodium as being major inorganic constituents of bovine saliva (Hungate, 1966).

Mesophilic pig waste digestion at 10.7 days RT operated at depressed pH (table III) due to the accumulation of

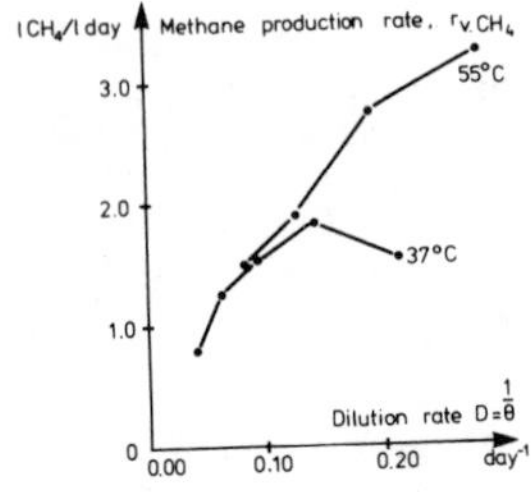

Fig. 3 Digestion of bovine waste, 8% VS.

volatile fatty acids, acetate being the dominant acid present (table II). A further decrease of the RT led to a cessation of digestion. Digestion appeared to require a minimum RT of 10.7 days. At this RT gas bubbles escaping from solution gave rise to scum formation. The odour of the effluent appears to be strong and very unpleasant at this RT.

Mesophilic digestion of bovine waste showed reduced VS destruction and methane yield at RTs shorter than 16 days (table III). Some accumulation of volatile fatty acids occurred at 10.7 days RT (table II).

Table II Concentrations of volatile fatty acids from steady state periods (mM).

Waste	bw	bw	bw	bw	bw	bw	pw
RT (days)	7.1	4.7	12.0	8.0	5.3	3.6	10.7
t (° C)	37	37	55	55	55	55	37
Acetate	11.1	5.2	2.9	3.4	1.7	5.4	28.6
Propionate	6.2	41.2	-	1.4	2.8	31.7	18.0
i-Butyrate	1.0	3.5	-	-	-	-	2.9
Butyrate	-	-	-	-	-	-	-
i-Valerate	-	4.3	-	-	-	2.5	11.8
Valerate	-	-	-	-	-	-	-

Values are means of two determinations of the steady state period.

bw = bovine waste, pw = pig waste.

- : concentration < 1 mM

An ultimate methane yield of waste can be estimated by plotting the methane yield versus the reciprocal of the RT and extrapolating to infinite RT (Chen et al., 1980). Hashimoto et al. (1981) demonstrated a reasonable agreement between this method and values obtained from batch fermentation of bovine waste.

The ultimate methane yield of pig waste is 0.441 ℓ methane g VS^{-1} (r = -0.989). The respective yields for meso- and thermophilic digestion of bovine waste are 0.295 ℓ methane g VS^{-1} (r = -0.987) and 0,250 ℓ methane g VS^{-1} (r = -0.971).

The ultimate methane yields of bovine waste are in the range between the yield of beef cattle fed high grain diet of 0.32 ℓ CH_4 g VS^{-1} (Hashimoto et al., 1981, Chen et al., 1980) and the yield of dairy cattle fed mainly roughage diet yielding 0.16 ℓ g VS^{-1} as estimated from the data of Bryant et al. (1977).

The ultimate methane yield of pig waste is in the lower range of yield (0.50 ± 0.05 ℓ g VS^{-1} fed) generally found as reported by Hashimoto (1983) for pig manures.

In order to compare methane yields of supplemented animal wastes with wastes digested seperately, an apparent methane yield Y ($\Delta CH_4/\Delta S_o$) of the waste supplemented is calculated. In this calculation the increase of methane production is attributed to the increase in VS content. Further it is assumed that the methane yield of sewage sludge organic matter is unaffected by waste supplementation. The results are given in table III. The apparent methane yields of pig waste supplemented are by a mean of 21% and 13% higher than the corresponding digestions at 8% VS at 1.5% and 3% VS supplementation respectively when results from digestion at different RTs are pooled. The corresponding values of bovine waste supplementation are 20% and 12% better methane yield. This observation may be an effect of the VS concentration as generally found (Varel et al., 1977, Summers & Bousfield, 1980, Schultze, 1958 and Wujcik & Jewell, 1980).

Digestion of bovine waste at loading rates as high as 17 g VS $\ell^{-1}d^{-1}$ at 37° C and 22.5 g VS $\ell^{-1}d^{-1}$ at 55° C were possible. At both temperatures the high loading rate effected accumulation of volatile fatty acids and caused a decrease of the methane yield (table III and IV). When the RT of the thermophilic digestion decreased below 8 days, propionic acid was the dominant acid present (table II). At 3.6 days RT the total concentration of acids sharply increased due to significant propionate accumulation. Similar propionate accumulation has been recorded in the digestion of beef cattle manure at 60° C of the same VS content (Varel et al., 1977). In spite of the acids accumulated, the pH was only slightly depressed, and titrimetric determination of acids showed unchanged total concentration during the period of measurement.

The accumulation of propionate at 4.7 days RT at 37° C was somewhat greater (table II), and this digestion was apparently stable too. The pH was somewhat depressed at this RT but was still higher than pH 7.

When mesophilic and thermophilic digestion of bovine waste are compared, highly increased production rates were experienced at 55° C at short RT, fig. 3. In this study the highest measured thermiphilic digestion rate of bovine waste surmounted the highest measured mesophilic rate with 76%. Fig. 3 shows that at a RT shorter than about 8 days the 55° C digestion is superior with respect to specific methane production rates. At RTs of more than 8 days, the thermophilic methane production rate is only slightly higher than the mesophilic methane production rate.

Pig waste digestion exhibited problems of instability. The phenomenon was only observed at RTs shorter than 16 days of the mesophilic digestion and shorter than 8 days at the thermophilic digestion. The phenomenon was most common in the thermophilic digestion. The phenomenon was characterised by a very rapid decline of gas production. This allowed the feeding to be stopped before pH was lowered beyond 7. Even if feeding was stopped as soon as gas production fell off, the gas production continued to decrease for some days, before it accelerated again, and feeding could be initiated. In these situations the normal load could not be given without causing a rapid decline of gas production. It was necessary to approach the loading rate gradually. The same problems were experienced when the RT was decreased from 8 to 5.3 days. Adjustment of the pH from a value of 6.7 to 7.3 with sodium hydroxide did not cause any immediate increase of gas production rate. The gas production rate increased gradually within 5 days from a minimum gas production of 2% of the normal. At the time of minimum gas production a total of 108 mmoles of volatile fatty acids had accumulated. Acetate and propionate accounted for 52% and 21% respectively of the acids on a molar basis. When gas production rate had reached 60% of the normal, the digesters were fed corresponding to 8 days RT. This effected a new period where gas production declined even though pH due to the prior base addition was still at 7.5.

A change of the RT of the mesophilic pig waste digester from 10.7 days to 7.1 days did not affect the gas production rate until 5 days later when gas production declined sharply. During the next 4 days the gas production rate gradually reached a value of 4-12% of the normal rate. The digester did not recover within 3 weeks though the pH of the digester had been raised from 6.5 to 7 with sodium hydroxide.

Due to the low level of ammonia nitrogen, the pig waste digestions in this study are expected to be free of instability problems that otherwise might be caused by high ammonia levels experienced when urine-faeces mixtures are digested (Hobson & Shaw, 1973, Kroeker et al., 1979, van Velsen et al., 1979). When inhibition of thermophilic pig waste digestion was attributed to ammonia inhibition, there was a pronounced effect even at long RTs (van Velsen et al., 1979, Hashimoto, 1983). In contrast, the thermophilic digestion at 12 days RT in this study operated stably. Since the total level of ammonia nitrogen and free ammonia of the pig waste digestion is lower than the corresponding levels in bovine waste digestion, important inhibition from this compound can be ruled out. The maximum possible mesophilic loading rate 7.5 g VS $\ell^{-1}d^{-1}$ in fact appears to be at least as high as the highest obtained rate from piggery waste digestion (Hashimoto, 1983, Summers & Bousfield, 1980) with significant VS-destruction. The specific methane production rate of 2.1 $\ell\ \ell^{-1}d^{-1}$ was much higher than earlier reported at high loading rates of piggery waste digestion (Fischer et al., 1979, Summers & Bousfield, 1980).

Propionate has been claimed inhibitory to digestion and has been reported as an inhibitor of M.formicicum (Hobson and Shaw, 1976). Propionate concentrations of inhibited pig waste digesters were in fact lower than the steady state concentrations of some of the bovine digesters operated in this study. Addition of acetic acid to bovine waste digesters caused a lower pH than found in pig waste digesters after cessation of the gas production without inhibition of methane production. Neither pH nor volatile fatty acids nor the combined action caused any immediate inhibitory effect on gas production. Consequently, volatile fatty acids do not appear to be the prime cause of instability in neither their dissociated nor their protonated form.

A hypothesis of the origin of the pig waste digester instability was formed in the course of this study. It was hypothesised either that the feed contained toxic compounds or the diurnal variation in nutritional or environmental conditions of different bacteria was too great to keep up a fermentation pattern leading to the ultimate formation of methane.

Experiments employing almost ideal continuous culture conditions were carried out in the thermophilic digestion of pig waste. During the first week after the initiation of feeding at a RT of 5.3 days, the average daily gas production was 7.2 ℓ gas $\ell^{-1}d^{-1}$. 35% of the gas was carbon dioxide. Gas production varied somewhat during the day and gas production rates as high as 9.5 $\ell\ \ell^{-1}d^{-1}$ were commonly measured. The pH of the digester declined slowly to pH 7.0. The concentration of volatile fatty acids was slowly increasing and reached a value of 11 mM acetate and 18 mM propionate. Butyrate, i-butyrate, valerate and i-valerate were also present in minute amounts. During the first 13 days after the start of the experiment, the gas production averaged 6.3 ℓ gas $\ell^{-1}d^{-1}$ (35% CO_2). At this time pH had declined to 6.7 and 25 mM acetate, 22 mM propionate, 1.4 mM butyrate and 1.4 mM i-butyrate were present. Hereafter the gas production declined more rapidly and attained a value of 3.9 ℓ gas $\ell^{-1}d^{-1}$ before feeding was stopped. Although pH was udjusted to pH = 7, gas production stopped. The accumulated volatile fatty acids reached a total concentration of 101 mM. 67

Table III	Retention time, days	A	B	C	D	E	F	G
Sewage sludge %VS (w/w)		3.0	3.0	3.0	3.0	3.0		
Pig wastes "			1.5	3.0			8.0	
Bovine wastes "					1.5	3.0		8.0
Total organic matter "		3.0	4.5	6.0	4.5	6.0	8.0	8.0
Total dry matter %TS		4.2	5.9	7.5	6.0	7.7	8.8	9.4
Gas prod.rate l gas/l day	7.1	1.25	2.593	3.207	1.90	2.46	-	2.67
	1o.7	0.827	1.648	2.432	1.32	1.75	3.346	2.22
	16.0	0.511	1.109	1.681	0.902	1.31	2.648	1.85
	24.0	0.420	0.792	1.17	0.675	0.918	1.77	1.19
Methane prod.rate lCH_4/l day	7.1	0.888	1.84	2.26	1.318	1.708	-	1.85
	1o.7	0.581	1.17	1.70	0.936	1.196	2.098	1.54
	16.0	0.360	0.880	1.17	0.633	0.891	1.785	1.26
	24.0	0.297	0.549	0.828	0.476	0.627	1.26	0.81
Gas yield l gas/gVS fed	7.1	0.296	0.410	0.380	0.300	0.291	-	0.237
	10.7	0.294	0.391	0.432	0.314	0.311	0.446	0.296
	16.0	0.273	0.394	0.448	0.321	0.350	0.530	0.371
	24.o	0.335	0.422	0.467	0.359	0.366	0.530	0.356
Δ gas yield l gas/gVS suppl.	7.1		0.637	0.464	0.307	0.286		
	1o.7		0.584	0.571	0.353	0.329		
	16.0		0.638	0.524	0.416	0.428		
	24.0		0.594	0.599	0.407	0.397		
Methane yield l CH_4/gVS fed	7.1	0.210	0.291	0.267	0.209	0.202	-	0.164
	10.7	0.206	0.276	0.302	0.222	0.212	0.303	0.205
	16.0	0.193	0.277	0.312	0.225	0.237	0.357	0.252
	24.0	0.237	0.293	0.326	0.253	0.250	0.377	0.242
Δ methane yield l CH_4/gVS suppl.	7.1		0.451	0.324	0.204	0.194		
	1o.7		0.415	0.397	0.252	0.219		
	16.0		0.555	0.433	0.291	0.283		
	24.0		0.402	0.415	0.286	0.264		
VS-reduction, % a)	7.1	42(3)	37(2)	40(6)	38(3)	37(9)		32(4)
	10.7	34(5)	38(4)	40(5)	36(4)	39(6)	49(6)	31(7)
	16.0	36(8)	40(4)	43(4)	38(4)	39(2)	54(13)	37(3)
	24.0	47(9)	48(2)	48(4)	43(3)	43(2)	58(5)	37(3)
pH b)	7.1	7.3	7.4	7.3	7.4	7.5	-	7.4
	1o.7	7.3	7.3	7.4	7.4	7.4	7.1	7.5
	16.0	7.2	7.3	7.4	7.3	7.4	7.4	7.4
	24.0	7.2	7.3	7.3	7.3	7.4	7.4	7.4
Alcalinity c) end point pH 4.5, mmolH^+/l	7.1	75(2)	92(2)	96(2)	95(2)	122(4)	-	128(2)
end point pH 5.5, "	"	64(2)	78(2)	79(3)	82(3)	104(4)	-	104(3)
end point pH 4.5, "	1o.7	71(2)	90(1)	104(2)	97(1)	120(1)	91(5)	142(3)
end point pH 5.5, "	"	62(2)	78(2)	91(5)	84(2)	105(1)	58(4)	125(6)
end point pH 4.5, "	16.0	67(2)	84(2)	99(2)	80(3)	125(3)	124(4)	151(4)
end point pH 5.5, "	"	59(2)	74(2)	87(2)	88(4)	110(3)	108(3)	131(3)
end point pH 4.5, "	24.0	72(3)	90(5)	104(8)	106(6)	138(3)	122(9)	162(2)
end point pH 5.5, "	"	-	-	-	-	-	-	-
Ammonia, (NH_3+NH_4)-N, mmol/l d)	7.1	71(2)	89(3)	95(2)	81(1)	98(5)	-	66(5)
	10.7	69(1)	86(2)	99(1)	83(1)	94(1)	48(2)	65(1)
	16.0	65.2	78(1)	92(5)	82(5)	96(2)	42(3)	66(1)
	24.0	71(2)	86(1)	90(2)	92(2)	103(3)	50(2)	77(3)

a) Mean of 5 determinations
b) " " 2 "
c) Mean of 4-5 determinations
d) " " 3-5 "

Standard deviation in paranthesis.

mM acetate and 26 mM propionate had accumulated. Another experiment was performed in which the RT of 5.3 days was approached gradually and pH was adjusted currently. The results of this experiment was similar to the previous one.

The continuous feeding experiment did not exhibit the characteristic abrupt cessation of gas production. Instead, failure occurred when pH was slowly depressed below the optium range of methanogenesis, probably due to the high loading rate used in these experiments.

Table IV Thermophilic digestions, 55°C

	Retention time, days	Bovine wastes 8 % VS	Pig wastes 8 % VS
Gas production rate l gas/l day	3.6 5.3 8.0 12.0	4.67 4.03 2.71 1.98	3.70
Methane production rate l CH_4/l day	3.6 5.3 8.0 12.0	3.25 2.77 1.90 1.50	2.54
Gas yield l gas/g VS fed	3.6 5.3 8.0 12.0	0.207 0.269 0.271 0.297	0.554
Methane yield l CH_4/ g VS fed	3.6 5.3 8.0 12.0	0.144 0.185 0.192 0.225	0.381
VS-reduction, %	3.6 5.3 8.0 12.0	21(12) 33(28) 32(10) 38(9)	57
pH	3.6 5.3 8.0 12.0	7.5 7.7 7.7 7.7	7.6
Alkalinity end point pH 5.5 end point pH 4.5 end point pH 3.5	 3.6 " "	 86(3) 124(5) 148(9)	
end point pH 5.5 end point pH 4.5 end point pH 3.5	5.3 " "	104(3) 125(3) 137(2)	
end point pH 5.5 end point pH 4.5 end point pH 3.5	8.0 " "	116(2) 138(1) 150(2)	
end point pH 5.5 end point pH 4.5 end point pH 3.5	12.0 " "	124(2) 144(6) 154(6)	108(14)
Ammonia (NH_3 + NH_4)-N mmol/l	3.6 5.3 8.0 12.0	69(5) 65(2) 71(1) 77(1)	48(1)

6. REFERENCES

American Public Health Association (1971). Standard methods for examination of water and wastewater. 13th ed. American Public Health Association Inc., Washington, DC.

Bryant, M.P., Varel, V.H., Frobish, R.A. & Isaacson, H.R. (1977). Biological potential of thermophilic methanogenesis from cattle wastes. In Schlegel, H.G. & J. Barnea (eds.): Microbial Energy Conversion, p. 347-359. E. Goltze KG. Göttingen.

Chen, Y.R., Varel, V.H. & A.G. Hashimoto (1980). Effect of temperature on methane fermentation kinetics of beef cattle manure. Biotechnology and Bioengineering Symposium, 10:325-39.

Hashimoto, A.G. (1983). Thermophilic and mesophilic anaerobic fermentation of swine manure. Agric. Wastes 6: 175-191.

Hashimoto, A.G., Varel, V.H. & Y.R. Chen (1981). Ultimate methane yield from beef cattle manure: Effect of temperature, ration constituents, antibiotics and manure age. Agricultural Wastes 3(4): 241-256.

Hobson, P.N. & Shaw, B.G. (1973). The bacterial population of piggery-waste anaerobic digesters. Water Research 8: 507-516.

Hobson, P.N. & Shaw, B.G. (1976). Inhibition of methane production by Methanobacterium formicicum. Water Research 10: 849-852.

Hungate, R.E. (1966). The rumen and its microbes. Academic Press, New York.

Jensen, K. (1973). Gas-solid adsorption chromatographic determination of short-chain fatty acids in rumen fluid. Acta vet. scand. 14: 335-337.

Kroeker, E.J., Schulte, D.D., Sparling, A.B. & Lapp, H.M. (1979). Anaerobic treatment process stability. J. Water Pollut. Control. Fed. 51:717-727.

Nørgaard, P. (1983). Note on the direct titrimetric determination of volatile fatty acids of anaerobic digesters. Environmental Engineering Laboratory, Department of Civil Engineering, University of Aalborg, Denmark.

Schulze, K.L. (1958). Studies on sludge digestion and methane fermentation. I. Sludge digestion at increased solids concentrations. Sewage Ind. Wastes 30(1): 28-45.

Summers, R. & Bousfield, S. (1980). A detailed study of piggery-waste anaerobic digestion. Agricultural Wastes 2: 61-78.

van Velsen, A.F.M., Lettinga, G. & den Ottelander, D. (1979). Anaerobic digestion of piggery waste. 3. Influence of temperature. Neth. J. agric. Sci. 27: 255-267.

Varel, V.H., Isaacson, H.R. & Bryant, M.P. (1977). Thermophilic methane production from cattle waste. Appl. Environ. Microbiol. 33(2): 289-307.

Wujcik, W.J. & Jewell, W.J. (1980). Dry anaerobic fermentation. Biotechnology and Bioengineering Symposium 10: 43-65.

Summary of the discussion of Session II :

PART 2 - ANAEROBIC DIGESTION

Chairman : H. NAVEAU
Rapporteur : A.D. WHEATLEY

Dr. Taldrup asked two questions:-

He had noted that Dr. Lettinga mentioned that suspended solids (SS) removal was better in anaerobic filtration (AF) than in sludge blanket reactors (USAB). Did Dr. Lettinga have any theories why this should be so?

Dr. Lettinga replied:

That his results showed that SS removal by AF was slightly better than USAB treating raw domestic sewage at 25-30°C. At lower temperatures (10-12°C) the efficiency was reversed with UASB system performing better than AF. Treating abbatoir waste (COD 1500-2000 80% insoluble) the AF performed much less satisfactorily than the UASB presumably because of clogging problems.

Dr. Taldrup's 2nd question:

Asked how a UASB would perform treating a separated pig slurry with 2% volatile solids (VS) and 50,000mgl^{-1} COD. Would recycling improve process stability?

Dr. Lettinga replied:

That UASB can be used to treat piggery wastes but the performance depends on the SS content. Recycling would normally be beneficial by balancing the load and volatile acid content (VFA) of the feed.

Dr. E. Colleran, University College, Galway

Asked if Dr. Lettinga had an explanation why small amounts of sucrose in the feed to UASB reactors improved start up as was mentioned in Dr. Lettinga's paper.

Dr. Lettinga replied:

That granulation of the sludge seemed to proceed more quickly in the presence of scurose. It had been observed that granulation followed on from the growth of filamentous organisms and the sucrose may accelerate the growth of these organisms.

Dr. Fernandes, Water Research Centre, U.K.

Asked what loading rates and temperatures were required to start a UASB treating domestic sewage and how important was the presence of an inert material in granulation.

Dr. Lettinga replied:

That it was not possible to give accurate recommendations regarding start up of all UASB systems treating sewage. The reactor should be inocculated with spetic tank sludge if available. 15-20°C was the minimum temperature and then performance should steadily improve. The sucrose and inert material discussed in the paper assisted granulation and attachment.

Dr. G.H. Kristensen, Denmark

Asked Dr. Lettinga if he knew of any investigations on the kinetics of the bacterial hydrolytic step for the treatment of industrial waste waters. This information might form the basis for selecting a single or a two stage process for anaerobic treatment.

Dr. Lettinga

Replied that there was little information available on the detail rates of this step it may or may not be the rate limiting step depending on the substrate. If the rate of hydrolysis was slow then an additional stage may be beneficial.

Dr. G. Petersen, Denmark

Asked Dr. Lettinga whether the use of a separate hydrolysing reactor would improve biogas yields and rates in biogas reactors treating waste waters containing suspended solids.

Dr. Lettinga replied:

That suspended solids were not well retianed by high rate reactors and consequently had little effect on gas yield. Settling then liquifaction of the solids would yield a soluble substrate for methanogensis.

Ms. D. Oakley, University of Birmingham

Ms Oakley had noted from Dr. Lettinga's presentation that UASB's seemed to start up slower the fixed films. Data from the University of Birmingham showed the opposite trend.

Dr. Lettinga

Replied that this was a matter of definitions, UASB reactors could start up more quickly than fixed films and this depended on how start up was judged.

Paper 2 Energy Recovery from Agro-Industrial Wastes Using Fixed Bed Anaerobic Reactors

by P. Sanna, M. Camilli and L. Degen

Dr. R. Campagna, Instituto Guido Donegani, Novara, Italy
had 3 questions for Dr. Sanna

1. What was the suspended solids concentration in the effluent waste?
2. What type and what were the main characteristics of the filling that was used on the full scale plant?
3. Was the loading rate and biogas production expressed in terms of the total volume of the reactor or to the void volume excluding the filling?

Dr. Sanna replied:

That normally the suspended solids were between 2-3grms/litre. The support media was a mineral aggregate with a voidage of about 50%, they were about 5cm in diameter, this was the best material available locally at that time. The loads are expressed in terms of the voidage or useful space in the reactor that is excluding the packing. This method of expressing results used by Dr. Sanna enabled easier comparison of efficiency relative to biomass, with other reactor types.

Dr. P. Weiland, Brauuschweig Deutchland
also asked 3 questions:-

1. based on the tests on different types of support materials, which was the best and which would be used in the next plant.
2. How much of the total biomass is attached to the support and
3. what is the optimum height of the reactor for COD reduction?

Dr. Sanna

Answered that the best material was a lignocellulose residue but it was impossible to use this at full scale because of the very large amount of support material required. Economically plastic was not as attractive as mineral media. For future work they would evaluate the costs of each media at each site but they would use stone again if it were the cheapest.

2. A significant portion of the biomass was not firmly attached to the support and this could be one of the reasons why high voidage plastic media did not give as good a performance at low HRT.

3. Most of the COD removal occurs in the base of the reactor and tall systems do not seem to be beneficial in terms of performance.

Dr. E. Colleran, University College Calway

Noted Dr. Sanna's work on laboratory scale anaerobic filters which had investigated different packing materials, and asked which support media was used in the final full scale plant and what were the reasons for the choice?

Dr. Sanna answered

That many packing materials could be used and the best results were obtained from waste lignins. This could not be supplied in sufficient quantity for the full scale plant and so stone had been used as the next most cheaply available material.

Paper 3 An investigation into energy recovery and waste treatment by anaerobic biofiltration

by A.D. Wheatley

Dr. E. Colleran, University College, Galway

Opened the discussion to Dr. Wheatley's paper with 2 questions:-

1. That the choice of media seemed to have been based on specific surface area and voidage but that work by Dahab and Young in the USA and Dr. Colleran's own work showed that pore size and packing arrangement together with surface characteristics were the most important media. parameters.
2. The $10m^3$ plant discussed in Dr. Wheatley's paper was very tall compared to its diameter, was this based on any laboratory work and had the COD removal profile been investigated through the reactor.

Dr. Wheatley replied:

1. That he agreed that the smallest pore size in the packing and the nature of the surface of the packing were the most important technical factors affecting the choice of media. His own work had shown that blockage problems most frequently occurred at media contact points where the pore size was smallest. Microbial adhesion was also better to mineral media than plastic (Wheatley 1981). However another factor to be considered was cost, and the media used in the $10m^3$ plant was an easily available cheap aerobic support media (Flocor ICI Ltd, £$30m^{-3}$). This was made from PVC which was better than Polypropylene for microbial adhesion. This material had not been available for the larger $22m^3$ plant and ETA PVC media at £$50m^{-3}$ had been used. It coincidently had a larger pore structure.

2. In reply to the 2nd question the aspect ratio of the reactor was 2 : 1 some of the photographs were deceptive because of the insulation. The site was an exposed hillside location at about 300m and subzero winter temperatures were normal. The reactor was insulated with 300mm of rockwool and the reactor was also on legs which made it look taller than it was. 2 : 1 aspect ratio vessels were the most common and therefore the cheapest design.

Dr. Campagna, Instituto Guide Donegani, Novara, Italy

Asked Dr. Wheatley what evidence he had that his reactors were infact plug flow and not completely mixed by the evolution of gas.

Dr. Wheatley replied:

That two operating parameters were regularly measured, as detailed in the full paper, which indicated continuing plug flow conditions existed within the reactor. Sampling down the side of the reactor showed more COD removal at the base of the reactor and this was similar to Dr. Sanna's results. Tracer studies with LiCl showed plug release of the tracer from the reactor. The upflow velocity was about 0.1m/hr.

Dr. E.J. Stanhuis, Dept. Chem. Eng., Ryksuniversiteit, Groninga, Netherlands

Asked 2 questions

1. Were ther any differences between the thickness of biofilm between the laboratory and pilot reactor?
2. Was there any evidence that the mass transfer from the bulk liquid into the biomass layer was limiting the biogas production rate.

Dr. Wheatley replied:

1. That no comparisons of film thickness between laboratory and pilot plant had been carried out. On inspection they seemed similar. There was a lagyer of biomass firmly bound to the media, a layer of thick biomass loosly held in the voidage and some unicellular bacteria. The accumulated biomass was often up to 50mm thick.

2. Yes, there was some evidence that mass transfer limited rates of COD removal although not necessarily gas production which took rather longer to respond to changes. Reactors of identical dimensions with and without support and with different amounts of biomass could produce similar removals of COD. Below a critical HRT COD removal was proportional to HRT and COD removal was also related to the nature of the substrate.

Dr. B. Ruggeri

Asked if the mesophilic operating temperature of digestion was an optimum from a net energy production point of view? The best working temperature was a balance between energy produced as biogas and energy consumed to keep the fermenter warm. This depended on volume, surface ratio, retention time, thermal characteristics, construction material, efficiency of heating and general layout of the plant.

Dr. Wheatley replied:

That many of the food, drinks and fermentation waste waters he had used were mesophilic and required no heating. His own $25m^3$ was only insulated. There was research in progress on other temperatures for digestion but these had not, as yet, produced a commercial plant. Dr. Wheatley agreed that where heating was necessary a proper economic estimate of benefit should be carried out. Normally in the UK 70% of the savings from using anaerobic treatment were to be derived from the reduction in discharge costs so heating digestion was still often economic.

Paper 4 Anaerobic treatment of olive processing wastes
by Dr. Staud

Dr. Rozzi, CNR, Italy

Had some comments about operational problems related to anaerobic digestion. Dr. Rozzi had investigated the use of UASB for the treatment of olive processing wastes and found that chemicals were absolutely necessary most of the time but particularly for long periods during start up. It was his conclusion that there was little hope that biological waste treatment could be used for olive processing waste waters. It had been proposed in the Apulice region of Italy to build centralised recovery by distillation to overcome the waste water problems. For small isolated mills 'switch on' 'switch off' plants were necessary and this would be extremely difficult.

Paper 5 Development of industrial biogas production units from fermentation of agicultural wastes and food processing wastes
by F. Bully

Dr. A. Ferretra, Orey, Portugal

What is the diameter of the evacuation pipe for solids which is at the bottom of the digester and how often is it necessary to service it?

Dr. Bully replied:

1. It seemed necessary after two years operations with the digester; the depths of temperature found on the digester indicated a bad caloric separation. Agitation of the digester, which broke up the calories, and encouraged fermentation had become ineffective.

2. The frequency of the service seems to be around twice per year. The diameter of the evacuation pipe is around 25 cm. A pump is attached to this pipe.

Paper 6 Thermophilic and mesophilic anaerobic digestion of abattoir wastes by K. Dahl and P. Nørgaard

Dr. R. Campagna, Instituto Guido Donegani, Novara, Italy

Concurred with the authors with regard to their conclusions about the ease mesophilic digestion of bovine and porcine wastes. The work in Italy now included a $2000m^3$ digester operation at 30^oC.

ANAEROBIC DIGESTION IN NORTH AMERICA

R.L. WENTWORTH

Dynatech R/D Company
99 Erie Street
Cambridge, Massachusetts 02139, USA

Summary

A survey is presented of the scope of utilization of anaerobic digestion in various regions of North America. Full-scale installations in Canada have been oriented principally toward pollution control. An active program of research and development in Canada has supported the adoption of advanced digester designs. At least 13 large-scale digesters are now in operation or proposed. Utilization of anaerobic digestion in the United States has had energy recovery as its prime motivation. In early 1984, the number of digesters of various capacities in the 50 states is estimated to be 84. At least 11 digesters have been installed in outlying areas of the U.S. such as Puerto Rico and the Pacific Islands. Of growing importance in the United States is recovery of the fuel gas produced by anaerobic digestion in landfills of municipal solid waste. There are 28 landfill gas recovery plants in operation. In Mexico and Central America about 150 digesters are in operation. Smaller installations are oriented toward meeting family energy needs in Central America; larger installations serve agricultural industry needs for disposal and fertilizer production. Throughout the continent scientific and engineering interest and activity is strong.

1. INTRODUCTION

In various regions of North America there are incentives for utilization of anaerobic digestion including all of its practical benefits. Examples can be cited of individual or family fuel gas generators, schemes for large scale gas generation, pollution control processes, and exploitation of landfills. This is supported by research and development work aimed at all of these areas.

The present status (1984) of the implementation of anaerobic digestion is one of conservative growth. Relief of pressure on energy prices, reduction in government support, and more realistic assessments of the economics and capabilities of anaerobic digestion have slowed its adaption. The scope of this technology may be summarized separately for Canada, the United States, and Mexico and Central America.

2. CANADA

Anaerobic digestion facilities in Canada include a number of manure digesters, as well as digesters dedicated to food processing wastes and wastes from pulp and paper manufacture. The designs vary from conventional stirred or unstirred tank digesters to advanced sludge blanket and anaerobic filter systems. The incentives for adoption of digestion in Canada have been principally pollution control, but methane utilization for heating and generation of electricity is in place.

Full scale plants for anaerobic digestion in Canada (1) are equally divided between animal waste applications and industrial waste treatment. These are located in the eastern provinces, as indicated in Table I.

Table I

DISTRIBUTION OF ANAEROBIC DIGESTION FACILITIES IN CANADA

LOCATION	NUMBER OF FACILITIES AGRICULTURAL	INDUSTRIAL
New Brunswick	-	1
Ontario	5	4
Prince Edward Island	1	-
Quebec	1	1
	7	6

Because Canada is a relatively energy-rich country, economic incentives have not developed for adoption of digestion technology in the western provinces, e.g., for digestion of plant residues or crop-grown substrates. As Figure 1 indicates, the present installations lie in the southernmost reaches of the country.

Three of the digesters listed in Table I are at present in the design stage. One of these, to be dedicated to the fermentation of sulphite liquor produced at a pulp mill, is projected to be more than $50,000 m^3$ in volume. It will be an unstirred vessel operated at a low loading level. Other digesters cited are applied to cattle and hog manure, vegetable wastes and milk and cheese waste. The total volume of these other digesters is estimated to be $12,000 m^3$.

At two landfills in Canada gas recovery systems are in place. One of these, at Kitchener, Ontario, delivers $2.6x10^6 m^3/yr$ dehydrated, medium energy content gas to an adjacent factory for steam generation. The second recovery system, in Mississauga, Ontario, is operated solely as a means of reducing the hazard from gas to nearby residents. The gas is flared on site.

In his review of the status of Canadian activity in 1983, Hall (1), pointed out that a vigorous program of research and development in Canada has advanced digestion technology there significantly. About twenty universities, government laboratories, and engineering firms are concerned with research and process development in the field. Notable emphasis has been placed on advancing the understanding of high rate digestion processes with the aim of tailoring performance to specific substrates. Currently, special attention is being given to the value of refeeding the residue from digestion of agricultural wastes and to the significant contribution anaerobic wastewater treatment can make to the success of new pulp and paper technologies so important to Canada. A notable activity is the development of the downflow stationary film (DSFF) reactor, based on work at the National Research Council in Ottawa.

The digesters cited in Table I include mixed and unmixed tank systems, DSFF digesters, upflow anaerobic film digesters, and upflow anaerobic sludge blanket (UASB) digesters. Some of the industrial waste installations have been supported financially, in part, by government funds, either national or provincial. Experience with these plants extends from 1980, and these publicly assisted systems have provided development sites for advancing the art and disseminating information. Hall (1) cites the evolution at a 4500 head beef feedlot (Roslyn Park) where centrifugally separated manure solids are sold for compost while centrate is digested, polymer added, and a 20% crude protein centrifuge cake produced as refeed supplement. Digester gas drives internal combustion engines which provide compressed air and hydraulic energy for plant operation. Waste heat is utilized in the plant. In another innovative commercial approach hybrid anaerobic filter/UASB reactors have been developed to improve start-up, alleviate plugging, and assist solids retention.

3. THE UNITED STATES

3.1 Anaerobic Digesters in the United States

The familiar application of anaerobic digestion to sewage treatment is widespread in the United States, but it is the rare instance that the fuel gas generated is utilized outside the sewage plant itself. However, technical appreciation of this application as well as the pioneering work of A.N. Buswell in the decade of the 1930's, provided background for serious investigation of the potential of anaerobic digestion for generation of significant domestically controlled fuel gas supplies in the early 1970's. Biomass resources in the U.S. suitable as raw material for such digestion were estimated at that time to have the potential for producing the following amounts of energy (2):

Substrate	Recoverable Energy Equivalent, joules
Animal Manures	1.0×10^{17}
Agricultural Residues	1.2×10^{18}
Forestry Wastes	2.1×10^{18}

Both private and government-financed development programs were undertaken to bring about realization of useful energy resources via anaerobic digestion. Substantial funds were expended by the U.S. Department of Energy in support of this field and much enthusiasm was generated by individuals, particularly in the agricultural community, about the prospects. This history emphasizes energy recovery by anaerobic digestion rather than pollution control. The abatement of nuisance, e.g., accumulation of manures at cattle feedlots or on dairy farms, was an attractive dividend, but the incentives in the U.S. until now for exploitation of anaerobic digestion have centered on energy production.

In assessing the scope of adoption of anaerobic digestion technology in the U.S. there is difficulty in allowing for the scale of individual units. There is a number of pilot-scale digesters which are of importance locally and nationally in providing experience and direction to the field. Similarly, some small farm installations have contributed importantly to the spread of information. Another factor of imprecision is lack of current knowledge of inactive digesters. The easing of price and availability of fossil fuels has removed some of the incentive for digestion technology among those who were motivated to construct digesters initially by the energy generation potential. Other influences affecting the discontinuation of digester operation are fluctuations in substrate availability, e.g., manure from beef cattle raising, and the completion of demonstration programs. Accordingly, the census data presented here count digesters of substantially varying sizes and some of uncertain status.

In each of the states and outlying areas of the United States offices have been established to direct, coordinate, and inform about energy and energy conservation. Information about the current status of utilization of anaerobic digestion was solicited from each of these sources. Responses were received from 86% of these offices. This information was supplemented with the review of U.S. digestion technology presented by Hashimoto (3) in 1983, and with personal reconnaissance and correspondence. On this basis, the estimates presented in Tables II and III show that there are presently about 84 digesters in the 50 states plus an additional 11 others is outlying areas of the U.S.. There are digester installations in 2/3 of the states, with notable concentrations in California, in a column of states near the Mississippi River, and in the northeastern states. This distribution arises because the principal substrates are manures from beef, pork or poultry raising or from dairy operation, and it is in these regions that these businesses are notably active. The concentration of activity is emphasized in a different way by noting that no digesters are reported in 18 states and that only one installation is reported in 12 states; thus activity is absent or minimal in 60% of the states. A visual appreciation of the way in which these digesters are distributed geographically is provided by Figure 2.

These digesters in the United States are principally mesophilic, stirred tank or plug flow digesters. A few anaerobic filters and other high rate digesters are in place, but these are a small minority at present. The total volume of these digesters is estimated to be about $90,000 m^3$ ($3.2 \times 10^6 ft^3$). Individual digesters range in size up to

Table II
NUMBERS OF ANAEROBIC DIGESTERS AND LANDFILL GAS RECOVERY SYSTEMS IN THE UNITED STATES

	ANAEROBIC DIGESTION FACILITIES (EXCLUSIVE OF SEWAGE TREATMENT)	LANDFILL GAS SYSTEMS	
		OPERATING	PLANNED
Alabama	0	1	0
Alaska	1		
Arizona	1		
Arkansas	5		
California	7	13	23
Colorado	2		
Connecticut	1		
Delaware	0	0	1
Florida	4	0	1
Georgia	2	1	0
Hawaii	4		
Idaho	1	1	0
Illinois	1	3	1
Indiana	0		
Iowa	8		
Kansas	0		
Kentucky	0	0	1
Louisiana	0		
Maine	3		
Maryland	3	0	4
Massachusetts	1	0	1
Michigan	4	0	4
Minnesota	2		
Mississippi	0		
Missouri	5	1	0
Montana	2		
Nebraska	1		
Nevada	0		
New Hampshire	3		
New Jersey	0	3	1
New Mexico	0		
New York	5	4	6
North Carolina	1	1	0
North Dakota	1		
Ohio	0	0	1
Oklahoma	1		
Oregon	0	0	2
Pennsylvania	3	0	2
Rhode Island	0	0	3
South Carolina	0		
South Dakota	0		
Tennessee	0		
Texas	3	0	2
Utah	0		
Vermont	3		
Virginia	1		
Washington	2		
West Virginia	1		
Wisconsin	2		
Wyoming	0		
TOTAL	84	28	53

Table III

CENSUS OF ANAEROBIC DIGESTERS
IN OUTLYING AREAS OF THE UNITED STATES
(Exclusive of Sewage Treatment Plants)

AREA	NUMBER OF FACILITIES
American Samoa	0
Guam	3
Commonwealth of Northern Marianas	5
Commonwealth of Puerto Rico	2
Trust Territory of the Pacific Islands	0
Virgin Islands	1
	11

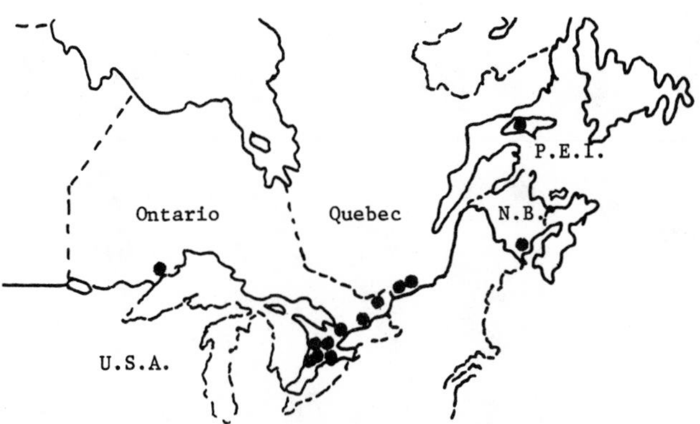

Figure 1. Location of Anaerobic Digesters in Canada.

$13{,}000m^3$ ($4.6 \times 10^5 ft^3$), but the majority have volumes in the range 250 - $750m^3$. In addition to manure processing, applications have been made to cannery waste, slaughter-house waste, whey, and distillery waste.

3.2 Landfills as Fuel Gas Sources in the United States

The practice of landfilling to dispose of municipal solid waste in the United States has created many hundreds of accumulations of digestible material. Indeed, like natural, plant-derived wastes, the contents of landfills spontaneously undergo anaerobic digestion. The composition of municipal waste in the U.S. generally is 50% by weight paper and paperboard, an ideal substrate for fermentation. Active exploitation of landfills for fuel gas is now under way. In a few cases the prime incentive for removing gas from landfills is the abatement of nuisance or safety hazard, but for the most part recovery of fuel gas is the prime objective.

The first undertaking of this kind was completed in 1975, at Palos Verdes, California. A census (4) of landfill gas recovery plants completed and in operation in early 1984, shows 29 gas recovery plants in operation. At the same time advanced planning or construction is under way on 53 additional sites. Many more than these are being considered for exploitation, but without firm intentions or contracts in place.

The location of landfill gas recovery projects is summarized in Table II. California clearly is the leading seat of this activity, the New York - New Jersey area being second. This distribution is shown graphically in Figure 3. The quantity and concentration of waste in these particular densely populated areas accounts for this leadership. The rate of fuel gas recovery from the 29 recovery systems in place is about 4×10^8 cubic meters annually (40 million cubic feet daily). These figures are cited in terms of raw gas containing more or less carbon dioxide and having an energy content of approximately 19 megajoules per cubic meter (500 Btu per cubic foot). The majority of these landfill gas recovery plants practice some kind of dehydration and removal of corrosive constituents, but only 6 of the present plants remove carbon dioxide to produce near-pure methane. Raw gas is entirely acceptable for local combustion, either heating or electricity generation.

In terms of pure methane the quantity of gas produced through landfill exploitation, about $200 \times 10^6 m^3$/year, is nearly 7 times the amount of methane estimated to be produced by the United States digesters surveyed above, i.e., $30 \times 10^6 m^3$/year. Taken together this amount of methane is

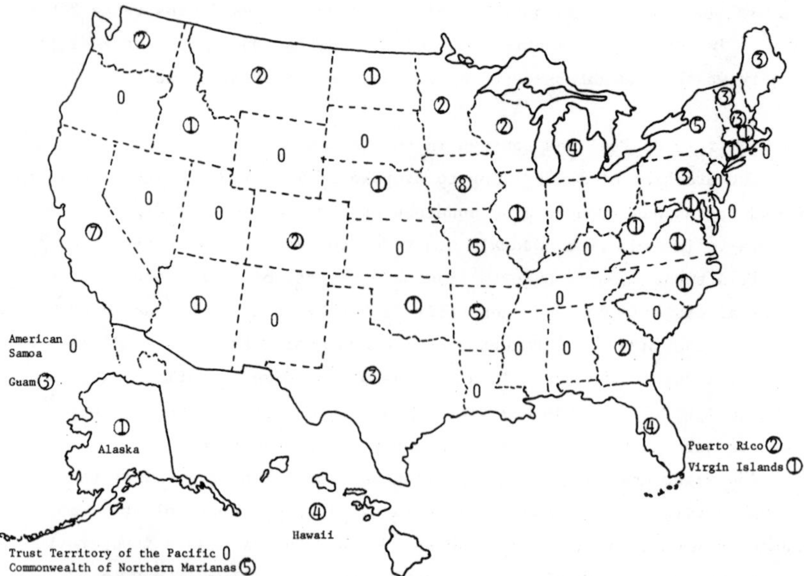

Figure 2. Distribution of Digesters in the United States and Outlying Areas.

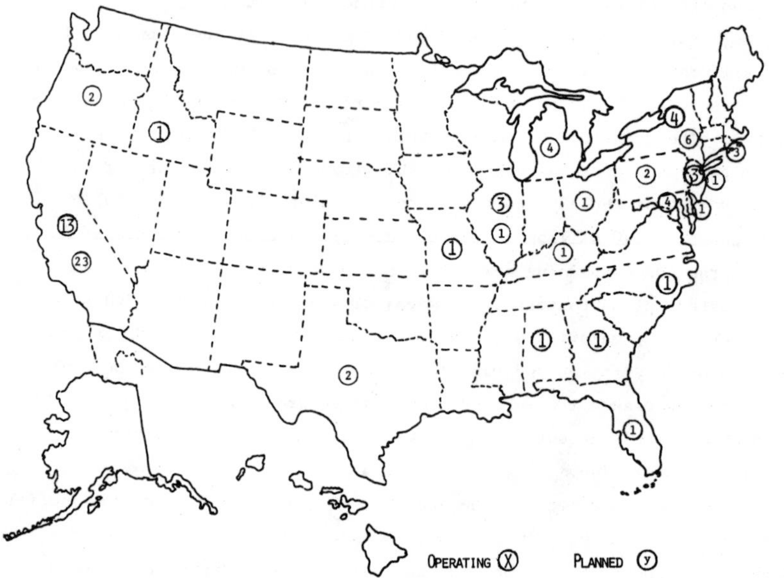

Figure 3. Locations of Landfill Gas Recovery Plants in the United States.

approximately 0.04% of the annual natural gas consumption in the United States. Clearly, this contribution to fuel supply is not determining, but growth is anticipated and nearly every individual production source is a financially significant operation to its managers.

It is notable that landfill gas recovery is sufficiently attractive economically that it has attracted private capital investment. Several firms are engaged actively in the acquisition of rights to landfill gas recovery. A reasonable estimate of the potential for such installations in the U.S. is 200. Given an estimate of 30 ± 10 years as the practical lifetime of gas recovery from a landfill it may be appreciated that this is going to be a vigorous industry. Improvements in the technology of landfill management may be expected to improve the economics and significance. Research and development now under way in the U.S. is aimed at optimizing moisture, nutrients, pH, and leachate circulation in landfills so that more complete gas generation at higher rates is obtained than is the case in untouched landfills. The prospects of "on-line" processing of municipal solid waste in this way for gas recovery are likely to insure many decades of useful gas production.

Academic support of research on anaerobic digestion in the U.S.A. continues strong with greatest emphasis currently in microbiological fundamentals. The over-all level of research and development on anaerobic digestion has diminished drastically in the past five years as financial support from the U.S. Department of Energy has been reduced. The reduction in this support, which has included the construction of large demonstration plants, has made the present level of development activity seem relatively slow. There remain in place a number of Federal and State financial incentives which are available to assist implementation of anaerobic digestion systems. Such incentives, an improved climate for private investment, and growing appreciation of the value of anaerobic treatment of industrial wastes will support steady growth in the utilization of this technology in the U.S.A.

4. CENTRAL AMERICA AND MEXICO

The potential of anaerobic digestion to produce supplemental energy supplies is the principal attraction in Central America and Mexico. The second level of interest concerns disposal of agricultural wastes. At least four research institutions are supporting original research as well as the promotion of practical use of anaerobic digestion.

The widespread dependence of individual families in Central America on indigenous fuels has stressed these resources severely, and supplemental energy sources that avoid recourse to petroleum-based fuels are needed greatly. Family or communal digesters ranging in capacity up to $40m^3$ and operated with agricultural residues, especially cow and pig manure, can make practical contributions to this need. Institutions in Mexico, Guatemala, and Costa Rica have contributed to the development and dissemination of such technology. Cattle raising and agricultural operations, notably the coffee industry, offer opportunities for larger scale application of digestion.

The total number of digesters in Central America and Mexico is estimated to be 150 (5,6). The majority of these, 90, are located in Guatemala, according to Calzada (6). A dozen of the 150 digesters are larger than $50m^3$. Ten of these, having volumes between 55 - $700m^3$ are in Guatemala. The remaining two are located in Honduras (2 x $30m^3$) and Costa Rica ($100m^3$). The designs utilized include flexible bags, rigid tops, half-bags, and so-called Chinese designs. High rate digesters have reached only the pilot scale in Central America so far.

The substrates for these digesters are cow and pig manures and vegetable wastes. In many cases batch digesters are operated for fertilizer and compost production alone with no gas collection. Promotion of the technology by agricultural extension agents is active. Scientists and engineers in Central America have participated actively in the world community of anaerobic digestion technology. The need coupled with this capability insure that steady development will occur.

5. REFERENCES

1. E.R. Hall, Proceedings of the Third International Symposium on Anaerobic Digestion, p. 393, Cambridge, MA, 1983.
2. D.L. Wise, R.L. Wentworth, and E. Ashare, Ind. Eng. Chem. Prod. Res. Dev. 18 (2) 150 (1979).
3. A.G. Hashimoto, Proceedings of the Third International Symposium on Anaerobic Digestion, p. 437, Cambridge, MA, 1983.
4. H.L. Hickman, editor, Landfill Gas Project Information. A Compilation of Information from Selected Landfill Projects in England, Canada, and the United States, Silver Springs, MD, Governmental Refuse Collection and Disposal Association, 1984.
5. A.M. Martinez and P. Mulas, Proceedings of the Third International Symposium on Anaerobic Digestion, p. 409, Cambridge, MA 1983.
6. J.F. Calzada, personal communication, 1984.

ANAEROBIC DIGESTION IN LATIN AMERICAN COUNTRIES

F. MONTEVERDE and E. J. OLGUIN

Energía y Técnica, S. C. and Instituto
Mexicano de Tecnologías Apropiadas, S. C.

Summary:

Most efforts concerned with anaerobic digestion in developing countries have been directed towards the rural sector. However, - recycling of urban and industrial waste has been only recently a research subject in these countries. This paper discusses two sorts of issues: on one hand the state of the currents research efforts and on the other hand, the potential of the anaerobic digestion in the industrial economy of latin american countries.

Regarding the first issue, the efforts done in Cuba, Mexico, Argentina and Guatemala are reviewed. The ICIDCA team at Cuba has shown that the digestion of stillage from sugar cane by means of the UASB process produces 0.32 m^3 of biogas/kg QOD when loading with 25 kg/m^3. day and running at 2 days retention time. The PROIMI group in Argentina has reported on the efficiency of three different sorts of processes, the anaerobic filter (AUF), the packed bed reactor (PBR) and the upflow anaerobic sludge bed (UASB) for treating also stillage from sugar cane. They found reduction of QOD around 55% to 85% depending more on the type of waste reather than on the loads.

Recycling through anaerobic digestion of the coffee industry waste has been investigated by ICAITI in Guatemala and INIREB in Mexico.

Concerning the potential of anaerobic digestion to treat urban and industrial waste in Latin America, two study cases were chosen. One is dealing with the sugar industry in Mexico and indicates that an annual biogas production around 10-12 millions cubic meters which is equivalent to 35 000-43 000 barrels of petrol, could be produced from the stillage of sugar cane.

The second case study chosen is the potential of biogas production from urban waste.

The main conclusion of the present study is that there is an urgent need to promote technological development and commercial operation of low -cost efficient anaerobic digestors to treat and recycle the urban and industrial waste in developing countries. The potential of these processes is enormous in these countries where waste recycling is not a current practice and energy deficit and environmental problems are huge.

2. INTRODUCTION

Anaerobic digestion in developing countries and more specifically, in Latin American countries, has been applied mainly to the treatment and recycling of animal waste in the rural areas. It has been only recently that this process is gaining attention as a tremendous tool for the treatment of industrial and urban waste and some interesting research projects are starting to be developed in few countries of the area.

This paper reviews most of the research which has been recently reported by Latin Americans Groups in order to provide a general view of the current state of the art in this field.

On the other hand, the potential for anaerobic digestion to treat industrial and urban waste is also presented. Data from the sugar industry and its waste in Mexico points out a clear need to introduce technology for biogas production utilizing the UASB process. The waste from the coffee industry is also shown as an attractive case. Finally, data from the biggest cities in the area, shows the potential of the urban waste as a substrate for anaerobic digestion.

RESEARCH PROJECTS IN VARIOUS COUNTRIES

I - ANAEROBIC DIGESTION OF STILLAGE FROM SUGARCANE

Sugarcane is one of the most important products of the area. The by-products of the sugar industry are processed into various other products. Thus, molasses, the liquid residue from the process of sugar crystallization can be fermented into ethanol with a subsequent production of stillage as residue. A flow diagram is shown below in which it is evident that stillage might be produced in considerable amounts.

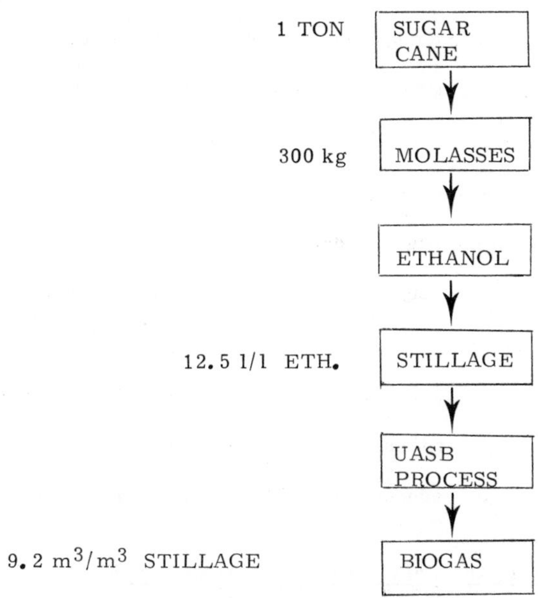

Stillage is a high-contaminant residue which can be effectivily treated by anaerobic digestion. Its composition varies depending on various factors but always containing a very high organic matter content.

The ICIDCA Institute (Cuban Institute for Fesearch on Sugar-Cane By-products) has undertaken a research project on anaerobic digestion of stillage by means of the "Upflow Anaerobic Sludge Planket" (UASB) process. They found(1) that a loading rate of 25 kg $QOD/m^3 \circ$ day and a retention time of 2 days resulted in a remotion of 70% of the original QOD. The methane yield was 0.33 m^3 CH_4/kg of removed QOD. The economic analysis for the Cuban conditions shows it feasible and no importation of technology is required.

Continuous anaerobic digestion has also been investigated by the PROIMI group in Argentina. The results reported by them[2] indicate that rather similar methane yields might be obtained either by means of an anaerobic filter, a packed bed reactor with recirculation or an UASB reactor. Differences should be atributed mainly to the stillage source, being the process more efficient with those non centrifugated and using a loading rate in the range of 18-30 kg QOD/m^3 ∘ day.

The mexican group at the Electric Research Institute (I. I. E.) has investigated the digestion of sugar cane stillage under batch operation[3]. They found that an addition of 40% of caw manure (diluted at 50% with water) was adecuate to maintain an estable system. The initial QOD was 15.8 g/l and it was reduced to 6.8 g/l after 60 days of operation.

II - COFFEE PROCESSING WASTE AS SUBSTRATE

The coffee industry is also relevant in Latin America. As shown in the flow diagram below, the washed coffee processing produces three different types of residues: pulp (40%), effluent from pulp elimination and effluent from washing (7 ton water/ton of coffee grain).

FLOW DIAGRAM FOR
WASHED COFFEE PROCESSING

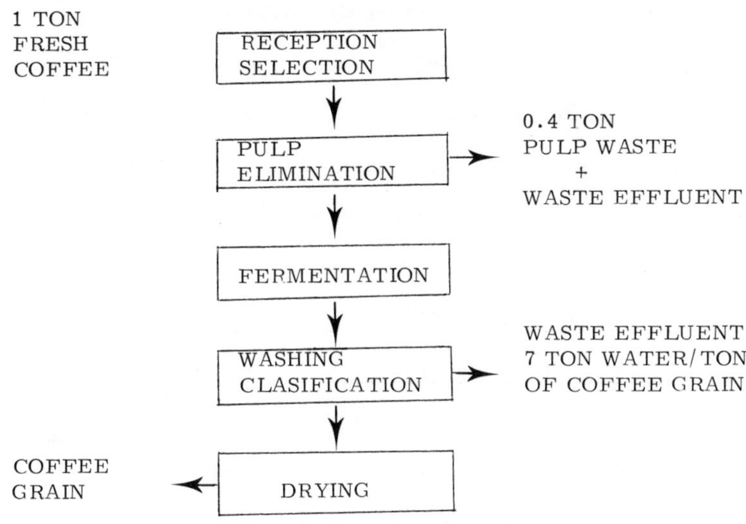

Only to provide some example, the mexican coffee industry has produced 240 x 10^3 tons of fresh coffee in 1981. According to this, the expected quantity of pulp waste was 96 x 10^3 tons.

The ICAITI group at Guatemala has been very active in the investigation of anaerobic digestion of the coffee processing residues. More recently[(4)], they have reported on the use of a continuous upflow reactor fitted with sponges to digest the juice of the coffee pulp. The use of these sponges enabled them to have methane production in the range of 2.5 to 3.1 VCH_4/V_o day at a retention time of 2-4 days and with a methane content in the biogas as high as 87-93%.

In Mexico, the INIREB group has investigated the use of semi-continuous reactors for the coffee pulp digestion[5]. They reported that in order to avoid long residence time and small loading rates as it was reported before for coffee pulp digestion without acidification, it is required to provide mixtures to the digestor containing 84% pulp and 16% manure. The use of these mixtures allows for retention times of 20 days and loading rates of 3.6 kg $VS_o/m^3 \circ$ day.

3. THE POTENTIAL OF ANAEROBIC DIGESTION FOR INDUSTRIAL AND URBAN WASTE IN THE AREA

Latin American countries are favoured by warm climates and tropical agriculture products are produced in vast amounts. Agro-industries such as those derived from sugar, coffee and fruits are relevant in the area and so are the amounts of residues suitable for anaerobic digestion. Only to provide a brief account of the availability of industrial and urban waste in the area, table 1 provides information concerning main agro-industries, percentage of urban population, number of cities above 500 000 habitants, annual growth rate and the input of agriculture and industry in the National Gross Product for some of the countries of the region.

Although a deep study of availability of wastes is required, figures from table 1 suggest that there is a potential for anaerobic digestion to treat the industrial and urban waste in the latin american region. Furthermore, the mexican sugar industry was chosen as a case study to provide a closer estimation of the potential for biogas production from stillage. Table 2 shows the annual production of sugar from 1979 to 1982. It also shows the actual ethanol production for those years and the estimated amount of stillage as residue. From these figures, and assuming a production of 9.2 m^3 biogas/m^3 stillage, annual biogas production was estimated. Finally, assuming that 1000 m^3 of biogas are equivalent to 553 liters of petrol, the equivalent to barrels of petrol is presented.

4. CONCLUSIONS

a) Latin America is a region with potential for anaerobic digestion of industrial and urban waste. The sugar and coffee industry residues are available in large quantities, besides other residues such as those coming from the canning and fruit industry. Concerning recycling of urban waste some countries such as Brasil Colombia, Venezuela, Perú and México have a large urban population and many cities above 500 000 inhabitants.

b) Regarding technology development in the area, there are few research groups working mainly with stillage from sugar cane and coffee processing residues. Their results suggest that economic systems based in continuous reactors such as the UASB process could be developed in the short term.

c) Although some countries in the region have qualified human resources and good industrial infraestructure, technology development

TABLE 1. RELEVANT DATA FROM SOME COUNTRIES OF THE REGION

COUNTRY	MAIN AGRO-INDUSTRIES	URBAN POPULATION %	CITIES ABOVE 500 000 HABITANTS	ANNUAL GROWTH RATE (%)	CONTRIBUTION TO THE GROSS NATIONAL PRODUCT
BRASIL	FIRST WORLD COFFEE PRODUCER. SUGAR - CANE TROPICAL FRUITS	65	14	2.6-3.1	AGRIC - 11% INDUSTRY - 38%
COLOMBIA	COFFEE SUGAR FRUITS	70	7	3.75	AGRIC - 29% INDUSTRY - 28%
PERU	COFFEE SUGAR FRUITS	67	1	3.0	AGRIC - 10% INDUSTRY - 43%
ECUADOR	COFFEE SUGAR FRUITS	45	2	3.3	AGRIC - 15% INDUSTRY - 37%
GUATEMALA	COFFEE SUGAR FRUITS	39	1	2.2	AGRIC - 26% INDUSTRY - 20%
PANAMA	COFFEE SUGAR FRUITS	54	1	3.1	AGRIC - 23% INDUSTRY - 21%

TABLE 2. ANNUAL PRODUCTION IN THE MEXICAN SUGAR INDUSTRY AND THE POTENTIAL OF BIOGAS

YEAR	TONS OF SUGAR X 10^6	LITERS OF (1) ETHANOL PRODUCED X 10^6	LITERS OF (2) STILLAGE PRODUCED X 10^6	BIOGAS (3) PRODUCED $m^3 \times 10^6$	EQUIVALENT (4) TO BARRELS OF PETROL
1979	2.88	106.89	1336.12	12.29	42 749.8
1980	2.6	87.72	1096.5	10.08	35 062.5
1981	2.37	88.35	1104.37	10.16	35 340.8
1982	2.68	80.59	1007.37	9.26	32 210.2

(1) Actual production, not estimated production
(2) Assuming 12.5 l of stillage/l of alcohol
(3) Assuming 9.2 m^3 biogas/m^3 stillage
(4) Assuming 1000 m^3 of biogas are equivalent to 553 l of petrol. According to Ref (6)

is not an easy task in the region, since many other problems occur. The major one is the lack of Technology Promotion Centers which are able to vinculate research groups to industry. Thus although research groups have interesting results at bench scale, pilot plants are generally missing and industrial enterprises do not risk in financing them.

d) A deep study of the actual availability and localization of indus trial and urban waste in the region, is required. Such study could indicate the real opportunities for technology transfer, either local technology or imported technology.

5. REFERENCES

(1) Valdez, E., A. Obaya y A. García Peña. 1982. El tratamiento de los Residuales en la Industria Alcoholera. En Simposio Internacional "Avances en Digestión Anaerobia". Organizado por ICAITI, UAM, IMETA. Méx. 25-27 Octubre.

(2) Siñeriz, F., H. F. Díaz, P. R. Córdoba y F. Sánchez Riera. 1982. Producción contínua de Metano a partir de Vinazas. En Simposio Internacional "Avances en Digestión Anaerobia". Organizado por ICAITI, UAM, IMETA. México, 25-27 Octubre.

(3) Arvizu, J. L. y A. Martínez. 1983. Efecto de la Concentración de Microorganismos en la digestión anaeróbica de la vinaza. Memorias de la VII Reunión Nacional de Energía Solar. 5-7 Octubre, Saltillo, Coahuila.

(4) Calzada, J. F. 1982. Biogas de Subproductos del Beneficio Húmedo de Café. En Simposio Internacional "Avances en Digestión Anaerobia" Organizado por ICAITI, UAM, IMETA. México 25-27 Octubre.

(5) Young, M. A. 1982. Biometización de las Pulpas de Café. Alimentación Alternada o Mezclada con Estiércol de Bovino. En Simposio Internacional "Avances en Digestión Anaerobia". Organizado por ICAITI, UAM, IMETA. México 25-27 Octubre.

(6) Mandujano, M. I., A. Felix and A. M. Martínez. 1981. Biogas, Energía y Fertilizantes a partir de desechos orgánicos. Manual para el Promotor de la Tecnología Olade.

ANAEROBIC CONVERSION OF AGRICULTURAL WASTES TO CHEMICALS OR GASES

P.N. HOBSON, W.G. REID and V.K. SHARMA

Rowett Research Institute, Bucksburn, Aberdeen, Scotland

Summary

Naturally-ocurring, mixed, undefined bacterial cultures can convert agricultural wastes to biogas in anaerobic digesters. From the mixed digester flora, anaerobic bacteria can be obtained which can be used in the production of chemicals from starches or fibres.

An amylolytic bacterium producing acetic, butyric and lactic acids, hydrogen and carbon dioxide was selected from about 140 mesophilic and thermophilic amylolytic bacteria isolated from anaerobic digesters. The paper describes experiments undertaken to obtain maximum production of volatile fatty acids, with maximum butyric acid content and minimum lactic acid. In stirred-tank continuous culture maximum butyric acid production was obtained at low growth rates. To try to obtain low bacterial growth rates and high hydraulic flow rates, various continuous cultures with immobilized bacteria were tested. A fixed-film, stirred-tank type of fermentor with pH control gave best results. Good results were also obtained with fed-batch cultures. Both these need to be optimised.

Some 400 cellulolytic bacterial isolates were made over a period from a mesophilic cattle-waste digester. Analysis showed that at least 13 different species and genera were present with varying abilities to degrade cellulose or straw fibres to mixtures of acids, alcohols and gases.

1. INTRODUCTION

Reserves of oil in the world will decrease and eventually be used up. Coal will probably last longer than oil, but again, is a diminishing asset. Nuclear power has so far not fulfilled its promise and nuclear power stations pose large environmental and other problems. Other sources are, or could be, available to diversify energy production and thus help to conserve fossil fuels in the short term. In the longer term these could become major contributors to the world's energy requirements, particularly in countries with no fossil fuel reserves.

Amongst these potential sources of energy is biomass, in the form of special crops and wastes from the processing of crops grown for human or animal foods, and animal and human excreta and wastes from animal processing for foods. Biomass can be converted to energy, or products that can be used as sources of energy, by chemical and physical means; or, in many circumstances more conveniently and better, by microbiological means.

Microorganisms can convert organic matter to microbial biomass for further processing, and/or to gases or non-gaseous fermentation products which can be used directly as fuels or as feedstocks for a chemical industry producing fuels or other chemicals.

With a naturally-occurring, mixed microbial culture and a suitable environment (an anaerobic digester) organic matter of many kinds can be converted to biogas, a mixture of methane and carbon dioxide. The biogas can be used directly as a fuel for burners or internal-combustion engines, or the methane can be purified for use as a fuel or chemical feedstock. This technology has, for many forms of biomass, been developed beyond the laboratory into full- scale plants (see e.g. 1). However, the laboratory work has shown that the production of biogas involves a number of steps and many bacteria with different properties. The first step in anaerobic digestion of solid wastes such as native vegetable matter and residues of vegetable matter in animal excreta or factory wastes is hydrolysis of polymeric plant structural or reserve polysaccharides to monomers or dimers which are then fermented to lower volatile fatty acids, hydrogen and carbon dioxide, the substrates for methanogenic bacteria. While the presence of the methanogenic bacteria 'guides' the fermentation of the many bacteria in an anaerobic digester towards a limited number of fermentation products, individually the bacteria show activities on many substrates and produce mixtures from a wide variety of fermentation products.

By suppression, for instance by chemical additives (2), of the methanogenic step of anaerobic digestion a mixture of volatile fatty acids may be obtained. This type of process uses the natural mixed culture of an anaerobic digester, and while it has advantages in being operable under non-sterile conditions there could be problems in control of the reactions and in maintenance of the desired mixed culture. For closer control of the reactions and production of defined acids or other fermentation products, use of pure or defined- mixed cultures is preferable.

Methane is almost insoluble in water and is easily removed from the digester contents. Ethanol is very soluble and its removal from fermentation liquids at present involves processes with high energy inputs. Various methods are theoretically possible for removal of organic acids from fermentation liquids, and these could be less energy-consuming than the distillation of alcohol. These include extraction as acids by organic solvents (e.g. 2), production and extraction of esters (3), precipitation of acids as salts, use of resins and electrodialysis, ultrafiltration and other physical methods. The acids or esters can be used as such in manufacturing processes or converted to alkanes, alkenes, ketones, ethers, etc.

Cellulosic structural polysaccharides of plants, in crops and wastes, represent a vast resource of potential feedstocks for micro-biological processing. On the other hand lignification and other structural complexities of fibres lead to low conversions to sugars and fermentation products unless the fibre is disrupted by chemical or physical means. Starch, however, whilst readily available in grains or tubers of tropical and temperate plants, or as a waste from food processing, can be dissolved in fermentation liquids and is almost completely fermentable.

The present work had two aspects. One investigation was designed to enumerate and characterise the fibre-degrading bacteria of an anaerobic digester as an aid to understanding of the process and to see the potential of these bacteria for fibre fermentations. The other was to investigate the possibilities of butyric acid production from starch, as butyric acid has, itself, manufacturing uses, and can be a precursor of other chemicals and fuels. Since butyric acid is one of mixed fermentation products the work involved the isolation of a suitable bacterium and the manipulation of the fermentation to maximise butyric acid production. Genetic manipulation to improve yields might be an extension of the work once the possibilities of the wild-type was explored but could not be attempted in a short-term investigation.

2. PRODUCTION OF BUTYRIC ACID FROM STARCH

Some of this work was described in a paper given at the EEC meeting in Broni (4).

Isolation of bacteria. Previous work (5) had suggested that amylolysis was a common property of digester bacteria although the digester feedstocks contained no starch. Some 75 isolates were made initially by culturing dilutions of fluids from pig- or cattle-waste mesophilic digesters using strictly anaerobic techniques based on those of Hungate (6). Of these isolates the majority produced acetic and lactic acids from starch; 7 produced lactic alone or plus a little acetic, and a minority produced acetic with one or more of the acids propionic, butyric, lactic, succinic. Two isolates were finally selected for ability to produce butyric acid with acetic and some lactic and to grow on ammonia as nitrogen source. These were designated A2012 (the taxonomic position is uncertain) and B211, a *Clostridium* sp. A2012 was selected after initial batch and continuous culture experiments as B211 produced more lactic acid under all conditions and did not grow as well on ammonia medium. Since then 50 more strains have been isolated but only a minority produced butyric acid and none seemed better than A2012. About 20 strains of thermophilic amylolytic bacteria have been isolated at intervals from a thermophilic pig-waste digester, but all have proved unstable on continued subculture and have died out. A2012 was a Gram-negative rod, 0.5 to 2-6 µm, giving rapid growth (μ_{max} 1.6h^{-1}) in a medium with maltose or starch plus Casitone and yeast extract under CO_2 and with cysteine as reducing agent. In a medium with only NH_4 as N source μ_{max} was 0.4h^{-1} and filamentous growth occurred; yeast extract increased growth in NH_4 medium. Urea was not utilised. It fermented starch, maltose, xylose, glucose, fructose, mannose, lactose, raffinose, sucrose, salicin, but not arabinose, glycerol and sorbitol. The extracellular α-amylase was induced by starch or maltose (*e.g.* specific activity 16 units in glucose, complex N, medium; 260 in maltose; 243 in starch). Optimum growth was at pH 6.7, there was little growth below 6.4 or above 7.2 and none at pH 5.4 or 7.7. Growth occurred at 25, 35, 39°C but not at 15, 45 or 60°C in 24h incubation. All experimental cultures were at 39°C. There was no growth under aerobic conditions, and a heavy inoculum was needed to ensure growth in non-reduced media under CO_2: growth could occur under N_2. In all experimental cultures hydrogen was produced.

Batch fermentations. The bacterium was not inhibited by high concentrations of substrates or acetic or butyric acids, so high starch concentrations could, theoretically, be fermented. However, starch solutions are very viscous and above 10% concentration become progressively stronger gels and this led to problems in handling media. So, in continuous culture particularly, starch concentration was usually limited to 2 or 4%, and maltose was used in some experiments.

Preliminary experiments showed that even with 1% maltose or starch, pH control would be needed to prevent limitation of growth by fall in pH and the results of some batch cultures with high substrate concentrations and pH controlled at 6.6 are shown in Table I.

Table I. Batch fermentations

Substrate	Fermentation products (mM)				C-recovery (%)† on	
	Acetic	Butyric	Lactic	But:acet	substrate used	substrate added
6% maltose*	65	117	6.6	1.8:1	78.9	76.7
8% maltose*	85	99	28.3	1.2:1	91.5	86.4
10% maltose*	109	215	34.7	2.0:1	102	89.7
6% maltose	50	145	35.4	2.9:1	65.0	64.3
10% maltose	106	260	16.6	2.5:1	80.4	74.6
10% starch*	166	249	29.9	1.5:1	72.1	70.6

* Complex N medium (casitone, yeast extract, NH_4), otherwise simple N (NH_4) medium

† Allowing for dilution of culture by NaOH added for pH control.

In other batch cultures with lower maltose or starch concentrations up to 4:1 ratio of butyrate to acetate was found in fermentaton products.

Continuous cultures. Some stirred-tank pH controlled, continuous cultures were run along with the batch cultures. These used an anaerobic culture apparatus with pH control and gas mixing (7). Cultures on maltose in complex and simple N media were run and some typical results of a run of over 1000 hours in the complex medium are shown in Table II.

Table II Continuous culture in complex N medium, 1% maltose, pH 6.6

$D(h^{-1})$	Molar proportions of acids			Maltose* recovered (%)	Maltose† used (%)
	Acetic	Butyric	Lactic		
0.11	20	32.6	0.0	83.2	91.4
0.165	12.2	22.3	0.5	53.2	90.6
0.18	12.1	21.1	3.2	54.5	91.4
0.22	14.7	27.0	4.8	69.1	92.0
0.28	17.0	24.0	6.0	77.9	93.1
0.31	15.4	24.5	6.4	82.3	87.8
0.42	10.6	18.7	12.3	68.5	88.4

* Recovered in fermentation products (including CO_2 and H_2) as % of maltose used.

† As % of maltose in medium.

Formation of bacterial extracellular polysaccharide caused problems of foaming in cultures, and recoveries from other cultures suggested that the maltose used and unaccounted for in fermentation products was converted to extracellular polysaccharide. An analysis of one preparation of this gelatinous material gave 9.8% N (dry wt.) glucose (88%) of carbohydrate), with rhamnose, mannose, fucose, arabinose and ribose.

The results of the continuous culture quoted and others on maltose and starch, including lower dilution rates and up to 4% maltose or starch, together with the results of batch cultures, showed that fermentation products varied with growth rate. Little, if any, lactate was formed below about $\mu 0.15h^{-1}$ and the proportion of butyrate increased with decrease in growth rate.

For controlling growth rate and for optimising acid production, some form of continuous culture would be required. On the other hand a low dilution rate is not compatible with maximum volumetric production rate and a small fermentor. In the next experiments various types of fermentors were tested to try to maximise medium hydraulic flow rates with low bacterial growth rates, and to try to overcome problems of bacterial polysaccharide formation and starch viscosity already becoming apparent. Some typical results are given below.

A U-tube fermentor. A type of sludge-blanket fermentor where the bacteria sedimented in the bottom of a U tube was tested, but starch conversions and acid productions at high flow rate were not good: e.g. Hydraulic D = 1.0h^{-1}; Acetic, butyric, lactic = 4,6,6.2. Conversion of maltose to products 23.7%. pH was uncontrolled.

A fixed-film, column fermentor. This fermentor was a water-jacketed, vertical, glass tube packed with unglazed pottery chips fitted with inlet and outlet valves and a sampling system at the bottom. There was no pH control system. The conversion of 2% starch in the simple NH_3-N medium varied with hydraulic dilution rate and was higher at the lower dilution rates. A maximum ratio of butyrate to acetate of 2.94:1 was obtained at a hydraulic dilution rate of 0.62 h^{-1}, but fermentation acids were only 14% of medium starch. Lactic acid was low.

Fixed-film, tank, fermentors. In the previous culture the medium was pumped in drops on to the top of the pottery-chip column and trickled down the column. The whole system was gassed with CO_2 to preserve anaerobic conditions. The column did not suffer from problems of blockage with bacterial polysaccharide, although a good growth of bacteria was attached to the chips. However, it was thought that although starch conversion was probably being limited by rate of passage of the medium and by fall in pH, channelling through the chips would have limited contact of medium and bacteria. So the next cultures used a wide-bore tube filled with pottery chips, but with inlet and outlet tubes arranged so that although flow of medium was downward or upward over the chips the chips were always completely covered in medium - an 'anaerobic filter' or form of tank fermentor with immobilised bacteria. However, although a maximum butyric-acetic ratio of 3.59:1 was obtained at a nominal hydraulic dilution rate of 0.10 h^{-1} with the 'upflow' fermentor, recoveries were low (17.1% of medium starch in this case) and fermentation appeared to be limited by pH fall. Build-up of bacterial polysaccharide after some hundreds of hours running stopped the cultures.

A retained-biomass fermentor seemed to show possibilities. However, pH control is almost impossible to achieve in a filter type of fermentor with granular support medium so another fermentor was constructed in which the bacteria were retained on a fine nylon mesh supported by a coarser metal mesh convoluted radially. A pH control electrode and NaOH inlet occupied the centre space of the support mesh and a medium feed tube and tube for inlet of CO_2 for gas stirring were also fitted. In these experiments the complex-N medium was used to try to obtain better growth and 10% maltose was used as substrate. Up to 95% of the added maltose was used and up to 80% of maltose used was accounted for in fermentation products (with the pH controlled at 6.6). In this case butyrate:acetate was 1.26:1 and lactic acid was 5.5% of acetic + butyric. The maximum ratio of butyrate to acetate was 1.94:1, but here C recovery in fermentation products was only 63% of maltose used.

Build-up of bacterial polysaccharide and associated foaming finally stopped these cultures after some hundred hours and this also made determination of culture liquid volumes and so hydraulic dilution rates difficult.

These cultures showed that high conversions of concentrated maltose (and so starch) could be obtained in a retained-biomass fermentor if pH were controlled, but that bacterial polysaccharide could cause problems. In general, the results suggested that while bacterial growth and substrate conversions were highest at a pH about 6.6, lower pH's possibly tended to increase the proportion of butyric acid in the fermentation products.

Fed-batch fermentors. Although the retained-biomass, tank fermentor showed promise, it seemed that a fed-batch type of fermentor could exploit the features of a batch fermentation while giving continuous production. This type of fermentor could also be better in that viscous starch medium could be pumped in rapidly through a large diameter tubed rather than very slowly as in the continuous cultures and that once the initial batch was fermented the high starch- concentration in the initial medium would be diluted by fermented medium making the whole less viscous.

After some preliminary cultures on a 25 ml scale an anaerobic fermentor was constructed in which a medium reservoir could be filled by a pump and the contents then rapidly ejected by gas pressure into a culture vessel with pH control and gas mixing. Samples of culture could also be taken anaerobically. The vessel was designed to take 200 ml of culture initially and this could be increased to 400 ml before half of the culture was removed.

The preliminary cultures with 1% starch and no pH control suggested that a culture with medium additions at 1-2 day intervals could be used and that the system would then come to a steady-stage. The complex-N medium gave better results than the simple NH_3-N medium, as in other cases.

Further experiments used 8% starch in the complex-N medium; pH was controlled at 6.6 and initial culture volume was 200 ml. Steady states were obtained in about 5-10 days and 25 or 50 ml medium could be added every 48 hours (with 5 ml removed for analysis) to give about 83% conversion of starch to fermentation products with a butyrate:acetate ratio of 1.5-1.6:1 and lactate 1.5-2% of the other acids (total about 510 mM). Daily additions resulted in about 60% fermentation of the starch. Stopping the feeding for an extra 2 days resulted in 100% starch fermentation. Foaming and bacterial polysaccharide production was not a large problem in these cultures.

The results show that although the hydraulic dilution rate is very small compared with even the slowest of the continuous cultures, the fed-batch cultures do give promising results for acid production.

3. CELLULOLYTIC, FERMENTATIVE BACTERIA FROM AN ANAEROBIC DIGESTER

These bacteria were isolated from a 150 l mesophilic, stirred-tank digester that had been processing cattle wastes for nearly 5 years (8) and at the time of the isolations it was treating separated dairy-cattle waste at a retention time of 21 days at 35°C. The majority of isolations were made from cultures inoculated with dilutions of digester contents taken on 11 days over a period of about 3 weeks. The cultures contained as substrates either filter-paper powder, native, ground barley-straw or delignified straw. Anaerobic, cellulolytic bacteria were found up to the 10^6 dilution of digester contents, although numbers varied from day to day. Of the isolates, 367 were tested for ability to hydrolyze filter-paper, and representative isolates then tested for straw hydrolysis and in some cases cotton-fibre hydrolysis. The properties of many isolates were investigated in detail and they were provisionally identified with known genera and species of bacteria, or in some cases new species names were suggested.

Further isolations were made from an anaerobic continuous culture in which medium flowed over cotton thread retained in the culture vessel. This was inoculated from a batch culture containing cotton thread and inoculated with digester contents. The culture was run for more than 1000 hours and actively digested the cotton.

The survey, the most comprehensive so far reported on digester cellulolytic bacteria, showed that the isolates had a wide range of abilities to degrade prepared cellulose (filter paper) but that degradation of this cellulose was not necessarily correlated with ability to degrade straw. Xylanolytic activity was also found in most strains. Many of the bacteria were spore-forming and a large proportion were classified as *Clostridium* spp. Their temperature and pH optima were suitable for growth in a digester, and the bacteria appeared to exist in a population in a dynamic equilibrium.

Loss in weight of straw in batch cultures ranged from near zero to about 50%. Major fermentation products varied and were various mixtures of formic, acetic, propionic, butyric, lactic, succinic acids; ethanol, butanol, hydrogen and carbon dioxide. The proportions of fermentation products might be altered, and the potential exists for use of these bacteria in direct conversions of cellulose or plant fibres to acids, alcohols and gases of use as chemical feedstocks, fuels or solvents. This work has been described in detail in a thesis (8) and will be published elsewhere.

4. CONCLUSIONS

The results obtained show that by using either retained-biomass tank cultures or fed-batch cultures a continuous fermentation process could be set up to produce volatile fatty acids from starch. High starch concentrations can be fermented and the acid concentrations obtained in these cultures compare well with those produced in other small-scale experiments reported in the literature: *e.g.* 4.5 and 10 to 15 or 20 g acetic acid l^{-1} in batch fermentations with glucose as substrate (9,10); up to 45 g acetic acid l^{-1} in batch culture and 30-34 g l^{-1} in continuous culture from ethanol (11,12); up to 304 mM mixed acids, 95% acetic, in batch culture with undefined, mixed, bacterial population from CO_2/H_2 (13); about 200-350 mM mixed acids in continuous or batch cultures with undefined, mixed bacteria, from cellulosic substrates (14). The proportion of butyric acid in the fermentation products has varied, but in the cultures producing most acid was rather more than 1.5 x acetic. Higher proportions of butyric than this were obtained and might be produced in the best fermentations by allowing a period for conversion of acetic to butyric acid in the absence of carbohydrate or adjusting the culture pH to slightly more acid values. Problems with bacterial polysaccharide production can occur, but can be overcome. Search for a naturally occurring mutant with no polysaccharide production has not so far been successful.

The recovery of acids at the concentrations produced here should be practically and economically feasible. Recovery of bacterial cells, amylase, and hydrogen from the H_2/CO_2 fermentation gases formed might also be possible.

The work on digester cellulolytic bacteria showed that it would be possible to convert cellulose directly to various acids, alcohols and gases and that it should be possible to alter the products by manipulation of the fermentations.

REFERENCES

1. HOBSON, P.N., BOUSFIELD, S. and SUMMERS, R. (1981). Methane Production from Agricultural and Domestic Wastes. Applied Science Publishers, London.
2. LEVY, P.F., SANDERSON, J.E. and WIRE, D.L. (1981). Development of a process for production of liquid fuels from biomass. Biotechnology and Bioengineering Symposium 11, 239-248.
3. DALTA, R. (1981). Production of organic acid esters from biomass-novel processes and concepts. Biotechnology and Bioengineering Symposium 11, 521-532.
4. HOBSON, P. N. and REID, W.G. (1983). The production of acids from starch by anaerobic bacteria. Fourth Specialized Meeting ANAEROBIC DIGESTION, Broni, 17-19 May, 1983.
5. HOBSON, P.N. and SHAW, B.G. (1974). The bacterial population of piggery-waste anaerobic digesters. Water Research 8, 507-516.
6. HUNGATE, R.E. (1950). The anaerobic mesophilic cellulolytic bacteria. Bacteriological Reviews 14, 1-49.
7. HOBSON, P.N. and SUMMERS, R. (1967). Continuous culture of anaerobic bacteria. Journal of General Microbiology 47, 53-65.
8. SHARMA, V.K. (1983). Isolation and Characterisation of Cellulolytic Bacteria from a Cattle Waste Digester. Ph.D. Thesis, Aberdeen.
9. SCHWARTZ, R.D. and KELLER, F.A. (1982). Isolation of a strain of *Clostridium thermoaceticum* capable of growth and acetic acid production at pH 4.5. Applied and Environmental Microbiology 43, 117-123.
10. SCHWARTZ, R.D. and KELLER, F.A. (1982). Acetic acid production by *Clostridium thermoaceticum* in pH-controlled batch fermentations at acidic pH. Applied and Environmental Microbiology 43, 1385-1392.

11. GHOMMIDH, C., NAVARRO, J.M. and DURAND, G. (1982). A study of acetic acid production by immobilized *Acetobacter* cells: oxygen transfer. Biotechnology and Bioengineering 24, 605-618.

12. GHOMMIDH, C. and NAVARRO, J.M. (1982). A study of acetic acid production by immobilized *Acetobacter* cells: product inhibition effects. Biotechnology and Bioengineering 24, 1991-1999.

13. LEVY, P.F., BARNARD, G.W., GARCIA-MARTINEZ, D.V., SANDERSON, J.E. and WISE, D.C. (1981). Biotechnology and Bioengineering 23, 2293-2306

14. PLAYNE, M.J. (1980). Microbial conversion of cereal straw and bran into volatile fatty acids - key intermediates in the production of liquid fuels. Food Technology in Australia 32, 451-456.

DEVELOPMENT OF AN AUTOMATIC EQUIPMENT FOR THE STUDY OF ACID-BASE EQUILIBRIA FOR THE CONTROL OF ANAEROBIC DIGESTION

F. COLIN
Scientific Manager, Institut de Recherches Hydrologiques
NANCY, France

SUMMARY

Search for rapid response methods enabling the early detection of disturbances in the functioning of methane fermentation plants. Selection of methods based on the study of acid-base equilibria in the aqueous phase of the medium. Modelling of simultaneous equilibria and development of experimental determination allowing to gain the knowledge of the concentrations which characterize the physicalchemical condition of the medium as to the danger for the fermentation to be stopped by an inhibition of toxicity action:

$[H^+]$, $[OH^-]$, $[CH_3COO^-]$, $[CH_3COOH]$, $[H_2CO_3]$, $[H\ CO_3^-]$, $p\ CO_2$, $[NH_4^+]$, $[NH_4OH]$

Demonstration of the validity of a method using the acid-base titration curves of the fermenting liquid. Automatisation of the method through the development of a titration apparatus with continuous flow and rapid response.

1. OBJECTIVE OF WORK

The high-rate anaerobic digestion of sludges, the quality of which varies with time, or which may contain toxic or inhibitory elements, requires to dispose on control and monitoring methods of the plants, in order to operate them in an optimum way in spite of the disturbances which may happen.

Within the limits of the European program "Recirculation of urban and industrial refuse", we have effected researches in order to develop control and monitoring methods better adapted than the present ones (control of pH-value, of the decrease of volatile matter, etc.) in view of :

- the early detection of lacks of balance, the amplification of which may lead to the fast irreversible blocking of the methane fermentation,
- the facility to represent and interpret the results for the operation of plants during particularly critical periods: starting, changes of substrate, various disturbances coming from within or outside the plants.

These works were conducted on various substrates, including urban sewage sludge. Applications may be expected in this field.

2. SELECTION OF THE TYPE OF METHOD TO BE STUDIED

A critical examination of the parameters which may at first sight be utilized (significance, ease of determination, etc.) led us to retain in priority those based on the use of acid-base equilibria in the interstitial liquid phase of the methane-fermenting medium.

The selection of such parameters is derived from the available knowledge of the innermost mechanisms of methane fermentation.

The global velocity of the methane fermentation process is indeed controlled by the speed of the slowest of its constituent stages (liquefaction, acidogenesis, acetogenesis, methanogenesis). The concerned stage is in fact methanogenesis, and the parameters which characterize the development of fermentation, with the fastest response, will be :

- either connected with the occurrence of fermentation end products. The concerned parameters will be pratically the flow and composition of the produced gas,
- or related to the accumulation of intermediate metabolites, and more specially those which may cause an acid-base lack of balance in the medium, this being detrimental to the correct development of fermentation.

The pH-value of the aqueous medium may indicate a lack of balance when it lies outside the range considered as normal (6.5 to 7.5). This lack of balance is reflected by an accumulation of volatile acids. These are neutralized by the alkalinity of the medium (presence of bicarbonates), as long as the buffering capacity of the latter is not exhausted. Free volatile acids can accumulate only when it is exhausted and as the pH-value decreases. It is necessary to follow the pH-value, but its lowering indicates a lack of balance too late for the only knowledge of this parameter be sufficient.

The volatile acid concentration is a very interesting parameter, as the accumulation of such acids may be either the cause or the consequence of a lack of balance.

In the aqueous medium, volatile acids are partly dissociated in the form of salts. Their toxicity seems to be due to undissociated molecules, the proportion of which depends on the pH-value of the medium.

The magnitude of the lack of balance is a function of the ratio between the total volatile acids in the medium and the buffering capacity of the latter, which enables to limit the pH decrease. A global value, total alkalinity, can be determined by an acid-base titration. This notion groups salts of volatile acids and bicarbonate, whereas the concept of total volatile acids covers salts of volatile acids an indissociated volatile acids. The difference between total alkalinity and total volatile acids expressed in a common unit (milliequivalents/litre) corresponds to the difference between bicarbonates and undissociated free volatile acids. It thus characterizes the buffering capacity of the medium.

Another form of toxicity may be caused by the presence of the undissociated ammonia molecule NH_4OH, whereas the ammonium ion, partly present in the medium as bicarbonate, contributes to the buffering capacity of the medium.

Finally, the parameters best adapted for the control and monitoring of methane fermentation plants, in view of an early detection of the disturbances which may appear in the functioning, are then: the compostition of the gas and values related to the accumulation of intermediate products, of substrates from the methanogenesis phase (total volatile acids), of substances toxic to methanogenesis (acidity, molecules not dissociated into ions, and ammonia), and to the buffering capacity of the medium (physical-chemical ability to withstand the accumulation of these compounds).

We have shown that all these parameters may be integrated into a unique model of the physical-chemical equilibria of the medium, the use of which is possible on the basis of a very limited number of values which can be measured and are described hereafter.

3. PRINCIPLE OF METHODS BASED ON ACID-BASE EQUILIBRIA

A medium, the methane fermentation of which is in progress, may be considered as constituted of a solid phase (insoluble substrate) and a liquid phase containing in solution dissolved substrates and intermediate or end products from the fermentation, namely:

3.1. Volatile acids (acetic, propionic, butyric...) in equilibrium with their salts. The dissociation of these acids of the AH-form, through the reaction: $AH \rightleftharpoons A^- + H^+$, follows the mass-action law: $\frac{[A^-]\,[H^+]}{[AH]} = K_a$

which, according to the acidity of the medium (concentration in H^+ ions, i.e. $[H^+]$), determines the proportions of undissociated acids AH and of anions A^- in presence.

The following table gives the acidity constants K_a for the principal volatile acids concerned in the methane fermentation.

Acid	K_a
Acetic, CH_3COOH	$1.76 \cdot 10^{-5}$
Propionic, CH_3CH_2COOH.	$1.34 \cdot 10^{-5}$
Butyric, $CH_3(CH_2)_2COOH$	$1.54 \cdot 10^{-5}$

As the ionization constants of the various volatile acids are very close one to another, everything happens, from the acid-base point of view, as if only one acid, with an ionization constant close to $1.5 \cdot 10^{-5}$, were present. In view of the prevailing character of acetic acid, this molecule shall be considered for the constitution of the simplified model.

3.2. Carbonic acid, which may be dissociated in two stages

$$1.\ CO_3H_2 \rightleftharpoons HCO_3^- + H^+ \quad (K_{aI} = 4,3.10^{-7})$$

$$2.\ HCO_3^- \rightleftharpoons CO_3^= + H^+ \quad (K_{aII} = 5,6.10^{-11})$$

As two acidic functions are neutralized about proportionally to their dissociation constants, and as the second acidity constant of carbonic acid is about 10,000 times lower than the first one, one is fully justified to neglect the second function of carbonic acid in the study of the acid-base equilibrium. (It is generally considered that for a pH-value lower than 8.3, the concentration of $CO_3^=$ ions is negligible.

Moreover, the concentration of carbonic acid $[H_2CO_3]$ is related to P_{CO_2}, the partial pressure of CO_2 in the fermentation gas, according to the law which governs the equilibrium of physical dissolution of CO_2 in water:

$$[H_2CO_3] = k\ P_{CO2}$$

3.3. Ammonium salts (end products from the degradation of proteins) in equilibrium with ammonia:

$$NH_4OH \rightleftharpoons NH_4^+ + OH^- \quad (K_b = 1,79.10^{-5})$$

As a first totally justified approximation, it is thus possible to represent the liquid phase of a medium, in course of methanization, by means of a simplified model where following equilibria are simultaneously established, in the presence of a number of inert salts:

- Ionization of water: $H_2O \rightleftharpoons H^+ + OH^-$
- Dissociation of a volatile acid: $CH_3COOH \rightleftharpoons CH_3COO^- + H^+$
- Physical dissolution of carbon dioxide in water, starting from the gas phase: $CO_2 + H_2O \rightleftharpoons H_2CO_3$
- Dissociation of carbonic acid: $H_2CO_3 \rightleftharpoons HCO_3^- + H^+$
- Dissociation of ammonia: $NH_4OH \rightleftharpoons NH_4^+ + OH^-$

That is to say that concentrations of ionic activities will be such that following relations will be simultaneously verified (application of the mass-action law):

$$[H^+]\ [OH^-] = K_e \qquad (1)$$

$$\frac{[CH_3COO^-]\ [H^+]}{[CH_3COOH]} = K_1 \qquad (2)$$

$$\frac{[H_2CO_3]}{p\ CO_2} = K_6 \qquad (3)$$

$$\frac{[HCO_3^-]\ [H^+]}{[H_2CO_3]} = K_2 \qquad (4)$$

$$\frac{[NH_4^+]\ [OH^-]}{[NH_4OH]} = \frac{[NH_4^+]}{[NH_4OH]\ [H^+]} = K_3 \qquad (5)$$

The 5 above relations together constitute a system with 9 unknown quantities. The various studied methods thus consisted to add 4 supplementary relations, derived from experimental determinations, in order to solve this system of equations, and so to gain knowledge of all concentrations which characterize the physical-chemical condition of the medium, in relation with the risk of blocking the fermentation through an action of inhibition or toxicity:

$[H^+]$, $[OH^-]$, $[CH_3COO^-]$, $[CH_3COOH]$, $[H_2CO_3]$, $[H\ CO_3^-]$, $p\ CO_2$, $[NH_4^+]$, $[NH_4OH]$

4. LABORATORY METHODS

4.1. Analytical method

This first method consists only to apply the techniques of classical analytical chemistry, for the determination of at least 4 concentrations which intervene in the above equation system, and to solve this system by informatic way.

To this end, we have selected following determinations:

- Measurement of pH-value (electrometric method)

  We obtain : $[H^+] = 10^{-pH}$ (6)

- Determination of total volatile acids (method of acid-base global determination)

  We obtain : $[CH_3COO^-] + [CH_3COOH] = A$ (7)

- Measurement of total ammonia (colorimetric method with Nessler's reagent)

  We obtain : $[NH_4^+] + [NH_4OH] = N$ (8)

- At will :

  . Determination of partial pressure of CO_2 in the gas phase (use of Orsat's apparatus or gas-chromatography with catharometric detection):

$$p\ CO_2 = C \tag{9}$$

  . Instrumental measurement of total dissolved inorganic carbon (vaporization and measurement of infra-red adsorption by the gas flow):

$$[H\ CO_3^-] + [H_2CO_3] = T \tag{10}$$

The first of both these options was preferentially selected because of its simplicity.

4.2. Method using acid-base titration curves

This method requires following experimental determinations:

- Measurement of the initial pH-value of the interstitial liquid phase of the sludge after separation of the latter,
- Establishment of titration curves presenting 3 parts:
  . acidification of the initial liquid phase up to pH = 2 with a titrated HCl-solution (part 1),
  . alkalinization of the previously obtained solution by addition of a titrated NaOH solution (part 3 or "return curve") up to pH $\geqslant$ 10,
  . alkalinization of the initial liquid phase of sludge with a titrated NaOH solution (part 2) up to pH $\geqslant$ 10.

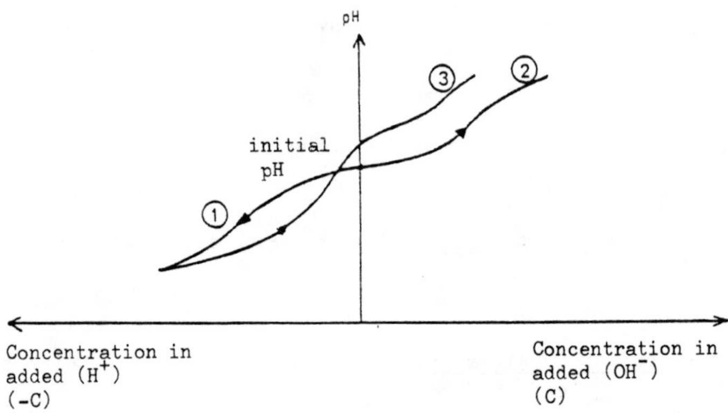

FIGURE 1

Theoretically, it should be possible to calculate directy A, T and N by solving a system of 3 equations with 3 unknown values (see above), by introducing into the system the pH-value of the initial liquid phase and 3 pairs of values (dose of reagent, resulting pH-value) derived from parts 1 and 2 of curves. A mini-program of data processing with BASIC language was developed to this end. Its application unfortunately reveals that the results are largely dependent on the selection of the 3 utilized experimental points.

In order to solve this problem, we were obliged to resort to a clearly more elaborate program working with successive approximations of A and T from n points. This program has given a very rapid convergency, and enabled to obtain a precision of $5 \cdot 10^{-3}$ M. Various improvements were then made, bearing on the selection of the experimental points (pH-values) to be considered for reaching the maximum of sensitivity, and on a relative balance of the various experimental points, taking into account the special form of the titration curves.

On the basis of the A, T and N values thus determined and of the initial pH-value integrated in the model for the physical-chemical equilibria, the various required ionic and molecular concentrations are calculated by means of following relations:

$$[HCO_3^-] = \frac{T}{1 + \frac{[H^+]}{K_1} + \frac{K_2}{[H^+]}} \qquad [CO_3^{2-}] = \frac{T}{1 + \frac{[H^+]}{K_2} + \frac{[H^+]^2}{K_1 K_2}}$$

$$[CH_3COO^-] = \frac{A}{1 + \frac{[H^+]}{K_a}} \qquad [NH_4^+] = \frac{N}{1 + \frac{Kn}{[H^+]}} \qquad [OH^-] = \frac{K_e}{[H^+]}$$

where : $K_1 = 10^{-6,3}$ $K_2 = 10^{-10,3}$

$K_a = 10^{-4,75}$ $Kn = 10^{-9,25}$ $K_e = 10^{-14}$

$[CH_3COOH]$, $[NH_3]$ and $[CO_2]$ may be determined by difference, as A, T, N and the above ionic concentrations are known.

4.3. Comparison of both laboratory methods

Both methods were systematically compared. Table I gives an extract of obtained results.

TABLE I

Sample No.	Comparison of concentrations of total inorganic carbon T (mole/litre)		Comparison of concentrations of total volatile acids A (mole/litre)	
	Method using analytical determinations	Method using titration curves	Method using analytical determinations	Method using titration curves
1	0.073	0.070	0.010	0.018
2	0.046	0.051	0.022	0.027
3	0.022	0.032	0.024	Indeterminable (*)
4	0.113	0.126	0.390	0.364
5	0.200	0.194	0.234	0.178
6	0.088	0.090	0.001	0.004
7	0.093	0.105	0.004	Indeterminable (*)
8	0.017	0.021	0.018	0.008
9	0.112	0.108	0.024	0.011
10	0.106	0.069	0.054	0.050
11	0.045	Indeterminable (*)	0.057	0.062
12	0.065	Indeterminable (*)	0.089	0.091
13	0.082	Indeterminable (*)	0.132	0.119

(*) Case where a too large scattering of experimental results does not allow a convergency of calculation or leads to an aberrant result (negative value).

One may generally observe a good agreement between the orders of magnitude determined with both methods. This agreement seems to be slightly better for total inorganic carbon than for total volatile acids. These performances are sufficient for an objective of rapid control and monitoring of functioning. In a restricted number of cases, the informatized mathematical method does not converge, because of the lack of precision of the introduced experimental values. In order to cope with this difficulty, it is necessary to measure the pH-value with the best possible precision.

Despite these defects, the method using titration curves seems thus to be founded, so that a tentative for automatisation - which practically cannot be contemplated for the method using the results of classical analytical determinations - shall be justified.

5. AUTOMATISATION OF THE ACID-BASE TITRATION METHOD

5.1. Conception and realization of the automatic titration apparatus

This apparatus is controlled by a microprocessor, which regulates its functioning by acting on the constituent organs, and which moreover is utilized for applying acid-basee quilibria to the calculation and presentation of results.

An original conception of a continuous or semi-continuous apparatus was defined according to a principle, which consists to mix a continuous flow of a liquid to be analyzed with a flow of titration reagent (acid or base) by varying the magnitude of the corresponding flows, in order to maintain permanently constant the signal from a detector (pH-recorder) branched on the resulting flow. In such conditions, the ratio of the values of both flows is a measure of the concentration of the liquid to be analyzed.

In practice, the solution to be analyzed is pumped at a constant rate, and the titrated acid solution is pumped at a variable rate by means of displacement pumps, the rotation speed of which is measured with precision. These pumps are driven by high torque motors, without reducing gear, and can work in a very large speed range, with a very high acceleration. Each elementary measurement consists to determine the ratio of the flows of solution to be analyzed and of titration solution, in the equilibrium conditions of the system.

The apparatus is built around a process multipurpose microprocessor (ANALOG DEVICES Macsym 150), which simultaneously effects all functions of data collection, control, regulation of data processing and display of results. Several working particulars of the apparatus, the above general principle of which is known, have been the object of an application for French patents.

Measurements are effected according to a sequential mode, whereas the data processing runs in parallel. The structure of the program allows a fast time of response, compatible with a utilization on a real time basis.

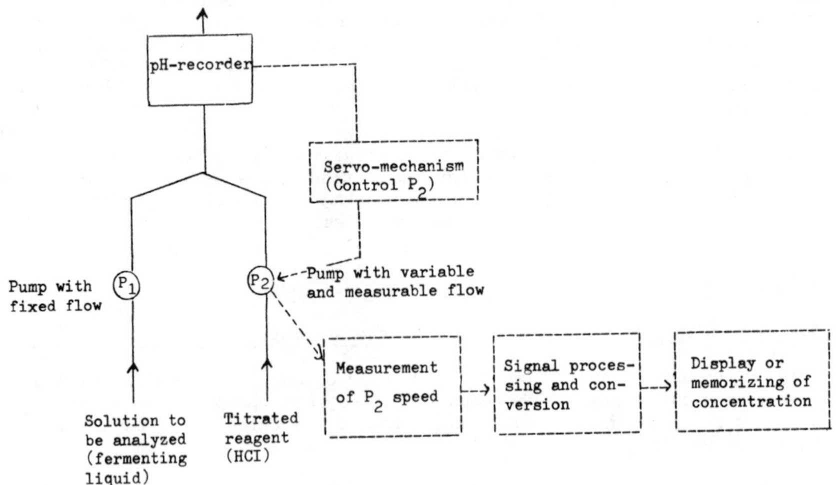

FIGURE 2 - Basic schema

5.2. Adaptation to the specific case of waste water sludge

The apparatus as described above is conceived for working on solutions.

When the determinations are made in the laboratory, the solution to be titrated is generally separated by centrifuging a sample of sludge tapped from the digester.

Such a separation connot be contemplated for the automatic feeding of a control apparatus with continuous or rapid sequential functioning.

We have studied the method consisting to flocculate the medium with a polyelectrolyte and to isolate the liquid phase by natural draining through a sieve.

From the point of view of the quality of the separated liquid to be analyzed, such a technique is equivalent to the addition, to the solution under analysis, of water and polyelectrolyte which constitute the flocculant solution.

We have compared the results obtained during the measurement, with the informatic method, of total inorganic carbon and total volatile acids:

- on the liquid phase separated by centrifugation (reference values),

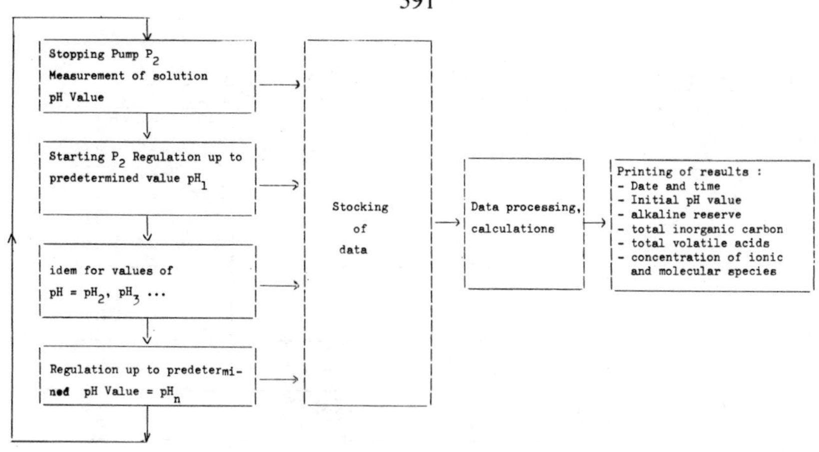

FIGURE 3 - Working operating chart

- on the liquid phase separated by natural drainage after flocculation of the fermenting medium with a solution of the polyelectrolyte which proved to be the most efficient after a previous experimental selection (10 % of a solution of Praestol 444 K at 1 g/l),

- on the liquid phase separated by centrifuging, after addition of a volume of demineralized water corresponding to the quantity of polyelectrolyte solution required to promote flocculation.

In the last two cases, the determined concentrations were corrected for taking into account the effected dilution, so that all values in the following tables can be directly compared one to the other, sample per sample.

Influence of the preparation mode of the solution on the measured concentration in total inorganic carbon (mole/litre of initial solution)

Preparation mode / Sample No.	Centrifugation	Addition of 10 % water and centrifugation (*)	Addition of 10 % of polyelectrolyte solution and natural drainage on sieve (*)
1	0.001	0.002	0.001
2	0.004	0.002	0.003
3	0.018	0.019	0.017
4	0.024	0.024	0.024
5	0.054	0.052	0.052
6	0.057	0.052	0.049
7	0.084	0.086	0.087
8	0.132	0.131	0.131

(*) Values calculated by taking into account the effected dilution

Influence of the preparation mode of the solution on the measured concentration in total volatile acids (mole/litre of initial solution)

Preparation mode / Sample N°	Centri-fugation	Addition of 10% water and cen-trifugation (*)	Addition of 10 % of polyelectro-lyte solution and natural drai nage on sieve (*)
1	0.088	0.090	0.087
2	0.094	0.093	0.092
3	0.021	0.022	0.021
4	0.112	0.112	0.112
5	0.106	0.104	0.104
6	0.045	0.044	0.044
7	0.065	0.064	0.064
8	0.082	0.086	0.086

(*) Values calculated by taking into account the effected dilution.

We observe that the separation technique with flocculation with a poly-electrolyte and natural drainage of the liquid phase does not disturb the obtained results, so that it is fully justified to select this technique for feeding a continuous titration apparatus.

Such a separation device can be easily conceived for continuous service (rotating drum).

6. INTERPRETATION OF EXPERIMENTAL RESULTS

When the values of the various ionic and molecular concentrations concerned in acid-base equilibria are known, it is possible:

. to drive therefrom :

- the risks of immediate toxicity to the methanogenesis microflora, by comparing the concentrations of undissociated CH_3COOH an NH_4OH molecules with the toxicity threshold values mentioned in the literature,
- the acid-base buffering capacity of the medium, which, in the pH range explored in practice (between 6.0 and 7.5), is essentially related to the concentration of HCO_3^- ion,

to verify

that the level of the above different parameters is situated, or not, in the range of values corresponding to a normal operation.

The established model enables to forecast the influence of additions of reagents which one may desire to introduce as cure or a preventive, in order to increase the buffering capacity of the medium.

Finally, in the point of view of the routine control of plants where the nitrogen content of the substrate is relatively low, one may be satisfied to determine, in complement to the measurement of the gas flow, the values of A and T, and to automatically effect control measurements at a frequency as high as required, and which could go up to a complete balance-sheet every 5 minutes if esteemed necessary.

A more precise interpretation of the results, in order to appreciate the quality of the development of methane fermentation, may resort to a method of graphic representation using the most significant parameters and enabling to differentiate zones of normal working, zones with risk of blocking and inhibition zones.

At the outset of our work, we have developed a simplified bidimensional representation method, which expresses the difference between total alkalinity and total volatile acids in function of the pH-value (see Fig.4).

One may observe that the points representing samples, the fermentation of which progresses satisfactorily, are situated in well determined zones of the plan, without any interpenetration with points représentative of media in unbalanced fermentation. This demonstrates that the selection of the parameters chosen for this representation is justified, and that the latter may be utilized to establish a global value judgement on the quality of methane fermentation.

The application of the developed equipment should enable to improve such methods, by taking into account a higher number of parameters and by interpreting and graphically representing the results by means of statistical multifactorial methods for the analysis of data (factorial analysis in principal constituents), which may be easily applied to the microprocessor utilized in the construction of the apparatus.

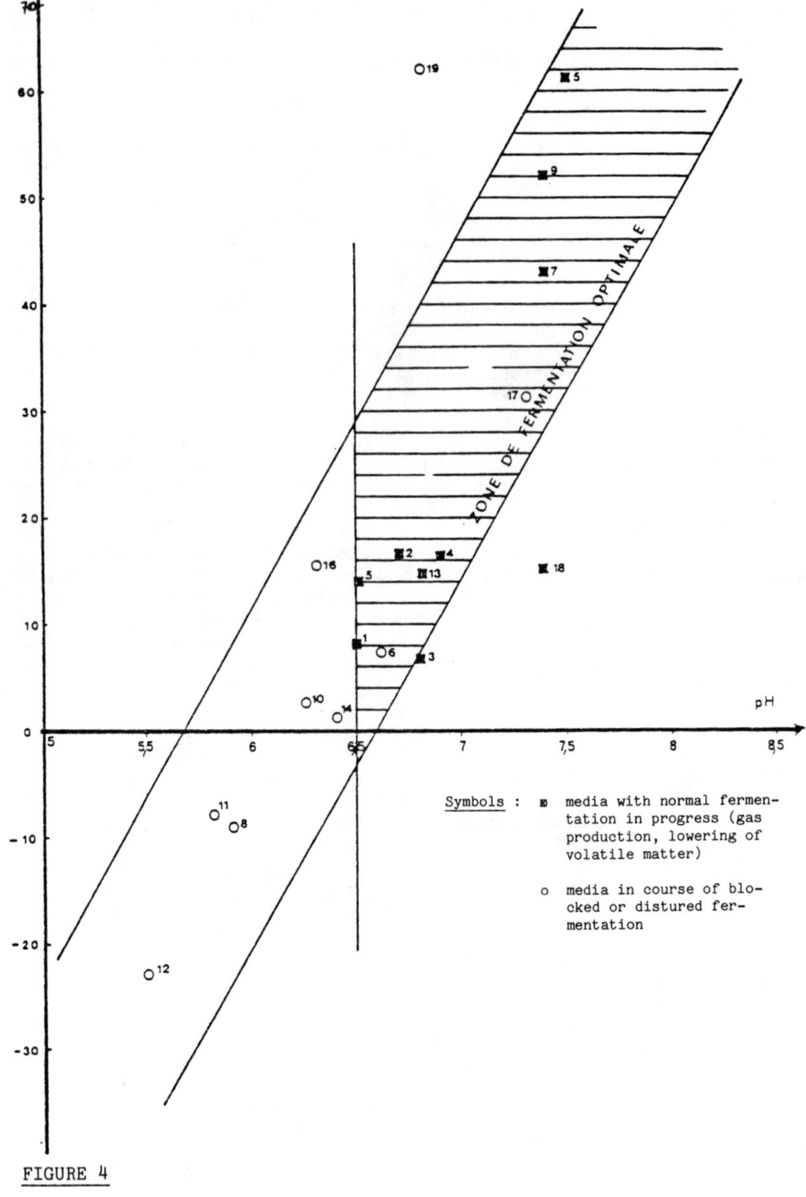
Total alkalinity - Total volatile acids (meq/l)
70
60
50
40
30
20
10
0
- 10
- 20
- 30
5
5,5
6
6,5
7
7,5
8
8,5
pH
ZONE DE FERMENTATION OPTIMALE
Symbols :
media with normal fermentation in progress (gas production, lowering of volatile matter)
o media in course of blocked or distured fermentation

FIGURE 4

METHANE PRODUCTION AND RECOVERY FROM HOUSEHOLD WASTE LANDFILLS

C. MOUTON
Agence nationale pour la récupération et l'élimination des déchets, France

Summary

The main aims of the research programme are to determine the kinetics of biogas formation in household waste landfills, to optimize biogas production and to devise various means of collecting the biogas. Only a few month after the first dumping of waste in experimental cells, the volatile fatty acids disappear from the leachates, active methanogenic organisms become established and the biogas composition stabilizes at a methane content of 50 to 55%. Methane production would be in the region of 30 to 50 m^3 per tonne of raw waste. Improvement of the energy yield from buried waste depends on a large number of parameters including waste density, composition, size consist and moisture content. The pragmatic approach adopted to the experiments demonstrated how difficult it is to manipulate these various parameters (on an operational scale), particularly as they are interdependent. Various types of equipment and devices for biogas collection were tested at a number of landfill sites and two approaches to gas recovery were worked out. The biogas recovery system for landfill sites where dumping has been completed consists of vertical PVC wells interconnected by a gastight horizontal collection network leading to a compressor and a flare. For sites where dumping is in progress, two systems may be considered : a vertical system consisting of perforated concrete pipe sections placed on top of each other or a horizontal system consisting either of tyres placed side by side or drainage pipes of very thick high-density polyethylene.

1. Objectives of the research programme

A landfill site is a constantly changing system : as soon as it is buried the waste is subject to intense biological activity, one of the consequences of which is the transformation of the organic matter (paper, food scraps etc.) into gaseous products (methane and carbon dioxide). The biogas so formed is regarded both as a nuisance - because of the nauseous odours associated with its emission and its adverse effects on nearby vegetation (root asphyxia) - and as a source of energy, since it consists in part of methane (with a net calorific value of 8.56 th/m^3).

In view of the above, it was decided to embark on a research programme with the aim, firstly, of improving the qualitative and quantitative information available on landfill gas and, secondly, of developing gas recovery techniques in order to eliminate the nuisance resulting from inopportune emission of this gas and to utilize its energy content.

A number of research topics were selected.

- Devising and testing of various types of equipment and systems for biogas collection.

- Optimization of methane production in household waste landfills without contemplating any radical change in established dumping practice for sanitary landfills.

Improvements in the final energy yield of a landfill site depend on the available knowledge of the complex microbiological and biochemical processes involved in the formation of methane from organic waste and on the control of the physical parameters affecting these processes.

Four research schedules were therefore drawn up for variation of the waste composition, size consist, density and moisture content. The second research topic was backed up by laboratory research into the microbiological and biochemical processes involved in methanogenesis.

- The study of possible uses for the biogas recovered.

Several landfill sites were chosen for the proposed experimental work, viz Arnouville-les-Mantes (Yvelynes), Roche-La-Molière (Loire), Vert-Le-Grand (Essonne), Fretin (Nord), Beuvry-La-Forêt (Nord), Hersin-Coupigny (Pas de Calais).

2. Results of the research programme

The objectives originally laid down in the research programme have so far not all been attained. In particular, work on biogas utilization will not get under way until 1984 under the second contract between the CEC and the ANRED. The research so far completed, however, has already yielded specific findings on certain problems arising with regard to the formation and recovery of methane at landfill sites.

2.1 Formation and production of biogas at landfill sites

Observation of the changing temperatures and concentrations of volatile fatty acids in the leachates and of the biogas composition reveals very early hydrolysis (depolymerization of the complex molecules) and solubilization of the organic matter in household waste landfills , beginning as soon as the first waste is buried. The hydrolisates are fermented to organic acids (acetate, propionate, butyrate, valerate, isovalerate, caproate) to hydrogen and to carbon dioxide. Hydrolysis and acidogenesis proceed at the same time as methanogenesis.

Only a few months after dumping has commenced, even before filling is completed and the refuse is covered with a gastight layer, the various bacterial populations (hydrolytic, acidogenic and methanogenic bacteria) are balanced in a medium which favours their development (leachate temperature 35°C, Ph between 7 and 7.5, no volatile fatty acids). This results in very marked stabilization of the composition of the biogas with a methane content of 50 to 55% and a carbon dioxide content of 50 to 45% (Figure 1).

At present, however, it is difficult to reach any conclusions with regard to degradation kinetics of organic matter in landfills since the local conditions (humidity etc.) vary greatly from site to site.

Average methane production from waste samples in laboratory tests of biodegradability is in the region of 130 1/kg TVM (Total volatile matter), only 43% of the organic matter being directly accessible to attack by the microorganisms.

One may conclude that one tonne of raw waste would produce about 30 to 50 M^3 of methane (depending on its TVM content) in the first years after tipping. This output would result from degradation of the readily fermentable proportion of the organic matter.

The 57% of organic matter of which is not readily degradable would be converted only very slowly over a much longer period which it is difficult to determine.

In the light of the foregoing, two questions arise :
what action can be taken to recover the biogas potential which is thus lost to the atmosphere and what parameters can be controlled at the landfill site to achieve a consistent improvement in the total methane yield?

2.2 Methods for biogas recovery

The experiments carried out at the Vert-Le-Grand, Arnouville, Fretin and Roche-La-Molière sites provide a basis for developing and offering for use two methods of biogas recovery, one for sites where dumping has been completed and the other for sites where dumping is still in progress.

2.2.1 Dumping completed (Figures I and III)

The biogas recovery system consists of vertical wells interconnected by a horizontal collection network. The wells are installed in boreholes drilled through the entire thickness of the deposited waste.

Drilling is carried out with a 300 mm auger and a tricone roller bit if necessary. A 150 mm pipe of polyvinyl chloride (PVC) is then inserted into the hole. The PVC pipes consist of solid-walled tubes at the top of the well and slotted tubes in the rest of the hole. The width of the slots in the plastic tubes is 4 mm. The annular space is filled with gravel (diameter 5 mm) and sealed at the top with a plug.

Whether the aim is to eliminate the odours associated with the biogas (pollution control) or the utilization of the gas (e.g. as fuel by an industrial user), the gas extracted by the various wells is collected by a horizontal network of gastight pipelines. These pipes must be buried at a sufficient depth (about 0.8 m) and protected by fine sand to eliminate any risk of piercing of the material.The material used for the collection pipes cannot be PVC, which is regarded as too fragile : only high-density polyethylene (HDPE) satisfies the requirement of flexibility and resistance to crushing. Condensation of the water vapour entrained by the biogas may cause blockages at dips in the pipelines, which must therefore be fitted with drain pots. It is of course preferable to minimize the number of dips by laying all the pipework at an even gradient.

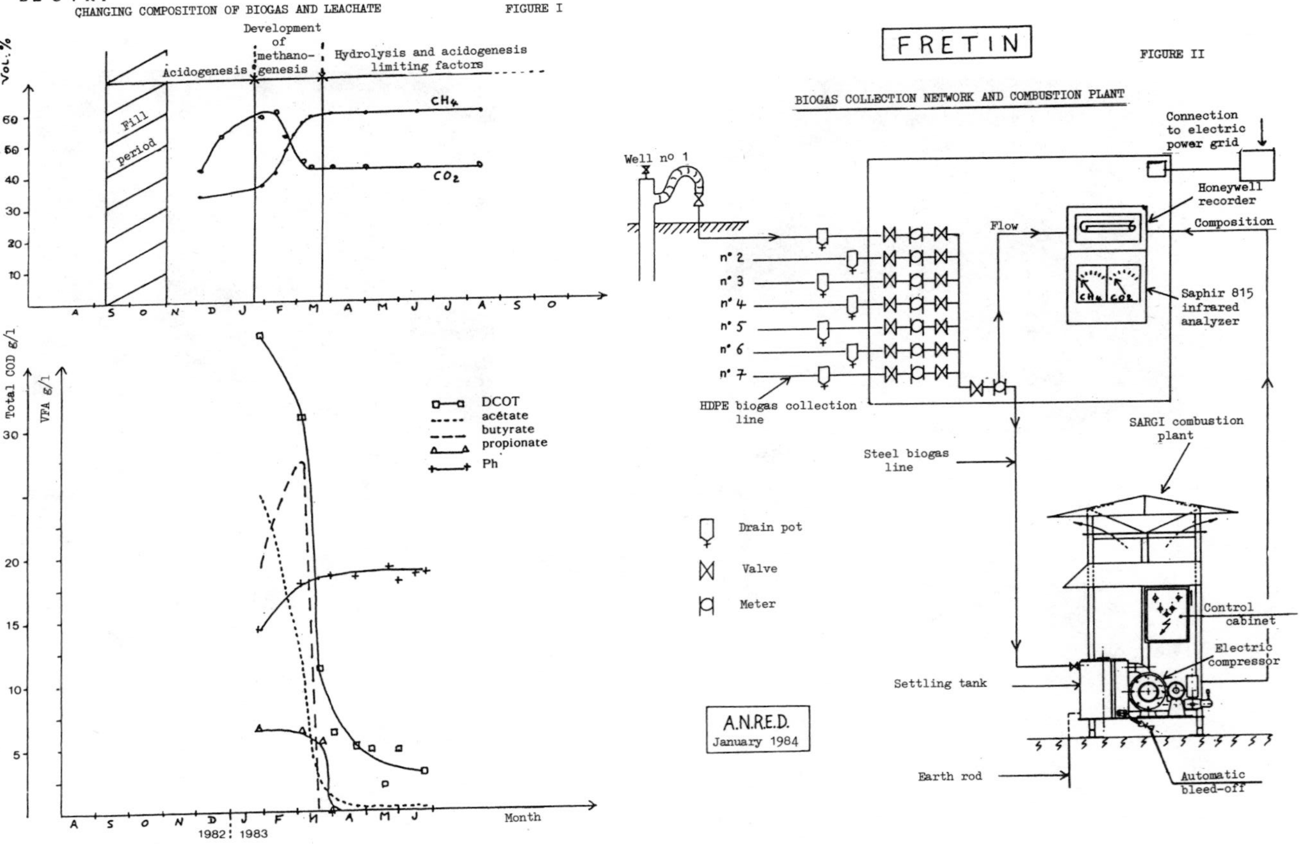
BE U V R Y
CHANGING COMPOSITION OF BIOGAS AND LEACHATE
FIGURE I
Vol. %
Fill period
Acidogenesis
Development of methano-genesis
Hydrolysis and acidogenesis limiting factors
CH4
CO2
60
50
40
30
20
10
A S O N D J F M A M J J A S O
Total COD g/l
VFA g/l
30
20
15
10
5
DCOT
acêtate
butyrate
propionate
Ph
A S O N D J F M A M J
1982 1983
Month
FRETIN
FIGURE II
BIOGAS COLLECTION NETWORK AND COMBUSTION PLANT
Connection to electric power grid
Well no 1
Honeywell recorder
Flow
Composition
n° 2
n° 3
n° 4
n° 5
n° 6
n° 7
Saphir 815 infrared analyzer
HDPE biogas collection line
SARGI combustion plant
Steel biogas line
Drain pot
Valve
Meter
Control cabinet
Electric compressor
Settling tank
A.N.R.E.D.
January 1984
Earth rod
Automatic bleed-off

FIGURE III

FRETIN

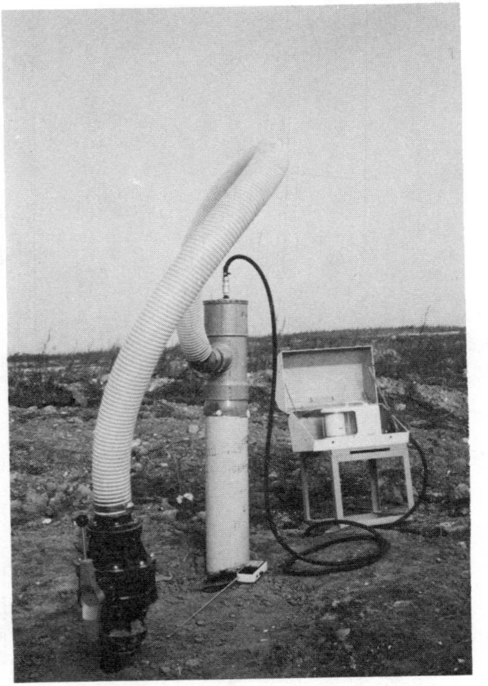

WELLHEAD

COMBUSTION UNIT

The low static pressure of the biogas (a few millibars) is not sufficient for normal feeding to a pipeline system. It is thus essential to install a compressor at the outlet from this network. The suction of the compressor must be sufficient to overcome the pressure drop across the network and must nonetheless be very low : if the suction were excessive, it would draw atmospheric air into the tip, thus making it possible for an explosive methane/air mixture to form.

2.2.2 Dumping in progress

At such sites the collection system is installed progressively as the site is filled. Two systems may be considered.

a) Vertical collection system

The wells are spaced at 80 to 100 m intervals and consist of perforated concrete pipe sections (diameter 1000 mm) placed one on top of the other as dumping progresses. A PVC pipe of 150 mm bore is placed in the centre of each well. The annular space must be filled with stones or gravel in order to preclude the formation of an explosive gas mixture. The wells can be interconnected once dumping has been completed. Individual flares at each wellhead make it possible to burn off the biogas until a collection network can be connected up.

b) Horizontal collection system

The advantage of the horizontal system over the vertical system is that the biogas can be collected and hence used as soon as production commences and throughout the entire period when refuse is being tipped.

The installation of horizontal drainage lines, however, poses two major problems : firstly, the pipes must withstand the tensile and compressive stresses to which they are subjected within the mass of buried refuse (PVC drain pipes of the type used in agriculture are therefore not suitable); secondly, if the pipes are not laid with sufficient care they act as drains not only for the biogas but also for the water seeping through the refuse.

The biogas may be recovered by means of drainage lines consisting of tyres placed side by side. These drains are installed in advance of the face of tipped household refuse. At one of the ends of each drain, a HDPE pipeline takes the gas to the pump station. It is also possible to use perforated polyethylene piping for the drainage line itself.

The best material at present appears to be high-density polyethylene 10 mm thick.
In order to avoid blockage of the pipes by water accumulation, however, they must be laid at an adequate gradient in the refuse.

Vertical wells are thus the standard biogas recovery system for a site where dumping operations have been completed.

At a site where dumping is still in progress, on the other hand, the choice of recovery system is more difficult and studies of the equipment to be used and the collection system as such are still in progress.

2.3 Determination of parameters for optimum methane production

The level of methane production per tonne of raw waste and the rate of utilization of the organic matter of a high or medium degree of degradability are directly related to the following factors : composition of the waste (percentage of biodegradable matter : papers, food scraps and vegetable waste etc.), moisture content of the material, Ph of the leachate, size consist of the waste, density (degree of compaction), permeability and porosity, temperature (optimum at 30-35° C for mesophile methanogenic bacteria). On site, only a few of these parameters can be manipulated : temperature, for example, seems difficult to control in such an environment.

The only parameters investigated were waste density, composition, size and moisture content.

The approach adopted was very pragmatic. There was no question of carrying out laboratory studies, as the method of tipping the waste is an essential element whose influence would be difficult to determine when extrapolating laboratory data to full-scale operation.

-Density

The studies carried out at Beuvry-La-Forêt and Hersin-Coupigny demonstrated how difficult it is to control this parameter, firstly because significant differences in density cannot be readily obtained with the equipment currently in use (compactors), and secondly because the waste 'density'as such is a function of size and moisture content of the waste. A significant economic factor which must also be taken into account is the fact that operators prefer to increase the density ($d > 1$) of the material rather than reduce it so that maximum use is made of the available space (quarries etc.).

-Waste composition

With regard to the parameter 'waste composition', it was proposed to bury sludge from an urban sewage treatment plant along with the household waste in order both to increase the total percentage of organic matter and moisture in the waste and to pre-seed the waste with methanogenic bacterial strains which were already well developed. These experiments could not be completed because the operator was unable to fit reliable gas meters in time to the collection systems serving the experimental units. Since data obtained on biogas production from other sources had shown that methanogenesis developed naturally in a very short time, initial seeding of the waste no longer seems a determining factor. Sewage works sludge of course yields greater quantities of methane than household waste : 400 to 500 1/kg of TVM with a methane content of 70%. If this sludge were deposited along with household waste it could only improve the output and quality of the biogas.

-Size consist of the waste

The experiments at Roche-La-Molière consisted in comparing the biogas output from two sealed cells of 300 m^3, one containing coarsely crushed household waste (Unit A), and the other uncrushed household waste (Unit B). It was observed that methane production occurred later in unit B than in unit A, but the difference was only about two weeks. Various problems connected with the excessively low temperature of the waste and gas leakage through the walls of the cells prevented any analysis of data on the quantities of biogas produced.

-Moisture content of the waste

Experiments to investigate the effect of the moisture content of the waste are now in progress. The method adopted was to sprinkle the surface of a unit in which dumping had been completed with a certain amount of water in order to increase the total moisture content by 10%.

At Hersin-Coupigny the quantity of water which it was originally intended to apply was not completely absorbed : it passed rapidly through the layer of waste and accumulated in the lowest 2 to 3 m of the refuse, threatening to undermine the walls around the unit.

The effect of this water was apparent in a substantial increase in hydrolytic activity (a temporary accumulation of volatile fatty acids).

It can, however, already be concluded that the application of very large quantities of water may involve severe risks for sites where the 'conventional' cell method is followed : the accumulation of water at the base of the units endangers the stability of the walls. Water spraying must therefore be adapted to the system of operation practised.

The studies of these four parameters have shown how difficult it is to control such a complex process as methanogenesis : manipulation of any particular factor may result in additional operating constraints and at the same time cause uncontrolled changes in other factors. In the light of the knowledge now acquired in respect both of gas production as such and in the technology of gas recovery, a more promising avenue of approach seems to be research into the actual utilization of the gas, especially as the energy potentially available in France is currently in the region of 50 000 toe.

3. Conclusions

This figure of 50 000 toe is of course a very rough approximation and should be regarded with a great deal of caution : it is not a single 'gas field' but consists of scores of small 'reservoirs', whose use must be considered case by case.

Several approaches seem possible :

- production of thermal energy : direct use of the biogas as a substitute fuel in industrial burners, for heating greenhouses etc ;
- production of electric energy : sypply to electricity generators ;
- production of mechanical energy : use as an automotive fuel after cleaning (CO_2, H_2S) and compression.

....

The heating of greenhouses seems to be the most flexible approach, which could be adjusted to the problems posed by the scale of tipping operations. A small landfill site -40 000 to 50 000 t/year- could heat about 4 000 m^2 of greenhouses, while a site accommodating 1000 t/year could heat about 10 000 m^2.

Since the technical studies of methane production and recovery are now virtually completed, the aim is to carry out one or more utilization projects under a second EEC-ANRED contact.

THE USE OF LANDFILL GAS AS AN AUTOMOTIVE FUEL

Reinhardt KNOP
Institut für Siedlungswasserbau,
Wassergüte- und Abfallwirtschaft,
University of Stuttgart

Aim of the project

The gas produced by anaerobic digestion processes at landfills represents, on the whole, a considerable potential source of energy. In many cases the gas is simply flared off, thus overcoming the problems of smell, danger of explosion and damage to vegetation, etc.

At some landfills generators are used to produce electricity, or the gas may be burned directly for heating. Where it is not possible to generate electricity or there is no demand for heat, waste gases may be used as fuel for site vehicles (refuse compactors, bulldozers, etc.) or for vehicles used outside (refuse collection trucks, lorries).

The gas may be available untreated or scrubbed and stored in pressurized canisters. For the purposes of the project it was necessary, first of all, to determine whether it is possible to remove the inert components of the waste gas, which is a mixture comprising about 40 % carbon dioxide and 60 % methane, simply by scrubbing and by applying a pressure swing adsorption process, and secondly to compress the gas to 300 bar in gas storage canisters with a view to fuelling a specially converted refuse compactor.

Gas treatment

The gas scrubber was loaned from the Buse company of Linz an Rhein and was installed at the end of 1982. After some teething troubles with the oilburner, and after certain adjustments had been made, it was possible to begin the gas scrubbing tests.

The gas required is passed through a gravel bed, which protects against explosion and acts as an initial condensated filter, before being fed by means of a suction blower through a flexible pipe to the gas scrubber. The waste gas, flowing in the opposite direction to the Alkazid, then passes through the first washing column and leaves the installation more or less free of carbon dioxide. The leach liquor, charged with carbon dioxide after leaving the washing column, is then fed into the desorption column mounted on the oil-fired boiler. The carbon dioxide is removed from the installation by heating to just under 130°C and by applying an overpressure of approximately 0.4 bar. The hot regenerated leach liquor passes through several heat exchangers, thereby transferring most of its heat to the saturated leach solution, and is then further cooled to approximately 40°C in a water cooler. The adsorption process is then repeated. The initial gas flowrate of approximately 50 Nm^3/h was reduced to approximately 20 Nm^3/h to achieve virtually complete removal of the carbon dioxide. The scrubber was fitted with an automatic pressure regulator on the desorption side to make desorption independent of flowrate.

After no decline in separation efficiency had been observed in tests extending over several days, the tests were repeated using a different scrubbing agent. The agent used was MEA (monoaethanolamine). The tests results were comparable with those obtained using Alkazid, but since MEA decomposes on contact with water and other trace elements in the landfill gas, preference should in future be given to Alkazid. Both scrubbing agents are very effective for the removal of hydrogen sulphide (H_2S).

The energy comsumption of the scrubber, apart from the low cost of operating the pump (approx. 0.15 DM/h), was 3.8 kg of oil per hour. In an industrial installation, untreated landfill gas can be used instead of oil, thereby reducing operating costs. If landfill gas had been used as fuel, 19 % of the total volume of gas would have been needed for desorption, and 81 % would have been available as purified methane. Since the test installation was not originally designed for landfill gas, it may be assumed that it would be possible to use only 15 % of the gas as fuel for the process.

The gas purification tests have ended, and it is planned to investigate - although not as part of this project - the extent to which other undesirable factors can be removed using this equipment, if necessary with different scrubbing agents.

Conversion of the refuse compactor

The refuse compactor made available for the tests by the Rems-Murr local authority was converted. The existing diesel motor was replaced by a gas motor from the firm of Klöckner-Humbolt-Deutz, which made it necessary to disassemble and extend the entire vehicle frame. A larger radiator and a pressure regulator for the gas supply were installed. To simplify the initial test runs, the metal tyres (rammer feet) were replaced by wheeled-loader tyres.

The gas was stored in pressurized canisters kindly provided by Hohenheim University. After evacuation of air, they were filled with landfill gas.

Compressor

Recent findings concerning the use of biogas have now resulted in the requirement that compressed gas must be dry. This is because carbon dioxide at a high pressure of 300 bar and in the presence of humidity forms an acid with a pH value of approx. 1.5. A dryer operating on the basis of the pressure swing adsoption technique was therefore loaned.

A four-stage hydrogen compressor was used. At the request of the project leader, a specially modified model was supplied. The modification concerned the incorporation of the dryer in the compression process.

The compressor has a power of 45 kW at 88 A and weighs over 2.5 t.

Power supply

It was found during the early stages of the tests that the power supply to Steinbach landfill is so low as to make it impossible to operate the gas compressor unit and the ground water pumps simultaneously. To overcome this problem, the local authorities acquired a second-hand 140 kVA emergency power unit, which is now being installed.

Gas recovery

In the middle of the year the local authorities, after lengthy delays due to a variety of factors, began the laying of gas collector mains. Connected to these are six drainage lines, three gas flues and a number of short probes. Each gas source is individually adjustable using slider valves.

By mid-October, after extensive sealing work, it was possible to begin the selective extraction tests. Unfortunately, the gas supply proved inadequated both in quality and quantity. The probable reason for this is that it was possible to connect up only some of the gas flues, and the enormous quantities of waste slurry and earth involved made it very difficult for the gas to find its way to the drainage system.

In order to be able to begin the test using the vehicle , the compressor was mounted onto a lorry trailer and connected up at the landfill at Burghof near Ludwigsburg. Thanks to a previous research project, the technical facilities there were relatively good.

Conclusion

Contrary to expectations, considerable problems were encountered during the project. These were caused by the conditions prevailing at the landfill and not by the methods applied. On the whole, however, the use of landfill gas as a fuel for motor vehicles was shown to be a viable proposition. At a later stage, when adequate technical conditions have been achieved, it is intended to investigate the liquefaction of methane.

AGRICULTURAL USES OF DIGESTED EFFLUENTS

M. DEMUYNCK, E.-J. NYNS and H. NAVEAU

Unit of Bioengineering, University of Louvain,

B-1348 Louvain-la-Neuve, Belgium

Summary

Whereas anaerobic digestion has no effect on the quantity of the waste treated, it has well an effect on its quality and consequently on its fertilizer value. Indeed, first of all, 30 to 40 per cent of the organic matter of the waste digested which can be either manure or sludge, are transformed into methane and the remaining organic matter is more stable. If the total nitrogen content of the waste remains more or less the same, the proportion of ammoniacal nitrogen increases (up to 70 %) and the proportion of organic nitrogen decreases.

The reduction of pathogens is good in theory for bacteria, viruses, parasitic cysts and plant pathogens but this effect is reduced in practice, especially in completely mixed digesters. The digested wastes are only free of pathogens after thermophilic treatment.

Concerning the spreading of the digested effluent on land, the effect of the biomethanation treatment is positive since first, the waste is deodorized and secondly, the waste is liquified, homogenized and less sticky and therefore disappears more quickly into the sward which avoids the smothering of the grass and hinders weeds to come out.

Since the availability of the nitrogen of the digested effluents is higher than for effluents which are raw or stabilized by other means, good yields of crops high demanding in nitrogen can be obtained. Neverthess precautions for avoiding ammonia losses must be taken. Therefore, recommendations for the utilization of digested effluents are proposed.

1. INTRODUCTION

If anaerobic digestion was already well developed in the past for the stabilization of sewage sludges, this process becomes at present currently applied for the treatment of agricultural and industrial wastes. Most of the anaerobically digested effluents are not further treated at all and are used in agriculture as fertilizer or soil conditioner.

This is especially the case for part of the sludges produced in the European Community and for the largest part of the agricultural wastes digested. If the farmers know well what is the fertilizer value of their untreated manures or other farm residues, there is at present no precise estimation of the fertilizer value of the digested agricultural wastes. If digested sewage sludges are already often used in agriculture, their potential users may not know exactly what type of fertilizer manure they have.

It may be asked if the anaerobic digestion process has an effect on the organic waste, if yes to which extent and consequently what it involves in practice. Does anaerobic digestion modify the composition of the waste ? Are the anaerobically digested wastes more free of pathogens than the raw ones ? Is their quality improved by the treatment ? Are the availabilities of their nutrients better or worst after digestion ? Can farmers obtain the same crop yields with the digested effluents than with the raw wastes ? Do they have to modify the methods and the times of waste application ? Can they use them on all types of crops ?

Based on a review of the literature and on known experimental results, a study has been realised attempting to answer these questions. This paper gives the essential conclusions of that study.

2. EFFECT OF THE ANAEROBIC DIGESTION ON THE WASTE COMPOSITION

A general consideration to be repeated here is that the anaerobic digestion or biomethanation process affects only the quality of the waste that is treated and that it does not change its quantity at all. On the contrary, sometimes the volumes of the wastes may be increased; this can be the case for semi-solid or solid wastes which are diluted before entering the digester.

The effect of the process on the quality of the waste will depend on several factors as the biodegradability of the waste, the retention time in the digester and the temperature of digestion. A high content in easily biodegradable matters of the waste, a long retention time and a high digestion temperature may lead to important modifications of the nutrient content of the digested effluent. At the contrary, if the waste is not easily degradable and if the digester is not well operating, there can be practically no modifications at all. It is so that the composition of sewage sludges is in general more affected by anaerobic digestion than the agricultural wastes, especially those containing high quantities of lignin, cellulose, those two organic substances being not or partly bioconverted.

In summary, anaerobic digestion has two effects on the nutrient content of the wastes : first, it decreases their organic content and consequently their carbon content and secondly, it transforms part of the organic nitrogen into ammonia nitrogen. The organic matter can be reduced by almost half for sewage sludges (1). For liquid and semi-solid manures, the organic matter reductions reported are of about 40 per cent (2) (3) whereas for manures with bedding, the reduction is less than 20 per cent (4). The remaining organic matter is more stable than originally.

The proportion of ammoniacal nitrogen related to the total nitrogen content increases from about 5 per cent up to as high as 70 per cent. The mineralizations of nitrogen are again more spectacular for sewage sludges and are more important under heated digestion than cold digestion. This mineralized or ammoniacal nitrogen is more readily available to plants than the organic one. This is thus an improvement of the quality of the treated waste. Nevertheless ammonia can be rapidly lost under certain circumstances; precautions for utilization of the digested effluents have so to be taken so as it will be seen further.

A direct consequence of these two effects is that the carbon to nitrogen ratio decreases after anaerobic digestion. For wastes having originally C/N ratios above 15 (raw sewage sludges, manures with bedding, crop residues), this is thus also an improvement. Indeed, generally after digestion, their C/N ratios decrease to around 10 and so there is no nitrogen immobilization once the waste is incorporated in the soil.

Concerning the other nutrients i.e. phosphorous, potassium and other macro nutrients, they appear to remain unaltered by anaerobic digestion. Nevertheless, since the dry matter content decreases in the digested effluents, their concentrations increase after digestion when expressed to the dry matter content.

The same observation is valid for the heavy metals and other trace elements' content. Although, these potentially toxic elements are precipitated during anaerobic digestion, it appears that once in the soil, these precipitates will oxidize and so have the same behaviour as those of raw wastes (5). Concerning the degradation of nonionic surfactants that may be present in sewage sludges, it appears that anaerobic digestion favours the production of 4-nonyphenol, a persistent toxic substance. Nevertheless, there is at present no case of intoxication reported. On the other hand, the process appears to favour the degradation of most of the insecticides (7). Although this has still to be verified for full-scale operation, this may be a point in favour of the anaerobic digestion.

3. REDUCTION OF PATHOGENS BY ANAEROBIC DIGESTION

Both sewage sludge and animal wastes may contain a variety of pathogens which present disease hazards to man and food animals or plants. They may be bacteria, viruses, helminths, fungi or protozoa.

At laboratory scale, mesophilic anaerobic digestion allows a significant reduction of those pathogens. Indeed, the relative reduction is good for plant pathogens and parasitic cysts, moderate to good for bacteria, moderate for viruses (12), and poor for helminths ova. This pathogens' reduction effect is very similar to conventional aerobic digestion. In practice, especially in completely mixed digesters (12) (14), this effect is reduced by short-circuiting and by simultaneous drawing off and feeding of the digester (15) (16). In that case, there is always a risk that the mixed liquor does not remain inside the reactor during the whole retention time; infectious organisms may so be discharged and recontaminate the rest of the effluent. To obtain at full-scale operation the same disinfection effect as at laboratory-scale, the waste should be treated in batch or plug-flow digestion systems. An effluent completely free of pathogens can only be obtained after digestion at thermophilic temperatures i.e. around 55 °C. Even for helminths eggs, there is a complete inhibition of normal egg development (17).

Although thermophilic anaerobic digestion is seldom employed until now, it should be recommended in the future to digest at 55 °C the most infected wastes as abattoir wastes for example.

4. QUALITY IMPROVEMENT OF THE WASTE BY ANAEROBIC DIGESTION

Besides its two effects on waste composition and disease survival, anaerobic digestion may also improve the quality of the waste. First of all, it reduces the odour of the waste. Indeed, both for sludge and for manure, smells during utilization (spreading) cause more complaints than any other aspect of the operation. These problems can be minimized and controlled by using digested effluents. It has for example been calculated that the emission from 1 ha spread with digested slurry is 10 per cent of the emission from 1 ha spread with untreated slurry (18). The reduction of smell could justify the fact that cattle prefer grazing on pastures spread with digested manures (19). Although the odour reduction is not as complete as the one obtained by continuous aeration, the odour reduction by anaerobic digestion is of higher long term value. Indeed for well anaerobically digested manure, odour remains unchanged even after 120 days of storage (19), and the concentrations of malodorous compounds may even be reduced during that time (20). Comparatively, well aerated manure, has been reported to smell again after two to three weeks (19).

As most of the waste treatments, anaerobic digestion reduces the viscosity of the wastes; consequently they become less sticky and more homogeneous (21) (22) (23). Their improved physical state leads to an easier handling and blockages in spreading equipment are avoided. The digested effluents disappear more quickly into the sward and do not form a mat as they dry which smothers new grass. Consequently, there is no or less opportunity for weeds to grow (22). Weed control occurs also in the anaerobic digester itself. Anaerobic digestion appears to have a quite significant fatal effect on weeds as rumex, millet, effect which should be superior to the one of storage or aeration (22). This weed control is nevertheless restricted to the wastes which undergo the treatment and has consequently limitations. It avoids at least the transmission of weeds by manure and other waste.

5. INFLUENCE OF DIGESTED EFFLUENTS ON THE SOIL

Organic waste may be used as soil conditioner so as to restore derelict and disturbed land e.g. mining spoil, landfill sites, unproductive land... Such soils have very low levels of nutrients and the lack of organic matter can render them more susceptible to physical damage, particularly compaction from heavy machinery resulting in poor root penetration, low infiltration rates with consequent risk of runoff, erosion and pollution (24).

By increasing the carbon content of the soil, organic wastes applications increase aggregation, decrease bulk density, increase water holding capacity and hydraulic conductivity (25). Furthermore organic matter and soil aggregation is inversely related to runoff volumes and sediment loss.

Digested organic wastes can be used to soil restoration as well than the organic wastes. Even it has been seen below point 2 that the organic matter content is reduced on average by 30-40 per cent, this is nevertheless largely compensated by the fact that the organic matter of digested effluents is more stable than the organic matter of raw organic wastes and will consequently degrade more slowly in the soil. The effect of digested effluents on soil physical conditions is consequently more permanent than for raw wastes. As for example, the aggregate stability of soil appears to be particularly improved by anaerobic wastes. By comparing the Water Stability Index after application of anaerobically digested sludge, aerobically digested sludge and compost of sludges with domestic refuse, Guidi et al., (26) observe that the stabilizing effect of the anaerobic sludge treatment is always higher and longer lasting than that of the other treatments. This phenomenon is attributed to the relative large amounts of stable organic compounds, such as lignin, cellulose, lipids and humic like materials which are not modified during the anaerobic digestion (27). These classes of compounds are highly reactive and can interact directly with soil surfaces, thus strenghtening the aggregates. Moreover, such chemical compounds are rather resistant to microbial degradation, and could so explain the lower rate of decay observed in Water Stability Index of the anaerobic sludge treated samples.

Nevertheless, so as to obtain measurable beneficial effects on soil physical conditions, applications rates as high as those used for farmyard manure i.e. about 30 to 50 tons dry solids/ha are required. Therefore, the digested effluents with the highest dry solids content are particularly indicated; these can be dry or dewatered digested sludges or digested manure with bedding or abattoir wastes...

Significant effects on soil improvements cannot be expected from application rates of liquid digested effluents (1-2 tons/ha) which are commonly in use. The small beneficial results may nevertheless accrue from repeated applications, at least if there are no problems of excessive contents of nitrogen or heavy metals in the liquid effluent.

6. AVAILABILITY OF THE NUTRIENTS OF THE DIGESTED EFFLUENTS

The biomethanation process affects essentially the nitrogen availability of the digested wastes but apparently does not influence the availability of the other nutrients. Since during anaerobic digestion part of the organic nitrogen is transformed into ammonia nitrogen and since ammonia-N is more readily available to the plant (assumed to be equal to the N-availability of N-inorganic fertilizer), the global nitrogen availability of digested effluents is higher than that of raw wastes. The total available nitrogen of digested effluents will essentially depend on the proportion of the ammonia nitrogen (in per cent of the total nitrogen) since it is in general assumed that ammoniacal nitrogen and organic nitrogen are respectively 100 and around 20 per cent available (28).

Compared to organic nitrogen, ammoniacal nitrogen presents more risks for volatilization or nitrification and may consequently be lost for the plant. The factors which affect the most the degree of NH_4^+-N losses are the post-treatment of the digested waste and its application conditions. Dewatering or drying are possible post-treatments mostly applied for sewage sludges. Much of the ammonium-N is lost after such treatment. The amount of nitrogen in liquid digested sludge is up to 3 times that in dried digested sludge (29) and about 20 per cent nitrogen availabilities are reported for such sludges (28). Storage is another possible post-treatment applied for both digested sludges and farm wastes. In such case, ammonia losses are not as high as for dewatering or drying. It appears in general that stored digested effluent has a lower nitrogen availability than freshly digested effluent, but it has nevertheless a higher efficiency than dewatered digested effluent. Although this would need more confirmation, it appears also that the losses are identical for untreated and digested wastes during their storage (30).

Among the application conditions having an effect upon the nitrogen effectiveness of digested effluents, we find :

- the timing of effluents application,
- the soil type, the soil cover, and its environmental conditions i.e., the rainfall and soil temperature conditions,
- the way of application.

Application of digested effluents over autumn and winter period can lead to high losses of nitrogen. Particularly, for autumn-early winter application, the soil may still be warm enough to allow nitrification of the ammoniacal-N and so it is possible that much of the nitrogen may be lost in drainage (29). In heavy textured soils, autumn and winter rainfalls maintain a high degree of moisture in the soil which leads again to ammoniacal-N nitrification with consequent losses. Nevertheless, it has been demonstrated in the United Kingdom, that winter applications of digested sludges do not lead to nitrogen losses, at least in hight textured soil (29). With equivalent application rates of digested sludges and fertilizer N winter applications of digested effluents may even lead to higher yields of grass or cereals than spring applications (28,30). The fact that most of the liquid digested effluents have a high water content is benefic for summer applications. It has been observed for example that liquid digested sludge applied to drying soil in summer gives a quicker growth than conventional N-fertilizer applied at the same time (28).

The weather conditions under which the waste is applied influence greatly the ammonia volatilization. It is mostly affected by high air temperature (31) and wind, particularly on bare soils. Consequently, digested effluents should preferably be applied on growing crops covering the soil and under rainy weather conditions.

The application method of the digested effluent is also of great importance for ammonia losses. Injection of the influent appears to be the best method for avoiding such losses, especially compared to surface application but is not very applied in practice. In surface application, losses can nevertheless be minimized by immediate incorporation into the soil following the spreading (32).

7. PLANT RESPONSE TO DIGESTED EFFLUENTS

As far as the response of the plant is considered, anaerobically digested effluents appear to have a good effect. Indeed, since anaerobic digestion produces effluents with lower carbon to nitrogen ratios (C/N = 10), the effect of nitrogen immobilization does not occur after the application of the digested waste whereas it can well occur with some of the untreated wastes (sewage sludges, farm wastes with a high carbon content). Since digested effluents contain a higher percentage of directly available nitrogen i.e. ammonia nitrogen, the time response of the plant is more immediate compared to fresh wastes. It has for example been observed that grass grows faster with digested slurry compared to untreated slurry and can so be harvested 10 days earlier (33).

A restriction has nevertheless been observed concerning the seedling emergence, especially in laboratory experiments. Due to free ammonia, there is a risk that seed germinates poorly if sown on cultivated ground which has just been treated with freshly digested effluents (28,34). Normally, most ammonia is absorbed in soil as the ammonium ion, but at high pH levels, particularly above 7.5 increasing proportions remain as free ammonia from which levels in excess of 2.5 mg/l are toxic to many plants (28). This effect dissappears within one week and has until now not been reported in field experiments.

Concerning the crop yields that can be observed, it is generally accepted that liquid digested wastes give increases in yield of crops similar to those produced by equivalent amounts of nitrogenous fertilizer (35,22). Crops high demanding in nitrogen so as ryegrass and maize are particularly indicated for the utilization of such wastes. Cereals would appear to profit less than other crops and there may be even an insufficiency in available nitrogen to carry the crop to maturity (28). In a 6 years field trial on ryegrass comparing digested manure to untreated and aerated manure, Besson (22) observes that grass yields obtained are more or less identical for the 3 differently treated manures. In fact, grass yields appear to be more function of the type of manure (bovine cattle or pig) and of the application rates. The feed value of the crop (grass or maize) expressed in digestible crude albumin appears to be unchanged for all the treatments (29). Both for the utilization of digested sludge or manure on mixed herbage, it has nevertheless been observed that clover tends to be suppressed, but this effect is less than that of ammonium nitrate (22,28).

8. CONCLUSIONS

Anaerobic digestion changes only the quality of the waste i.e. compared to raw waste, digested effluents have a lower content of organic matter, a higher content of ammonia nitrogen, are more easily usable and are more disinfected although not completely.

Dried digested effluents or solid digested wastes should preferably be used for restoration of unfertile soils so as to improve their physical conditions. Concerning the recommendations that may be given for the utilization of liquid digested effluents, the essential point is to take the maximum advantage of the ammonia nitrogen that contain digested waste and therefore to avoid ammonia losses. Therefore the digested effluents should never be spread under dry and sunny weather. In case of use on bare soils, surface application should immediately be followed by the ploughing in of the waste. An ideal cultural practice is for example to combine the incorporation of liquid digested effluents with that of crop residues rich in carbon e.g. stubble.

Defining the optimal times for digested effluents utilization is rather difficult and depends in fact on several factors as climates, types of crops, types of soils... On a general rule, the wastes should be applied when crops need nutrients for growing, otherwise, there is always a risk for nitrogen leaching especially on soils having a low organic matter content. Winter applications can be allowed only on light textured and levelled soils and when temperatures are low. In any case, winter spreading on slopes should be forbidden because there may be risks for superficial runoff.

The typical way for the utilization of digested effluents on grass is applying them firstly before the growth of the new shoots and than after each cut. Indeed spreading on new grass is dangerous and can lead to the necrosis of the new shoots and well developed grass may be easily burned when too high a quantity of ammonia is applied in once. Concerning maize growth, the best season of application appears to be a first time before sowing and a second time, somewhat 2 months after (22). For cereals although it is not recommended, the best yields appear to be obtained with January and February applications.

This study has been supported by grant ECI-1012-B7210-83-B from the Commission of the European Communities.

REFERENCES

1. HALL, J.E., 1983. Predicting the nitrogen values of sewage sludges. In Characterization, Treatment and Use of Sewage Sludge, Proceedings of the third European Symposium, Brighton, 26-30 September 1983. Commission of the European Communities, L'Hermite P. and Ott H. eds., Reidel D., Publishing Company, Dordrecht, Holland, in the press.

2. DEMUYNCK, M., NYNS, E.-J., 1984. Biogas plants in Europe - practical handbook. In Energy from Biomass, Série E, Vol. 6., Grassi G. and Palz, W., eds., D. Reidel Publishing Company, Dordrecht, Holland, in the press.

3. BESSON, J.-M., LEHMANN, V., ROULET, M., WELLINGER, A., 1982. Influence de la méthanisation sur la composition des lisiers, Revue Suisse Agric. 14 (3), pp. 143-151.

4. CHAUSSOD, R., SANCHEZ, C., DUMET, M.C., CATROUX, G., 1983. Influence d'une digestion anaérobie (méthanisation) préalable sur l'évolution dans le sol du carbone et de l'azote des déchets organiques. D.G.R.S.T. - Action concertée "Valorisation énergétique des déchets agricoles" - Aide n° 78-7-2909. Rapport de l'I.N.R.A. - Laboratoire de Microbiologie des sols - BV 1540-21034 Dijon - Cedex.

5. BALBWIN, A., BROWN, T.A., BECKETT, P.H.T., ELLIOTT, G.E.P., 1983. The forms of combination of Cu and Zn in digested sewage sludge. Water. Res., 17 (12), pp. 1935-1944.

6. SCHAFFNER, C., BRUNNER, P.H. and GIGER, W., 1983. 4-NONYLPHENOL, a highly concentrated degradation product of nonionic surfactants in sewage sludge. In Characterization, Treatment and Use of Sewage Sludge, Proceedings 3d European Symposium, Brighton, 26-30 September 1983. Commission of the European Communities, L'Hermite P. and Ott H. eds, Reidel D. Publishing Company, Dordrecht, Holland, in the press.

7. LESTER, J.N., 1983. Presence of organic micropollutants in sewage sludges, in Characterization, Treatment and Use of Sewage Sludge, Proceedings third European Symposium, Brighton, 26-30 September 1983. Commission of the European Communities, L'Hermite, P. and Ott, H. eds., Reidel D. Publishing Company, Dordrecht, Holland, in the press.

8. TURNER, J., STAFFORD, D.A., HUGHES, D.E., CLARKSON, J., 1983. The reduction of three plant pathogens (*Fusarium*, *Corynebacterium* and *Globodera*) in anaerobic digesters. Agricultural Wastes, 6, pp. 1-11.

9. CARRINGTON, E.G., 1980. The fate of pathogenic micro-organisms during wastewater treatment and disposal. Technical report TR 128, Stevenage Laboratory, Water Research Centre, 58 pp.

10. PIKE, E.B., 1980. The control of Salmonellosis in the use of sewage sludge on agricultural land. In Characterization, Treatment and Use of Sewage Sludge, Proceedings of the 2nd European Symposium, Vienna, October 21-23, 1980; P. L'Hermite and H. Ott eds, D. Reidel Publishing Company, Dordrecht, Holland, pp. 315-327.

11. WILLINGER, H., THIEMANN, G., 1983. Survival of resident and artificially added bacteria in slurries to be digested anaerobically. In Proceedings of a joint workshop of expert groups of the Commission of the European Communities, German Veterinary Medical Society (DVG) and Food and Agricultural Organisation; D. Strauch ed., Institute for Animal Medicine and Animal Hygiene, University of Hohenheim, Stuttgart, Federal Republic of Germany, pp. 210-217.

12. LUND, E., LYDHOLM, B. and NIELSEN, A.L., 1982. The fate of viruses during sludge stabilization, especially during thermophilic digestion. In Disinfection of Sewage Sludge : technical, economic and microbiological aspects, Proceedings of a Workshop, Zurich, May 11-13, 1982. A.M. Bruce, A.H. Havelaar, P. L'Hermite, eds., D. Reidel Publishing Company, Dordrecht, Holland, pp. 115-124.

13. CARRINGTON, E.G., HARMAN, S.A., 1983. The effect of anaerobic digestion temperature and retention period on the survival of *Salmonella* and *Ascaris* ova, presented at WRC Conference on stabilisation and disinfection of sewage sludge, Paper 19, Librarian, WRC Processes, Stevenage, 13 pp.

14. BLACK, M.I., SCARPINO, P.V., O'DONNELL, C.J., MEYER, K.B., JONES, J.V., KANESHIRO, E.S., 1982. Survival rates of parasite eggs in sludge during aerobic and anaerobic digestion. Applied and Environmental Microbiology, 44 (5), pp. 1138-1143.

15. GOLUEKE, C.G., 1983. Epidemiological aspects of sludge handling and management. Biocycle, 24, (3), 52-58.

16. MUNCH, B., SCHLUNDT, J., 1983. The reduction of pathogenic and indicator bacteria in animal slurry and sewage sludge subjected to anaerobic digestion or chemical disinfection. In Proceedings of a joint workshop of expert groups of the Commission of the European Communities, German Veterinary Medical Society (DVG) and Food and Agricultural Organization; D. Strauch ed., Institute for Animal Medicine and Animal Hygiene, University of Hohenheim, Stuttgart, Federal Republic of Germany, pp. 130-149.

17. KIFF, R.J., LEWIS-JONES, R., 1983. Factors that govern the survival of selected parasites in sewage sludges. Pollution Research Unit Umist, Manchester.

18. KLARENBEEK, J.V., 1982. Odour measurements in Dutch agriculture : current results and techniques. Research Report 82-2, Institute of Agricultural Engineering, Wageningen, The Netherlands.

19. WELLINGER, A., 1983. Anaerobic digestion - the number one in manure treatment with respect to energy cost ? In Proceedings of a joint workshop of expert groups of the Commission of the European Communities, German Veterinary Medical Society (DVG) and Food and Agricultural Organization D. Strauch, ed., Institute for Animal Medicine and Animal Hygiene, University of Hohenheim, Stuttgart, Federal Republic of Germany, pp. 163-183.

20. VAN VELSEN, A.F.M., 1981. Anaerobic digestion of piggery waste. Ph. D. Thesis of the Agric. University, Wageningen, 103 p.

21. Anon., 1983. Practical guidelines for the farmer in the EC with respect to utilization of animal manures.

22. BESSON, J.M., 1983. Personal Communication.

23. PIKE, E.B., DAVIS, R.D., 1983. Stabilisation and disinfection - their relevance to agricultural utilisation of sludge, presented at WRC Conference on stabilisation and disinfection of sewage sludge, Paper 3, Session 1, Librarian, WRC Processes, Stevenage, 30 pp.

24. HALL, J.E., VIGERUST, E., 1983. The use of sewage sludge in restoring disturbed and derelict land to agriculture. In Utilization of Sewage Sludge on Land : Rates of Application and Long-term Effects of Metals, Proceedings of a Seminar, Uppsala, June 7-9, S. Berglund, R.D. Davis and P. L'Hermite eds., D. Reidel Publishing Co., Dordrecht, Holland pp. 91-102.

25. KHALLEL, R., REDDY, K.R., OVERCASH, M.R., 1981. Changes in soil physical properties due to organic waste applications : a review Journ. of Environmental Quality, 10, pp. 133-141.

26. GUIDI, G., PAGLIAI, M., GIACHETTI, M. 1981. Modifications of some physical and chemical soil properties following sludge and compost applications. In The Influence of Sewage Sludge Application on Physical and Biological Properties of soils, Proceedings of a seminar organized jointly by the Commission of the European Communities, Directorate - General for Science, Research and Development and the Bayerische Landesanstalt für Bodenkultur und pflanzenbau, Munich, Federal Republic of Germany, held in Munich, June 23-24, pp. 122-130.

27. PAGLIAI, M., GUIDI, G., LAMARCA, M., GIACHETTI, M., LUCHAMANTE, G., 1981. Effects of sewage sludges and composts on soil porosity and aggregation. J. Environ. Qual. 10, pp. 556-561.

28. COKER, E.G., 1978. The utilization of liquid digested sludge. Paper 7 in WRC Conference, "Utilization of Sewage Sludge on Land". Water Research Centre, 1979 (Stevenage, Herts, SG1 1TH).

29. TOUSSAINT, B., MARCIN, C., 1981. Comparaison de lisiers de bovins avant et après méthanisation. Internal report of the "Laboratoire d'Ecologie des Prairies", Prof. J. Lambert, UCL, B-6654 Michamps (Longville).

30. HALL, J.E., WILLIAMS, J.H., 1983. The use of sewage sludge on arable and grassland. In Utilisation of Sewage Sludge on Land Rates of Application and Long-term Effects of Metals, Proceedings of a seminar, Uppsala, June, 7-9, S. Berglund, R.D. Davis and P. L'Hermite eds., D. Reidel Publishing Co., Dordrecht, Holland, pp. 22-35.

31. BEAUCHAMP, E.G., KIDD, G.E., THURTELL, G., 1978. Ammonia volatilization from sewage sludge applied in the field. J. Environ. Qual., 7, pp. 141-146.

32. RYAN, J.A., KEENEY, D.R., WALSH, L.M., 1973. Nitrogen transformations and availability of an anaerobically digested sewage sludge in soil. J. Environ. Quality. 2, pp. 489-492.

33. STEPHAN, B. 1983. Personal communication.

34. GUIDI, G., HALL, J.E., 1983. Effects of sewage sludge on the physical and chemical properties of soils, in Characterization, Treatment and Use of Sewage Sludge, Proceedings, 3d European Symposium, Brighton, 26-30 September 1983. Commission of the European Communities, L'Hermite P. and Ott H. eds., Reidel D. Publishing company, Dordrecht, Holland, in the press.

35. WILLIAMS, J.H., 1979. Utilisation of sewage sludge and other organic manures on agricultural land. In Treatment and Use of Sewage Sludge, Proceedings of the 1st European Symposium, Cadarache, 13-15 February 1979, D. Alexandre & H. Ott eds, Commission of the European Communities, pp. 227-242.

Summary of the discussion of Session II :

PART 3 - ANAEROBIC DIGESTION

Chairman : M. GERLETTI

Rapporteur : P.N. HOBSON

In reply to a question from Mr. J. Field of Wageningen about his paper on American digesters, Mr. Wentworth said that so far as he knew only about 4, or perhaps 6, of the 84 full-scale digesters in America were of the immobilised-biomass type, the majority were continuous-flow stirred tank or 'plug-flow' designs.

Mr. N. S. Kang from Birmingham asked for some details of the proposed 57000 m^3 digester. Mr. Wentworth said that it was a proprietary design of ADI Ltd., Nova Scotia, and was called a 'Bulk Volume Fermentor', which suggested a static system, but he had no details. Mr. Kang also asked about a pig-farm digester in Alaska. Mr. Wentworth said that this was a mesophilic digester: he had no knowledge of its economics in the cold climate, but he would assume that the digester would be worthwhile solely because it converted the piggery waste to an effluent that could be more safely disposed of in winter to a land ecological system that was very finely balanced.

Mr. Karim Ali Bhal from Kenya asked Ms. Olguin a question about her paper with Mr. Monteverde. Replying, Ms. Olguin said that the figures she gave were only for the theoretical potential production of biogas from wastes from distilleries. However, a survey had shown that this biogas could be profitably used in the distilleries in place of the present petroleum fuels and it would not be sold for external use.

Mr. Hobson replied to Mr. A. M. Breure of Amsterdam that in the fermentation of starch they had not seen any relationship between VFA concentrations and acetic:butyric ratio, the ratio seemed to depend more on the rate of growth of the bacteria. In fermentations where total acid production was low and the butyric:acetic ratio was high, rates of growth of the bacteria and formation of acids were also low.

In reply to Mr. E. Senior of Strathcylde, Mr. Hobson said that they did not monitor for acetogenesis from CO_2 and H_2 as this would need isotope labelling, but from fermentation balances there was no evidence for acetic acid formation by this pathway.

Ms. D. Oakley from Birmingham said that in aerobic systems bacterial polymer production was influenced by C:N ratio in the medium, and could this be used as a method of control in the anaerobic systems. In reply Mr. Hobson said that in general polymer (polysaccharide) production took place when cultures were nitrogen limited and growth rate was low. In the present experiments they had tried to obtain the lowest ratio of N to C which would give maximum fermentation of substrate to acids without polysaccharide formation. It was difficult to envisage polysaccharide formation as a control method.

Mr. F. Edeline asked Mr. Colin a question about temperature equilibria and gas production in the titration cell of his apparatus. Mr. Colin said that temperatures in the digester and the titration cell had to be taken into account, but that the titration produced some reaction heat which offset the cooling of the digester sample. The titration cell was fitted with a gas-release valve.

Replying to similar questions from Mr. Hobson and Mr. P. Weiland from Braunschweig, Mr. Colin said that the apparatus was now used only for monitoring the fermentor; no computer program had yet been written to enable the fermentation to be controlled on the basis of the measurements made by the analytical system.

In reply to Mr. A. Pauss, Louvain, Mr. Colin said that the pH values measured agreed with pH values calculated from the constituents of the digester fluids. Mr. P. Nørgaard, Allborg, was told that the reliability of the VFA determinations had been checked by comparison with results obtained by Standard Methods and that the difference was no more than about 5%.

Mr. A. Rozzi, Bari, asked about fouling of electrodes, a point also raised by a previous questioner. Mr. Colin said that the only feedstock that had been used was sugar juices which contained little nitrogenous material and fouling was not a problem and proteins did not contribute to buffering. On a further question Mr. Rozzi was told that the price of the apparatus was about FFr 50000, but by using a home computer this could be reduced to 20-30000.

Mr. K. Dahl, Lund, asked a question about the use in the calculations of concentrations rather than activities. Mr. Colin said that strictly one should use activities rather than concentrations, but they did not do this as there seemed to be no reasonable way of calculating activities. Tests showed that activities calculated by correcting values for dissociation constants from the literature used considerable computer time and were only approximations. Concentrations seemed sufficiently accurate for the present purposes.

Mr. Knop told Mr. Wentworth that disposal of land-fill gas by flaring is unlikely to be permitted in Germany, except possibly on a very small scale, because it could cause air pollution in excess of that permitted by regulations. He had already seen problems of tree death on land-fill sites caused by burning of gas with its H_2S, chloride and cyanide impurities; carbon monoxide from incomplete combustion might also be a factor.

Mr. S. Picchiolutto, Modena, asked whether it would be possible to use other methods of gas purification and so lose less than the 19% of gas quoted for the method described in the paper. Mr. Knop replied that it would be possible in a number of ways. The easiest system would be water-scrubbing, but this left the problem of the disposal of contaminated water.

In reply to other questions Mr. Knop said that according to the manufacturers of the cylinders the gas should be dried to the equivalent of a dew-point of -20°C before being compressed into cylinders, and a dual-fuel, methane-diesel, system for vehicle engines would need two fuel tanks with attendant problems, it would be possible for stationary engines.

Mr. Pronost, SOBEA, asked whether the isohumic coefficient of organic matter is modified after anaerobic digestion as it is after aerobic composting. Mr. H. Naveau, who had read the paper by Mr. Demuynck, said that he did not know the answer but pointed out that the processes differ in that anaerobic digestion treats slurries whereas composting is carried out on dry, solid material.

Mr. Wentworth commented that Dr. Marchaim, in Israel, had found washed solid residues from animal-waste digesters to be exceptional in promoting plant growth and that he spoke of growth factors being present. Mr. Navean said that he had no experience of this but that plant toxins are not present in digested sludges. Mr. Wentworth said, further, that Marchaim had been interested in producing a substitute for expensive peat when he noted the enhanced plant growth caused by washed digester solids. The explanation for this effect was at present obscure, Marchaim believed that growth factors were introduced during passage of digesta through the animal.

POSTER SESSION

BIOTECHNICAL CONVERSION OF XYLOSE TO XYLOTOL BY PACHYSOLEN TANNOPHILUS

P. Thonart, X. Gomez, J. Didelez and M. Paquot

Département de Technologie
Faculté des Sciences Agronomiques
5800 Gembloux (Belgium)

SUMMARY

Biotechnical production of xylitol was investigated using Pachysolen tannophilus. The xylose reductase and xylitol dehydrogenase activities proved to be inducible by xylose, but the influence of the xylose concentration seems slight, although xylitol production is optimum at a concentration of 6%. Xylitol production and excretion do not seem to depend on the concentration of these enzymes but the hypothesis is formulated that it depends on the NADPH concentration. A study of the mutants involved in xylose assimilation does not disprove this hypothesis.

INTRODUCTION

In order to utilize paper mill wastes, it is essential to find uses for the sugars which they contain. The possible approaches involve the production of single-cell proteins, ethanol or metabolite of greater added value.

As a first stage we are investigating the biotechnical production of xylitol, with single-cell proteins as a by-product. Pachysolen tannophilus will be used as a model organism with which we shall seek to investigate physiological or genetic means of dissociating the growth of the microorganism from xylitol production with a view to designing an immobilized-cell reactor with regeneration of the cofactor (NADPH).

APPARATUS AND METHODS

The strain used was Pachysolen tannophilus ATCC 32691. The methods of measuring the xylose reductase and xylitol dehydrogenase activities have already been described (1)(2)(3), as has the mutagenesis technique (4). The SC medium consisted of Yeast Nitrogen Base (Difco) 1.75 g/l, ammonium sulphate 5 g/l with the carbon source added at different concentrations.

RESULTS AND DISCUSSION

Effect of the carbon source on the growth of Pachysolen tannophilus

Pachysolen tannophilus was cultured on a SC minimal medium with various carbon sources added. The generation times were as follows: 2h45 on glucose at 2%; 4h15 on xylose at 2%. This strain does not grow on the SC medium with 2% xylitol added. When cultured under these conditions, it is thus incapable of metabolizing a product which it could have secreted. Cells induced on xylose are also incapable of using xylitol. However, in a SC medium with added xylitol, the ammonium sulphate being replaced by nitrate, the generation time was 5h35. When the xylitol concentration was increased to 5% in the SC medium, slight growth was observed.

Study of xylitol dehydrogenase and xylulose reductase

The technique used for the determination of these activities has already been described. The maximum xylitol dehydrogenase activity is attained at a temperature of 40°C and a pH of 6.6 - 6.8, whereas optimum xylose reductase activity is observed at 40°C and a pH of 8.5 - 9.

The two activities are inducible by xylose but not by glucose. Moreover, the level of the activity studied is particularly high in the stationary phase and the xylose concentration is not the determinant factor. Measurement of xylose and xylitol by HPLC, however, shows that xylitol production is nil at a low xylose concentration (1%) and increases markedly to reach 0.860 g/100ml at 6% xylose.

Xylitol production does not depend on the xylitol dehydrogenase activity in the cell. It is probably more affected by the NADPH concentration.

Selection of mutants involved in xylose assimilation

Mutants involved in xylose assimilation were selected after the ethylmethanesulphonate reaction as previously described (4). The xylose$^-$ mutants exhibit reduced D-xylose reductase and xylitol dehydrogenase activities.

CONCLUSIONS

The factor determining xylitol production by Pachysolen tannophilus is neither xylose reductase nor xylitol dehydrogenase activity. Xylitol production is probably rather a means of regulating the quantity of NADPH in the cell in relation to the phosphate pentose cycle.

REFERENCES

(1) THONART, P. and GOMEZ, X., 1984
VIIth International Biotechnology Symposium, New Delhi, India

(2) SMILEY, K.L. and BOLAN, P.L., 1982
Biotechnology Letters, 4, 607-610

(3) SUIKKO, M., SOUMALAIEN, I. and ENARI, T.M., 1983
Biotechnology Letters, 5, 525-530

(4) THONART, P., BECHET, J., HILGER, F. and BURNY, A., 1976
J. Bacteriol., 125, 25-32.

THE PRODUCTION OF CHEMICAL AND FERMENTATION FEEDSTOCKS FROM LIGNOCELLULOSIC MATERIAL

A. J. BEARDSMORE
Imperial Chemical Industries PLC, Agricultural Division
Research and Development Department, United Kingdom

Summary

Novel chemical catalysts are being developed by ICI which rapidly hydrolyse cellulose and hemicellulose to monosaccharides. The catalysts can be either lithium chloride, zinc chloride or calcium chloride in combination with hydrochloric acid. The ICI catalytic reaction is complete within several minutes with high yields (>75%). The temperature of the reaction is less than 90°C and the pressure (developed internally) is no greater than 150 psig. The catalyst system has been shown to rapidly hydrolyse a large number of lignocellulosic materials, including straw, newsprint, cardboard and wood. Because hemicellulosic sugars have been shown to degrade more rapidly than glucose to furans, a pre-treatment has been successfully used to remove these sugars prior to cellulose hydrolysis. Microbiological studies have shown that the sugars derived can be used as substrates for fermentation.

1 INTRODUCTION

Lignocellulosic material is a renewable resource derived by the fixation of CO_2 into green plants by light energy. It consists of three major components - cellulose, hemicellulose and lignin. Cellulose and hemicellulose are polysaccharides which when hydrolysed yield sugars for fermentation. The hydrolysis process has taxed the minds of academics and industrialists alike because it is extremely difficult to achieve rapid breakdown of lignocellulosic components. This is due to the complex interwoven structure of the material in its native state.

Many processes have been attempted; these can be crudely separated into two categories: 1.1 chemical hydrolysis 1.2 enzymic hydrolysis.

1.1 Chemical Hydrolysis

Chemical breakdown of cellulose and hemicellulose can be achieved using acid. The Madison Scholler (1) process involving dilute H_2SO_4 was used during 1933-1945 to produce sugars for fermentation. It suffered from low yields (<50% sugars) and high by-products (furfural and hydroxymethylfurfural).

The temperature and pressures involved were high and the time of reaction was 3 hours. The plants were shown to be uneconomic in the later 1940's and were closed.

Further research on acidic hydrolysis has examined alternative catalysts including hydrochloric acid (dilute, concentrated and anhydrous (2,3)), hydrogen fluoride (4), trichloracetic acid (5) and other studies on H_2SO_4 (dilute and concentrated (6)).

The recovery and recycle of the catalyst has appeared to limit their commercialisation.

1.2 Enzymic Hydrolysis

The complex close-knit structure of lignocellulose precludes rapid enzyme hydrolysis. This can be illustrated by the length of time required for a piece of wood to degrade. The enzymes involved cannot readily attack the cellulosic substrate. In all cases a pre-treatment is required which will effectively open out the structure to allow subsequent enzymic hydrolysis. Pre-treatments include steam explosion (7), liquid NH_3 explosion (8) and UV irradiation (9). Following these treatments hydrolysis is enhanced and only takes several days to produce 80% yields of glucose.

Research aimed at achieving effective lignocellulosic hydrolysis was started at ICI in collaboration with Professor S A Barker and his colleagues at the University of Birmingham. EEC sponsorship has led to the results described in this paper.

2 CATALYST DESCRIPTION

Lithium chloride, zinc chloride and calcium chloride in combination with hydrochloric acid hydrolyse cellulose at an extremely rapid rate. The reaction will occur over a range of experimental temperatures but is optimum between 50°C and 90°C. A typical reaction is shown in Figure 1. Although the substrate for these results was 10% w/w filter paper (ie pure cellulose) a number of 'real' substrates have been tested. Some of these are shown in Table I using $CaCl_2$/HCl as the catalyst. The reaction temperature was 70°C and the yields are based on the cellulose content being converted to glucose. The hemicellulose content of these substrates was also hydrolysed to monosaccharides, however the lignin appeared to be unaffected.

3 PROCESS OPTIMISATION

It was realised that the catalyst degraded the hemicellulosic content of various substrates more rapidly than cellulose. To obtain maximum sugar concentrations a dilute acid pre-treatment is now used to catalyse the reaction. Typical conditions are shown in Table II. In excess of 90% of these sugars are recovered from the substrate using this technique. Less than 2% of the sugars are further degraded to hydroxymethylfurfural and furfural.

Solids derived from the treatment outlined above are then used as the substrate for the metal halide and hydrochloric reaction. Using paper or wheat straw as the substrate results comparable to those shown in Table I have been obtained. Hydroxymethylfurfural accounts for less than 2% of the substrate added to the catalyst.

Lignin has been recovered from these two reactions as a dark brown solid. Initial analysis suggests that it has not been affected by the reaction and is free of carbohydrate.

4 CATALYST RECOVERY

As mentioned earlier catalyst recovery is of importance in obtaining an economic process. HCl is difficult to recover from water because it forms a constant boiling azeotropic mixture at 21% HCl. However it has been shown that various metal salts can break this azeotrope. ICI has shown that one of the best common salts to achieve this is $CaCl_2$. Dependent of the $CaCl_2$ concentration in the catalyst all the HCl can be

FIGURE I

The hydrolysis of cellulose using $CaCl_2$ or LiCl in combination with HCl. The results are expressed in terms of glucose production. The concentrations of catalyst used were 6M and 7M HCl, and 1.64M $CaCl_2$ and 6M HCl. The reaction temperature was 90°C.

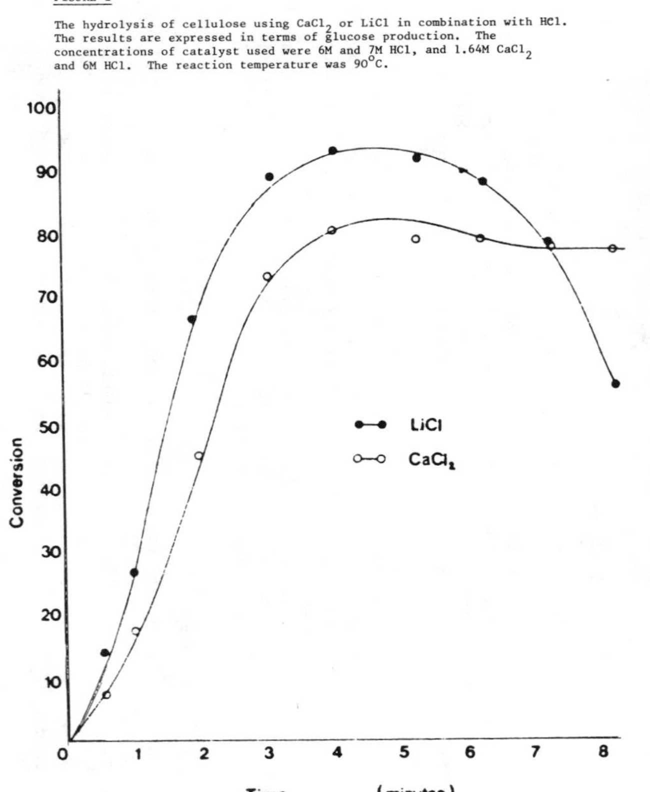

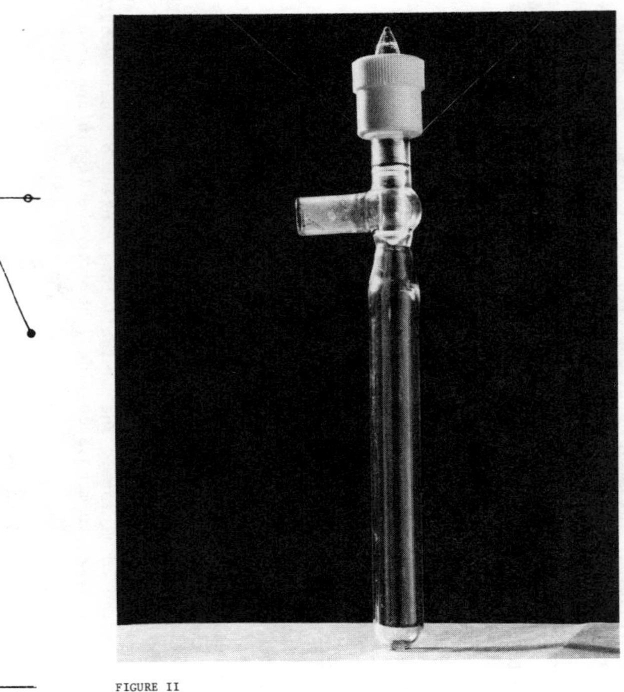

FIGURE II

Glass reactor used in laboratory experiments on lignocellulose hydrolysis. The tubes were purpose-built and can withstand up to 200 psig internal pressure.

TABLE I

THE HYDROLYSIS OF LIGNOCELLULOSIC MATERIALS USING THE ICI CATALYST

(Results all expressed in terms of 10% w/w solids concentration glucose concentration)

Substrate	Cellulose Content	% Conversion
Straw (ground)	32	85.6
Wood shavings	32	89.7
Cardboard	55	88.7
Bleached white paper	70	72
Glossy coloured paper	40	86
Newsprint	42	93.3
Brewing grain waste	14	100

TABLE II

Conditions used to remove hemicellulosic sugars from a range of lignocellulosic material

Temperature = 100-150°C

Time = 5-60 minutes

Pressure (developed internally) = ~60 psig

Acid concentration = 0.5-1.5% w/w

recovered and recycled. In addition $CaCl_2$ can be easily recovered from glucose using a variety of techniques including solvent extraction, differential crystallisation or electrodialysis.

5 SCALE OF OPERATION

Until November 1983 the experimental programme consisted of laboratory work using purpose built glass reactors as illustrated in Figure II. Studies have now been extended to a 2.5 l pressure vessel in which it has been much easier to study the effect of mixing on the reaction rate. Also a pilot rig is now being built which when completed will carry out the whole operation taking 0.1-0.2 tonnes of substrate through to the three major products - glucose, hemicellulosic sugars and lignin. Some of this work has already been published (10).

REFERENCES

1 Harris, E. E. and Beglinger, E. (1946). Madison Wood Sugar Process Ind. Eng. Chem., 38, 890-895.

2 Goldstein, I. S., (1980). Hydrolysis of wood. Ann. Meet. Assoc. Pulp Inds. p413-416.

3 Goldstein, I. S., Pereira, H., Pittman, J. L., Strausse, B. A., and Scaringelli, F. P. (1983) Hydrolysis of lignocellulosic material using concentrated and anhydrous HCl. froc 5th syrup Biotech Fuels Chem (in press).

4 Franz, R., Erckel, R., Riehm, T., Woernle, R. and Deger, H. M. (1983) Lignocellulosic Saccharification by HF. Energy from Biomass 2nd EC Conference p873-878.

5 Fengel, D. and Wegener, G. (1979) Hydrolysis of Polysaccharides with Influoracetic acid and its Application to Rapid Wood and Pulp Analysis Am Chem Soc 145-157.

6 Grethlein, H. E., (1978) Acid Hydrolysis of Cellulose Biomass 2nd Ann Fuel Biomass Syrup p461-479.

7 Fieber, C. A., Roberts, R. S., Faass, G. S., Muzzy, J. D., Colchord, A. R. and Bery, M. K. (1982) Continuous Steam Hydrolysis of Tulip Poplar Proc Intersoc Energy Convers, Eng. Conf. 17th 1, 271-275.

8 Dale, B. E., and Moreira, M. J. (1982) A Freeze Explosion Technique for Increasing Cellulose Hydrolysis Proc 4th Syrup Biotech Fuels Chem p72-81.

9 Kaetsu, I., Kumakwa, M., Fujimwa, T., Yoshila, F., Kojinea, T. and Tamada, M. (1981) Utilisation of Radiation Technique for the Saccharification and Fermentation of Biomass Radiat Phys Chem 18, 827-835.

10 Beardsmore, A. J., (1983) The Production of Chemical and Fermentation Feedstocks from Lignocellulosic Material. Raw Materials for Ind Material Forum, Metal Society, London (in press).

ENZYMATIC HYDROLYSIS OF CELLULOSIC WASTES

M. CANTARELLA, A. GALLIFUOCO, L. PEZZULLO, F. ALFANI
Istituto di Principi di Ingegneria Chimica, Università di Napoli, P.le Tecchio, 80125, Italy

Summary

In the framework of the last ECC "Energy from Biomass" Project a research was done aiming to acquire basic knowledges on the saccharification of the cellulose present in agricultural residues and to specially study the hydrolysis of olive husks, one of the most abundant wastes in Mediterranean regions. To accomplish this, the enzymatic hydrolysis of pure amorphous and crystalline cellulose and the selected biomass was investigated in standard batch and ultrafiltration membrane reactors.
In this brief communication references are given of the major results achieved, namely the enzymatic reaction pattern, the thermal stability of the cellulase components and a biomass chemical pretreatment. However, the discussion is mainly focused on the use of two different cellulase complexes in parallel for increasing glucose yield and selectivity during the hydrolysis of native cellulose and on some preliminary tests dealing with the behaviour of immobilized cellulase in a membrane reactor.

1. Introduction

Enzymatic saccharification is one of the most interesting process for obtaining chemicals and fuels from biomass and several academic and industrial research groups are tackling the problems that still remain to be solved for the design of full scale plants.

Ultrafiltration membrane reactors possess some peculiarities which are interesting on a laboratory scale for this study, since they offer the possibility to measure separately the action of the enzymic components of the cellulase complex, their thermal stability and the deactivation induced by product inhibition (1). Furthermore a membrane reactor configuration could also represent on a full scale a valid solution, for achieving high yield of enzyme recovery and for preventing large glucose inhibition.

Firstly the enzymatic hydrolysis of two well characterized substrates, Avicel and CMC, was performed in both batch and UF-membrane reactors. Several commercial cellulases were tested and the synergistic attack of endo- and exoglucanases on the crystalline and amorphous regions of the cellulose molecules was confirmed (2). During the investigation, the body of experimental results indicated that in commercial preparations the activities of the three major components, C_1, C_x and β-glucosidase, largely vary, dependind on the source, but are poorly balanced. To overcome this limitation, the possibility is usually suggested to extract and to purify single components and then to properly combine them in a new preparation. Since this procedure is costly, it can be adopted in those processes where a very limited amount of enzyme is required and a large added value product is formed. This is not the case in upgrading of cellulosic wastes and therefore in this study an attempt was made of simultaneously using two crude preparations of cellulase from different sources at different percentages.

Two further results were achieved during the first phase of investigation. Thermal stability of β-glucosidase is generally poor as compared with those of the other components and its activity is significantly curtailed by the presence of low glucose concentration, less than 1 g/l, according to a non competitive inhibition mechanism (3). Since a satisfactory thermal stabilization is usually got working with immobilized enzymes, experiments were started with an immobilized cellulase from A. niger, rich in the β-glucosidase component. This enzymic crude preparation was preferred to other commercial purified enzymes since most of them are not specific for cellobiose conversion. The endo- and exo-glucanase components of the immobilized cellulase are strongly limited in their attack towards the insoluble fractions but a good performance of a UF-membrane reactor in connection with a two phase saccharification process can be expected. An initial breakdown of the cellulose in soluble phase, catalyzed by a cellulase rich in C_1 and C_x, is followed by a final stage catalyzed by an immobilized cellulase, rich in β-glucosidase.

Preliminary results dealing with the use of immobilized A. niger in a gel layer of poly-humanserumalbumin (pHSA) are afterwards discussed.

2. Cellulose hydrolysis by enzymic mixtures from different sources

During cellulose hydrolysis it is important to obtain high yield of saccharification together with good selectivity to glucose since the higher molecular weight reducing sugars cannot be fermented to ethanol. Glucose production mainly depends on biomass pretreatment and on the specific activities of the enzymic components in the cellulase complex. The experiments have been carried out with cellulase from T. viride (Miles, USA) and from A. niger (Sigma, USA) and olive husks chemically pretreated to remove lignin. In the raw material 42% by weight of total sugars and 40% lignin are present, while the remaining part are different materials, the characterization of which is in progress. This latter are almost completely removed during an initial acid reaction with H_2SO_4 1% w/v at 90°C for 3 hours together with a large fraction of hemicellulose. The resulting solid residue is submitted to an alkaline pretreatment with NaOH 1N at T = 80°C for 6 hours and in the final material cellulose content is 50-55% in a form which is accessible to enzymatic attack. In fact the lignin barrier is destroyed and the degree of polymerization of the cellulose is reduced in comparison with that in raw biomass.

Samples of the reacted residues were submitted to enzymatic hydrolysis tests in a batch reactor at T = 45°C, biomass concentration 5 mg/ml, enzyme concentration 0.03 mg/ml, and cellulose conversion and glucose selectivity were measured after 40 hours of reaction. The data are plotted in Fig.1 versus weight percentage of A. niger in the enzymic mixture. Cellulose conversion in the biomass depends on the composition of the cellulase mixture more markedly than glucose selectivity. The highest yield of saccha-

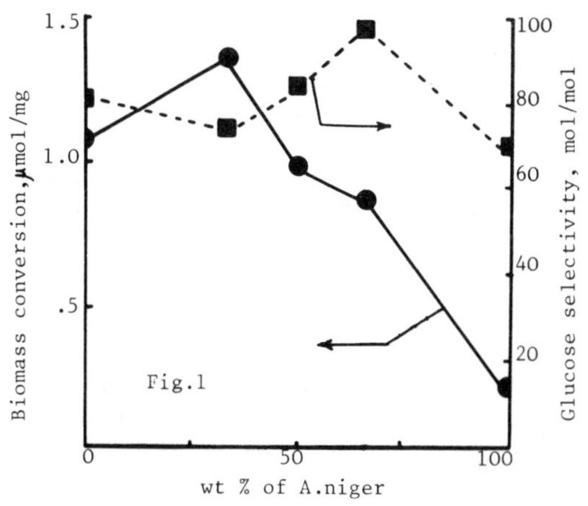

rification is achieved at 2 : 1 ratio between T. viride and A. niger, and in spite of the relatively low glucose selectivity, the highest glucose production per unit weight of biocatalyst is reached.

The maximun in the curve of reducing sugar production makes further on evident the presence in the biomass of both amorphous and crystalline forms of cellulose, the rates of degradation of which largely differs, the synergistic mode of action of C_1 and C_x, the amounts of which vary in the two sources of cellulases employed in this study and finally the unsatisfactory balanced activity of the enzymic components in the cellulase complex and the advantages to use mixtures of different preparations in the hydrolysis of native cellulose.

3. Immobilized cellulase

The experiments were focused to determine the effect induced by immobilization on the β-glucosidase component. Tests were performed using a UF-membrane reactor, Amicon ultrafiltration cell equipped with PM 10 or YM 10 flat membrane, and the enzyme, cellulase from A. niger was immobilized onto the active surface of the membrane in a gel layer of poly-humanserumalbumin ≃0.02mm thickness, according to the method described in (4).

Once enzyme immobilization at refrigerator temperature (+5°C) and in buffer stream (0.05 M Sodium Acetate/ Acetic Acid pH 4.8) was completed, a buffered solution of cellobiose, at saturating concentration (10 mM), was fed into the reactor from a N_2 pressurized reservoir and the bath temperature raised to 45°C. The rate of glucose production is constant for a prolonged time while experiments performed at equal operational conditions, but with A. niger in soluble form, have shown a progressive decay of enzyme activity with reaction time. According to the analysis reported in (1), thermal deactivation of β-glucosidase component in a cellulase from A.niger follows an exponential decay mechanism and at 45°C, for the preparation used in this study, an half life equal to 102 hours was determined.

The apparent reaction rate of the immobilized enzyme is half that observed for the soluble enzyme, since mass transfer resistance of reactant and product controls reaction mechanism. However the longer interval of significant enzyme activity largely exceeds the production loss during the first period of process.

An other point deserves a final consideration. It has been shown in (3) that β-glucosidase activity of a soluble cellulase is highly reduced by small concentration of glucose, Cg less than 2 mM, according to a non competitive inhibition mechanism. This behaviour is not always reported in the Literature, since, as shown in Fig.2, after an initial very rapid decline which last for Cg ≃ 2 mM, an almost stationary regime follows. Therefore it is difficult to detect an inhibitory glucose effect in experiments performed at medium and high product concentration.

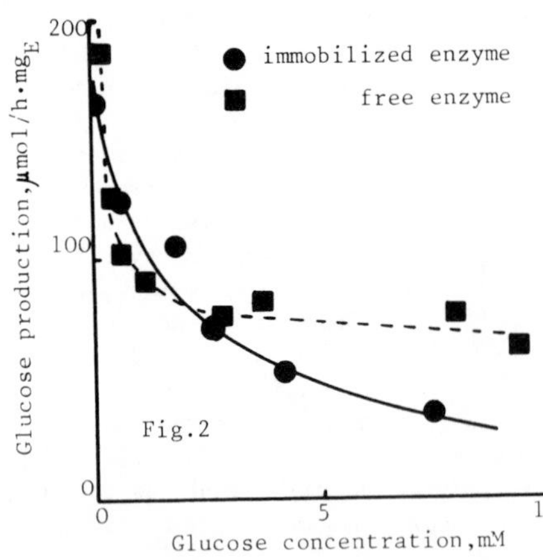

This problem can be overcome improving the selection and the culture of strains from which glucose resistant cellulases can be extracted. Nevertheless it is important to determine the process conditions under which product inhibition on the commercially available preparations can be controlled. In previous studies it was discussed the influence of residence time on product inhibition, in this communication the effect of enzyme immobilization is presented. The preliminary results seem to indicate that the activity decay with this method of immobilization is not reduced. Experiments are in progress to check if the mechanism for the immobilized and free enzyme is the same and to explore the phenomenon in a wider temperature range. In fact the data reported in Fig. 2 refer to 25°C, sufficiently low to avoid thermal enzyme deactivation, but it is interesting to study at higher temperature the effects of a combined enzyme activity decay caused by product inhibition and thermal deactivation.

4. References

1. Alfani, F., Cantarella, M. and Scardi, V. (1983). Use of membrane reactors for studying enzymatic hydrolysis of cellulose. J. Membrane Sci. 16, 407-416
2. Alfani, F., Cantarella, M. and Scardi, V. (1982). Cellulose saccharification : reaction kinetics and cellulase characterization. Energy from Biomass Series E, 3, 253-259. Grassi, G. and Palz, W. eds.
3. Alfani, F., Cantarella, M. and Scardi, V. (1983). Enzymatic saccharification of native cellulose : effect of product inhibiton and biomass pretreatment. Energy from Biomass Series E, 5, 336-343. Palz, W. and Pirriwtz D. eds.
4. Scardi, V., Cantarella, M., Gianfreda, L., Palescandolo, R., and Alfani, F., Greco, G. (1980). Enzyme immobilization by means of ultrafiltration techniques. Biochimie, 62, 635-643

5. Acknowledgment : This study is part of the research project ESE-R-041-I granted by DGXII Solar Energy Programme "Energy from Biomass".

Studies of soluble by-products resulting from organosolv-pulping

J. Feehl

MD Verwaltungsgesellschaft Nicolaus GmbH & Co.KG

Summary:

During the MD-Organosolv-Pulping process, in two extraction stages lignin and partially hemicelluloses are removed from wood. The resulting weakly acidic and alcaline extracts can be analytically fractionated by some precipitation steps.The fractions are characterized by IR spectroscopy, gel chromotography and degradation experiments. The results can be compared with literature data of carefully isolated products.

The MD-Organosolv-Process is primarly a pulp production process (1). The other components of wood are dissolved in water/methanol mixtures. To develop possible application for this parts,lignin and hemicelluloses, together at least 50 % of the raw material, the extracts must be characterized analytically.

The extracts can be fractionated by a few successiv precipitation steps (1,2). The process produces two extracts. Extract one is weakly acidic and contains as solvents only methanol and water. Extract two contains additional few percent of sodium hydroxide. This makes the fractionation more complicated. The salt can be removed by ion exchange or by electrolysis.

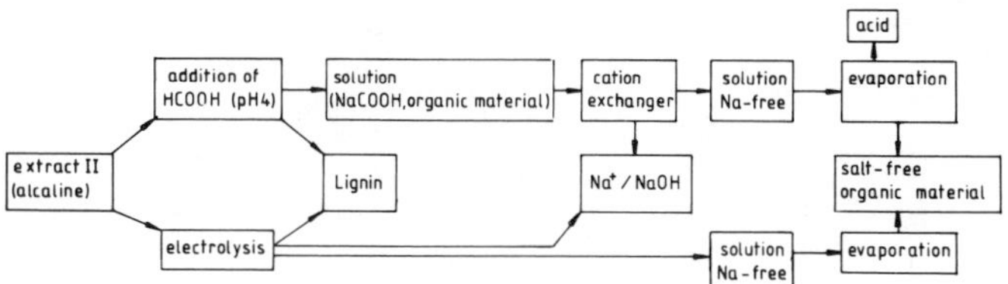

Principles of salt removement of extract II

Ion exchange should be the more careful method, but the products of the electrolytical treatment should be compared with it because of the technical importance of this method.

Some examples of characteristic products of spruce wood are shown. From the infrared spectra we see, that fractions remaining after precipitation of lignins and polysaccharides contain only small amounts of lignin, but perhaps lignin degradation products.

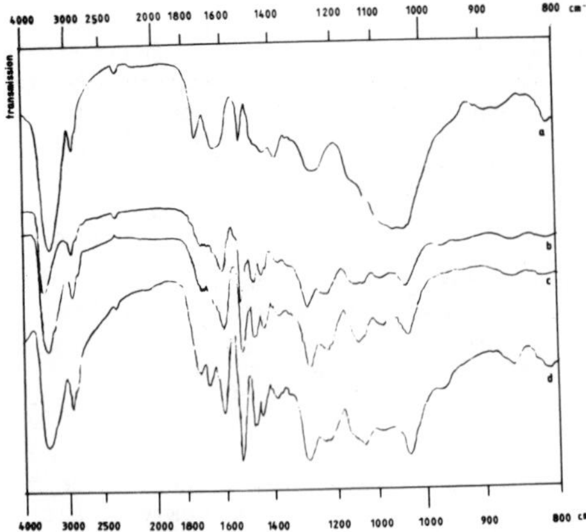

Infrared spectra of some fractions of organosolv-extract;

a) extract I after precipitation of lignin and polyoses

b) lignin of extract II, acid precipitation

c) lignin of extract II, electrolytical precipitation

d) lignin of extract I

The precipitated lignin differs only slightly from published spectra of milled wood lignins (3).
In the hydrolysates of lignin only small amounts of sugars can be found.

TABLE I

sugar content % of the hydrolysates of some fractions of organosolv-extract

	extract I	extract II *	lignin I	lignin II
rhamnose	0,3	-	-	-
glucose	5,1	0,6	0,3	0,1
mannose	12,8	0,7	0,1	-
galactose	6,3	1,0	0,1	0,1
arabinose	0,3	1,1	0,1	-
xylose	7,3	1,0	0,1	0,1
total	32,1	4,4	0,7	0,3

* after electrolytical precipitation of lignin

The molecular weight of the lignin is low compared with the values published on milled wood lignins (4).

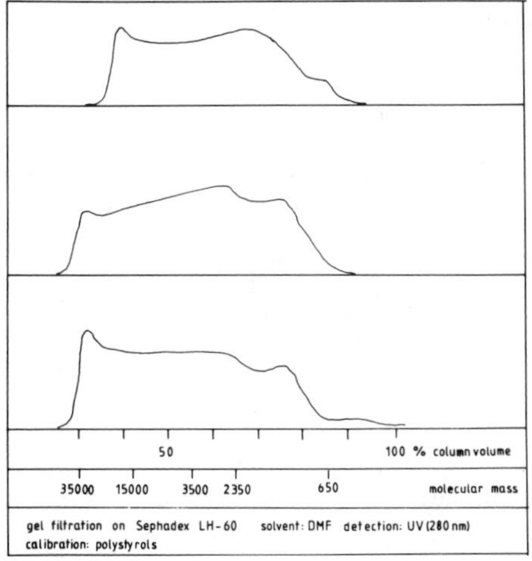

Molecular weight distribution of some samples of organosolv-lignin

Only 10 to 20 % of the carbohydrates can be precipitated, probably the higher molecular parts. Oligosaccharides seem to be soluble in alcohols containing 20 % of water. The sugar analysis shows (table I), that galactoglucomannan is the predominant component.

Reterences:

1. Edel, E., Feckl, J.: Studies of soluble by-products from organosolv-pulping (first report)

2. Edel, E.: Das MD-Organosolv-Zellstoffverfahren, Deutsche Papierwirtschaft, 1/1981 (in press)

3. Fengel, D., Wegener,G., Feckl, J.: Beitrag zur Charakterisierung analytischer und technischer Lignine II Holzforschung 35 111-118 (1981)

4. Fengel, D., Wegener, G.: Wood, Ultrastructure, Chemistry, Reactions; Berlin 1983

ANAEROBIC DIGESTION OF CRUSHED RAW URBAN WASTES
DEVELOPMENT OF A SEMI-CONTINUOUS, THERMOPHILIC, SOLID STATE PROCESS

B. MARTY - GERME S.A. - MARSEILLE - FRANCE

Summary

Experiences reported show that anaerobic digestion can be easily done in a semicontinuous way at elevated temperature (55°C) with raw domestic wastes. One important feature of the process is that it works at a high level of total solids (TS). The higher biogas production has been obtained at 35 % total solids concentration and 10 days retention time (θ). Other concentrations tested were 25 %, 30 % and 40 %. When TS = 40 % and θ = 10 days biogas production is still relatively important. Running of the digester is more steady when TS = 35 % or 40 %, underlying the stabilizing influence of bacteria fixation on inert materials and the buffering power brought by the latter. Intermittent mixing was done by an helicoidal device allowing a complete homogenization of the working content (60 liters) within a very short time (1 minute). The process could be easily scaled up on a technical point of view but needs more important economical evaluation before.

1. INTRODUCTION

The main features of the process and the digester design have been determined in order to enhance its efficiency in different ways. High temperature would allow to destroy all pathogenic organisms, to increase solubilization and degradation of organic solids and to improve the production of biogas. High level of total solids (TS) has been choosen because urban or agricultural wastes are commonly produced at total solids concentrations largely higher than 20 %.

Digestion of such wastes in conventional systems would require a previous dilution to lower TS concentration down to about 10 %. This would require larger installations for the treatment of a given bulk of wastes involving the handling and elimination of big volumes of waste water. Meanwhile, equipment and techniques adapted to the handling of semi-solid or solid wastes will be identical to those utilized for raw urban wastes or solid agricultural wastes.

This would allow the utilization of existing engines and would not require a complstly new know-how when installing such digestion unit within a treatment plant for city wastes.

2. MATERIAL AND METHODS

Total volume of each digester was 100 liters with 60 liters of working content. The mixing device was built of contra-rotatory helicoidal blades turning at 38 rpm. Biogas production was measured by domestic gas meters. Four total solids concentrations (25 %, 30 %, 35 %, 40 %) and four lengths of retention time (30, 15, 10, 8 days) have been investigated. All experiments were done at 55°C.

3. RESULTS

Thermophilic anaerobic digestion of semi-solid and solid wastes revealed itself as a process easy to handle. Previous experiences have been done by some authors (1 - 3) but there were batch experiments. Presently the digesters were fed once or twice a day and TS concentrations equal to 35 % and 40 % were the more favourable for digesters steadiness. Some minor decreases in biogas production have been observed, when TS = 25 %, resulting from a too rapid production of volatil fatty acids (VFA). At TS concentrations higher than 25 %, VFA and ammonia levels were fairly important but it seems that the buffering power of the wastes was enough to maintain the pH in a suitable range (pH $\simeq$ 7,8). Minimum retention times (θ) observed were 8 days when TS = 25 % and 10 days when TS = 40 %. The best biogas production was obtained for TS = 35 % and θ = 10 days where it reached more than 300 liters per day (see table).

% TS / θ (days)	25		30		35		40	
	(a)	(b)	(a)	(b)	(a)	(b)	(a)	(b)
30	1,6	0,305	2,2	0,321	2,6	0,290	2,8	0,268
15	2,4	0,230	3,3	0,236	4,4	0,246	3,3	0,158
10	2,9	0,185	3,8	0,180	5,6	0,208	4	0,128
8	3,2	0,163	3,5	0,133	4,8	0,143	-	-

(a) = Biogas $m^3.m^{-3}.d^{-1}$ (b) = YCH_4 $m^3.m^{-3}$.Kg VS

Attemps have been made to check other parameters such as ATP concentration and DCO. Until now the values that have been obtained during these assays show poor reliability (because of interferring ions and molecules present in the raw wastes) and need more investigation.

4. CONCLUSION

Thermophilic anaerobic digestion of raw urban wastes can be conducted in a continuous manner and in a semi-solid or solid state. It has proven to be a reliable process which can be monitored by controlling a small set of parameters. A technical and economical evaluation of a pilot-scale installation would be very useful because one can expect that such a process will require less investment in engineering, know-how and money and can be implanted in existing installations treating solid urban refuse.

REFERENCES

1. COONEY C.L., LINDSEY E.E., KIRK R.S. and OYEWOLE S. (1974) "Fuel gas from solid waste" Dynatech report n° 1213.

2. WUJCIK W.J. and JEWELL W.J. (1980) "Dry anaerobic fermentations" Biotechnol. Bioeng. Symp. 10 : 43-65.

3. SHELEF G., KIMCHIE S. and GRYNBERG H. (1980) "High-rate thermophilic anaerobic digestion of agricultural wastes" Biotechnol. Bioeng. Symp. 10 : 341-351.

ANAEROBIC DIGESTION OF M.S.W. IMPROVED BY OTHER ORGANIC WASTES IN A LARGE PILOT PLANT

M. GERLETTI
E. Bi. A. Coop. MILANO - BRONI (ITALY)

Summary

Two different kinds of sorted organic matter from MSW (a fresh compost and a simply comminuted organic fraction) were treated, in mixture with zootechnical wastes (cattle and swine) and municipal wastewater treat= ment sludges with industrial contributions, in a full size plant (dige= ster useful volume 1800 m^3).
Biogas was utilized by a total energy dual fuel group. Full laboratory service allowed the collection of an extended set of data on the beha= viour of the different mixtures. The energy balance in different seasons and the plant reliability were evaluated.

1. ORIGIN OF THE RAW MATERIALS

1.1 Municipal solid wastes.

a) Pistoia municipality. MSW from the entire town (120.000 inhabitants) in amounts of 100 t/d about (6 days per week) and 10 t/d about of slaughter house residues, sawdust and other residues are introduced in a rotating aero bic biostabilizer as received. After 50-70 h, with temperature reaching 60-65°C, the materials are screened and a fine fraction containing organic mat ter plus amounts of glass, plastics and metals is subjected to ballistic se paration. About 42 t/d of organic matter (fresh compost), containing very little glass, plastics and metals, are obtained. Fresh compost was transpor ted to the anaerobic digester by trucks (28 t per trip).

b) Busto Arsizio plant. MSW from 21 municipalities (400.000 inhabi= tants) in amounts of 200 t/d about are submitted to comminution by hammer mills and then sieved. The finer fraction, rich in organic matter, is sub= mitted to ballistic and magnetic separations for the removal of glass, hard plastics and metals. This organic fraction was utilized for anaerobic dige= stion.

1.2 Agricultural wastes.

a) Cattle excreta. Two types were obtained from a dairy farm (400 heads) and a fattening bull breeding/120 heads). In the first case the housing is in open sheds. The feeding with hay or silomais is "at libitum" while the concentrates are distributed by a computer driven system in pro= portion to milk production. Excreta are scraped and sent to a storage tank. In the second case the bulls are confinedin a closed shed. The feeding is mainly composed of silomais and concentrates distributed twice per day. Excreta are collected under slotted floor and flow by gravity to a tank; no washing water is used. Holding time in both cases is 2-4 months according to season.

b) Swine excreta. They were obtained from a complete cycle breeding (from piglets to 120 kg bacon pigs, 2000 heads). The housing is in closed sheds with separated defecation areas. The excreta are collected through slotted floors and flow to collecting tanks emptied twice per month. Little water is used for washing.

1.3 Wastewater sludges.

Thickened sludges from different aerobic wastewater treatment plants and one chemical-physical plant were obtained. In general they consisted of primary and active sludges from municipal plants with variable industrial contributions. In order to reduce transportation costs the sludges were thickened to 15-25% T.S. by Al or Fe salts, polyelectrolytes and belt presses (Ash 22-30% of T.S.).

2. PLANT DESCRIPTION

The whole plant is composed by four sections:

a) Storage of raw materials and preparation of the feed to anaerobic digester. Four concrete tanks are available (50+70+160+160 = 440 m^3). Relatively dry wastes (organic MSW or thickened sludges) are diluted to 10% TS with diluted wastes (piggery) and recycled water and homogenized with a submersed pump with cutting blades. The 10% TS slurry is passed through a vertical shaft rotary cutter for a further size reduction. The discharge of the rotary cutter is sent to one of the two larger tanks and diluted to 5-7% with diluted wastes and recycled water and stored for digester feeding. The content of the tanks is continuously agitated by submerged stirrers. A metering pump feeds continuously the digester.

b) Anaerobic digestion. The digester sidewall and bottom are in reinforced concrete; the cover is an insulated steel floating gasholder. Mixing is obtained by biogas recycling and gaslifter. Thermostatting at 35 °C is by an internal heat exchanger. Useful volume is 1000 m^3.

c) Electricity and hot water cogeneration. A dual fuel diesel total energy group (6 cylinders, 1500 rpm) consumes a maximum of 1180 MJ/h (85% as biogas, 43-47 m^3/h according to calorific value, 15% as diesel oil, 4.1 kg/h) and delivers 400 MJ/h (112 kWh 380 V) and 570 MJ/h as 65 °C hot water, with a total yield of 82%. Hot water is used for heating the digester to 35 °C. Electric heating coils can supply extra heat. A 330 MJ/h biogas fueled boiler supplies hot water during emergencies.

d) Sludge dewatering and final water treatment. Digested sludge is discharged continuously, added by polyelectrolite and aluminium salt and centrifuged. The dewatered sludge 20-25% TS is stored and used as an agricultural soil amender. Water is partly recycled as process water (60-100 m^3/d), partly treated by a rotating biological contactor (10.000 m^2 useful surface) for BOD polishing and ammonia nitrification.

3. RUNNING CONDITIONS

The digester was started during March 1983 by 205 m^3 of 1% TS swine manure, 180 m^3 of 9% TS bovine manure and 120 m^3 8.7% TS waste brewery yeast and 1300 m^3 of water (29 t TS, 22 t VS, 5.6 t VFA, alkalinity 3300 ppm, pH 7.0). After one week 1.2 t of lime were added due to an initial acidification (alkalinity to 4500 ppm). No seeding was purposely used. Gas evolution started on 47 th day. Up to 120 th day the feeding was discontinuous due to lacking official authorizations (only 8 m^3/d in average). Regular loading started thereafter. Loading was continuous at a mean rate of 3 kg VS $m^3/m^3/d$. Temperature of stored feed was very near to air temperature and as a mean 5 °C minimum Winter (30 days), 8 °C average Winter (120 days), 15 °C average Spring-Autumn (215 days). Recycling of process water (60-100 m^3/d at 25-30 °C allowed significant energy savings (5-10 GJ/d). H.R.T. was 9-14 days and no solid was recycled. Mixing was continuous (theoretical rate 3 tank turnovers per hour, P ≑ 9 kW).

Table I. Daily undiluted feed composition t/d.

	VS	TS	H_2O
Organic MSW	2.5 - 3.5	3.3 - 4.6	3.1 - 5.8
Wastewater sludges	1.0 - 1.6	1.5 - 2.7	6.7 - 9.8
Zootechnical wastes	0.9 - 1.4	1.0 - 1.2	20 - 64
Total	5.0 - 5.8	7.0 - 7.2	32.9 - 80.5

4. PLANT RELIABILITY

During 8 months of continuous operation (August 1983 - March 1983) only minor failures occured: a) pieces of glass filled the pre-chamber of metering pumps; b) strings of textiles wound around the shaft of stirrers; c) water condensated in critical sections of gas lines during winter; d) heavy foaming during the treatment of chemical-physical sludge; e) low biogas production when treating biological old sludges (stored for months).

Table II. Feed and untreated effluent composition (n.d. = not determinable)

	feed	effluent
TS %	3.5 - 5.6	2.5 - 3.7
VS %	2.5 - 4.5	1.4 - 2.2
VS removal %	-	43.2 - 53.8
Ammonia nitrogen ppm	-	260 - 1250
Volatile fatty acids ppm	300 - 3000	100 - 430
Acetic acid	250 - 1700	n.d. - 300
Propionic acid	n.d. - 900	n.d. - 100
pH	5.5 - 7.0	7.1 - 7.7
Total alkalinity ppm	-	3900 - 5400

Table III. Gas composition (% volume) and production.

Specific gas production	m^3/m^3 reactor/d	0.95 - 1.28
Methane yield	m^3/kg VS	0.18 - 0.23
CH_4 53-58	CO_2 40 - 45	N_2 1 - 2

Table IV. Average daily energy balance in different seasonal periods.

	Minimim Winter 30 d	Average Winter 120 d	Spring Autumn 215 d
Biogas production m^3/d	2050	2000	1900
Biogas energy available GJ/d	+ 46	+ 45	+ 43
Total energy group consumptions:			
Biogas GJ/d	- 25.5	- 25.5	- 25.5
Diesel oil GJ/d	- 4.2	- 4.2	- 4.2
	- 29.7	- 29.7	- 29.7
Electricity production GJ/d	9.6	9.6	9.6
Thermal energy recovered GJ/d	13.6	13.6	13.6
	23.2	23.2	23.2
Wasted heat GJ/d	6.5	6.5	6.5
Plant consumptions:			
Electricity GJ/d	- 3.2	- 3.2	- 3.2
Digester heat losses GJ/d	- 7.1	- 6.4	- 4.2
Raw sludge heating GJ/d	- 20.4	- 18.2	- 13.4
Heat recovery (water recycle) GJ/d	+ 7	+ 8	+ 7
Diesel oil equivalent GJ/d	- 4.2	- 4.2	- 4.2
Cogenerator heat losses GJ/d	- 6.5	- 6.5	- 6.5
	- 34.3	- 30.5	- 24.5
Energy available GJ/d	+ 11.7	+ 14.5	+ 18.5
Biogas equivalent m^3/d	520	640	820

ETHYLENE GLYCOL-WATER PULPING. KINETICS OF DELIGNIFICATION

D. GAST and J. PULS

Federal Research Centre of Forestry and Forest Products
Institute of Wood Chemistry and Chemical Technology of Wood
Leuschnerstraße 91 D-2050 Hamburg 80

Summary

In ethylene glycol-water pulping of birchwood at 180°C, 190°C and 200°C kinetics of delignification were determined. Obtained data indicate that delignification occurs in two distinct phases which can be described by first order reactions. Transition from the first phase to the second phase takes place after about 25-45 minutes. With increasing temperature the transition point shifts towards lower lignin levels. The activation energies for the first and the second delignification phase were found to be 81.2 kJ/g · mole and 152 kJ/g · mole, respectively.

1. INTRODUCTION

Recently we reported about a series of cooks with various low - and high boiling alcohols without and in mixture with water (1). A considerable delignification effect was found only when the alcohols were used with water. The pulping efficiency of the low boiling alcohols increased with increasing molecular weight, whereas in the case of high boiling alcohols in general contrary observations were made. Among the high boiling alcohols ethylene glycol seemed to be the most efficient delignifying agent. Thus a kinetic study with the ethylene glycol-water system was performed in order to investigate time and temperature effects.

1.1 Materials and Methods

In all cooks birchwood chips of 0.3 - 0.5 mm particle size were used. The wood was filled into small baskets, fitting into the boring of 50 ml autoclaves, 30 ml of liquor (ethylene glycol : water = 50 : 50 by volume) was added. In order to keep heating-up times short, the autoclaves were first dipped into an oilbath of 220°C until the desired reaction temperature was attained and then transferred into a second oilbath of cooking temperature. For exact timing, temperature inside the autoclaves were controlled by Ni-Cr/Ni-Al thermocouples. After cooling the baskets were put into soxhlet extractors to remove dissolved lignin and carbohydrates with ethanol : water = 80 : 20. The fibre materials were dried and treated with H_2SO_4 for total hydrolysis. The sugars in the hydrolysates were determined quantitatively and qualitatively according to (1). The lignin contents in the fibres were determined as hydrolysis residues.

2. Results and Discussion

Fig. 1 shows the results from examining the quantities of removed lignin. Depending on the temperature different levels of delignification can be achieved. 92 % lignin removal requires a cooking time of 50 minutes at 200°C. This value corresponds to 4.1 % residual lignin in fibres or 1.8 % lignin based on wood. At lower temperatures lignin removal is reduced. In any case delignification is accompanied by carbohydrate dissolution. The amount of hemicelluloses is reduced to values less than 2 % compared to the hemicellulose content in the original wood. At reaction temperatures of 180°C and 190°C and residence times of 45 minutes the cellulose portion starts to be attacked. This point corresponds to a lignin content of 10 % (180°C) and 5 % (190°C) in fibres respectively. At 200°C however cellulose starts to be attacked after cooking times of 30 minutes. These pulps have lignin contents of 6 %.

Experimental data in ethylene glycol-water pulping show two first order reaction curves for each temperature. This is in accordance with previous findings of other authors (2,3,4). Fig. 3 shows residual lignin versus cooking time. Inital fast delignification is partly ascribed to the breakdown of lignin-carbohydrate linkages and following extraction of the dissolved lignin molecules. The sharp decrease in delignification in the second stage may be explained by lower accessibility of the residual lignin, and by condensation and recondensation reactions of dissolved lignin molecules in the wood (2,5).

The rate of delignification is expressed by a first order equation of the linear form: $\ln L = \ln L_o - kt$. where L is the content of lignin in the fibre material, L_o is the content of lignin in the original wood an k is the rate constant in min^{-1}. The activation energies and the frequency factors can be calculated using the Arrhenius equation. Table I presents the values of these constants in ethylene glycol-water pulping. The values of E found for the initial delignification step (I) and for the residual delignification step (II) were 81 kJ/g · mole and 152 kJ/g · mole respectively. Kleinert (6) reported a value of 117.6 kJ/g · mole for the slow part of delignification in ethanol-water pulping with spruce sawdust, whereas Nguyen (7) found values of 75.4 kJ/g · mole for the inital step and 171.6 kJ/g · mole for the slow step using an ethanol-water/$Fe_2(SO_4)_3$ system and prehydrolysed sweetgum.

3. CONCLUSION

In the first stage of our investigations delignification efficiencies of various alcohols were determined. Ethyleneglycol showed promising delignification power. The second step involved a kinetic study of the ethylene glycol-water system. The results prove this alcohol to be a good delignifying agent. The third step will include production of larger amounts of cellulose, hemicellulose and lignin to test the practical applicability.

REFERENCES

1. GAST, D.; AYLA, C. and PULS, J. in "Energy from Biomass", A. Strub, P. Chartier and G. Schleser ed., Appl.Science Publ., London and New York (1983), 879
2. KLEINERT, T.N., Tappi 57, 8 (1974), 2381
3. APRIL, G.C.; KAMAL, M.M.; REDDY, J.A.; BOWERS, G.H. and HANSEN, S.M., Tappi 62, 5 (1979), 83
4. CHANG, P.C.; PASZNER, L. and BOHNENKAMP, G., Paper, Tappi Forest Biology/Wood Chemistry Symposium, June 20-22 (1977), Madison, Wisconsin
5. WAYMAN, M. and LORA, J.H., Tappi 61, 6 (1978), 55
6. KLEINERT, T.N., Tappi 58, 8 (1975), 170
7. Nguyen, X.N., Thesis, North Carolina State University, Raleigh (1980).

Table I

RATE CONSTANTS AND ACTIVATION ENERGIES IN ETHYLENE GLYCOL-WATER PULPING

	Temp. °C	I	II
Rate const. $\times 10^{-3}$ min^{-1}	180 190 200	26,2 47,8 69,7	2,4 7,2 15,2
Frequency factor A_0		$1{,}85 \times 10^{7}$	$1{,}18 \times 10^{13}$
Activ. Energy E kJ/gm-mole		81,2	135,5

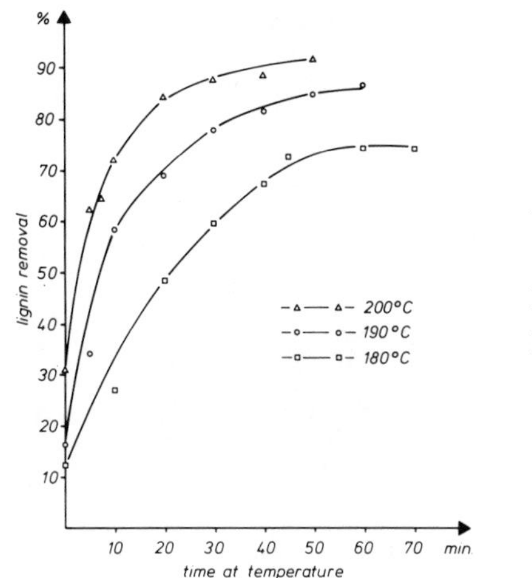

Fig. 1

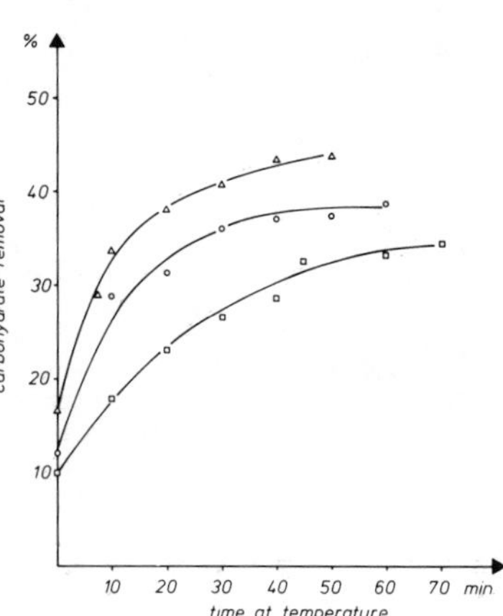

Fig. 2

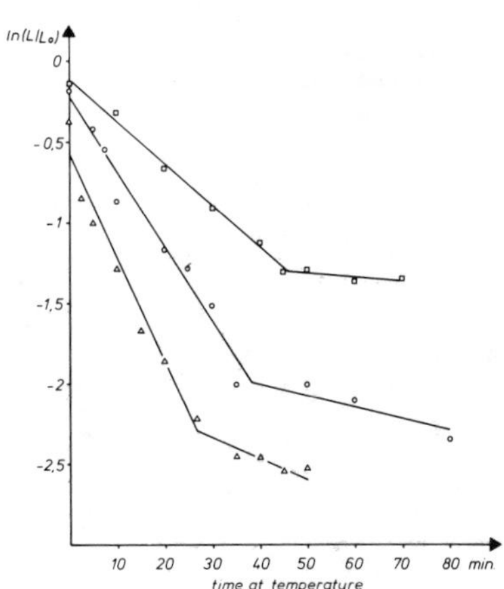

Fig.3

Fig. 1 Rate of delignification in ethylene glycol-water system

Fig. 2 Rate of carbohydrate removal in ethylene glycol-water system

Fig. 3 Logarithm of residual lignin in fibre material

DEVELOPMENT OF A PLUG-FLOW DIGESTER FOR SEWAGE SLUDGE

T.J. CASEY AND J. POWER

Department of Civil Engineering, University College, Dublin.

Summary

The technical feasibility of using plug-flow reactors for the anaerobic digestion of sewage sludge was established through an experimental programme of laboratory batch-testing and a complementary pilot plant continuous digestion study using primary sludge from the Ringsend Works of Dublin Corporation. The batch tests showed that stable digestion could be achieved with a digested: raw sludge volumetric mix ratio in the range 2:I to I.25:I. The pilot plant study established that a stable high-rate plug-flow digestion process could be operated in the retention time test range I2.5 to 20 days. These findings provide a basis for the design of plug-flow reactors as an alternative to the completely mixed reactors conventionally used for the digestion of sewage sludge.

I. INTRODUCTION

Developments in anaerobic digestion technology in the sewage sludge field have been almost exclusively based on the "completely mixed" reactor design concept. In this study attention was focussed on the alternative "plug-flow" reactor concept, which has potentially superior kinetic characteristics (I) and also allows considerable flexibility in the choice of reactor shape. The project was initiated as a technical feasibility study to determine if it would be possible to convert old redundant sedimentation tanks at the Ringsend Works of Dublin Corporation to anaerobic digestion use, to process some 600 m^3/d of primary sludge from a new wastewater treatment plant on an adjacent site. The old sedimentation tanks, which are about 29m square by 2.4m deep, are geometrically unsuited for conversion to conventional completely mixed reactors. However, by installation of an internal round-the-end baffle wall system, they could be readily converted to plug-flow reactors at a relatively modest cost.

The feasibility of designing a plug-flow reactor for this application was established through laboratory and pilot plant studies, using primary sludge from the new plant, which has been operating on part-load for some time.

2. EXPERIMENTAL STUDIES

When using a plug-flow anaerobic digester for sewage sludge stabilisation, it is essential to recycle digested sludge to the reactor inlet to provide microbial seed and a pH buffer for the raw sludge. The required proportion of digested: raw sludge in the reactor feed was determined in the laboratory using six batch digestion units. Each unit included a 4-litre cylindrical digester, a gas-recirculation pump for digester mixing and a gas-collection vessel. The six digesters were maintained at a temperature of 35°C in a common water bath. The input of mixing was intermittent, the gas pump being automatically controlled in in respect of frequency and duration of operation.

The pilot plant study was undertaken with the general objective of examining the feasibility of operating a continuous digestion process in a nominally plug-flow reactor or more accurately in a tank in which the horizontal flow length between inlet and outlet was very much larger than the dimensions of the flow cross-section, as would be the case in the envisaged adaptation of the old sedimentation tanks, to which reference has already been made.

The pilot digester was a closed steel tank with a longitudinal dividing baffle, providing a flow cross-section 0.78m deep by 0.7Im wide, with a centreline flow length of 6.6m, giving a digesting liquor volume of 3.Im^3 and a supernatant gas volume of I.7 m^3. The digester was lagged externally and heated by an internal hot water heat exchanger. The feed sludge was also pre-heated by a hot water pipe loop in an inlet chamber. The throughput was controlled by a volume-batching feed system, automatically regulated to provide the required retention time. In conjunction with each batch of feed sludge, a set volume of digested sludge was recycled from the digester outlet to the digester inlet. Intermittent mixing was provided by a gas-recirculation system, which pumped biogas from the head-space of the digester through a pipe manifold located near floor level. The operation of the pilot plant was fully automated, using a timer-based pneumatic control system.

3. EXPERIMENTAL RESULTS

Some 80 batch tests, in which the volumetric ratio of digested: raw sludge in the batch mix varied from 2:I to I.25:I, were carried out. A lower value of this ratio was not tested due to the inability of the laboratory scale mixing system to cope with higher proportions of raw sludge in the batch mix.

The test results showed that stable digestion could be achieved over the full sludge mixture range evaluated. It seems probable that sludge mixtures having a still higher proportion of raw sludge in the mix would also support a stable digestion process, provided mixing was adequate to prevent segregation of solids.

In the batch tests the biogas output typically rose to a peak level after 4 to 6 days digestion and thereafter declined to virtually zero value after about I0 to I2 days. The average yield of biogas in the batch tests was I.04 m^3 per kg of volatile solids removed.

The pilot plant was operated at hydraulic retention times in the range I2.5 to 20 days. The test results confirmed that horizontal flow tanks can be made into highly efficient anaerobic digesters, provided that appropriate design measures, particularly those related to sludge recycle and digester mixing, are employed. The following is a summary of process performance at retention times of 20 and I2.5 days:

Parameter	Retention Time		
	20 days	I2.5 days	
Raw sludge volatile solids	8I.2	78.4	%
Extent of conversion of vol. solids	54.4	54.7	%
Yield of biogas per kg. solids removed	I.03	I.0I	m^3
Yield of methane per kg. solids removed	0.72	0.69	m^3
Volatile fatty acid conc. in digested sludge	I75	287	mg/l

The foregoing are average parameter values for the respective test runs. It is noteworthy that there was very little difference in overall performance at these two quite different retention times. The feed material was primary sewage sludge with a solids concentration in the range 4 to 6%

by weight.

4. CONCLUSIONS

I. Laboratory tests have shown that stable and reproducible rates of digestion of primary sewage sludge can be obtained in batch mixtures where the volumetric ratio of digested: raw sludge was in the range 2:I to I.25:I. Lower ratios may also be satisfactory but were not tested. It was concluded that recycle: feed ratios in the batch-proven range should provide stable digestion in large-scale plug-flow digesters;

2. Continuous digestion studies, carried out in a pilot plant of the horizontal flow (plug-flow) type, verified that stable highrate digestion was feasible in this type of reactor. The system, which used intermittent mixing by gas recycling, was operated at retention times in the range I2.5 to 20 days.

5. ACKNOWLEDGEMENT

The project was supported by grants from Dublin Corporation and the National Board for Science and Technology.

6. REFERENCES

CASEY, T.J. Anaerobic digestion; some comments on process design. COST 68, EEC Concerted Action on Treatment and Use of Sewage Sludge, Working Party I Paper WPSP 30.

METHANOGENIC FERMENTATION OF SOLID AGRICULTURAL RESIDUES

J.L. ROUSTAN, A. AUMAITRE, A. MOUNIER and Y. PEIGNIER
I.N.R.A. Pig Husbandry Department

SUMMARY

Anaerobic degradation of agricultural residues was studied with batch on continuous process. A laboratory scale digester, designed to handle substrates without any mechanical pretreatment, showed a maximum CH_4 production rate per unit volume $r_{V\ CH4}$ of 1 l x l^{-1} x d^{-1}, for grass clippings or pig solid manure volumetric loading rates ($B_{V\ O.M.}$) of 4.1. kg^{-1} x m^{-3} x d^{-1} or 5.1 kg^{-1} x m^{-3} x d^{-1} respectively. Solubilization of the substrate, rather than lignin content appeared to be the limiting factor. With wheat straw or solid manure, the rate of cellulose degradation did not exceed 0.8 g x l^{-1} x d^{-1}. Tentatives to realize the two steps of liquefaction-methanization in separate reactors were unsuccessful, since only a limited percentage of substrate CH_4 potential was recovered in the second reactor. Batch or continuous fermentation achieved, in most case, the same gas yield over the R.T. used (22 to 26 days). In order to maximise $r_{v\ CH4}$ the substrate must be supplied in the highest concentration compatible with the mixing and heating system in the digester.

1. INTRODUCTION AND EXPERIMENTAL PROCEDURE

Anaerobic fermentation of dry substrates containing 15-30% dry-matter (D.M.) might be applied to recover energy as methane from numerous agricultural residues. In France the use of straw bedding in rearing operations produces yearly 13 millions tons of D.M.. Slurrying such a material for its use in a conventional digester requires water and energy supply to produce a pumpable substrate. A digester was designed (Fig.1) to study the anaerobic fermentation of solid manure at the laboratory scale. Without any mecanical pretreatment the substrate was handled in a container (4 litres capacity) and immersed in a liquid phase where fermentation was proceeded. The retention time (R.T.) is defined by frequency of addition-removal of containers. In order to determine the methane potential of the substrate a batch experiment was used as reference. Chemical analysis were performed to characterize the degradation of the various components of D.M.. The experiments were performed at 35°C temperature, with various feeding materials : straw, solid manures and grass clippings. The working volume refers to the total volume of the containers overlayed by the gaz collection system, i.e, 88 litres.

Our previous report (5) showed that an anaerobic fermentation can be accomplished at a relatively high D.M. content, without a complete immersion, thanks to a daily recycling of a limited amount of liquid. A comparative experiment was also conducted to measure the respective role of solid and liquid phase in gas production. Solid manure was put in a digester

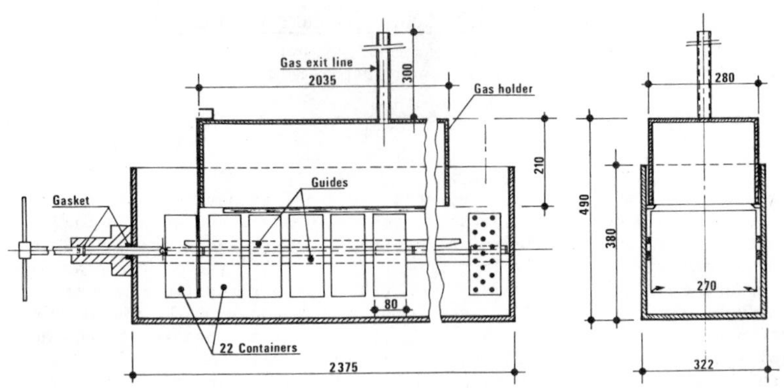

FIG.1 : LABORATORY SCALE METHANE DIGESTION PILOT FOR "DRY" SUBSTRATES

maintained at 35°C ; once a day the system has been completely filled with liquid and after a contact period of 30 minutes, the liquid has been drained out. A separate reactor, running on the continuous mode at 35°C and 10 days R.T., was used to exhaust methane precursors produced by the previous maceration and available in the liquid. Every day effluent from this reactor has been saved for the next refilling-emptying step, insuring a stock of liquid with constant characteristics.

2. RESULTS

2.1. Gas production :

Operating conditions and performance of the continuous digester are summarized in Table I. The higher methane production rate per unit volume ($r_{v\ CH4}$ = 1 l x l^{-1} x d^{-1}) was achieved with grass clippings and pig solid manure i.e, respectively the more degradable and the more concentrated substrate. Distribution of gas production along the four compartments of the gas holder varied according to the different substrates, and for a given substrate, methane percentage in the gas was increasing according to the rank of successive compartments (Table II). Average percentage are 58.6 for grass, 63.4 for pig solid manure, 64.6 for cattle solid manure, 52.8 and 49.4 for straw at 26 day-R.T. and 11 day-R.T. respectively.

TABLE 1 : OPERATING CONDITIONS AND DIGESTER PERFORMANCE ACCORDING TO THE SUBSTRATE

SUBSTRATE	RETENTION TIME (d)	VOLUMETRIC LOADING RATE (B_v) kg^{-1} O.M. x m^{-3} x d^{-1}	SUBSTRATE DENSITY kg O.M. x m^{-3}	CH_4 PRODUCTION RATE VOLUME $r_{v\ CH4}$ l x l^{-1} x d^{-1}	METHANE YIELD m^3 x kg^{-1} introduced O.M.		
					Continuous system	Batch same R.T	Batch (ultimate)
Wheat Straw	11	3.7	40.6	0.368	0.10	0.08	0.216
	26	1.6	40.6	0.283	0.18	0.15	0.216
Cattle solid manure	26	3.3	85.5	0.53	0.16	0.145	0.258
Pig solid manure	22	5.1	111.5	1.04	0.204	0.195	0.320
Grass clippings (1)	11	4.1	45.5	1.0	0.243	0.112	0.307

(1) fresh cuttings of green, mainly ray grass leaves

TABLE II : DAILY GAS PRODUCTION IN THE VARIOUS GAZ HOLDER COMPONENTS, l $CH_4 \times d^{-1}$ (% CH_4)

COMPARTMENT / SUBSTRATE	1st	2nd	3rd	4rd
Straw 11 day-R.T.	6.8 (44.6)	9.7 (49.4)	7.6 (52.0)	8.3 (52.0)
Straw 26 day-R.T.	5.6 (47.2)	8.1 (53.2)	6.1 (56.0)	5.2 (55.2)
Cattle solid manure	6.3 (59.7)	15.5 (63.7)	14.3 (65.7)	11.0 (67.8)
Pig solid manure	18.0 (56.3)	21.8 (61.4)	26.9 (66.2)	24.9 (68.0)
Grass clippings	25.8 (48.8)	21.3 (61.2)	21.9 (65.3)	19.7 (65.5)

2.2. Comparison with the batch system :

On the basis of results presentation in Table I, the fraction of the methane potential recovered with the continuous system can be calculated. Values ranged from 50% for straw at 11 day-R.T. to 80% for grass at 11 day R.T. or straw at 26 day-R.T.. Recovery appeared only slightly faster with the continuous system, excepted for grass where instantaneous accumulation of V.F.A. produced a lag phase at the start of the fermentation.

2.3. Changes in the major components of the O.M.

Variation in the amount of O.M. cellulose, hemicellulose and lignin in each container according to the R.T., is presented in Fig.2 for pig solid manure. Lignin remained unaffected while cellulose and hemicellulose, determinated by the Van Soest method, are degraded nearly at the same rate. Progressive exhaustion of the readily available fraction from these components was associated with a decrease in the rate of solubilization. Similar trends were observed with the other substrates ; the mean values of cellulose degradation rate according to the retention time are presented in Table III. With grass clippings some materials escaped through the holes of containers during the fermentation, precluding the set up of mass balances.

TABLE III : CELLULOSE UTILISATION WITH VARIOUS SUSBTRATES

SUBSTRATE	R.T. DAYS	RATE OF CELLULOSE DEGRADATION $g \times l^{-1} \times d^{-1}$	VOLUMETRIC LOADING RATE $B_{v\ cell:kg} \times m^{-3} \times d^{-1}$	% OF CELLULOSE ABATEMENT
Wheat straw	11	0.62	1.73	36.3
	26	0.37	0.64	58.8
Cattle solid manure	28	0.79	1.56	50.6
Pig solid manure	22	0.78	1.39	59.6

Similarity in the degradation of the two classes of structural polysaccharides was confirmed by G.L.C. of the individual sugars obtained after hydrolysis of pig solid manure D.M.. In the feed material arabinose, xylose and glucose represented respectively 13.2, 32.5 and 50.6 of the total sugars ; after a 22 day-R.T. the values obtained were 11.0, 33.3 and 50.4%. In a parallel experiment after 48 days of batch fermentation, the values were respectively 10.0, 36.4 and 53.6% suggesting a stable percentage of pentoses (from hemicellulose) and glucose (mainly from cellulose) relati--ly to the carbohydrate fraction of the substrate during the treatment.

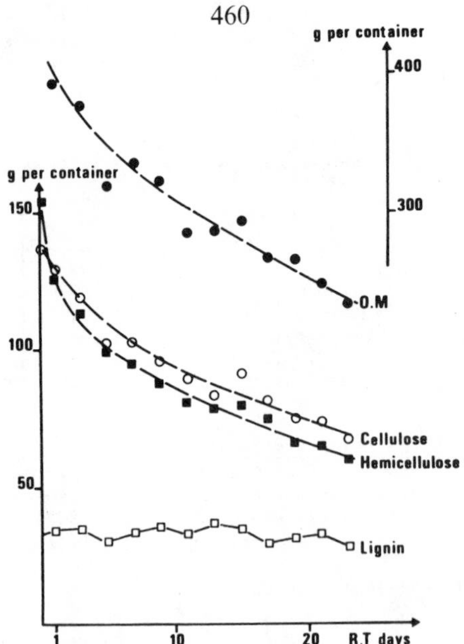

FIG.2 : VARIATIONS IN THE AMOUNT OF VARIOUS D.M. COMPONENTS ACCORDING TO THE R.T. (vertical axis : g. per container)

2.4. Pattern of the V.F.A. :

At the beginning of the process V.F.A. could be only detected in the liquid draining from pig solid manure and grass clippings. They dissapeared after day 16 in the case of pig manure, whereas with grass some V.F.A. remained at the end of the 11 day-R.T. (Fig.3). In this last case the concentration of acids present in the liquid draining from grass was equivalent to a C.O.D of 8 g x 1^{-1}. For the two containers used daily, on an average loss on drying of 2 x 1000 g, this C.O.D concentration represent a maximum wastage of 5.6 1 CH_4, while the production of methane of the reactor is 88.4 1 x d^{-1}.

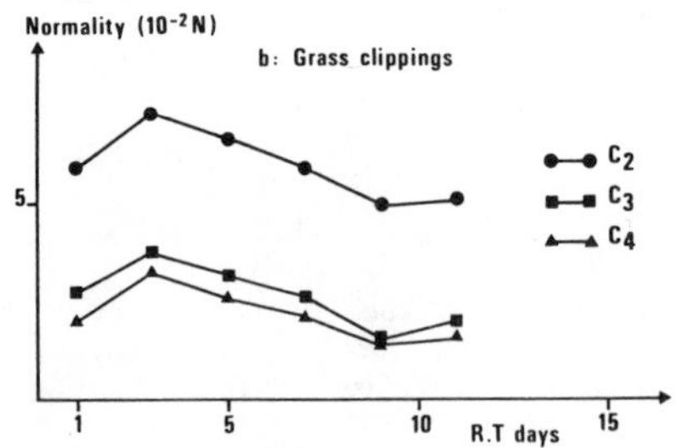

FIG.3 : VARIATIONS OF THE V.F.A. IN THE DRAINING LIQUID ACCORDING TO THE R.T.

2.5. Comparison with the "dry" system :

Comparative results obtained with pig solid manure as substrate are presented on Fig.4 . In the case of dry system, two steps can be clearly identified : in the first period methane was rapidly produced in the "liquid digester" while after 4-5 days the gas production settled in the "dry reactor" and decreased to stop at day 12 in the liquid one. Final production for each O.M. kg introduced was 42.8 l CH_4 from the liquid reactor (i.e 20% of the total) and 180.5 l CH_4 from the solid one. The dashed line in Fig.4 indicates the theorical production of a continuous system, and parallels the straight part of the cumulative curve of gas production by the dry system. This reflects the same efficiency of the two processes.

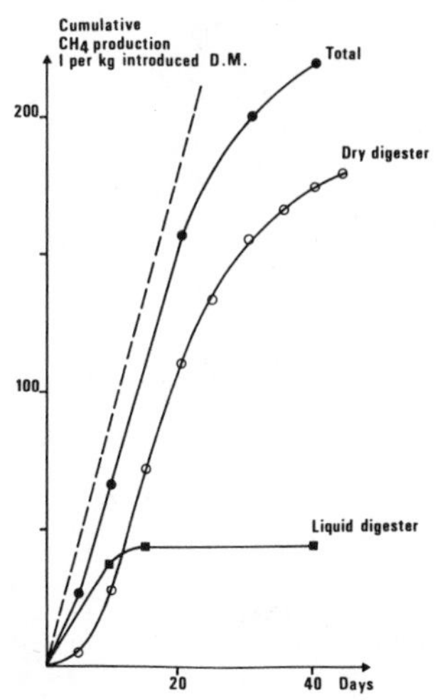

FIG. 4 : ESTIMATION OF THE RESPECTIVE ROLE OF SOLID AND LIQUID PHASE IN METHANE PRODUCTION

3. CONCLUSIVE REMARKS

3.1. Process stability :

Experiments performed with the digester designed on Fig.1 showed that a balanced fermentation can be maintained with high D.M. content substrates. Utilization of straw required an addition of ammonium carbonate at the level of 20 g per kg of fresh weight to reach stable performance. Other additives i.e other carbonates or other ammonium salts should be furter tested in order to evaluate the exact requirement of the system.

3.2. Process performance :

Methane percentage in the gas phase reflects the kind of the substrate ; a decrease is associated with an increase in the cellulose content of the substrate. It is also influenced by the R.T., straw achieved less than 50% CH_4 in the gas phase of a short R.T. of 11 days.

Volumetric loading rates ($B_{v\ O.M.}$) : in these experiments they were of the same range as those encountered in the case of other agricultural wastes. JEWELL (1) reported with dairy cattle manure 8 and 4.2 kg $^{-1}$ x m^3 x d^{-1} for 16 and 28 days R.T. respectively. The high D.M. content of the substrate did not influence this parameter. Our previous experiment (5) with pig solid manure processed in batch system with 3 m^3 working capacity allowed a concentration of 150 kg O.M. x m^{-3} for which a B_v O.M. of 6.8 kg^{-1} x m^{-3} x d^{-1} may be expected, i.e 25% higher than the maximum value obtained in this experiment.

Methane production per unit volume ($r_{v\,CH4}$) was greatly influenced by the origin of the substrate, staw being the less productive and also presented the lower ultimate gas yield. Differences observed between cattle and pig solid manure may reflect variations in the duration of the initial storage, as well as in the amount of straw used. $r_v\,CH_4$ obtained with pig solid manure and grass favourably compared with those reported, at the same $B_{v\ O.M.}$ for liquid manure. The compartments in the gas holder allowed an evaluation of the $r_{v\ CH4}$ achieved with a shorter R.T.. For pig solid manure the last compartment, i.e, 25% of the reactor volume, recovered 24,9 l CH_3 x d^{-1} i.e 27% of the daily methane production. In this case decreasing in the R.T. from 22 to 16 days did not change $r_{v\,CH4}$. A shorter R.T. may be considered beneficial only in the case of grass as shown in Table II, but with a risk of a wash-out for the slow-growing micro-organisms acting in the hydrolysis phase. In the digester used in our experiments a large amount of liquid was present, and mainly with grass, material escaping the containers could generate an active sludge, increasing artificially the stability of the reactor.

Comparison with the batch system :

Except with grass for the reason already discussed, batch experiments developed over the same duration, lead to the same methane yield. The batch system after a slower setting up overtaked the performance of the continuous one.

3.3. Biochemical aspects :

Methanisation settled rapidly whatever the substrate may be, reflecting a rapid colonisation of the substrate by the microorganism. In the case of grass, the initial acidification observed in the batch system was avoided by the continuous one . V.F.A. were associated with the two higher $B_{v\ O.M.}$; their levels were relatively constant, and in the case of pig solid manure they disappeared quickly after day 16. There was no evidence of propionic acid accumulation before its utilization, as routinely high for grass as substrate, whatever the fermentation process used. Variations in V.F.A. can suggest an equilibrium between solubilization of the substrate and methanization, and also reflects the poor diffusibility of these metabolites, hampered by the spongy structure of these substrates. Tentative measurements of cellulolytic and xylanolytic specific activities extracted from the pig solid manure displayed constant values along the reactor, after a rapid initial increase. Rate and extent of cellulose degradation, calculated with data obtained by the Van Soest procedure, appeared independant from the lignin content of the D.M. of the feed material (between 5.9 for pig and 12.9 % for cattle manure). Accessibility of the substrate was not the limiting factor during the process. Maximum rate of cellulose degradation i.e 0.8 g x l^{-1} x d^{-1} compares favourably with values reported for cellulolysis in the mesophilic range. KHAN et al. (2) with a mixed culture isolated from sewage sludge obtained a rate of 4g cellulose l^{-1} week $^{-1}$; SCHARER et MOO-YOUNG (6) reported an upper limit of cellulose hydrolysis of 1.3 g x l^{-1} x d^{-1}, but gas production was oscillatory and "the process does not appear to reach a steady in the traditional sense". Pure cultures of mesophilic Cloestridia utilize purified cellulose at the rate of 1.3 to 1.4 g x l^{-1} x d^{-1} (3).

3.4. Opportunity of phase separation :

Liquefaction of the solid substrate in a separate reactor became attractive according to the technological improvement in the methanization of soluble metabolites offered for example by reactors with a retention of active biomass. This concept has been discussed in the case of agricultural residues by RIJKENS (4). Our own results rule out the interest of phase separation for solid manure. If the dry fermenter is running at 35°C only 20% of the methane potential may be recovered in the liquid phase whatever the reactor used. The tentative of solubilization at laboratory temperature (18-20°C) to avoid active methanogenesis in the "dry" digester remained ineffective, solubilizing activity being too slow. Another limitation was associated with the poor diffusibility of metabolites inside the solid manure : field analysis demonstrated clearly that if the manure pile is sufficiently tight, limited changes would occur in the composition and gas potential of the substrate.

3.5. Practical implications

Improvement in the digester performance can be achieved by increasing the volumetric loading rate and running with the higher substrate permissible density. Further technical problems are associated with the maintenance of homogeneous temperature conditions and metabolite concentrations in the digester. A R.T. of 20 days appears favourable for a dense substrate such as solid manure, with a recovery of at least 60% of the CH_4 potential of the substrate. A shorter R.T. may be advisable if the substrate is readily degradable.

REFERENCES

1. JEWELL, W.J. et al. (1980). Anaerobic fermentation of agricultural residue potential for improvement and implementation. Cornell University ITHACCA New-York.
2. KHAN, A.W. et al. (1979). Nutrient requirement for the degradation of cellulose to methane by a mixed population of anaerobes. Journal of general microbiology, 112, p.365-372.
3. LESCHINE, S.B. and CANALE-PAROLA, E. (1983). Mesophilic cellulolytic Clostridia from fresh water environments. Applied and Environmental microbiology, 46, 3 p.728-737.
4. RIJKENS, A. (1982). Methane and comport. from straw. Commission of the European Communities. Contract N°665-78-1. Final report.
5. ROUSTAN,J.L., AUMAITRE, A. (1982) Progress report presented at the C.E.E. meeting on anaerobic digestion WAGENINGEN 2-3 March.
6. SCHARER,J.M. , MOO-YOUNG, M. (1979). Methane generation by anaerobic digestion of cellulose containing wastes. Adv. Biochemistry. Bioengineering, 11, p.85-101.

VALORIZATION OF HEMICELLULOSES AND LIGNIN : SYNTHESIS OF NEW POLYMERS. (CEE n° BOS-049-F).

M. DELMAS et A. GASET

Laboratoire de Chimie Organique et d'Agrochimie
Ecole Nationale Supérieure de Chimie, I.N.P.,
118, route de Narbonne - 31077 TOULOUSE Cédex - France

A. GANDINI

Laboratoire de Chimie des Polymères - Ecole Française de Papeterie -
BP 65 - F-38402 SAINT-MARTIN-D'HERES - France

G. ROUX

Division Physique des Matériaux - Centre Scientifique et Technique du Bâtiment - 24, rue Joseph Fourier
F-38420 SAINT-MARTIN-D'HERES - France

SUMMARY

The condensation reaction between aromatic and heteroaromatic aldehydes was carried out using a solid-liquid transfer process in the presence of an excess of potassium carbonate. 2-vinylfurans were obtained selectively and quantitatively. The polymerization of vinylfuran (VF) and methylvinylfuran (MVF) has been investigated with two types of initiators for polyaddition : free radical ones and coordination ones.

A new synthetic route has been developed for the preparation of 2-alkenylfurans in high yields and simple experimental conditions. Essentially, it is an extension and modification of the classical Wittig reaction. Typically, 2-furancarboxyaldehyde (or one of its homologues) is made to react with the appropriate triphenylalkylphosphonium bromide in an organic solvent in the presence of a solid base which is slightly hydrated. Within a few hours at about 100°C almost quantitative yields of the corresponding 2-alkenylfuran are obtained, particularly if potassium carbonate-water is the hydrated base used and dioxane the organic solvent 1-3).

In order to optimize the conditions for the specific synthesis of vinyl compounds (i.e. monomers of the furan series), namely, 2-vinylfuran (VF) and 5-methyl-2-vinylfuran (MVF), such parameters as temperature, water-to solid-base ratio and type of solvent were varied until very high conversions could be achieved :

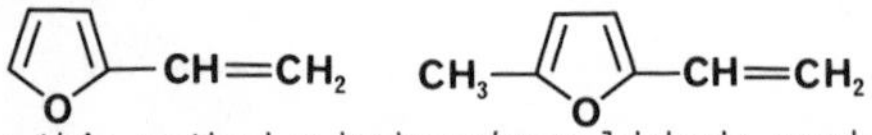

O —CH=CH₂   CH₃— O —CH=CH₂

According to this method p.hydroxybenzaldehyde produced by the oxidative degradation of lignin gives for the first time the p.hydroxystyrene with good yields without protecting the phenol group (2,4).

This new synthesis is allowed by the ability of potassium carbonate in solid-liquid transfer process to pull away an hydrogen ion from the methylene group α to the phosphorus of the methyltriphenylphosphonium bromide without interaction with the phenol group. The p-methoxystyrene from anisaldehyde was also obtained with quantitative yields.

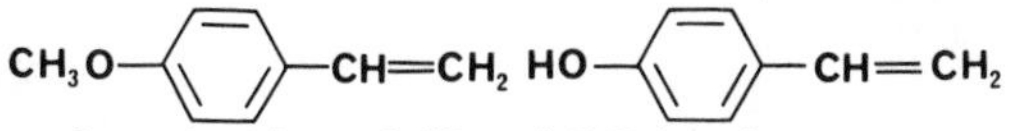

$$CH_3O-C_6H_4-CH=CH_2 \qquad HO-C_6H_4-CH=CH_2$$

The polymerization of VF and MVF has been investigated with two types of initiators for polyaddition : free-radical ones and coordination ones. The free radical processes in solution and in bulk gave poor yields and low molecular weights. Turning to emulsion techniques proved much more successful in that high conversions were readily obtained with persulfates at 50-70°C using various surfactants and the DP's of the products were ranged from 10^3 to 10^4. Interestingly, the polymers did not seem to suffer important cross-linking reactions on standing and remained soluble in appropriate organic solvents for months. The spectroscopic characterisation of these macromolecules revealed the expected vinylic structure with a remarkable degree of regularity. The products are being now tested for thermal properties, i.e., glass transition temperature (T_g), decomposition threshold, type of thermal degradation (loss of volatile materials and/or cross-linking, and for resistance to photooxidation in typical simulated atmospheric conditions (variable light intensity and spectral distribution in the uv-visible region and variable humidity). The synthesis of copolymers of VF and MVF with current monomers is also under way using the same emulsion procedures. These copolymers will be fully characterised and then submitted to chemical modification reactions based on the reactivity of the pendant furan rings. Preliminary results indicate that one can transform these moieties into carboxylic groups by treatment with ruthenium tetroxide or into reversible Diels-Alder adducts with such dienophiles as maleic anhydride and propiolic acid (and its esters). This allows the preparation of gels which return to soluble polymers by heating.

The use of coordination catalysts of the Ziegler-Natta type showed that "strong" catalytic combinations as $TiCl_4$-Et_3Al in heptane or toluene at 40-70°C lead to lightly cross-linked products and are therefore inadequate for the preparation of thermoplastics. Turning to less agressive initiators by replacing the transition metal halide with, e.g. oxyhalides like $VOCl_3$, and/or the organometallic compounds with milder reagents like iBu_3Al or $iBuOAlEt_2$, gave soluble products. The optimisation of these systems is now under way to improve both yields and molecular weights without falling back into cross-linking side reactions. The copolymerisation of VF and MVF with such monomers as styrene or propylene will then be studied.

The copolymerisation of MVF with isobutene using various Lewis acids (cationic initiation) has also been studied under a variety of experimental conditions. The furan monomer is incorporated into the chains more rapidly that isobutene owing to its higher nucleophilicity. At the same time some transfer to MVF has been detected. The copolymers possess a regular structure indicating that the insertion of each monomer was not marred by side reactions. The modification of the furanic structure by oxidation lead to a new material, i.e. a copolymer of isobutene with acrylic acid, which cannot be prepared by direct synthesis. This products exhibited properties typical of reversible ionomeric gels in the presence of aluminium ions and the study opens the way to a new interesting field which will be explored further with less nucleophilic alkenyl furans.

The polymerisation of vinyl phenols is presently being explored. Little information is available on the behaviour of these monomers and for the moment the reactions reported in the literature are being repeated to check their reliability.

REFERENCES

1 - Y. LE BIGOT, M. DELMAS et A. GASET,
Brevet Européen, 81.1782.5.

2 - Y. LE BIGOT, M. DELMAS et A. GASET
Brevet Européen, 82.90072.6.

3 - J. AREKION, M. DELMAS et A. GASET
Biomass, 3, 59, 1983.

4 - Y. LE BIGOT, M. DELMAS et A. GASET
Tetrahedron Letters, 24, 193, 1983.

5 - R. ALVAREZ, A. GANDINI et M. MARTINEZ
Makromol. Chem., 183, 2399, 1982.

6 - A. GANDINI, R. MARTINEZ
Makromol. Chem., 184, 1189, 1983.

MICROBIOLOGICAL TRANSFORMATION OF TECHNICAL ORGANOSOLV LIGNINS

A. HAARS, A. BAUER and A. HÜTTERMANN
Forstbotanisches Institut der Universität Göttingen

Summary

Organosolv lignins from spruce and beech supported growth for several types of fungi and induced extracellular laccase production up to 70 U/ml. The water-soluble part of the lignins was increased or decreased depending on type of lignin and fungus. Degree of enzymatic polymerization was dependent on the glucose concentration in the medium. When the water-solubility of organosolv lignins was increased by mild sulfonation they showed the capacity to bind pairs of wooden boards in combination with laccase and lignosulfonate.

1. INTRODUCTION

Separating lignin from wood by the organosolv process (1), which now is being performed in pilot plant scale, causes less environmental problems than the sulfite and sulfate pulping method. On the other hand, the energy costs of this process are rather high, so that this pulping method can only become technically and economically feasible if, besides the cellulose, the other main components such as lignin and hemicelluloses are converted to useful chemicals. In contrast to lignosulfonate and kraft lignin there are no studies yet on the biotransformation of technical organosolv lignin. It is to be expected, however, that this lignin is more accessible for microorganisms because it is less condensed, purer than the other technical by-product lignins and has a rather low molecular weight.

The aim of our work was (i) to get basic information about the type of transformations caused by different microorganisms and (ii) to find out whether organosolv lignins are suitable as two-component wood adhesives in combination with polymerizing enzymes according to the method developed earlier (2).

2. MATERIALS AND METHODS

Beech and spruce lignin from different organosolv pulping stages (SI, SI+II, BI, BI+II) were kindly supplied by Dr.Feckl, MD Nicolaus, Munich. Fungi were cultivated on B-medium (3) containing 1% organosolv lignin (OL) in the presence or absence of 1% glucose. For laccase production F-medium also was used (4). After the incubation period of 3 weeks the cultures were harvested using a complex schedule including successive filtration, acid precipitation, centrifugation, ether extraction, dioxane-, dioxane:H_2O-, and NaOH-dissolution, in order to separate water-soluble lignin (WSL), aromatic substances (AS), water-insoluble lignin (WIL) and mycelium-bound lignin (ML). Uninoculated lignin-containing cultures (C_1) as well as inoculated cultures without lignin (C_2) served as controls.

The different lignin fractions were examined by UV-spectrophotometry (5) and Sephadex LH20 and -LH60 gelchromatography using doxane:water=7:3 as solvent (6). Aromatics were analyzed by TLC on precoated Silicagel F-254 plates using benzene:dioxane:acetic acid (90:25:4 (v/v)) as solvent system. Aromatics with and without phenolic hydroxyl groups were detected by UV-radiation and 0.1% p-nitrobenzenediazonium fluoroborate (NBDF)(7). Laccase activity was determined using 2,6-dimethoxyphenol as substrate (8). Gluing capacity was tested in the system described by Schorning and Stegmann (9). Mild sulfonation was performed using the method of Boettger et al.(10).

3. RESULTS AND DISCUSSION

Growth: The following fungi were grown on 4 types of OL: white-rot fungi: *Heterobasidion annosum* (*H.a.*), *Sporotrichum pulverulentum* (*S.p.*), *Schizophyllum commune* (*S.c.*), *Pleurotus florida* (*P.f.*), *Polyporus versicolor* (*P.v.*); brown-rot fungus: *Gloeophyllum trabeum* (*G.t.*); mycorrhizal fungi: *Trichoderma aurantium* (*T.a.*), *Cenococcum geophilum* (*C.g.*); ubiquitous fungus: *Botrytis cinerea* (*B.c.*). Most of them, especially the white-rot fungi, produced more biomass on lignin than in controls without lignin.

Laccase production: All types of lignin induced extracellular laccase in *P.v.*-, *P.f.*- and *H.a.*-cultures, *S.p.*, *S.c.* and *B.c.* produced only negligible amounts. The highest induction rate was obtained in F-medium containing SI+II:

Fungus	type of lignin	laccase activity [U/ml]	induction
P.f.	SI+II in B	8	30 fold
H.a.	SI+II in B	23	5 fold
P.v.	SI+II in F	71	3 fold

So, organosolv lignins can serve as substrate for edible fungi (*P.f.*) and as laccase-inducing agent for the binding system according to Haars and Hüttermann (2).

Solubilization of lignin: Organosolv lignins are alkalisoluble, therefore only 10% of OL was dissolved in the culture media at pH 6.5. The amount of water-soluble, acid-precipitable (pH 2,8) spruce lignins (WSL), measured as dry weight, was increased 40% by *H.a.*. In cultures of *S.p.* and *B.c.*, however, the amount of WS decreased dependent on the type of lignin: in BI-cultures over 90% of WS was either metabolized by the fungi or transformed, so that it was no longer acid-precipitable.

Aromatics: The amount of ether-extractable aromatics decreased in all cultures due to metabolization or polymerization in cultures of laccase-producing fungi. Phenolic substances in C_1-cultures were also consumed by laccase-⊕ - and ⊖ -fungi as was detected by NBDF on TLC.

Changes in molecular weight distribution: Fig. 1 shows that WSL is polymerized by H.a.. The presence of 1% glucose inhibited the polymerization rate. The other white-rot fungi exhibited similar patterns. The water-insoluble part of the lignin remained unchanged.

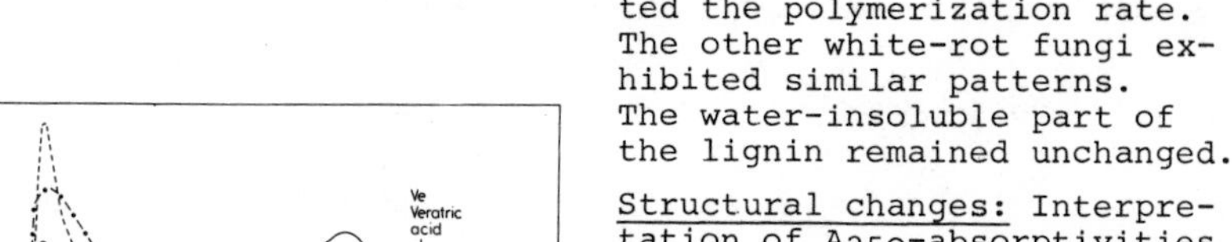

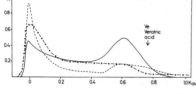

Fig. 1: Sephadex LH60 gelchromatogram of WSL before (—) and after attack by H.a. in presence (•-•-•) and absence (----) of glucose

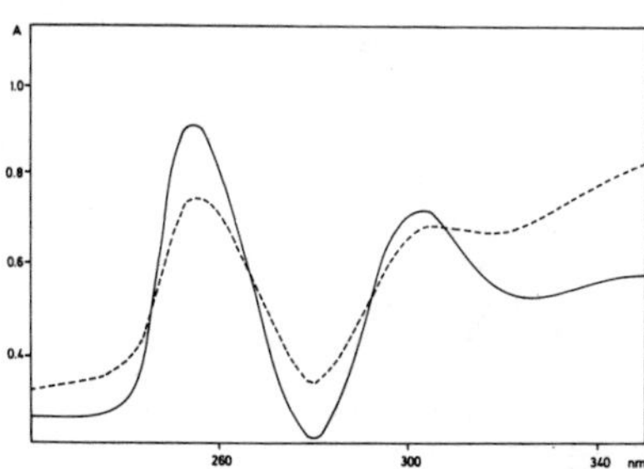

Fig. 2: Difference spectrograms of SI+II before (—) and after (---) attack by H.a.

Structural changes: Interpretation of A_{250}-absorptivities in difference spectrograms of SI+II before and after attack by H.a. suggested a decrease in the phenolic hydroxyl content, whereas the increased absorptivity of H.a.-WSL at wavelengths above 320 nm (Fig. 2) indicated an increase of carbonyl groups, carbon-carbon double bonds or biphenyl groups, etc.(11).

Gluing capacity: All types of organosolv lignins gave negative results in combination with different laccase preparations from P.f. and P.v.. Increasing the water-solubility by mild sulfonation had a positive effect on the versal tensile strength.

Starting from the results presented above we plan for the future fractionation of OL according to their molecular weight, increasing the water-solubility, and screening of demethylating organisms.

4. REFERENCES

1. KLEINERT, T.N., Tappi 58(8),170-171 (1975
2. HAARS, A., HÜTTERMANN, A., DBP 3037922 (1980)
3. HAIDER, K., TROJANOWSKI, J., Arch.Microbiol.105,33-41(1975)
4. FAHRAEUS, G. et al.,Physiol.Plant.11,631-643 (1978)
5. WEXLER, A.S., Anal.Chem.36,213-221 (1964)
6. CONCIN, R. et al., Holzforschung 35,279-282 (1981)
7. BOLLAG,J.-M. et al.,Soil Biol.Biochem.14,157-163 (1982)
8. HAARS, A. et al.,Eur.J.Forest Pathol.11,67-76 (1981)
9. SCHORNING, P., STEGMANN, G.,Holz Roh-Werkst.30,329-332 (1972)
10. BOETTGER, J. et al., Holzforschung 35,241-246 (1981
11. SARKANEN, K.Y., LUDWIG, C.H., "Lignins", 1971

BIOGAS PRODUCTION FROM FARM WASTE (BELGIUM)

A.I.D.R.
International Association for Rural Development

Summary

The International Association for Rural Development (A.I.D.R.) has directed its Energy programme towards the pratical application of biomethanisation of farm wastes. Since 1980, 23 digesters have been installed by the A.I.D.R. in Africa and in Belgium.
The operating principle of four digesters set up in Belgium with the support of C.E.C. and the results obtained are presented in this paper.

INTRODUCTION

Founded in 1964, the International Association for Rural Development (A.I.D.R.) has orientated its activities towards the rural development in the developing countries but also in Belgium.

The A.I.D.R. programme is specially orientated to the pratical application of biomethanisation of organic wastes, mainly farm wastes. Since 1980, the A.I.D.R. has built TWENTY-THREE digesters : eight in Rwanda, five in Burundi, four in Upper-Volta and six in Belgium.

The financial support of the Commission of the European Communities allowed the A.I.D.R. to set up four full-scale digesters on poultry and pig farms in Belgium.

FLEXIBLE PLASTIC DIGESTER FED WITH POULTRY MANURE

In 1980 a 20 m^3 plug-flow digester was installed on a farm comprising 850 battery hens and producing 4,000 chickens per year.

The digester consists of a flexible plastic, cylindrical tank. This semi-interred tank is covered by an insulated roof. The substrate is pumped into one end and is discharged via an overflow at the opposite end. The temperature of the mixed liquor is maintained at 35°C by a submerged heat exchanger.
The gas produced is collected in the upper half (20 m^3) of the digester and is discharged from the balloon to the control room where it is scrubbed, compressed and stored at a pressure of 4 bars in a 200 l steel tank. From there the gas expanded is burned in a boiler used to heat the digester, in infra-red heaters for the hen-house, in convectors and a cooker in the farmhouse.
The sludge pumped out of the digester every 2 months and the effluent are stored in a 12 m^3 tank and are periodically spread on crops.

The load reaches 60 g TSo kg^{-1} and the mean retention time is about 40 days.

The mean gas production rate reaches 1 m^3 gas m_{ML}^{-3} d^{-1}, 35 % of the gas produced is used for the maintenance of the digester (digester heating, pumps and compressor). The annual net energy production reaches 111,000 MJ.

STEEL DIGESTER FED WITH PIG MANURE

A 90 m^3 completely mixed digester was installed in 1981 on a farm raising 80 sows and 350 porkers.

The digester consists of a vertical, cylindrical steel tank covered with an insulating material. The fresh manure is introduced from overhead. After digestion it overflows into a 360 m^3 storage tank. The manure is collected from the conical base of the digester and reintroduced into the upper part at regular intervals to ensure homogenization of the mixed liquor and to avoid the formation of a scum on the surface of the liquid. A heat exchanger maintains the temperature of the manure at 35°C, and the biogas produced is stored at atmospheric pressure in a 100 m^3 flexible plastic tank housed in an aside building.
After scrubbing, the biogas is used to run a 15 kW total energy generating set which covers the electricity needs of the whole farm and heats the digester, the piggery and a part of the farmhouse using the water from the motor's cooling system.
The digested manure is spread on crops.

The digester is fed with fresh manure that has an average concentration in total solids of 65 g kg^{-1}. During the winter the mean retention time reaches 22.5 days and during the summer, about 36 days.

The mean gas production rate reaches 0.9 m^3 gas m_{ML}^{-3} d^{-1}. The total output of the generating set is equal to 0.84.
28 % of the gross energy production are consumed by the digester. Thus the annual net production is equal to 384,000 MJ.

FLEXIBLE PLASTIC DIGESTER FED WITH PIG MANURE

In 1983 a 270 m^3 digester was installed on a farm with 1,800 pigs.

The digester consists of a flexible plastic, semi-interred tank. It is covered by an insulated floor which in turn is covered by a 150 m^3 flexible tank. The whole system is protected by a hangar.
The fresh manure flows out of the piggeries to an influent-effluent heat exchanger where it is pumped into the digester. After digestion it overflows via the influent-effluent heat exchanger into a 2,000 m^3 lagoon. Three submerged mixers ensure the homogenization of the mixed liquor. Another heat exchanger maintains the temperature at 35°C.
The biogas produced is scrubbed and stored at atmospheric pressure in the 150 m^3 balloon. During the summer the gas is used to run a 16 kW generating set which covers the electricity needs of the farm and heats the digester using the water from the motor's cooling system. During the winter the gas is burned in two boilers used to heat both the digester and a piggery.
The effluent is spread twice a year on crops.

The digester is fed with fresh manure at a concentration in total solids of 60 g kg^{-1} and with a mean retention time of 42 days. This long mean retention time is caused by the reduction in livestock, for economic reasons, from 1,800 to 900 porkers.

Current production is 150 m^3 gas d^{-1} and the net electricity produced is 135 kWh d^{-1}. The depollution expressed in COD is about 70 %.

FLEXIBLE PLASTIC DIGESTER FED WITH POULTRY MANURE

A 300 m^3 digester was installed on a poultry farm in 1983. The farm comprises 48,000 chickens.

The digester consists of a flexible plastic tank connected to a 300 m^3 gas storage balloon..The substrate is diluted in a metallic tank with the liquid fractional part of the effluent and pumped into the digester. After biomethanisation it is pumped out of the digester. The effluent is poured through a filter. The resulting liquid is used to dilute the influent and the solid is used as fertilizer. The homogenization, the heating system and the gas storage are identical with those used in the flexible digester fed with pig manure. The gas is used in a boiler to heat the digester and in infra-red heaters for the hen-houses.

The digester is in its starting period.

ACKNOWLEDGEMENTS

The A.I.D.R. wishes to thank the Commission of the European Communities for supporting the present project and the European Development Fund (Brussels, Belgium) for supporting with the C.E.C. the project in Africa. The collaboration of the Unit of Bioengineering (University of Louvain, Louvain-La-Neuve, Belgium) and the Laboratory of Microbial Ecology (State University of Gent, Gent) is gratefully acknowledged.

BIOGAS FROM ORGANICS SORTED OUT FROM MUNICIPAL SOLID WASTES

SO.GE.IN. S.p.A. - SO.RA.IN. CECCHINI S.p.A. - SECIT S.p.A.
Project Leader: G.M.Baruchello

1.) INTRODUCTION

The purpose of the research carried out at the municipal solid waste (MSW) disposal plants in Rome was to check the feasibility of an anaerobic digestion process for this waste.
Particular care was paid to reproducing the working conditions already tested and amply described in specialized literature, based on the conditions already existing in the two Rome recycling plants.
This involved a series of on-site and laboratory operations for analysing the following aspects:

a) - quality of the potential substratum of an anaerobic digester;
b) - analysis of each potential substratum;
c) - laboratory test of operational systema for the digester;
d) - assessment of yields.

2.) DESCRIPTION

A major part of the work was dedicated, as already mentioned in previous progress-reports, to an assessment of the product features of the outputs of the Roman recycling plants by grading the different materials both on site and after laboratory tests.
The substrata analysed were the following:

(1) "medium" Rocca Cencia *
(2) "fine" Rocca Cencia
(3) "paper pulp" Rocca Cencia
(4) "fine" Ponte Malnome *

"Medium" is defined as the non-refined organic fraction of MSW, whilst "fine" is the selected and refined fraction.
"Paper pulp" instead is the salvaged product of one of the plant selection lines, whith a high cellulose content.
The analyses on product features were only carried out on-site on the "medium" because it was more representative than the other fractions. In fact a mean value of 80% was found in the organic substance + paper fraction.

---

* Rocca Cencia and Ponte Malnome: wastes disposal and recycling plants

The on-site tests were repeated in the laboratory on smaller but equally representative quantities: the aforesaid values were much the same even though, due to more careful selection, the paper percentage was higher. The laboratory analyses provided more detailed indications on the composition of the two "fine" and "paper pulp" fractions. At the same time both static and dynamic digestors (5 litre capacity) were sept up with different concentrations of MSW organic substance and urban sludge mixtures.

Analyses were carried out on these mixtures to determine their features, especially as regards the heavy metal contents and C/N/P ratio, which were found to be within the range of allowable values. Subsequently tests were carried out on the operation of systems for mechanical stirring, initial inoculation, variation of volatile acid, including all those parameters which could inhibit the process.

In synthesis, it was found that a mechanical stirring system is only feasible with a mixture which has a low dry percentage, independently of its biological kinetics. Therefore it is necessary to change the form of the digesters for materials with a high dry content.

As regards the production of biogas, the tests were carried out in parallel on the 4 aforesaid fractions, using 5 litre digesters with mixtures which had a low percentage of dry material (4.5% of ST), with a working load of 3 kg SV/m^3, in mesophyll conditions (33°C) and with a water retention time of about 12 days.

After analysis, before placing it in the digesters, the Ponte Malnome "fine" was defined as an excellent substratum both for its SV content (approx. 57%) and for its volatile acidity values which favour the anaerobic digestion process (Table 1).

TABLE 1

Chemical-physical features of substrata	Rocca Cencia "medium" compost	Rocca Cencia "fine" compost	Rocca Cencia "paper pulp"	Ponte Malnome "fine" compost
U%	34.9	36.9	59.4	25.9
ST%	65.1	63.1	40.6	74.1
SV%	51.8	40.2	35.5	55.9
Ashes %	13.3	22.9	5.1	18.2
pH	7.8	6.7	7.6	6.25
Alkalinity (mg/kg $CaCo_2$)	19,200	27,500	6,530	27,000
Volatile acid (mg/kg CH_3COOH)	2,400	10,800	2,650	17,650
Nitrogen Ammoniac (mg/kg)	250	375	n.d.	770

3.) RESULTS

As production values, the "medium" compost and the "paper pulp" have the greatest values in m^3/kg VS but the CH_4 percentage in the "paper" pulp is less; however, the values of the four samples are very similar, as can be seen in Table 2 and Table 3.

TABLE 2

Biogas production	Rocca Cencia "medium" compost	Rocca Cencia "fine" compost	Rocca Cencia "paper pulp"	Ponte Malnome "fine" compost
Mean production ml/d	4.798	4.563	5.335	4.668
Conversion m^3/kg VS	0,40	0,37	0,40	0,38

TABLE 3

Biogas production	Rocca Cencia "medium" compost	Rocca Cencia "fine" compost	Rocca Cencia "paper pulp"	Ponte Malnome "fine" compost
CH_4 % vol.	56,0	56,2	53,0	55,8
CO_2 % vol.	42,4	42,1	45,3	42,5
N_2 % vol.	1,6	1,7	1,7	1,7

Still to be defined is the optimal stirring system which, together with a concentration of total solids which does not reduce, if not inhibit, the biological reaction kinetics, allows the maximum exploitation of the biomass.

MONITORING ANAEROBIC DIGESTERS ON COMMERCIAL FARMS

R FRIMAN
Agricultural Development & Advisory Service
Ministry of Agriculture, Fisheries & Food
Great Britain

Summary

A monitoring programme was developed in 1980 in order to:
establish the performance of farm plants in terms of net energy production and utilisation;
determine the effects of digestion on slurry with regard to pollution reduction and fertiliser value.
Monitoring was accomplished using a mobile laboratory equipped to visit each site in turn for periods up to 8 weeks. The mobile laboratory was used effectively to assess the performance of 4 prototype farm digesters. Monitoring showed on full scale plants that pig slurry achieved a specific gross biogas production of $0.5m^3$ biogas/m^3 digester volume.day and dairy cow slurry $0.8m^3$ biogas/m^3 digester volume.day. A pilot plant operating on separated pig slurry at a retention time of 8.3 days produced $1.8m^3$ biogas/m^3 digester volume.day. The biogas yield from the total solids in pig slurry fed ranged from 0.28 to $0.37m^3$ biogas/kg TS fed. The yield from dairy cow slurry was $0.12m^3$ biogas/kg TS fed.

1. INTRODUCTION

A flexible monitoring programme was developed in 1980 to monitor the performance of existing anaerobic digesters on commercial farms. The monitoring programme had the following objectives: to establish the performance of the farm plants in terms of net energy production; to determine the effects of digestion on slurry with regard to pollution and fertiliser value; to identify indicators of digester performance which can be used to assist the operator on the farm to manage the plant effectively; to note the efficiency of energy utilisation on the farm.

1.1 Methods

Monitoring was undertaken at the farmers or manufacturers request when the plant operating parameters of hydraulic loading, feed quality and temperature had remained as constant as digester management would allow for approximately 4 retention times. Monitoring equipment was installed on site for 6-8 weeks. A data logger recorded hourly readings from meters measuring biogas production, biogas calorific value, electricity production, input feed volume and digester temperature. Samples representative of slurry entering and leaving the digester were taken daily and were analysed for total solids, volatile solids, total-N, NH_4-N, P_2O_5, K_2O, Na, Ca, Mg, Zn and Cu as previously described (1). Biogas was analysed on site for CO_2 and H_2S. On one pig farm slurry was sampled before and after digestion to assess the reduction in slurry odour offensiveness (1).

2. RESULTS

Site		A	B		C	D	
			Run 1	Run 2		Run 1	Run 2
Digester capacity	m^3	400	324		300	11	
Feed		Separated pig slurry	Whole dairy cow slurry		Whole pig slurry	Separated pig slurry	
Retention time	days	12.3	15.3	19.9	13.3	12.8	8.3
Slurry total solids	kg/m^3	23	105	114	22	59	40
Temperature	^oC	35.6	37.0	33.2	33.8	36.1	33.8
Organic loading	kg VS/m^3 digester .day	1.1	5.6	4.4	1.1	3.2	3.4
Daily biogas	m^3	215	252	217	163	17	20
Biogas yield	m^3/kg VS fed	0.50	0.14	0.15	0.47	0.49	0.52
	m^3/kg TS fed	0.28	0.11	0.12	0.32	0.33	0.37
Specific gross biogas production	m^3 biogas/m^3 /m^3 digester volume.day	0.5	0.8	0.7	0.5	1.6	1.8
Biogas calorific value	MJ/m^3	25.0	20.1	20.1	25.8	23.7	26.2
Hydrogen sulphide in biogas	%	0.6	1.0	1.3	0.5	0.5	0.7
Total solids reduction	%	21.0	29.5	17.0	34.0	33.0	13.0
Volatile solids reduction	%	29.0	33.0	19.0	43.0	37.0	15.5
COD reduction	%	41.0	34.0	36.0	40.0	-	-
Acetic + propionic acid reduction	%	88.0	96.0	66.0	88.0	-	56.0

3. DISCUSSION

3.1 Biogas production

The specific gross biogas production of the continuous stirred tank reactor type digesters A and B was 0.5m^3 biogas/m^3 digester volume.day which is lower than has been found experimentally (2) and on other full scale plants (3). The biogas yields of 0.28 and 0.32m^3 biogas/kg TS for digesters A and C respectively were as expected and the low specific gross biogas production was due to the low slurry feed total solids of 23kg/m^3 and 22kg/m^3.

Digesting separated pig slurry with a total solids content of 59kg/m^3 in the pilot scale continuous stirred tank reactor D gave an increased specific gross biogas production of 1.6m^3 biogas/m^3 digester volume.day. This was further improved to 1.8 by reducing the slurry total solids to 40kg/m^3 and reducing the retention time to 8.3 days by increasing feed throughput.

High dry matter cow slurry produced poor biogas yields of 0.11 and 0.12m^3/kg TS fed in the plug flow type digester B which were partly attributable to low retention times and low and variable temperatures in the digester. This resulted in a low specific gross biogas production of 0.8 and 0.7m^3 biogas/m^3 digester volume.day.

3.2 Slurry pollution reduction

The total solids concentration of the slurry was reduced by between 13 and 34% and the volatile solids concentration was reduced by between 15 and 43%. The least reduction occurred with low retention times of 8.3 days. The chemical oxygen demand of the slurry was reduced by between 34 and 41% in the large scale plants which was not considered significant in terms of water pollution control.

A single trial using an odour panel to estimate the effect on slurry odour by plant C showed that the odour offensiveness ratings were consistently lower for digested slurry than for the slurry feed to the digester.

4. CONCLUSIONS

Low slurry feed solids resulted in poor biogas production from full scale plants on pig farms. Reducing the retention time to 8.3 days and digesting separated pig slurry with a feed total solids of 40kg/m^3 in a pilot scale digester resulted in a good specific gross biogas production of 1.8m^3 biogas/m^3 digester volume day. The specific gross biogas production achieved by the full scale plants on pig farms showed that the digesters were not economic as producers of energy. The potential for producing biogas is being improved by developments in digester design. However the major problem is to obtain a consistent and high solids slurry feed for the digester which ultimately involves re-designing pig buildings to exclude sources of dilution water.

More work is needed to assess the environmental effects of anaerobically digesting animal manures.

REFERENCES

1. FRIMAN, R. Monitoring anaerobic digesters on farms. J. Ag. Eng. Res. To be published.
2. SUMMERS, R. and BOUSFIELD, S. (1980). A detailed study of piggery waste anaerobic digestion. Agricultural Wastes 2 (1) 61-78.
3. FISCHER, J. R. et al. (1979). Design and operation of a farm anaerobic digester for swine manure. Trans of the ASAE 2 (5) 1129-1136, 1144.

TWO-STEP ANAEROBIC DIGESTION OF SOLID WASTES

B.A. RIJKENS and J.W. VOETBERG

Institute for Storage and Processing of Agricultural Produce (IBVL)
Wageningen the Netherlands

SUMMARY

The two-step anaerobic digestion process for solid wastes, as developed by the IBVL, is a novel alternative for the processing of moist and perishable wastes into biogas and compost, without any environmental burden. The solid matter, after being chopped or crushed, is digested (liquefied) in a reactor by a population of acid-forming bacteria under percolation with water. The effluent from this reactor (R1) is treated in a UASB reactor. (R2), after which the water is recycled to the R1. Within 2 or 3 weeks the organic solid matter is digested, generally for more than 50 %, and has become free of odors. After discharging the drip dry residue is after-composted for about 2 months, resulting in a compost, identical to the products produced in the usual way. The biogas produced represents 40 to 80 % of the caloric value of the raw material. The process has been developed through laboratory and semi-technical stages for various kinds of raw waste. The final stage consisted of a system of 5 reactors (R1) of 60 m^3 each, connected to one R2 (USAB) of 10 m^3. Experience exists now with the organic fraction of MSW, solid wastes from sugarbeet and potato processing industry, waste tomato plants, waste onions and grass. Areasonable insight has been obtained in the technological and micro-biological aspects of the process and its costs under various circumstances.

1. INTRODUCTION

This new approach to the fermentation of solid waste material was initiated by the need to utilize renewable resources as good as possible. For wet and perishable solid wastes composting will economically be an alternative, but the dissipated heat generally will be wasted. An aerobic digestion will retrieve a major part of the caloric value, but has the disadvantage of the many expensive unit operations which are necessary (grinding or milling,slurrying etc.).Another disadvantage is that the "technology" of the one- or two-step slurry digestion process is poor, since the slowly growing methane flora never can be built up in higher concentrations due to continuous losses through the discharge. The novel IBVL -process separates the two main phases in the microbial break-down process, so both of these steps can be optimised separately.

The principle of the process is that chopped or crushed solid matter is attacked by an anaerobic flora, breaking down solid matter into volatile organic compounds. However, stiffling of the process is prevented by a continuous percolation with water. The percolate is pumped to a second reactor in which the water is treated under the production of biogas, where upon the treated water is recirculated to the first(liquefaction) reactor. The principle of the process is shown in figure 1. Since the second reactor needs a more or less constant loading rate, it has to be linked to a number of primary reactors, each of them in a different stage of break-down.

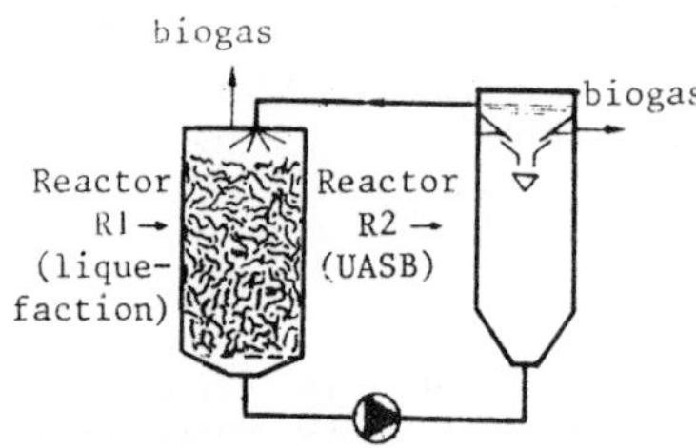

Fig.1: the principle of the two-step process for the fermentation of solid wastes

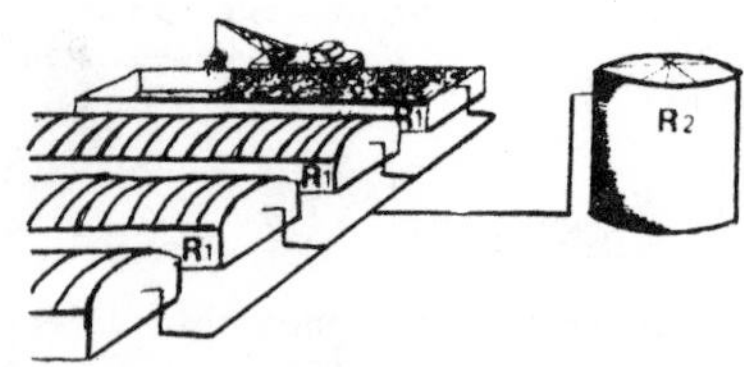

Fig.2: A possible concept for a full-scale plant

In two to three weeks the solid waste is sufficiently decayed and free of odours, so the residue can be discharged and submitted to an aerobic after-composting process on windrows during 2 to 3 months.

2. THE REACTOR R1

Of the liquefaction process in the reactor R1 no proper knowledge did exist, so an extensive laboratory programm had to be carried out to collect the microbiological and technological information, necessary to run this part of the process and to develop the reactor design.
It was found that the liquefaction proceeds relatively rapidly and that working in a fully submerged condition generally had to be preferred to the "sprayed" conditions, in which the percolation water is trickled through the solid mass. A basic requirement is that the water can be percolated and that clogging not occurs. The percolation rate amounts 1 to 2 volumes of the reactor R1 per day, which is usually easily obtainable. The break-down of the solid matter is increased by mechanical damage of the structure.

3. THE REACTOR R2

It was found that the percolate from the R1 contains mainly volatile fatty acids (vfa) and hardly any suspended matter; the liquid is a nearly clear, slightly brown solution that can be digested rapidly by the sludge blanket of methane bacteria in the R2. In nearly all cases a granular sludge with a capacity of at least 10 kg COD/m^3.day and sometimes upto 40 kg COD/m^3.day will develop. Since in the recirculating water some soluble components accumulate, attention has to be paid to inhibiting or toxic effects of i.e. ammonia (in combination with a higher pH),hydrogen sulfide, and "salts". No adverse effects of salts have been met in practice, but sometimes calcium can cause severe problems, because of scale or the production of a "heavy sludge" or "heavy granules" which fail to keep suspended.

4. THE PRODUCTION OF BIOGAS

Since there is hardly any loss of energy in the system, the biogas produced is equivalent to the decayed organic matter in respect to its energy content. The advantage of this process is that, even from a very moist or wet waste, 40 to 80% of the caloric value of the organic dry matter can be obtained as biogas.

5. THE COMPOST

The residue discharged from the R1 generally contains 20 to 60% of the original organic dry matter. Strongly lignified material will decay slowly and leave a higher percentage of residue, since lignin is not attacked under anaerobic conditions. When placed in windrows, the generally coarse residue will easily give access to air, so the aerobic after-composting starts and proceeds rapidly.
Within a week, temperatures of 60°C are obtained, resulting in a reduction of pathogens and seeds.

6. THE PILOT PLANT OPERATION

The final concept of the pilot-plant, consisting of five reactors R1 of 60 m^3 each and one reactor R2 (UASB) of 10 m^3, was run during nearly two years with wastes like waste tomato plants, sugar-beet wastes and municipal solid waste and has provided a lot of experience in the behaviour of the organic matter during the liquefaction process.

7. DISCUSSION

The two-step process turns out to be a useful alternative in the various ways of treatment or disposal of organic wastes. Every process has its own assembly of pros and cons, which enables us to chose the process, that fits best the situation and the requirements. Though this two-step process has not yet been developed completely and still shows child diseases, we believe that it is on the verge of application in practice now.
A feasibility study has been made for the processing of the organic fraction separated from municipal solid waste in the Netherlands and the implementation of this process is under consideration.

ANAEROBIC DIGESTION EXPERIENCES IN BOLOGNA NEARBY

G. VERONESI

Group of Physics Technologies for Agriculture
Institute of Genio Rurale
University of Bologna - Italy

SUMMARY

We describe the results of experiences done in an anaerobic plant for biogas production from swine waste, constructed with the C.E.E. contribution, near Bologna.

On the basis of this experience and for the good operative results obtained, we formulate a new project for the construction of a plant using the septic tanks already existing.

The instruction for the construction of a biogas plant are illustrated in a manual edited with the C.E.E. and Camera di Commercio di Bologna contribution.

A methane producing facility was built in the town of Monte S. Pietro with the funds contributed by the CEE (contract n. RUW-044-I) and the contribution of the Camera di Commercio I.A.A. of Bologna. This plant treats wastes from a pig farm, which is a dairy annex. This type of production unit is typical in the Parmigiano Reggiano production area.

The system consists of 12 re-enforced concrete cubical tubs (2.5 m x 2.5 m) with inverted pyramidal bottoms which are 1 m high. Each tub has a gross capacity of 18 cm, the total volume of the system is 216 cm; and the working volume is 190 cm. The heating system utilizes the heat from the fumes of the dairy's water heater.

All of the movements for the pumping, agitation and drainage of the wastes are done with a single pump, which is connected to the single tubs through a series of servo commanded valves. The anaerobic treatment cycle lasts 16 days and includes the progressive passage of the waste into the tubs which are place in a progressive series: therefore the system can be considered multi staged.

The transfer operations from one tub to another (which also substitute the agitation processes) are done every three hours and transfer about 12 cm of waste per day.

The reaction temperature is maintained at about 20°C in order to limit energy consumption for heating. Therefore the anaerobic activity is attributed to psycrophile bacteria.

Thus we were able to investigate the productivity and the reliability of the use of this type of bacteria which, until now, has been studied very little.

The output of the plant is discharged into a series of three small basins of the facoltative type, which total volume is 700 cm. The basins are able to hold their contents for about 60 days.

At the end of their cycle, the liquids are spread directly on fields or, before being filtered with sand, are recycled, being used for washing the stalls of the pig farm.

The gas produced is initially stored in large plastic bags placed on the reaction tubs. It is then sent by small compressor to a metallic gasometer which has a capacity of about 6 working hours. The biogas produced is used both to heat the dairy farmer's house and for the needs of the dairy.

In the dairy's water heater, a biogas burner was installed next to the fuel oil burner. A stainless steel burner was used to avoid possible corrosion due to the presence of sulphidric acid.

With this method, all of the gas produced daily is consumed, avoiding the need for costly storage.

In the three years of the plant's functioning, the mean daily biogas production was about 80 cm, 70 % of which was methane. The yield is equal to 0.4 - 0.5 cm of biogas/kg of volatile solids introduced into the system.

The decrease in polluting factors was large, even if not sufficient in terms of the law; on the average we obtained a 65 % reduction of C.O.D. and a 85 % reduction of B.O.D.; at the same time, however, we noted an imbalance in the composition of the ouput, prevalently in the C/N rations, which advise against further depuration using aerobic cycles with active slurries.

Given the positive results reached and the reliability of the system which did not present particular problems in handling nor in maintenance, an enlargement of the plant is underway, which will allow for the treatment of the waste produced from the entire pig farm (the above described project treated only about 50 %.

The problems relating to the polluting of the ouput and to the methods of spreading it will be dealt with in the near future.

As many of the breeding farms in the region already have accumulation tubs (septic tanks) which, given the experiences described above, could be easily utilized for the production of biogas; an executive project is being formulated for the construction of a plant at another farm.

This will also receive EEC funds. This project is designed to include the experimental use of a mixture of pig and cattle wastes.

The operational cost estimates, guarantee a profitable return on investment, as the layout is modest.

Considering that there are more than 1.000 dairies in the region, which are in condition similar to the above mentioned, the operational effectiveness of the plant is of indisputable interest.

The results of the experiences already acquired lead to the conviction that, wherever gas can be directly used as fuel without the need for costly storage and processing, current technological expertise enables such facilities to be built. These facilities, because of their low cost, management and reliability, fully respond to the needs of modern livestock farms. The main factor limiting the widespread use of anaerobic facilities, especially in the pig and dairy sectors, is the hight cost of adapting industrial technology to these plants.

The construction and operation of these low profit margin facilities (like the experimental ones) are usually dependent on small, local manufacturers and artisans. However the technical expertise of such personnel is often lacking.

To overcome this situation, a manual has been draften, entitled "Riciclo di rifiuti animali: guida alla costruzione di un impianto biogas" (Recycling animal wastes: a guide to the construction of a biogas plant), it describes the techniques and calculations necessary for the planning, construction and management of a plant treating the slurry from a pig breeding farm. This manual was concieved in order to make the knowledge acquired from the experimental plant available to the artisans and professionals working in this field in rural areas.

The most widespread types of biogas production plants are examined in the manual. However, the construction methods of those plants which most fully respond to the need for simplicity in conception and management and which are low cost are explained in detail.

The proposed projects are developed in such a way as to encourage the farmers to construct, furnish and manage the facilities themselves, so that they can be as self-sufficient as possible.

The last part of the manual deals with the present laws in Italy: on one hand the codes regulating the containment and storage of energy products and, on the other, the anti pollution codes.

ADSORPTION OF THE TRICHODERMA REESEI QM9414 CELLULASES WITH REFERENCE TO THE SYNERGISM OF THE CELLULOLYTIC COMPLEX

P. DESMONS*, P. THONART** and M. PAQUOT**

* IRSIA scholar

** Départment de Technologie, Faculté des Sciences Agronomiques, 5800 Gembloux (Belgium)

Enzymatic hydrolysis of cellulose necessarily involves the adsorption of the cellulases on the substrate. It is difficult to quantify this phenominon precisely because of the various reactions which are superimposed: adsorption as such, disappearance of the substrate as it is hydrolysed, enzyme denaturation (thermal or other) in the course of the reaction, changes in the liquid phase (viscosity, transfer phenomena, appearance of sugars in the medium).

Two methods have been employed to make these observations: measurement of the residual activities in the medium (1) by analysis of the sugars produced (2) and ion-exchange chromatography of the enzymes by HPLC (1 and 3).

Table 1 shows the changes in the residual activities in the liquid phase after five additions with subsequent separation of 2% of fresh cellulose. A cellulose/cellulases contact time of 30 minutes was chosen since it had previously been observed that most adsorption occurs in the first few minutes of the hydrolysis process (1, 4, 5 and 6).

Figure 1 shows the change in the level of the residual proteins as determined by the HPLC method described by Bisset (1979).

Measurements of activity and separation by HPLC show that different activities are adsorbed to different degrees. The exocellulase activity seems to undergo the greatest proportional adsorption.

Endocellulase activity does not seem to be a limiting factor: over 25% of the initial activity remains in the medium. Xylanase activity remains adequate (over 50%). There are only very slight changes in the -glucosidase and β-xylosidase activities.

In order to confirm these findings, we also tried to add a number of cellulase fractions purified by various chromatographic methods to the residual enzymes after adsorption. The purification techniques employed were gel permeation, ion-exchange, affinity chromatography and electrofocusing. The results, which have not been reported, show that the Trichoderma reesei QM9414 cellulolytic complex has a fairly good balance of the different activities required for the hydrolysis of crystalline cellulose.

		ENZYME ACTIVITIES				
Sample		FPase (%)	CMCase (%)(VIS)	β-glucosidase (%)	Xylanase (%)	β-xylosidase (%)
Control	Time 0	100	100	100	100	100
1st	After 15'	81	84	96	95	107
addn	After 30'	83	74	98	94	104 104
2nd	After 15'	50	53	94	83	104
addn	After 30'	49	48	98	78	104
3rd	After 15'	19	36	96	65	102
addn	After 30'	24	41	89	77	100
4th	After 15'	8	33	89	56	100
addn	After 30'	8	25	88	65	89
5th	After 15'	7	27	91	64	96
addn	After 30'	3	27	89	50	84

Table I : Whatman No 1 cellulose - Residual activities as percentages of the original activity detected in samples taken as a function of the number of substrate additions (2% on each occasion) and time (15 or 30 min incubation)

REFERENCES

1. DESMONS, P.; THONART, Ph.; PAQUOT, M. Pasteur Biosciences, 1983. Poster 49.

2. MILLER,G.L. Analytical Chemistry, 31, 426 - 428, 1959.

3. BISSET, F.H. J. Chromatogr., 178, 515 - 523, 1979.

4. PEITERSEN, N.; MEDEIROS, J.; MANDELS, M. Biotechnol. Bioeng., 19, 1091 - 1094, 1975.

5. CASTANON, M.; WILKE, L.R. Biotechnol. Bioeng., 21, 1037 - 1053, 1980.

6. LEE, S.B.; SHIN, H.S.; RYU, D.Y.; MANDELS, M. Biotechnol. Bioeng., 24, 2137 - 2153, 1982.

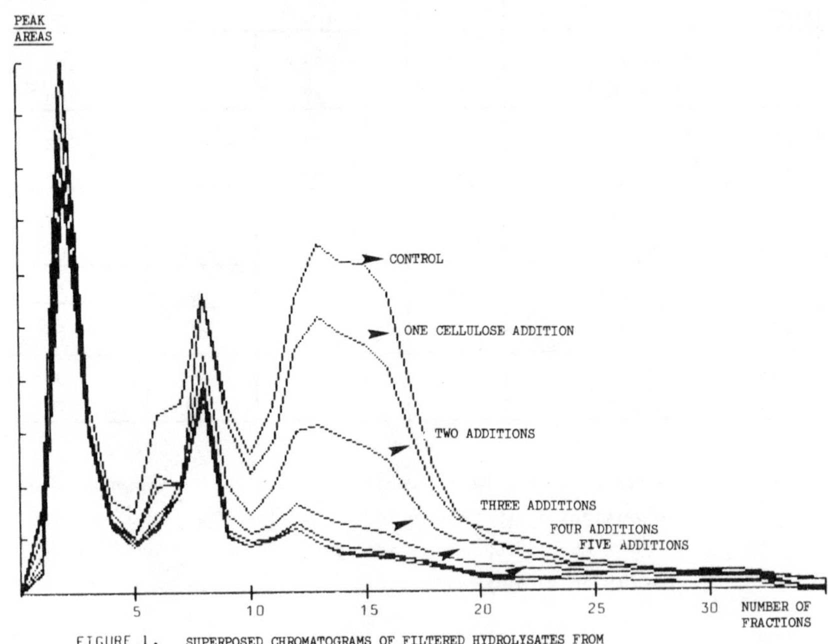

FIGURE 1. SUPERPOSED CHROMATOGRAMS OF FILTERED HYDROLYSATES FROM WHATMAN NO 1 CELLULOSE

BIOGAS RECOVERY FROM OLIVE-OIL MILL WASTE WATER BY ANAEROBIC DIGESTION

A. AVENI
Istituto Ricerche Breda - Bari -

Summary

The high organic content of the olive-oil waste water makes it, in theory, suitable to be treated by the anaerobic digestion process. However, the waste water composition causes some difficulties in maintaining the anaerobic process efficiency at high level.

Experimental tests on progress show that the anaerobic treatment of diluted olive oil waste water at organic loads ≤ 4 kg (COD) $m^{-3} \cdot d^{-1}$ gives high organic removal efficiency ($80 \div 85\%$) with biogas production $\simeq 550$ $l \cdot kg^{-1}$ (COD removed).

At high COD concentration, the control of organic acids and organic-inorganic (k^{+}) inhibition is difficult.

1. INTRODUCTION

The olive oil production gives waste waters with an high organic content ($\simeq 10\%$): mainly glucose and sucrose, nitrogenous compounds, fats, organic acids and small quantities of polyalcohols, polyphenols, pectins ecc. Inorganic compounds are also present ($\simeq 2\%$): mainly potassium and phosphorus salts.

The pollution effect of these waste waters is very strong. A study carried out in Puglia, in the southern Italy, where are operating about 2000 oil-mills, has shown that the COD values of the waste waters vary in the range: $\sim 30 \div \sim 200$ kg $\cdot$ m^{-3}. This high organic content makes, in theory, the waste water suitable to be treated by the anaerobic process with twofold advantage: to produce biogas and to reduce the organic polluting load. However, the waste water composition causes difficulties in maintaining the process efficiency at high level.

Aim of the study, carried out by laboratory and pilot tests, is to identify suitable operating conditions to perform the anaerobic process, both with diluted and concentrated olive-oil waste waters.

2. LABORATORY AND PILOT TESTS

Tests have been performed, at 35°C, with an anaerobic seed, produced by olive-oil waste water diluted with synthetic sewage.

The main chemical characteristics of the olive-oil waste water used in the experimental tests are:

- pH		=	5,4	
- Total Solids		=	107	kg . m^{-3}
- Organic Nitrogen	(N)	=	0,8	"
- Phosphorus	(P)	=	0,5	"
- COD	(O_2)	=	150	"
- Sodium	(Na)	=	0,35	"
- Potassium	(K)	=	10	"

The COD values of the waste water fed to the anaerobic process have been changed, by dilution, to point out the behaviour of the process in a large range of waste water polluting load.

To neutralize the acidity and to furnish the nitrogen for the cellular biosynthesis in the biodegradation processes sodium hydroxide and ammonia were added.

In the laboratory tests, excluding an operating period of low efficiency, caused by an anomaly in the thermostatic system, the plant ran properly, in the experimented volumetric load range: $0,2 \div 8,5$ Kg (COD_0) · m^{-3} · d^{-1}, with COD concentration of feed water $\simeq 5 \div 90$ kg · m^{-3}. After the starting period, the gas yield reached about 60% of a goal value, estimated considering that only 78% of the eliminated COD is transformed into biogas with CH_4 concentration = 60%. These values have been obtained experimentally when the anaerobic process developed properly. Afterwards, the efficiency of the anaerobic process increased: gas yield of $90 \div 100\%$ of the goal value and COD reduction of $85 \div 95\%$ have been recorded. The organic acid values kept at low levels in all the tests, excluding the tests at organic load of about 8,5 Kg (COD_0)· m^{-3} · d^{-1}. In these tests the process efficiency has been maintained at good levels by increasing the alcalinity.

The laboratory results have been used for programming pilot tests by a purposely built anaerobic digestor (fig. 1), that operated at organic loads = $1,5 \div 4,0$ kg (COD_0)· m^{-3} · d^{-1} and COD concentrations of feed water = $25 \div 70$ kg (COD) m^{-3}.

Fig. 2 shows the results of the tests.

3. FINAL REMARKS

The pilot tests on progress show that the anaerobic treatment of diluted olive oil waste waters, at organic loads ≤ 4 kg (COD) · m^{-3} · d^{-1}, gives high COD removal efficiency ($80 \div 85\%$) and biogas production = 550 l · kg^{-1} (COD removed), with CH_4 concentration = $50 \div 70\%$.

At high COD concentration the control of organic acids and organic-inorganic (k^+) inhibition effect is difficult.

Alkalinity and ammonia addition improves steady state process operation.

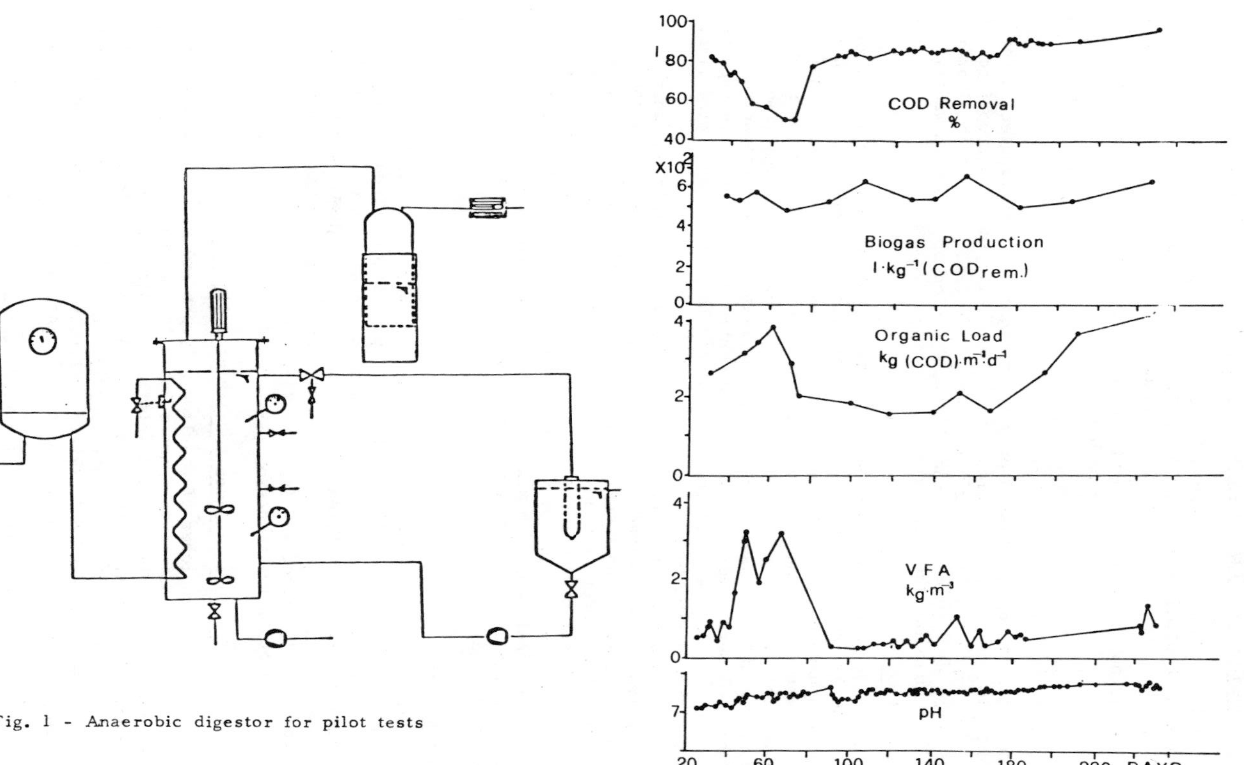

Fig. 1 - Anaerobic digestor for pilot tests

Fig. 2 - Results of the pilot tests.

CHEMICALLY MODIFIED LIGNIN FOR THE USE IN POLYMER BLENDS

A.H.A. TINNEMANS[1] and P.J. GREIDANUS[2]
[1]Institute of Applied Chemistry TNO, P.O. Box 5009, 3502 JA Utrecht The Netherlands
[2]Plastics and Rubber Research Institute TNO, P.O. Box 71, 2600 AB Delft, The Netherlands

Summary

The utilization of acylated pine kraft lignin is investigated in the formation of homogeneous blends by solvent alloying of appropriate resins, viz. copolymers of maleic anhydride, and as compatibilizer in mixtures of polymers from the plastic fraction of household waste.

1. Introduction

Polyblends such as the blends of copolymers of styrene and maleic anhydride (PSMA) with a homopolymer such as cellulosics, acrylates and polyvinyl esters have been used as release matrix, because of their potential to control water sorption and sorption rate by simply choosing the right mixing ration(1). These blends, called Aqualloy polymers, are thermoplastically processable. We have investigated the feasibility of preparing water -swellable polymers comparable with those mentioned above in which type of polymer the polyvinyl esters etc. are replaced in part by kraft pine lignin or its acylated derivatives.

Another possible bulk outlet for chemically modified lignin in the field of polymers is examined, viz. the utilization as compatibilizer in mixtures of polymers from the plastic fraction of household waste. This plastic fraction cannot be used in the plastic product manufacturing circuit, because after processing the mixture appears to be largely heterogeneous resulting in very low elongations-at-break (brittleness). To reduce the heterogeneity of such mixtures compatibilizing agents have been used functioning like "an adhesive" in providing a link between semicompatible resins (2).

2. Experimental

Acylated lignin derivatives were prepared by the action of an appropriate acyl chloride on a solution of kraft pine lignin (Indulin AT, Westvaco) in a mixture of pyridine/dioxane (3). A high molecular weight alternating copolymer of styrene and maleic anhydride (PSMA) was made by solution polymerization (1) in butanone (Mw 350,000). The maleic anhydride groups were partly protolyzed (approx. 35%), resulting in a lower critical solution temperature for PSMA/PVAc (1:1) of 135°C.

3.1. Aqualloy alloying techniques for acylated lignins

The alloying techniques of Aqualloy polymers are only applicable if, firstly, the dissimilar polymers are soluble in a mutual solvent. Secondly, a strong association between the dissimilar polymers has to exist which can be accomplished by dipole interactions or polar bonds, hydrogen bonds and polarizability. For instance, a stable polyblend can be formed from partly hydrolyzed PSMA with polyvinyl acetate (PVAc), because both polymers are soluble in acetone or butanone, and strongly interact each other by hydrogen bonding.

The degree of solubility of pine kraft lignin in a number of solvents with known solubility parameters (δ_r, δ_h) was determined by Roberts (4). In accordance with the solubility concept (5) only strong proton attracting solvents like dimethyl sulfoxide would solve pine kraft lignin. This behaviour might be influenced by derivation of the kraft lignin, e.g. esterification, resulting in lower values for δ_r and δ_h.

To determine the compatibility of the lignin derivatives with PSMA, films were cast from an acetone solution of PSMA and an acylated lignin in various weight ratios. After drying the mechanical and visual appearance, and the swellability in 1% NH_4OH and distilled water were determined.Using the same procedure also films with acetyl lignin and polymethyl vinylether maleic anhydride (PMVMA) were cast. The characteristics of the dry films are given in Table. From the swelling behaviour providing a good indication of the structural homogeneity of the blend matrix it appears that acetyl lignin is compatible with maleic anhydride copolymers by using the Aqualloy alloying technique. Depending on the weight ratio maleic anhydride copolymer/acetyl lignin the alloy swells 30-50 times its dry weight. However, these alloys are brittle. The brittleness can be reduced by adding another compatible polymer to the mixture, which has a relatively low glass-transition. Such ternary mixtures still have a swellability of 3000% and are rather flexible.

Further optimization of these ternary mixtures is necessary especially focused on thermoplastic processability and flexibility. In the future, swell characteristics such as swell ratio, swell rate and swell behaviour depending on pH and ion strength will be determined in order to evaluate potential matrices for controlled-release systems.

3.2. Acylated lignins as compatibilizer for the plastic fraction of household waste.

When the plastic fraction of household waste is reused and processed in new products the mechanical properties as tensile strength and strain-at-break are inferior. Therefore, it is difficult to find applications for such recycled plastic fractions. By applying a floating-separation technique to the plastic fraction of household waste a top fraction is obtained which is rich in low density polyethene (LDPE). Adding of only 5 parts of stearoyl lignin to this household top fraction the strain-at-break increased up to 50%. Adding either stearoyl lignin or oleolyl lignin to LDPE/polypropene- and LDPE/PS-blends relatively good values of tensile strength and strain-at-break are obtained. Moreover, these esters did also lubricate the mixture strongly as appeared from the torque values. All remaining lignin/polymer mixtures exhibit values for strain-at-break that were less than the mixtures without lignin or its acylated derivatives.

Although the absolute values of tensile strength and strain-at-break are far too low to be of practical relevance, the results indicate some compatibilizing effects of the stearoyl and oleolyl esters in the plastic fraction of actual household waste.

Further experiments will be focused on optimizing the compatibilizing effects observed. Special consideration will be given to household top fraction/oleoyl lignin because the latter component might be cross-linked advantageously by e.g. peroxides. Such cross-linking could lead to improvement of the mechanical properties.

Characteristics of dry films casted from a solution of maleic anhydride copolymers and acylated lignin

Lignin/copolymer	Ratio	Mechanical appearance	Swelling in % 1% NH_4OH	H_2O (dist.)
Acetyl lignin/PSMA	80/20	very brittle	160	200
	60/40	very brittle	160	2000
	50/50	very brittle	>100	
	40/60	brittle	200	5000/fragm.
Heptanoyl lignin/PSMA	50/50	very brittle	20-60	
Stearoyl lignin/PSMA	50/50	brittle	nihil	
Benzoyl lignin/PSMA	50/50	tacky	nihil	
Acetyl lignin/PMVMA	50/50	brittle	2000	5000/fragm.
Acetyl lignin/PMVMA/ PVAc	33/33/33		3000	3000
Acetyl lignin/PMVMA/ PVAc/VL	33/33/33	flexible but tacky	700	800

PSMA Polystyrene maleic anhydride

PMVMA Polymethyl vinylether maleic anhydride

PVAc Polyvinyl acetate

PVAc/VL Polyvinyl acetate/laurate (80/20)

References

1. A. Heslinga and C.A. van der Willigen, J.Appl.Polym.Sci., Appl.Polym. Symp., 35, 589 (1979).
2. C.E. Locke and D.R. Paul, Pol.Eng.Sci., 13, 308 (1973).
3. H.F. Lewis, F.E. Brauns, M.A. Buchanan and E.B. Brookbank, Ind.Eng. Chem., 35, 1113 (1943).
4. S.C. Roberts, Ph.D. Thesis, Univ. of Washington, 1974.
5. e.g. J.H. Hildebrand and R.L. Scott in "Regular Solutions", Prentice-Hall, Inc., Englewood Cliff, N.J. (1962); C.M. Hansen, J.Paint Techn., 39, 104, 505, 511 (1967); S.A. Chen, J.Appl.Pol.Sci., 15, 1247 (1971).

CONCLUDING SESSION

DISCUSSION ON THE PROPOSALS FOR THE DEFINITION OF PARAMETERS AND ANALYTICAL MEASUREMENTS APPLICABLE TO ANAEROBIC DIGESTION AND CARBOHYDRATE HYDROLYSIS PROCESSES

Reported by H. NAVEAU.

Unit of Bioengineering
Catholic University of Louvain,
B-1348 Louvain-la-Neuve, Belgium

1. INTRODUCTION : THE PROPOSALS

Reporting of experiments and results in the scientific literature in the fields of anaerobic digestion and carbohydrate hydrolysis is often difficult to understand. Indeed, symbols signification vary widely from paper to paper, units are often poorly defined and substrate materials, process and reactor are not clearly described.

The Contact-Coordination Group "Anaerobic Digestion and Carbohydrate Hydrolysis" of the Programme "Recycling of Urban and Industrial Wastes", felt the usefulness to have reference papers on how to describe experiments and to report results in these fields, and decided to prepare such papers. These have been published both for anaerobic digestion (1) and carbohydrate hydrolysis (2). They contain the following paragraphs, each with advices on data and description to give so that the reader has the right amount of information to understand and interpret the experiment :

- origin and description of the raw materials, treatments for substrate preparation
- sampling
- process description : reaction, reactor, running conditions
- description of feed, reactor content and products (effluent) with a list of the minimum data necessary for description, a list of additional usefull data and a list of other (specific) data, and possible methods

of analysis

- expression of results. This last paragraph is quite important and symbols and units are recommended for concentrations (substrate, active biomass, enzymes), rates (volumetric loading rate, production rate,...) and yields (conversion, methane yield,...).

2. DISCUSSION

It is suggested to give the elemental analysis of the substrates, or the oxidation state of carbon, that allows to calculate the theoretical potential yield of methane. However, this analysis is not possible for most laboratories because the necessary instrumentation is not available, so that a reference method could not be proposed.

The chemical oxygen demand (COD) could also be used because it can be linked with methane production. However, the potential methane yield can also be assessed by a batch digestion with a long period of digestion. Results in continuous digestors may then be compared with this practical potential methane yield.

The chemical oxygen demand (COD) is often used to express a yield of methane per g of COD removed. The COD of methane is a constant value and every gramme of COD removed corresponds and can correspond only to 0,395 liter methane at 35 °C ($Y_{CH4/CODr}$ = 0,395 l CH_4/g CODr). Any value that differ from the theoretical value can only mean that there is an error somewhere in the analytical procedure. This may be due to a wrong gas volume or a wrong methane content, to hydrogen sulfide content, substances refractory to the COD determination, bad sampling, volatile solids accumulation or loss in the digester or only to solubility of methane in the liquid at short retention time and low feed concentration.

A similar conclusion may be drawn when the calorific power of the methane produced is higher than the calorific power of the substrate !

The necessity to agitate continuously fed completely mixed digestors during feeding seems obvious; without agitation, the mixed liquor obtained is not representative of the digestor content because some sedimentation occurs very rapidly and the analytical results would have no meaning as concerns the digestor's content. However, is it meaningful to feed digestors on 5 days mean retention time once every day, 5 days a week ? This is a practical problem.

It is asked wether terms like efficiency and capacity would still have a use if the proposal is adopted. Efficiency and capacity lead to misunderstanding because they may have different meaning : the use of a clear symbol with well defined units would allow a better understanding. Capacity can be related to a unit of volume of reactor (volumetric capacity), e.g. capacity to treat a given amount of substrate (kg) per m^3 of digester, or it can express the total capacity of a given installation, but then it should be clearly specified that the total capacity is concerned and the working volume of the installation should be given.

Concerning processes using various stages or phases or steps, the following rules may be recommended :

- phases should be used for processes when different reactions take place in different reactors, e.g. two phases biomethanation means an acidification phase in one reactor and a methanogenesis phase in an other digester.
- stades should be used when the same process or reaction takes place in two consecutive reactors : e.g. biomethanation of wastewater sludges is often performed in two stades, that is two methane producing digesters, differing only e.g. by their mean retention time and temperature.
- steps should be avoided in this context; it is used to describe the various steps of a reaction, considering e.g. the mechanism of a chemical reaction.

As a practical rule for typing subscripts, one should type only one line of subscripts and never type a subscript to a subscript; for instance type CH_4 for methane but $r_{V.CH4}$ for methane production rate or $Y_{CH4/VSo}$ for methane yield (based on added volatile solids).

The use of these proposals in the scientific community could be promoted through the editors of journals or through the patronage of an international scientific organization like IUPAC (International Union of Pure and Applied Chemistry). The aim of the Contact Coordination group of the CEC was only to promote such use on a voluntary basis within the European scientific community. But sometimes, it is difficult to get CEC contractors to use it !

However, it does not seem that individual editors of journal could oblige their authors to use the symbols but only advise them to do so.

3. REFERENCES

(1) Proposal for the Definition of Parameters and Analytical Measurements Applicable to the Anaerobic Digestion Processes,

F. COLIN, G.L. FERRERO, M. GERLETTI, P. HOBSON, P. L'HERMITE, H.P. NAVEAU and E.-J. NYNS.

Agricultural Wastes, 7, 183-193, (1983).

(2) Proposal for the definition of parameters and analytical measurements applicable to carbohydrate hydrolysis processes.

J. BELAICH, G.L. FERRERO, M.P. L'HERMITE, H. NAVEAU, PH. THONART and T.M. WOOD.

Process Biochemistry, 19 (1), 2-5 (1984).

CONCLUSIONS AND RECOMMENDATIONS

J.P. BELAICH, CNRS, Marseille
J.M. LYNCH, Glasshouse Crops Research Institute, Littlehampton
W. VERSTRAETE, Rijkuniversiteit, Gent
H. NAVEAU, Université Catholique de Louvain

1. PRELIMINARY CONSIDERATIONS

1.1. The Community has sponsored R&D shared cost contracts since 1981 in the area of carbohydrate hydrolysis and anaerobic digestion. The response of the different countries and the results presented in this conference clearly show that the action of the community was appropriate and has given rise to valuable research.

1.2. It is clear that the EEC countries have obtained a leading position in the field of carbohydrate hydrolysis and anaerobic digestion. A good deal of work has been done. Yet a variety of questions remain. From the reports presented it is evident that a deeper knowledge of the composition and ultrastructure of lignocellulose as well as of the reacting micro-organisms has become imperative. If we are not one day to be submerged by agricultural, urban and industrial wastes, efficient methods must be developed for transforming them into chemicals and energy. This is far from being a "fashionable" field of research, such as molecular biology which makes headlines in the scientific journals, but it is nevertheless absolutely indispensable. Given that research requires a certain continuity over a minimum of about a decade to make fully effective use of the acquired hard and software, it is important to profit from the current momentum and consequently for the EEC to continue its support in this area, irrespective of the current policies of the individual countries.

1.3. The programme should continue to operate under the heading "Recycling". Indeed the programme covers in the first place research to transform secondary materials into useful end-products. The programme has also some important side-effects because it deals with environmental problems and contributes to the recovery of suitable amounts of energy for the EEC countries. Yet, it is felt that in view of the integrated approach which is necessary in the field of recycling, research efforts and coordination activities on "composting" and "chemicals" should be considered in the framework of the research areas on carbohydrate hydrolysis and anaerobic digestion within the "Recycling" programme.

2. CARBOHYDRATE HYDROLYSIS : CONCLUSIONS AND RECOMMENDATIONS

2.1. The present state of knowledge

There is little use of enzymatic reactions on an industrial scale at present. Some assessments indicate that even with zero enzyme cost it would be more expensive to produce alcohol from enzymatic routes than from acid hydrolysis. However, it is felt there is insufficient information available to make such assessments. For example, modern biotechnological techniques could greatly increase enzyme activity in micro-organisms.

2.2. Needs for study

There is a need for more information to be accumulated on fundamental aspects on the microbiology, biochemistry and genetics leading to the enzymatic hydrolysis of carbohydrates. Among the specific areas which need to be studied are :

- Establishment of culture collection of relevant cellulotytic micro-organisms including bacteria and fungi.
- Genetic manipulation of cellulase and hemicellulase positive micro-organisms.
- Physiological studies to optimise cellulase activities.
- Mixed culture studies.

One of the major targets must be to establish the most relevant secondary pathways from the primary hydrolysis which yields simple sugars from the cellulosic substrates. The alcohol fermentation may be a much less satisfactory target than the production of chemicals, biochemicals or microbial biomass, and there should be exploration for high-value end-products.

2.3. The future

There is little immediate scope for the construction of demonstration units in this field. However, this must be a target for the next 3-5 years. This will need the coordination of activities between biologists and engineers, especially those currently involved in acid hydrolysis. Within Europe, there is a large surplus capacity in the fermentation industry and where possible this should be used for demonstrations, thus minimizing capital investment costs. However, low technology systems should also be considered. For example, the successful European mushroom industry is based on a simple non-axenic but controlled solid substrate fermentation of lignocelluloses.
The use of lignocellulose fermentations involving hydrogen donors for action in agricultural and other environment systems, especially those concerned with mineral recycling, should receive particular attention. It should also be pointed out that cellulolysis is often the limiting factor in anaerobic digestion.

3. ANAEROBIC DIGESTION CONCLUSIONS AND RECOMMENDATIONS

3.1. Fundamental research

This is of utmost importance in order to further advance applications of biomethanation. The following topics are specifically relevant for future investigations :

3.1.1. Characterization of obligate proton reducing and acetogenic bacteria capable of metabolizing not only conventional fatty acids but also aromatic compounds in association with H_2-consuming bacteria. Potential applications of anaerobic detoxification should also be considered.

3.1.2. Study on the physiology of sulfate reducing bacteria and mechanisms to control them in anaerobic digestion, first to remove sulfur from sulfates and also for their role, important although not well understood, in the methane production mechanisms.

3.1.3. In depth analysis of the various mechanisms by which anaerobic bacteria adhere to support matrixes and to one another.

3.2. Applied research

3.2.1. Domestic wastewaters

The possibility of direct anaerobic treatment or cold, raw domestic wastewaters should be explored in the form of pilot and also of demonstration plants. The integration of such an anaerobic pretreatment into a total treatment scheme should be emphasized. Particular attention should be given to the evaluation of overall treatment efficiency with respect to the investment and operation costs of such combined anaerobic - aerobic wastewater treatment systems. The same attention should also be given to cold industrial wastewaters which have similar characteristics.

3.2.2. Industrial wastes and wastewaters

A variety of troublesome wastewaters can possibly be made less obnoxious through an anaerobic pretreatment. Methods to detoxify, decolorize, deodorize certain types of wastewaters deserve therefore to be explored and possibly tried out on demonstration plants.

3.2.3. Agricultural wastes and substrates

A relatively large number of digesters have been constructed on farms in the EEC member countries and much experience has been gained. It is of paramount importance that this experience be evaluated in detail. The programme should therefore maintain a limited number of activities covering the follow up and further improvement of biogas production from animal wastes on the farm. There remains the problem of the lack of a good process to digest bedding manure.

The possibility of recycling agricultural surpluses by means of an integrated anaerobic technology deserves special attention, in view of the agricultural overproduction capacity in the EEC. In areas with potential overproduction of certain crops, the possibility of constructing such an integrated recovery and recycling system should be tried out at a demonstration level.

3.2.4. Domestic refuse

In view of the quantities of domestic refuse produced, it appears necessary to continue efforts in the field of anaerobic digestion of these wastes. Of particular interest should be methods which combine efficiently with post-treatment in the form of composting. Particular attention should be given to anaerobic pretreatments which result in an improvement of the quality of the subsequent compost, e.g. lower level of heavy metals. A demonstration plant trial seems worthwhile.

4. DISCUSSION

4.1. Carbohydrate hydrolysis

As regards enzymatic hydrolysis, it would appear premature to build demonstration plants within the next 3-5 years, although pilot plants remain a possibility. Cooperation between scientists should be increased : for instance, exchange of cultures could result in better and/or faster results. Coordination activities in this field could be reinforced.

Acid hydrolysis is more advanced and processes are available at pilot scale. The next step should be demonstration plants with promising techniques, such as short-time hydrolysis and a process using diluted acid.

More emphasis could be given to the production of substances with high added values; that aspect should be stressed by using the specific nature of the chemical structure of ligno-cellulosic materials, and not use as substrate glucose or starch.

Fermentation on solid substrates to produce enzymes or other products should be supported. Production of animal feed is already being carried out but should be emphasized.

4.2. Anaerobic digestion

Research is already under way in some places on priority topics indicated in anaerobic digestion (§ 3). Although it is often difficult to obtain relevant information on ongoing activities in this field, the goal of the EEC R&D coordination programme should nevertheless be to disseminate information collected in the member states and results obtained in the framework of the Recycling programme. However, coordination of national activities is beyond the purpose of the Commission.

The conclusions and recommendations given in this paper mean that these fields of research are important and should start or continue, e.g. in a future EEC programme. One of the aims of this conference was to obtain suggestions for future work and the conclusions will be taken into account when preparing future programmes.

LIST OF PARTICIPANTS

AHRING, B.
Manager Scientist
Institute of Thallophytes
University of Copenhagen
Ø. Farimagsgade 2D
DK - 1353 COPENHAGEN

ALBUQUERQUE D'OREY, J.L.
Engineer
Sociedade Comercial Crey
Antunes Sarl
R. de Sacramento A Lapa 72 3° D10
P - 1200 LISBON

ALCOBIA FERMEIRA, V.M.
Engineer
Orby Technica Naval
Industria, Lda.
R. Remolares 12, 4°
P - 1200 LISBON

ALFANI, F.
University Professor
University of Naples
Istituto Principi Ingegneria Chimica
Piazzale Tecchio
I - 80125 NAPOLI

ALIBHAI, K.
Research Associate
University of Birmingham
Department of Civil Engineering
P.O. Box 363
GB - BIRMINGHAM B15 2TT

ANDERSEN, H.M.
Hans Møller Andersen ApS
Falkonér Allé 7,1
DK - 2000 COPENHAGEN F

ATARHOUCH, T.
Université Libre de Bruxelles
24, rue General Thys
B - 1050 BRUXELLES

ATHANASOPOULOS, N.
Environmental Protection Department
Coton Manufacturing Co. Inc.
Piraiki-Patraiki
GR - KPIA ITEON, PATRAS

AUBART, C.
Centre de Recherche de la S.C.P.A.
Aspach-le-Bas
F - 68700 CERNAY

AUMAITRE, A.
I.N.R.A.
Station de Recherches
sur l'Elevage des Porcs
Juint-Gilles
F - 35590

AVENI, A.
Istituto di Ricerche Breda SpA
Viale Earca 335
I - MILANO

BARUCHELLO, G.M.
SOGEIN, SpA
Via Giorgione 61
I - 000147 ROMA

BEARDSMORE, A.
Research ICI
Agricultural Division
P.O.Box 1
GB - BILLINGHAM

BEAUDRY, T.
Director
Bio-Productions-Lubex S.A.
40, rue du Tournoi
B - 1190 BRUXELLES

BELAICH, J.
Professeur
C.N.R.S.
Laboratoire de Chimie Bactérienne
31, Chemin Joseph Aiguier
F - 13274 MARSEILLE CEDEX 2

BOCCIARELLI, M.
Tecnico
Viale Piacenza 4
I - 43100 PARMA

BORGHANS, L.
Research Scientist
CSM Suiker Bv
Nienoord 13
NL - DIEMEN

BOURDEAU, P.
Director
Commission of the European
Communities
Directorate-General "Science,
Research and Development"
200, rue de la Loi
B - 1049 BRUSSELS

BOURLET, P.
Ingénieur IRIS
362, rue J. Guesde
F - 59650 VILLENEUVE D'ASCQ

BRETSCHER,
European Energy and
Environmental Engineering + Co. Ltd.
Oberdorfstrasse 32
CH - 8953 DIETIKON/ZH

BREURE, A.M.
Microbiologist
Laboratorie van Microbiologie
Universiteit van Amsterdam
Nieuwe Achtergracht 127
NL - 1018 WS AMSTERDAM

BRICTEUX, J.
Chef de Département
INIEX
200, rue du Chéa
B - 4000 LIEGE

BROWN, M.
Microbiologist
QMC Industrial
Research Ltd.
229, Mile End Road
GB - LONDON E 1

BRUXELMANE, M.
Professeur
Faculté Polytechnique de Mons
Laboratoire de Génie Chimique
57, rue de l'Epargne
B - 7000 MONS

BULLY, F.
Centre de Recherches de la
S.C.P.A. Aspach le bas
F - 68700 CERNAY

BUREL, C.
Stagiaire
CREATE
Boulevard Louis Seguin
F - 92700 COLOMBES

BURNS, R.G.
Biological Laboratory
University of Kent
Canterbury, Kent
GB - CANTERBURY

BURTON, J.
Development Technologist
Dairy Crest, Research
and Development Division
MMB Crudgington
GB - TELFORD, Shropshire TF6 6HY

BUVET, R.
Professeur
Université Paris-Val de Marne
23, Allée de la Toison d'Or
F - 94000 CRETEIL

CADONICA, R.
Biologist SIDI s.r.l.
Viale Mentana 92
I - 43100 PARMA

CAMPAGNA, R.
Researcher
Istituto Guido Donegani
Centro Ricerche - NOVARA
Via Fauser 4
I - 28100 NOVARA

CANNAZZA, S.
European Energy and
Environmental Engineering + Co. Ltd.
Oberdorfstrasse 32
CH - 8953 DIETIKON/ZH

CANTARELLA, M.
University Researcher
University of Naples
Istituto Principi Ingegneria Chimica
Piazzale Tecchio
I - 80125 NAPOLI

CARBONE, D.
Ricercatore
CONPHOEBUS
Via G. Leopardi 60
I - 95127 CATANIA

CECCHI, D.
Engineer
Lombardia Risorse
Via Dante 12
I - 20121 MILANO

CHAPMAN, J.
Agricultural Engineer
National Institute for
Research in Dairying
Shinfield
GB - READING, BERKSHIRE, RG2 9AT

CHASSAING, P.
European Energy and
Environmental Engineering + Co. Ltd.
Oberdorfstrasse 32
CH - 8953 DIETIKON/ZH

COLARDEAU, J.
Agence Nationale pour la Récupération
et l'Elimination des Déchêts (ANRED)
2, Square Lafayette - B.P. 306
F - 49004 ANGERS CEDEX

COLIN, F.
Directeur Scientifique
Institut de Recherches
Hydrologiques
10, rue Ernest Bichat
F - 54000 NANCY

COLLERAN, E.
University Lecturer
University of Galway
Department of Microbiology
University College
IR - GALWAY

COUGHLAN, M.
University Professor
Biochemistry Department
University College
IR - GALWAY

COUSSEMENT, I.
Head of Energy Department
AIDR International Association
for Rural Development
20, rue du Commerce
B - 1040 BRUXELLES

CREYF, H.
R & D
Recticel
Division of NV PRB
Damstraat, 2
B - 9000 WETTEREN

DAHL, K.
Assistant Professor
University of Aalborg
Sohngaardsholmvej 57
DK - 9000 AALBORG

DAHL, A.
Manager Scientific
Chemical Engineer
Enerchem AB
Box 7030
S - 22007 LUND

DAVID-GEUSKENS, C.
Université Libre de Bruxelles
Faculté des Sciences,
Chimie Macromoléculaire
Campus Plaine C.P. 206/1
Boulevard du Triomphe
B - 1050 BRUXELLES

DE LEENHEER, L.
Research Scientist
Artois Breweries
Vaartstraat 94
B - 3000 LEUVEN

DE POLI, F.
Researcher
Italian Commission for
Nuclear and Alternative Energy
Casaccia
P.O. Box 2400
I - 00100 ROMA

DEL CERRO, C.
Lecturer
Imperial College
Department of Chemical Engineering
Prince Consort Road
GB - LONDON SW7

DELMAS, M.
Professeur
Institut National Polytechnique
ENSC
118, route de Narbonne
F - 31077 TOULOUSE

DEMUYNCK, M.
Searcher
Unit of Bioengineering
University of Louvain
119, Place de la Croix Sud
B - 1348 LOUVAIN-LA-NEUVE

DIDIER, A.F.
Engineer
L'école Nationale des
Travaux Public de l'Etat
1, rue Maurile Audim
F - 69120 VAULX-EN-VELIN

DIERICKX, L.
Attaché - Recherches
CNPEM
32, Boulevard de la Constitution
B - 4020 LIEGE

EDELINE, F.
Directeur
CEBEDEAU A.S.B.L.
2, rue A. Stévart
B - 4000 LIEGE

FARGET, M.A.
Délégué à l'Energie
INRA
B.P. 47
F - 38040 GRENOBLE CEDEX

FAUL, W.
Diplom Ingenieur
Kernforschungsanlage
Jülich
Box 1913
D - JUELICH

FECKL, J.
MD Verwaltungsgesellschaft
Nicolaus GmbH & Co.KG
Zellstoffprojekt
Planegger Strasse 8 B
D - 8000 MUENCHEN 60

FELGEN, E.
Ingénieur diplomé EPFZ
Freylinger & Felgen
Ingénieurs Conseils
46, rue du Cimetière
L - 1011 LUXEMBOURG

FERNANDES, X.
Wastewater Scientist
Water Research Centre
Stevenage Laboratory
Elder Way
GB - STEVENAGE, Hertfordshire SG1 1TH

FERRANTI, M.P.
Commission of the European Communities
Directorate-General "Science,
Research and Development"
200, rue de la Loi
B - 1049 BRUSSELS

FERRERO, G.L.
Commission of the European Communities
Directorate-General "Energy"
200, rue de la Loi
B - 1049 BRUSSELS

FIELD, J.
Research Fellow
Department Water Pollution Control
Wageningen University
De Dreijen 12
NL - WAGENINGEN

FIESTAS ROS DE URSINOS, J.A.
Professor de Investigacion
Instituto de la Grasa
y sus Derivados
Avenida P.P. Garcia Tejero 4
Aptdo 1078
E - SEVILLA 12

FOND, O.
Etudiant 3ème cycle
Université de Nancy I
Faculté des Sciences
Laboratoire de Chimie Biologique I
B.P. 239
F - 54506 VANDOEUVRE LES NANCY

FREY, M.
Engineer
L'école Nationale des
Travaux Public de l'Etat
1, rue Maurile Audim
F - 69120 VAULX-EN-VELIN

FRIMAN, R.
Farm Advisor (Wastes)
Ministry of Agriculture
Block & Coley Park
GB - READING, BERKSHIRE

GALLARATI, E.
Dottore Ingegnere
S.I.D.I. Srl
Viale Mentana 92
I - 43100 PARMA

GARCIA, J.L.
Research
ORSTOM - France
Université de Provence
3, Place Victor-Hugo
F - 13331 MARSEILLE Cedex 3

GAST, D.
Institut für Holzchemie und
Chemische Technologie des Holzes
Leuschnerstrasse 91
D - 2050 HAMBURG 80

GAUDILLIERE, R.
Responsable Cellule Energie
Environnement
Agence Française pour
la Maitrise de l'Energie
27, rue Louis Vicat - "Le Béarn"
F - 75015 PARIS

GAY, R.
Professeur
Laboratoire de Chimie Biologique 1
Faculté des Sciences
Université de Nancy-I
B.P. 239
F - 54506 VANDOEUVRE-LES-NANCY CEDEX

GAY, C.
Université de Nancy I
Faculté des Sciences
Laboratoire de Chimie Biologique I
B.P. 239
F - 54506 VANDOEUVRE LES NANCY CEDEX

GEOGHEGAN, M.
Department of Industrial Microbiology
University College Belfield
IR - DUBLIN 4

GEORGACAKIS, D.
Lecturer Age. Enge (PhD)
Agricultural College of Athens
Laboratory of Agricultural Structures
GR - ATHENS 11855 (Iera Odos 75)L

GERLETTI, M.
E.Bi.A. COOP. a.r.l.
Via Bergamo 21
I - 20135 MILANO

GIJZEN, H.J.
Microbiologist
Catholic University of Nijmegen
Laboratory for Microbiology, K.U.N.
Toernooiveld
NL - 6525 ED NIJMEGEN

GLAUSER, M.
Biologiste
Université Neuchâtel
Laboratoire de Microbiologie
22, Chantemerle
CH - 2000 NEUCHATEL 7

GRETHLEIN, H.
Professor of Engineering
Thayer School of Engineering
Dartmouth College
6, Dunster Drive
USA - HANOVER NH 03755

HAARS, A.
Scientist
Institut für Forstbotanik
der Universität Göttingen
Büsgenweg 2
D - 3400 GOETTINGEN

HACK, P.
Environmental Engineer
Paques BV
Postbus 52
NL - 8560 AB BALK

HARTMANN, L.
Professor
Universität Karlsruhe
Institut für Miljoteknik
Am Fasanengarten
D - 7500 KARLSRUHE 1

HASENBOEHLER, A.
Dr. rer.Nat.
Sulzer, Wasser- und
Abwassertechnik
Postfach 380
D - 6308 BUTZBACH

HEIERS, W.
Commission of the European Communities
Directorate-General "Information
Market and Innovation"
P.O.B. 1907
L - 2029 LUXEMBOURG

HOBSON, P.N.
Rowett Research Institute
Technology Group
Microbiology Department
Greenburn Road
Bucksburn
GB - ABERDEEN AG2 9SB

HUETTERMANN, A.
Universität Goettingen
Forstbotanisches Institut
Bisgenweg 2
D - 3400 GOETTINGEN

ISIK, H.
Process Engineer
Energy & Waste Systems Ltd
Bridge House
Station Road
GB - WESTBURY Wilts BA13 4HS

JONES, T.
Commission of the European Communities
Directorate-General "Information
Market and Innovation"
P.O.B. 1907
L - 2920 LUXEMBOURG

JØRGENSEN, M.H.
Chemical Scientist
DTH Denmark
Anker Engelundsvej
DK - 2800 LYNGBY

KANG, N.S.
University of Birmingham
Deptartment of Chemical Engineering
P.O. Box 363
GB - BIRMINGHAM B15 27T

KARAGEORGOS, A.
Member of ACPM in Recycling
Hellenic Republic Ministry
for Research & Technology
2, Ermou Str.
GR - 105 63 ATHENS

KARPER, R.
P.O. Box 17
NL - 8200 AA LELYSTAD

KELLER, P.
Head of Department
The Danish Testing
Station for Biomass-Plants
Statens Jordbrugstekniske
Forsøg, Bygholm
DK - 8700 HORSENS

KIELY, P.V.
An Foras Taluntais
Johnstown Castle Research Centre
IR - WEXFORD

KJAERGAARD, L.
Head of Section
De Danske Sukkerfabrikker
Biotechnology Section
P.O. Box 17
DK - 1001 COPENHAGEN

KLEIN, K.
Deputy Head of Division
Commission of the European
Communities, Directorate-General
"Environment, Consumer Protection
and Nuclear Safety"
200, rue de la Loi
B - 1049 BRUSSELS

KNOP, R.
Institut für Siedlungswasserbau,
Wassergütewirtschaft und Abfallwirtschaft
der Universität Stuttgart
Bandtäle, 1
D - 7000 STUTTGART 80 (Büsnau)

KOERNER, H.U.
Institut für Holzchemie und
Chemische Technologie des Holzes
Leuschnerstrasse 91
D - 2050 HAMBURG 80

KOFOD, B.
Civil Engineer
Nordisk Triclair
A/S Hotaco
P.O. Box 49
DK - 4300 HOLBAEK

KRAEMER, S.
Engineer
CARL BAD A/S
Granskoven 8
DK - 2600 GLOSTRUP

KRISTENSEN, G.H.
Research Engineer
Krüger
Gladsaxeves 363
DK - 2860 SØBORG

KRUIJDENBERG, H.
Projectbureau
Energieonderzoek
P.O. Box 342
NL - 7300 AM APELDOORN

KUNTZIGER, E.
Manager Department Entreprises
SIDMAR N.V.
51, John Kennedylaan
B - 9020 GENT

LANG, J.
Pollution Scientist
Ministry of Agriculture
Fisheries and Food
Great Westminster House
Horseferry Road
GB - LONDON SW1P 2AB

LAWSON, V.
Managing Director
WMC (Resource Recovery) Ltd.
2, Eaton Crescent
Clifton
GB - BRISTOL BS8 2EJ

LEDERGERBER, E.
Diplom Physiker
Intercon H. Strittmatter AG
Suhrerstrasse 24
CH - 5036 OBEIENTFELDEN

LE ROUX, N.
Research Scientist/Biotechnologist
Warren Spring Laboratory
Gunnels Wood Road
GB - STEVENAGE Herts SG1 2BX

LENZEN, C.
Chef de travaux
C.N.P.E.M.
32, boulevard de la Constitution
B - 4020 LIEGE

LESTY, Y.
Ingénieur Recherche
S.L.E.E.
Laboratoire Central
38, rue du Président Wilson
F - 78230 LE PECQ

LETTINGA, G.
Agricultural University Wageningen
Associate Professor
De Dreyen 12
NL - WAGENINGEN

LEVERT, J.M.
Professeur
Faculté Polytechnique de Mons
Institut de Chimie et de Métallurgie
Rue de l'Epargne
B - 7000 MONS

LINDSAY, J.
Microbiologist
Research and Development Department
Gist-Brocades
P.O. Box 1
NL - 2600 MA DELFT

LINKO, M.
Professor
VTT Biotechnical
Laboratory
Tietotie 2
SF - 02150 ESPOO 15

LINNE, M.
Manager Scientist Chemical Engineering
Enerchem AB
Box 7030
S - 22007 LUND

LIPS, S.
Technical Engineer
IBVL
Bornesteeg 59
NL - WAGENINGEN

LONGIN, R.
Ingénieur de Recherche
Institut Pasteur
28, rue du Dr. Roux
F - 75015 PARIS

LUCARELLI, M.T.
Responsabile Settore Ricerche
SOGEIN SpA
Via del Giorgione 63
I - ROMA

LYNCH, J.M.
Microbiologist
Plant Pathology &
Microbiological Deptartment
Glasshouse Crops Research Institute
Worthing Road, Rustington
GB - LITTLEHAMPTON, West
Sussex BN16 3PU

MAESTROJUAN SAEZ DE SAUREGUI, G.
Becario
Instituto de la Grasa
y sus Derivados
Avenida P.P. Garcia Tejero
Aptdo 1078
E - SEVILLA 12

MACRIS, B.J.
Biology Division
N.R.C. "DEMOCRITOS"
Aghia Paraskevi
GR - AGHIA PARASKEVI

MALLIDIS, S.
Ingénieur
THARAKI S.A.
Panepistimioy Str. 13
GR - ATHEN

MARTY, B.
GERME
27, Boulevard Charles Moretti
F - 13014 MARSEILLE

MEDICI, F.
Researcher
Institute Applied Chemistry
University of L'Aquila
Facoltà di Ingegneria
I - 67100 MONTELUCO (AQ)

MEMBREZ, Y.
Ingénieur
EREP S.A.
La Rosaire, Zone Industrielle
CH - 1111 ACLENS

MEYER, F.
Chef de Laboratoire
IRCHA
B.P. 1
F - 91710 VERT LE PETIT

MOLLE, J.-F.
Head of Project Biomass/Energy
Commission of the European
Communities, Directorate-General
"Science, Research and Development"
200, rue de la Loi
B - 1049 BRUSSELS

MONTIES, B.
CCMGP
MST Ministère Industrie Recherche
1, rue Descartes
F - 75005 PARIS CEDEX

MOOIJ, H.P.
General Manager, Refcom
Waste Management, Inc.
1341, S.W. 25th AVE.
USA - BOYNTON BEACH, FL. 33435

MOSS, D.A.
Group Leader
Fermentation Technology
Sturge Biochemicals
Denison Road
GB - SELBY, N.YORKSHIRE YO8 8EF

MOUTON, C.
Ingénieur
ANRED
2, Square Lafayette
F - 49 ANGERS

MØLLER, A.
Agri Contact
Torupvejen 97
DK - 3390 HUNDESTED

McNERNEY, M.
Microbiologist
Harper and Fay
The Jones Group
Beechill, Clonskeagh
IR - DUBLIN 4

NAEHLE, C.
Wissenschaftlicher Mitarbeiter
Süddeutsche Zucker-AG
Zentral-Laboratorium
Postfach 11 27
D - GRUENSTADT

NAVEAU, H.
Professor
Unit of Bioengineering
Catholic University of Louvain
1/9, place Croix du Sud
B - 1348 LOUVAIN-LA-NEUVE

NEKREP, F.V.
Professor of Microbiology
Biotechnology Faculty
Zootechnology Department
Groblje 3
YU - 61230 DOMZALE

NORRMAN, J.
Biosystem Engineering Pte Ltd
70, Senoko Road
SINGAPORE - 2775 SINGAPORE

NØRGAARD, P.
Research Worker, Msc.
University of Aalborg
Sohngaardsholmvej 57
DK - 9000 AALBORG

NTALIS, D.
Institut für Oekologie
und Taxonomy
Universität Athen
Panepistimiopolis
GR - ATHEN

NYNS,
Professeur
Université Catholique de Louvain
1/9, Place Croix du Sud
B - 1348 LOUVAIN-LA-NEUVE

OAKLEY, D.
Research Fellow
University of Birmingham
Department of Chemical Engineering
Edgbaston
GB - BIRMINGHAM BI5 2TT

O'GORMAN, V.
National Board for Science
and Technology
Shelbourne House
Shelbourne Road
IR - DUBLIN

O'SHEA, J.
Biochemist
Ceimici Teoranta
Fitzwilton House, Wilton Place
IR - DUBLIN 2

OESTERGAARD, N.
Manager Scientist (Chemical Engineer)
The Technological Institute
of Denmark
Department of Chemical Technology
DK - TAASTRUP

OLGUIN, E.
Doctor in Biotechnology
Instituto Mexicano de
Technologias Apropriadas
IMETA
Appd. Postal 63-257
MEXICO - 02000 MEXICO D.F.

OLIVER, B.
Chemical Engineer
National Institute of
Agricultural Engineering
West Park
GB - SILSOE, Beds MK45 4HS

PAIN, B.
Agricultural Scientist
National Institute for
Research in Dairying
Shinfield
GB - READING RG2 9AT

PAQUOT, M.
1er Assistant
Faculté des Sciences Agronomiques
2, Passage des Déportés
B - 5800 GEMBLOUX

PARISI, F.
Professore
Università di Genova
Via All'Opera Pia
I - GENOVA

PARRY, J.
Research Fellow
University of Warwick
Deptartment of Environmental Science
GB - COVENTRY CV4 7AL

PAUSS, A.
Searcher
Unit of Bioengineering
University of Louvain
1/9, place Croix du Sud
B - 1348 LOUVAIN-LA-NEUVE

PETERSEN, G.
M. Sc. (Chem. Eng.)
The Technological Institute
of Denmark
Department of Chemical Technology
DK - 2630 TAASTRUP

PHILLIPS, V.R.
Chemical Engineer
National Institute of
Agricultural Engineering
West Park
GB - SILSOE, Beds MK45 4HS

PICCHIOLUTTO, S.
Ingegniere
Comune di Modena
Piazza Grande
I - 41100 MODENA

PICCININI, S.
Researcher
Centro Richerche
Produzioni Animali
Via Crispi 3
I - 42100 REGGIO EMILIA

PICCININI, N.
Professore
Politecnico di Torino
Departimento Scienza dei
Materiali e Ingegneria Chimica
Corso Duca Degli Abruzzi 24
I - 10100 TORINO

POLLAK,
Schmidt Reuter
Ingenieursgesellschaft
Abteilung Forschung und Entwicklung
Graeffstrasse, 5
D - KOELN 30

POURQUIE, J.
Institut Français du Pétrole
Centre de Documentation
B.P. 311
F - 92506 RUEIL MALMAISON CEDEX

PRONOST, M.
Ingénieur Agricole
SOBEA - M/ENV
280, Avenue Napoleon Bonaparte
F - RUEIL MALMAISON

PROST, C.
Professeur
Laboratoire des Sciences
du Génie Chimique
CNRS - ENSIC
1, rue Grandville
F - 54042 NANCY CEDEX

PULS, J.
Bundesforschungsanstalt für
Forst- und Holzwirtschaft
Leuscherstrasse
D - 2050 HAMBURG 80

QUICKENDEN, J.
Microbiologist
Hamworthy Engineering Ltd.
(Pump & Compressor Division)
Fleets Corner
Poole
GB - DORSET BH17 7LA

REID, W.G.
Microbiologist
Rowett Research Institute
Green Burn Road
Bucksburn
GB - ABERDEEN AB2 9SB

RICHARDS, K.M.
Technologist
Energy Technology Support
Unit , Ukaga
E.T.S.U. A.E.R.E.
Harwell
GB - OXON OX11 ORA

RIEHM, T.
Diplom-Chemiker
Bothestrasse 30
D - 6900 HEIDELBERG

RIERADEVALL-PONS, J.
Quimico
Servicio de Agricultura
Diputacion de Barcelona
Urgell, 187
E - BARCELONA 36

RIJKENS, B.A.
Research Officer/ Project Leader
Institute for Storage and
Processing of Agricultural
Products (IBVL)
P.O. Box 18
NL - 6700AA WAGENINGEN

ROTONDO', P.P.
Commission of the European
Communities, Directorate-General
"Information Market and Innovation"
P.O.B. 1907
L - 2920 LUXEMBOURG

ROUSTAN, J.L.
I.N.R.A.
Domaine de la Prise
St. Gilles
F - 35590 L'HERMITAGE

ROUX, G.
Centre Scientifique
et Technique du Bâtiment
Division Physique des Matériaux
24, rue Jospeh Fournier
F - 38400 SAINT MARTIN D'HERES

ROZZI, A.
Research Scientist
IRSA - CNR
Via de Blasio 5
I - 70123 BARI

RUGGERI, B.
Ricercatore
Politecnico di Torino
Duca Degli Abruzzi 24
I - 10124 TORINO

RULKENS, W.H.
T.N.O. - Organisatie vor Toegepaste
Natuurwetenschappelijk Onderzoek
Postbus 342
NL - 7300 AH APELDOORN

SANNA, P.
Research
ASSORENI
Via Ercole Ramarini, 32
I - 00015 MONTEROTONDO (Roma)

SARKAR, J.
Research Scientist
University of Kent
at Canterbury Kent
Biological Laboratory
GB - CANTERBURY, KENT CT2 7NJ

SCHELLER, W.A.
Professor of Chemical Engineering
University of Nebraska
Deptartment of Chemical Engineering
USA - LINCOLN, NEBRASKA 68588 0126

SCHILEO, G.
Assistant to the President
ANSALDO
Viale Pilsudski 92
I - 00197 ROMA

SCHUBERT-KLEMPNAUER, H.
Diplom-Geologe
Messerschmitt-Bölkow-Blohm
Space Division
P.O. Box 80 11 69
D - 8000 MUENCHEN 80

SENIOR, E.
Lecturer in Applied Microbiology
University of Strathclyde
Royal College
204, George Street
GB - GLASGOW G1 1XW

SLACK, J.C.
Director
Energy and Waste Systems Ltd
Bridge House
Station Road
GB - WESTBURY, WILTSHIRE BA13 4HS

SLATER, J.H.
Department of Environmental Sciences
University of Warwick
GB - COVENTRY, WARWICKSHIRE CV4 7AL

SLEAT, R.
Microbiologist
Biotechnica Ltd
5, Chiltern Close
GB - CARDIFF CF4 5DL

SMITHER, R.
Government Scientist
Biotechnology Unit
Laboratory of the Government Chemistry
Deptartment of Trade & Industry
Cornwall House, Stamford Str.
GB - LONDON SE1 9NQ

SPEECE, R.
Professor
Drexel University
3100, Chestnut
USA - PHILADELPHIA, PA 19104

STAMHUIS, E.J.
Rijksuniversiteit Groningen
Department of Chemical Engineering
Nijenborgh 16
NL - 9747 AG GRONINGEN

STANEFF, Th.
Dornier System GmbH
Postfach 1360
D - 7990 FRIEDRICHSHAFEN 1

STAUD, R.
Research Scientist
Institut für Ingenieurbiologie
Universität Karlsruhe
Am Fasanengarten
D - 7500 KARLSRUHE

STELLOY, V.
Agricultural Engineer
Ministry of Agriculture
60, Lomvardoy St.
GR - ATHENS (701)

STRUBE, R.
Consultant
Danish Fermentation
Industry Ltd.
Abriksparken 58
DK - 2600 GLOSTRUP

SUTTER, K.
Chemist
Biogas Projekt
Eidg. Forschungsantalt für
Betriebswirtschaft und Landtechnik
CH - 8355 TANIKON

SVENSSON, T.
Head of Section
National Swedish Board
For Technical Development
Box 43200
S - 10072 STOCKHOLM

TAFDRUP, S.
Laboratory Manager
Hojbogaard Biogas Plants Ltd.
Industrivej 9
DK - 5580 NORRE AABY

TER MEULEN, B.P.
Chemisch Ingenieur
Institute of Applied
Chemistry TNO
P.O. Box 108
NL - 3700 AC ZEIST

THONART, P.
Chargé de Cours
Faculté des Sciences
Agronomiques
2, Passage des Déportés
B - 5800 GEMBLOUX

TINNEMANS, A.
Institute of Applied
Chemistry TNO
P.O. Box 5009
NL - 3502 JA UTRECHT

TRANCART, J.L.
Ingénieur
CREATE
Boulevard Louis Seguin
F - 92700 COLOMBES

TROESCH, W.
Fraunhofer-Institut für Grenzfiechen
und Bioverfahrenstechnik (IGB)
Nobelstrasse 12
D - 7000 STUTTGART 80

TSAO, G.T.
Professor of Chemical Engineering
Purdue University
West Lafayette
USA - INDIANA 47907

VAN ANDEL, J.G.
Microbiologist
University of Amsterdam
Nieuwe Achtergracht 127
NL - 1018 WS AMSTERDAM

VAN DER VLUGT, A.J.
MT TNO
P.O. Box 342
NL - 7300 AH APELDOORN

VANDERBEKE, E.
Research
AVV - Molens Boerenbond
6, E. Meeusstraat
B - 2060 MERKSEM

VANLANDUYT, G.
Biochemical Research Manager
ACEC Charleroi
B - 6000 CHARLEROI

VERONESI, G.
University Professor
lst. Genio Rurale
Università Bologna
Via Filippo 4
I - BOLOGNA

VEROUGSTRAETE, A.
\gricultural Engineer
Catholic University of Louvain
Unité de Génie Biologique
1, place Croix du Sud
B - 1348 LOUVAIN LA NEUVE

VERRIER, D.
Ingénieur
I.N.R.A.
Certia Bp. 39
F - 59651 VILLENEUVE D'ASCQ CEDEX

VERSTRAETE, W.
Professor
University of Gent
Laboratory of Microbial Ecology
Coupure L 653
B - 9000 GENT

VETTER, L.R.
Director of Research
A.O. Smith Harvestore Products, Inc.
550, West Algonquin Road
USA - ARLINGTON HEIGHTS, ILLINOIS 60005

VIAL, V.
Research Administrator
An Foras Taluntais
19, Sandymount Ave.
IR - DUBLIN 4

VISSCHER, K.
Ministerie VROM
P.O. Box 450
NL - 2260 MB LEIDSCHENDAM

VOGEL, M.
Wissentschaftlicher Mitarbeiter
Süddeutsche Zucker-AG
Zentral-Laboratorium
Postfach 11 27
D - GRUENSTADT

WEBB, L.J.
Unit Head
Pira
Randalls Road
GB - LEATHERHEAD, SURREY HT22 7RN

WEILAND, P.
Engineer
Bundesforschungsanstalt für
Landwirtschaft, Institut
für Technologie
Bundesalle 50
D - 3300 BRAUNSCHWEIG

WENTWORTH, R.L.
Chemical Engineer
Dynatech R/D Company
99, Erie Street
USA - CAMBRIDGE, Massachusetts 02139

WESTERMANN, P.
Manager Scientist
Institute of Thallophytes
University of Copenhagen
Ø. Farimagsgade 2D
DK - 1353 COPENHAGEN

WHEATLEY, A.D.
The Environmental Biotechnology Group
Chemical Engineering Department
University of Manchester
Institute of Science & Technology
P.O. Box 88
GB - MANCHESTER M60 1QD

WHITMORE, T.N.
Research Miocrobiologist
Department of Microbiology
University College, Cardiff
Newport Road
GB - CARDIFF CF2 1TA

WIND, E.
CSM Suiker B.V.
Nienoord 13
NL - DIEMEN

WOOD, T.M.
Biochemist
Rowett Research Institute
Department of Microbial Biochemistry
Bucksburn
GB - ABERDEEN

YADAV, K.S.
Department of Microbiology
Agricultural University Wageningen
414 1AC, Lawickse Allee II
NL - 6701 AN WAGENINGEN

ZEEMAN, G.
Research Worker
Agricultural University Department
Water Pollution Control
De Dreyen 12
NL - 6703 BC WAGENINGEN

ZEEVALKINK, J.A.
Specialist
Heidemij Adviesbureau BV
P.O. Box 264
Nl - 6800 AG ARNHEM

ZOGLIA, M.
Mechanical Engineer
Daneco - Danieli Ecologia SpA.
Via Nazionale 85
I - 30048 SAN GIOVANNI AL NAT.(UD)

ZUBR, J.
Senior Fellow
Department of Crop Husbandry
and Plant Breeding
Thorvaldsensvej 40
DK - 1871 V COPENHAGEN